On the Road to Precision Medicine: Magnetic Systems for Tissue Regeneration, Drug Delivery, Imaging, and Theranostics

On the Road to Precision Medicine: Magnetic Systems for Tissue Regeneration, Drug Delivery, Imaging, and Theranostics

Editors

Bogdan Parakhonskiy
Francesca Garello
Miriam Filippi
Yulia I. Svenskaya

MDPI • Basel • Beijing • Wuhan • Barcelona • Belgrade • Manchester • Tokyo • Cluj • Tianjin

Editors

Bogdan Parakhonskiy
Department of Biotechnology
Ghent University
Ghent
Belgium

Francesca Garello
Department of Molecular
Biotechnology and Health
Sciences
University of Torino
Turin
Italy

Miriam Filippi
Soft Robotics Lab
ETH
Zurich
Switzerland

Yulia I. Svenskaya
Scientific Medical Center
Saratov State Unversity
Saratov
Russia

Editorial Office
MDPI
St. Alban-Anlage 66
4052 Basel, Switzerland

This is a reprint of articles from the Special Issue published online in the open access journal *Pharmaceutics* (ISSN 1999-4923) (available at: www.mdpi.com/journal/pharmaceutics/special_issues/magnetic_systems).

For citation purposes, cite each article independently as indicated on the article page online and as indicated below:

LastName, A.A.; LastName, B.B.; LastName, C.C. Article Title. *Journal Name* **Year**, *Volume Number*, Page Range.

ISBN 978-3-0365-8433-1 (Hbk)
ISBN 978-3-0365-8432-4 (PDF)

Cover image courtesy of Bogdan Parakhonskiy

© 2023 by the authors. Articles in this book are Open Access and distributed under the Creative Commons Attribution (CC BY) license, which allows users to download, copy and build upon published articles, as long as the author and publisher are properly credited, which ensures maximum dissemination and a wider impact of our publications.
The book as a whole is distributed by MDPI under the terms and conditions of the Creative Commons license CC BY-NC-ND.

Contents

Francesca Garello, Yulia Svenskaya, Bogdan Parakhonskiy and Miriam Filippi
On the Road to Precision Medicine: Magnetic Systems for Tissue Regeneration, Drug Delivery, Imaging, and Theranostics
Reprinted from: *Pharmaceutics* **2023**, 15, 1812, doi:10.3390/pharmaceutics15071812 1

Vera L. Kovalenko, Elena N. Komedchikova, Anna S. Sogomonyan, Ekaterina D. Tereshina, Olga A. Kolesnikova and Aziz B. Mirkasymov et al.
Lectin-Modified Magnetic Nano-PLGA for Photodynamic Therapy In Vivo
Reprinted from: *Pharmaceutics* **2022**, 15, 92, doi:10.3390/pharmaceutics15010092 5

Olga Yu. Griaznova, Iaroslav B. Belyaev, Anna S. Sogomonyan, Ivan V. Zelepukin, Gleb V. Tikhonowski and Anton A. Popov et al.
Laser Synthesized Core-Satellite Fe-Au Nanoparticles for Multimodal In Vivo Imaging and In Vitro Photothermal Therapy
Reprinted from: *Pharmaceutics* **2022**, 14, 994, doi:10.3390/pharmaceutics14050994 23

Cristian Iacoviță, Ionel Fizeșan, Stefan Nitica, Adrian Florea, Lucian Barbu-Tudoran and Roxana Dudric et al.
Silica Coating of Ferromagnetic Iron Oxide Magnetic Nanoparticles Significantly Enhances Their Hyperthermia Performances for Efficiently Inducing Cancer Cells Death In Vitro
Reprinted from: *Pharmaceutics* **2021**, 13, 2026, doi:10.3390/pharmaceutics13122026 43

Guilherme A. Soares, Gabriele M. Pereira, Guilherme R. Romualdo, Gabriel G. A. Biasotti, Erick G. Stoppa and Andris F. Bakuzis et al.
Biodistribution Profile of Magnetic Nanoparticles in Cirrhosis-Associated Hepatocarcinogenesis in Rats by AC Biosusceptometry
Reprinted from: *Pharmaceutics* **2022**, 14, 1907, doi:10.3390/pharmaceutics14091907 75

Guan-Ling Lu, Ya-Chi Lin, Ping-Ching Wu and Yen-Chin Liu
The Surface Amine Group of Ultrasmall Magnetic Iron Oxide Nanoparticles Produce Analgesia in the Spinal Cord and Decrease Long-Term Potentiation
Reprinted from: *Pharmaceutics* **2022**, 14, 366, doi:10.3390/pharmaceutics14020366 95

Sandor I. Bernad, Vlad Socoliuc, Daniela Susan-Resiga, Izabell Crăciunescu, Rodica Turcu and Etelka Tombácz et al.
Magnetoresponsive Functionalized Nanocomposite Aggregation Kinetics and Chain Formation at the Targeted Site during Magnetic Targeting
Reprinted from: *Pharmaceutics* **2022**, 14, 1923, doi:10.3390/pharmaceutics14091923 109

Irena Pashkunova-Martic, Rositsa Kukeva, Radostina Stoyanova, Ivayla Pantcheva, Peter Dorkov and Joachim Friske et al.
Novel Salinomycin-Based Paramagnetic Complexes—First Evaluation of Their Potential Theranostic Properties
Reprinted from: *Pharmaceutics* **2022**, 14, 2319, doi:10.3390/pharmaceutics14112319 131

Stefan H. Bossmann, Macy M. Payne, Mausam Kalita, Reece M. D. Bristow, Ayda Afshar and Ayomi S. Perera
Iron-Based Magnetic Nanosystems for Diagnostic Imaging and Drug Delivery: Towards Transformative Biomedical Applications
Reprinted from: *Pharmaceutics* **2022**, 14, 2093, doi:10.3390/pharmaceutics14102093 153

Mónica Cerqueira, Efres Belmonte-Reche, Juan Gallo, Fátima Baltazar and Manuel Bañobre-López
Magnetic Solid Nanoparticles and Their Counterparts: Recent Advances towards Cancer Theranostics
Reprinted from: *Pharmaceutics* **2022**, *14*, 506, doi:10.3390/pharmaceutics14030506 . . . . **179**

Francesca Garello, Yulia Svenskaya, Bogdan Parakhonskiy and Miriam Filippi
Micro/Nanosystems for Magnetic Targeted Delivery of Bioagents
Reprinted from: *Pharmaceutics* **2022**, *14*, 1132, doi:10.3390/pharmaceutics14061132 . . . . **211**

Editorial

On the Road to Precision Medicine: Magnetic Systems for Tissue Regeneration, Drug Delivery, Imaging, and Theranostics

Francesca Garello [1,*], Yulia Svenskaya [2], Bogdan Parakhonskiy [3] and Miriam Filippi [4,*]

1. Molecular and Preclinical Imaging Centers, Department of Molecular Biotechnology and Health Sciences, University of Torino, Via Nizza 52, 10126 Torino, Italy
2. Science Medical Center, Saratov State University, 410012 Saratov, Russia; yulia_svenskaya@mail.ru
3. Faculty of Bioscience Engineering, Ghent University, Coupure Links 653, B-9000 Ghent, Belgium; bogdan.parakhonskiy@ugent.be
4. Soft Robotics Laboratory, Department of Mechanical and Process Engineering, ETH Zurich, 8092 Zurich, Switzerland
* Correspondence: francesca.garello@unito.it (F.G.); miriam.filippi@srl.ethz.ch (M.F.)

Citation: Garello, F.; Svenskaya, Y.; Parakhonskiy, B.; Filippi, M. On the Road to Precision Medicine: Magnetic Systems for Tissue Regeneration, Drug Delivery, Imaging, and Theranostics. *Pharmaceutics* **2023**, *15*, 1812. https://doi.org/10.3390/pharmaceutics15071812

Received: 5 June 2023
Accepted: 20 June 2023
Published: 24 June 2023

Copyright: © 2023 by the authors. Licensee MDPI, Basel, Switzerland. This article is an open access article distributed under the terms and conditions of the Creative Commons Attribution (CC BY) license (https://creativecommons.org/licenses/by/4.0/).

Magnetic systems have always been considered as attractive due to their remarkable versatility. As proof of this, the number of articles concerning magnetic systems has dramatically increased over the past 20 years (Figure 1). In the pharmaceutical industry, magnetic systems are widely investigated both clinically and preclinically. Their strength lies in the ability to use them for imaging, therapy, cell stimulation, or guidance purposes. To target selected cells or pathological areas, magnetic systems can be functionalized with antibodies, peptides, or molecules that recognize specific molecular markers. In addition, these systems can be also loaded with drugs, thus resulting in platforms for simultaneous therapy and diagnosis (i.e., theranostics). This set of features makes magnetic systems attractive tools for personalized and precision medicine, in which therapy selection is tailored to a specific individual. As personalized medicine is expected to make modern medical methods more accessible, improve the control of personal health data, and drive the economic development of health technologies, magnetic systems will render future healthcare more equitable and efficient. Nevertheless, even if the impact of these magnetic tools on future personalized medical approaches is expected to be notable, this relation has not been extolled and carefully analyzed yet.

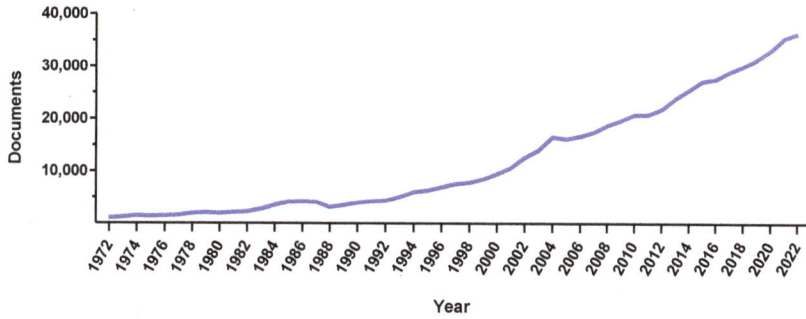

Figure 1. Increasing interest in magnetic systems. Number of documents published in the last 50 years that contain "magnetic systems" as a keyword (Source SCOPUS, accessed on 29 May 2023).

This Special Issue brings together magnetic systems and personalized medicine for the first time. Here, seven research articles and three comprehensive literature reviews were published. The first paper, authored by V.L. Kovalenko et al., reported on the synthesis and validation of multifunctional, biodegradable, and biocompatible magnetic nanoparticles with anti-cancer properties [1]. The nanoparticles were based on Poly Lactic-co-Glycolic Acid (PLGA), which was loaded with magnetite to quantitatively assess its accumulation in various organs and enable magnetic-assisted delivery. IR775 dye was then added to enable fluorescence in vivo imaging and Photodynamic Therapy (PDT), a promising strategy for the treatment of aggressive tumors. Finally, to selectively recognize the glycosylation profile of breast cancer cells, the nanoparticles were functionalized with lectin concanavalin A. Under external light irradiation, the resulting hybrid nanoparticles fully inhibited allograft solid tumor growth in murine models, thus demonstrating a great potential for combined magnetically assisted targeted delivery to tumor sites, tumor bioimaging, and treatment. Similarly, O. Y. Griaznova et al. reported on hybrid multimodal nanoparticles for the noninvasive imaging and therapy of cancer [2]. These researchers prepared bimetallic polyacrylic acid-coated Fe-Au core-satellite nanosystems to combine the magnetic and plasmonic properties of iron and gold nanoparticles in one core–shell. Then, they tested these systems as dual contrast agents for magnetic resonance imaging (MRI) and computed tomography (CT), and as sensitizers for the laser-induced hyperthermia of cancer cells. The particles were synthesized via pulsed laser ablation using the liquids (PLAL) method, which provided them with contaminated-free surfaces. While in the absence of photostimulation, the nanoparticles did not cause cytotoxicity on the cancer cell lines, and under laser irradiation, they induced a 100% cell death. When tested in vivo, the nanoparticles enabled the visualization of tumor boundaries in T_2-weighted MR images and CT scans, thus proving the theranostic potential of this bimetallic system.

C. Iacoviță et al. investigated coating ferromagnetic iron oxide magnetic nanoparticles with silica layers of various thicknesses by adding different amounts of tetraethyl orthosilicate via a reverse microemulsion method [3]. The silica coating improved the nanosystems' colloidal stability without affecting their magnetic properties and increased their biocompatibility and cellular uptake, enhancing their magnetic heating performance. A critical silica layer thickness was found, beyond which the nanoparticles' colloidal stability decreased. The most stable formulation of coated nanoparticles presented an enhanced magnetic hyperthermia performance in water, and their specific absorption rate values increased by almost 1000 W/g_{Fe} compared to bare Fe_3O_4 nanoparticles. Finally, intracellular magnetic hyperthermia experiments revealed that the malignant cells were more sensitive to magnetic hyperthermia treatment compared to the normal ones. It was concluded that the controlled silica coating of ferromagnetic iron oxide nanoparticles enhanced their hyperthermia performance, cellular uptake, and destructive action on cancer cells, thus suggesting another design parameter that could be used to tune the performance of magnetic nanoparticles in biomedical applications.

Three additional research articles reporting on magnetic systems designed for applications to specific pathological scenarios were collected. In more detail, the study by G. A. Soares et al. investigated how cirrhosis-associated hepatocarcinogenesis can alter the biodistribution of hepatic magnetic nanoparticles [4]. The researchers used a multichannel alternate current biosusceptometry system to real-time image the biodistribution of magnetic nanoparticles in the blood circulation and liver. Another customized sensor was used for ex vivo quantification of the MNPs that accumulated in the various organs. The investigated particles were manganese ferrite nanoparticles coated with citrate, which showed pharmacokinetic profiles that were remarkably affected by the pathophysiological factors induced by a cirrhosis state. Since the number of liver monocytes and macrophages did not vary, an altered hepatic blood flow likely caused an abnormal biodistribution and accumulation of nanoparticles injected into the cirrhotic animals. This study draws attention to the host pathophysiological state as a crucial parameter to be considered when designing in vivo nanosystems for optimized interactions between therapeutic agents and the injured target tissue. G.L. Lu et al. instead focused on using iron oxide nanoparticles for pain management [5]. A form of amine-terminated ultrasmall superparamagnetic iron oxide was found to have analgesic effects, even in the absence of any conjugated pain-relieving drug. Nevertheless, the cytotoxic risks of this nanoformulation were unclear. G.L. Lu and coworkers studied the effect of the nanoparticles' toxicity on hippocampal long-term potentiation, revealing a double-side action of the nanosystems. These particles could relieve inflammatory pain in the spinal cord but also induce neurotoxicity in the central brain. The localized administration routes (e.g., intrathecal or intraplantar administration) of the nanoformulation did not elicit a toxicity response during the experiments performed by the authors; however, if the particles could leak to the brain and accumulate at high concentrations, they would have impaired hippocampal long-term potentiation. Finally, S.I. Bernad et al. focused on guiding magnetic nanoparticles through a stented artery model [6]. By applying external magnetic fields with a precisely positioned permanent magnet, polyethylene glycol-coated magnetic nanoparticles were remotely driven to a stented artery region, in which they aggregated and formed chains within and around the implanted stent. The performance of these magnetized, controllable nanoclusters as drug-loaded vectors for stented arteries was assessed. This contribution provided a new tool for site-specific drug delivery approaches based on the uniform-field-induced magnetization effect.

Moving to paramagnetic systems, I. Pashkunova-Martic et al. designed two salinomycin (Sal)-based theranostic paramagnetic probes, comprising either gadolinium(III) or manganese(II) ions [7]. Salinomycin is a natural polyether antibiotic and a highly selective cytotoxic agent that forms complexes with metal(II) ions such as Zn^{2+}, Cu^{2+}, Co^{2+}, and Ni^{2+}. In this work, Sal was conjugated to Gd(III) or Mn(II) to obtain an MRI contrast agent and, to overcome the water insolubility of the two Sal complexes, they loaded them into empty bacterial ghost cells acting as transport vectors. The relaxivity values of the resulting systems were similar to those of the clinical MRI contrast agents, thus indicating that these probes can act as performant bioimaging agents. Considering such a contrast enhancement ability combined with strong cytotoxic activity, cancer theranostics could be the ideal applicative arena for these bio-hybrid nanotools.

In addition to original primary research, this Special Issue includes three comprehensive review articles that discuss the use of magnetic systems for personalized medicine from different points of view. S.H. Bossmann et al. described the current trends in applying iron-incorporated nanosystems to various areas of diagnostic imaging and drug delivery, focusing on cancer treatment, diagnosis, and wound care [8]. M. Cerqueira et al. provided an extensive overview of cancer theranostics with magnetic solid lipid nanoparticles being used in clinical trials [9], thus depicting the emergence of next-gen theranostic nanomedicines with potential for the selective, controlled, and safe delivery of chemotherapy. F. Garello et al. discussed micro and nanosystems for a magnetically targeted delivery of bioagents by describing the different classes of magnetic carriers that can serve as drug delivery platforms and their use in the magnetic guidance and delivery of bio-active agents (e.g., genes, drugs, and cells, etc.) [10]. Moreover, the authors highlighted the emergent applications of magnetic nano/microsystems of a synthetic or bio-hybrid composition (i.e., labeled cells) in regenerative medicine and tissue engineering, describing the magnetized bioagents that can be localized to specific target tissue following systemic or local administration and, remotely, noninvasively stimulated in vivo to promote tissue regenerative processes. Finally, the review presented the latest advances in combining magnetic targeting and imaging technologies to assist with drug delivery.

With the study collection presented in this Special Issue, we hope to guide scientists, clinicians, and other interested parties in designing magnetic systems for precision medicine and theranostics. Additional research based on creative, multidisciplinary collaboration will hopefully accelerate the progress in precision theranostics, aiming to use controllable magnetic nanotools for a better understanding and the personalized treatment of diseases. Hoping that readers will enjoy this Special Issue, we expect that our research content selection will catalyze debate and shape the future development of precision nanomedicine.

Author Contributions: Writing—original draft preparation, F.G.; writing—review and editing, F.G., Y.S., B.P. and M.F. All authors have read and agreed to the published version of the manuscript.

Institutional Review Board Statement: Not applicable.

Informed Consent Statement: Not applicable.

Data Availability Statement: Not applicable.

Conflicts of Interest: The authors declare no conflict of interest.

References

1. Kovalenko, V.L.; Komedchikova, E.N.; Sogomonyan, A.S.; Tereshina, E.D.; Kolesnikova, O.A.; Mirkasymov, A.B.; Iureva, A.M.; Zvyagin, A.V.; Nikitin, P.I.; Shipunova, V.O. Lectin-Modified Magnetic Nano-PLGA for Photodynamic Therapy In Vivo. *Pharmaceutics* **2022**, *15*, 92. [CrossRef] [PubMed]
2. Griaznova, O.Y.; Belyaev, I.B.; Sogomonyan, A.S.; Zelepukin, I.V.; Tikhonowski, G.V.; Popov, A.A.; Komlev, A.S.; Nikitin, P.I.; Gorin, D.A.; Kabashin, A.V.; et al. Laser Synthesized Core-Satellite Fe-Au Nanoparticles for Multimodal In Vivo Imaging and In Vitro Photothermal Therapy. *Pharmaceutics* **2022**, *14*, 994. [CrossRef] [PubMed]
3. Iacoviță, C.; Fizeșan, I.; Nitica, S.; Florea, A.; Barbu-Tudoran, L.; Dudric, R.; Pop, A.; Vedeanu, N.; Crisan, O.; Tetean, R.; et al. Silica Coating of Ferromagnetic Iron Oxide Magnetic Nanoparticles Significantly Enhances Their Hyperthermia Performances for Efficiently Inducing Cancer Cells Death In Vitro. *Pharmaceutics* **2021**, *13*, 2026. [CrossRef] [PubMed]
4. Soares, G.A.; Pereira, G.M.; Romualdo, G.R.; Biasotti, G.G.A.; Stoppa, E.G.; Bakuzis, A.F.; Baffa, O.; Barbisan, L.F.; Miranda, J.R.A. Biodistribution Profile of Magnetic Nanoparticles in Cirrhosis-Associated Hepatocarcinogenesis in Rats by AC Biosusceptometry. *Pharmaceutics* **2022**, *14*, 1907. [CrossRef] [PubMed]
5. Lu, G.-L.; Lin, Y.-C.; Wu, P.-C.; Liu, Y.-C. The Surface Amine Group of Ultrasmall Magnetic Iron Oxide Nanoparticles Produce Analgesia in the Spinal Cord and Decrease Long-Term Potentiation. *Pharmaceutics* **2022**, *14*, 366. [CrossRef] [PubMed]
6. Bernad, S.I.; Socoliuc, V.; Susan-Resiga, D.; Crăciunescu, I.; Turcu, R.; Tombácz, E.; Vékás, L.; Ioncica, M.C.; Bernad, E.S. Magnetoresponsive Functionalized Nanocomposite Aggregation Kinetics and Chain Formation at the Targeted Site during Magnetic Targeting. *Pharmaceutics* **2022**, *14*, 1923. [CrossRef] [PubMed]
7. Pashkunova-Martic, I.; Kukeva, R.; Stoyanova, R.; Pantcheva, I.; Dorkov, P.; Friske, J.; Hejl, M.; Jakupec, M.; Hohagen, M.; Legin, A.; et al. Novel Salinomycin-Based Paramagnetic Complexes—First Evaluation of Their Potential Theranostic Properties. *Pharmaceutics* **2022**, *14*, 2319. [CrossRef] [PubMed]
8. Bossmann, S.H.; Payne, M.M.; Kalita, M.; Bristow, R.M.D.; Afshar, A.; Perera, A.S. Iron-Based Magnetic Nanosystems for Diagnostic Imaging and Drug Delivery: Towards Transformative Biomedical Applications. *Pharmaceutics* **2022**, *14*, 2093. [CrossRef] [PubMed]
9. Cerqueira, M.; Belmonte-Reche, E.; Gallo, J.; Baltazar, F.; Bañobre-López, M. Magnetic Solid Nanoparticles and Their Counterparts: Recent Advances towards Cancer Theranostics. *Pharmaceutics* **2022**, *14*, 506. [CrossRef] [PubMed]
10. Garello, F.; Svenskaya, Y.; Parakhonskiy, B.; Filippi, M. Micro/Nanosystems for Magnetic Targeted Delivery of Bioagents. *Pharmaceutics* **2022**, *14*, 1132. [CrossRef] [PubMed]

Disclaimer/Publisher's Note: The statements, opinions and data contained in all publications are solely those of the individual author(s) and contributor(s) and not of MDPI and/or the editor(s). MDPI and/or the editor(s) disclaim responsibility for any injury to people or property resulting from any ideas, methods, instructions or products referred to in the content.

Article

Lectin-Modified Magnetic Nano-PLGA for Photodynamic Therapy In Vivo

Vera L. Kovalenko [1,†], Elena N. Komedchikova [1,†], Anna S. Sogomonyan [2], Ekaterina D. Tereshina [1], Olga A. Kolesnikova [1], Aziz B. Mirkasymov [2], Anna M. Iureva [1], Andrei V. Zvyagin [2], Petr I. Nikitin [3] and Victoria O. Shipunova [1,2,4,*]

1. Moscow Institute of Physics and Technology, 9 Institutskiy Per., 141701 Dolgoprudny, Russia
2. Shemyakin-Ovchinnikov Institute of Bioorganic Chemistry, Russian Academy of Sciences, 16/10 Miklukho-Maklaya St., 117997 Moscow, Russia
3. Prokhorov General Physics Institute, Russian Academy of Sciences, 38 Vavilov Street, 119991 Moscow, Russia
4. Nanobiomedicine Division, Sirius University of Science and Technology, 1 Olympic Ave., 354340 Sochi, Russia
* Correspondence: viktoriya.shipunova@phystech.edu
† These authors contributed equally to this work.

Abstract: The extreme aggressiveness and lethality of many cancer types appeal to the problem of the development of new-generation treatment strategies based on smart materials with a mechanism of action that differs from standard treatment approaches. The targeted delivery of nanoparticles to specific cancer cell receptors is believed to be such a strategy; however, there are no targeted nano-drugs that have successfully completed clinical trials to date. To meet the challenge, we designed an alternative way to eliminate tumors in vivo. Here, we show for the first time that the targeting of lectin-equipped polymer nanoparticles to the glycosylation profile of cancer cells, followed by photodynamic therapy (PDT), is a promising strategy for the treatment of aggressive tumors. We synthesized polymer nanoparticles loaded with magnetite and a PDT agent, IR775 dye (mPLGA/IR775). The magnetite incorporation into the PLGA particle structure allows for the quantitative tracking of their accumulation in different organs and the performing of magnetic-assisted delivery, while IR775 makes fluorescent in vivo bioimaging as well as light-induced PDT possible, thus realizing the theranostics concept. To equip PLGA nanoparticles with targeting modality, the particles were conjugated with lectins of different origins, and the flow cytometry screening revealed that the most effective candidate for breast cancer cell labeling is ConA, a lectin from *Canavalia ensiformis*. In vivo experiments showed that after i.v. administration, mPLGA/IR775–ConA nanoparticles efficiently accumulated in the allograft tumors under the external magnetic field; produced a bright fluorescent signal for in vivo bioimaging; and led to 100% tumor growth inhibition after the single session of PDT, even for large solid tumors of more than 200 mm^3 in BALB/c mice. The obtained results indicate that the mPLGA/IR775 nanostructure has great potential to become a highly effective oncotheranostic agent.

Keywords: lectin; PLGA; ConA; magnetic polymer nanoparticles; MPQ; allografts; photodynamic therapy; IR775; image-guided therapy

1. Introduction

Nanobiotechnology opens up new possibilities in the diagnostics and treatment of socially significant diseases, including oncological ones [1–4]. A number of drugs based on nanoparticles have already been approved for clinical applications [5]. For example, liposomal forms of chemotherapeutic drugs, such as doxorubicin, penetrate the tumor through the EPR effect (enhanced permeability and retention effect) [6]. However, for the majority of human solid tumors, in contrast to laboratory rodents, the EPR effect is much less pronounced, and passive delivery is not as efficient as expected [7].

To solve this problem, different methods of targeted delivery are now being developed, e.g., targeted drug delivery, magnetic delivery, or combination delivery strategies, thus implementing a "magic bullet" concept. Namely, magnetically enhanced nanoparticle delivery is one of the mainstream directions of modern nanobiotechnology. Magnetite incorporation into nanoparticle structure allows for non-invasively visualizing different pathologies within the organism with different MRI regimens [8,9], quantitatively assessing the nanoparticle accumulations in different tissues in fundamental applications [10], and performing a magnetically-guided drug delivery using an external source of electromagnetic field near the region of interest, e.g., the tumor site [11,12]. Different magnetite-based nanoparticles are already used in clinical applications as MRI contrasting agents, thus proving the effectiveness of this kind of nanoparticle for biomedicine (ferucarbotran for hepatocellular carcinoma or ferumoxide for the imaging of mononuclear phagocyte systems) [13].

Another actively developing approach is the targeted delivery strategy of different drugs to the tumor site. To impart cancer-cell-targeting modality to the nanoparticle, the particle surface is equipped with different recognizing molecules. Different proteins are traditionally used for the targeted delivery of nanoparticles to cancer cells: antibodies and their derivatives [14,15]; scaffold polypeptides of various natures, such as DARPins or affibodies [16–20]; transferrin [21]; lactoferrin [22]; and various peptides [23]. However, despite the variety of tools for targeted drug delivery and the number of fundamental works devoted to this topic, there are currently no targeted nanomedications approved for human administration.

This is due both to the difficulty of marketing new compounds and the insufficient efficiency of existing candidate nanoformulations. The described problems require the development of new smart nanosystems with a fundamentally different mechanism of action that selectively affects cancer cells. In particular, we believe that it is necessary to develop nanoformulations that would target the tumor with greater efficiency and in which the anti-cancer agent would have maximum cytotoxicity, but being activated only in the tumor area in order to reduce side effects: hepatotoxicity and cardiotoxicity.

One alternative way to affect cancer cells is to target their glycosylation profile that differs from that of normal ones [24–29]. Aberrant glycosylation is one of the significant hallmarks of cancer cells that has contributed to cancer progression, angiogenesis, and metastasis, and therefore is a promising target for drug delivery [30]. In this regard, proteins that specifically bind carbohydrate residues in the composition of cell membrane proteins present a promising alternative to existing targeting molecules [31,32]. In particular, various lectins, including those of plant origin, are able to specifically and reversibly bind carbohydrate residues in other biomolecules. We do believe that lectin-modified nanoparticles combined with magnetically guided delivery present a promising alternative to the existing cell targeting strategies for the implementation of different types of cancer therapy: chemotherapy, photothermal therapy, photodynamic therapy, and so on.

In particular, one of the effective methods of fighting tumors is photodynamic therapy (PDT) based on the conversion of external electromagnetic radiation into reactive oxygen species (ROS) that are harmful to cancer cells. The main advantage of PDT in comparison to, e.g., chemotherapy or surgery, is the full non-invasiveness and the activation only with light at the site of action, thus making undesirable side effects negligible.

We believe the polymer PLGA nanoparticles to be the most effective matrixes for the PDT sensitizers incorporation, since the copolymer of lactic and glycolic acids is fully biocompatible, biodegradable, and makes possible slow release of PDT sensitizer from the nanoparticle at the tumor site. PLGA-based nanoformulations are already approved for human use and confirmed their efficacy for biomedicine [5,33].

Currently, PLGA nanoparticles conjugated with lectins were shown to significantly improve the delivery and cytotoxic efficacy of chemotherapy in vitro and in vivo. Namely, wheat-germ-agglutinin-conjugated PLGA nanoparticles loaded with paclitaxel were found to be more effective at cell cycle arrest of A549 cells compared to free paclitaxel in vitro and had greater tumor doubling time (25 vs. 11 days) in vivo [34]. Despite these promising results, currently, there are only a few studies on the use of lectin-conjugated PLGA nanoparticles for the delivery of chemotherapeutic drugs, and there are no data on the use of such promising lectin-conjugated PLGA nanoparticle formulations as hybrid magnetic-PLGA nanoparticles enabling MRI tracking and nanoparticles for photodynamic drug delivery, which allow for a non-invasive tumor treatment.

Here, we demonstrate the rational design of lectin-modified nano-PLGA loaded with PDT sensitizer (IR775) and magnetite, namely, mPLGA/IR775-lectin. The developed nanoparticles realized a combined targeted delivery strategy, namely, the combination of active targeting mediated by lectin and magnetically guided targeting mediated by magnetite loaded inside nanoparticles and an external magnetic field. The anti-cancer efficacy is mediated by the loading of the photosensitizer, IR775 dye, into PLGA nanoparticle structure. The heptamethine cyanine derivative IR775 is one of the most effective photosensitizers, especially inside polymer nanoparticles, due to its hydrophobic nature; under external light irradiation in the near-infrared window in biological tissue, it produces reactive oxygen species (ROS) leading to cancer cell death, thus affecting only the specific tissue site only on demand under light exposure [35–37].

The synthesized particles were shown to be effective theranostic agents, realizing effective magnetically assisted targeted delivery, tumor bioimaging, and treatment under external light irradiation. The in vitro and in vivo functionality of these nanoparticles were thoroughly tested, and 100% inhibition of allograft solid tumor growth was shown, thus confirming the great potential of the developed nanoformulation for bioimaging and PDT.

2. Materials and Methods

2.1. Magnetite Synthesis

The magnetite cores incorporated into PLGA were synthesized as described by us previously with some modifications [9,38,39]. A total of 0.86 g $FeCl_2 \cdot 4H_2O$ and 2.36 g $FeCl_3 \cdot 6H_2O$ were dissolved in 40 mL of Milli-Q water, then 3 mL of 25% NH_4OH was added, and the resulting mixture was rapidly stirred. The mixture was then incubated in a water bath at 80 °C for 2 h with subsequent cooling to room temperature. The resulting magnetic fraction was washed using the magnetic separation and then 10 mL of 0.5 M HNO_3 was added, with the mixture incubated for 5 min and magnetically separated through the magnet application for 5 min. The supernatant was then removed, and 10 mL of H_2O was added to the magnetic fraction. Then, this fraction was sonicated and magnetically separated again. The procedure was repeated thrice, and 3 magnetic fractions were collected.

Next, 600 µL of the third magnetite fraction was mixed with 2 mL of oleic acid and 5 mL of chloroform and sonicated for 3 min. Next, 15 mL of chloroform was added, and the mixture was sonicated again for 2 min. After that, 300 µL of 2M NaOH was added, and the mixture was sonicated again for 1 min. The oleic-acid-stabilized magnetite was then centrifuged for 1 h at 20,000× g at 20 °C, and the resulting mixture was resuspended in 3 mL of dichloromethane.

2.2. mPLGA/IR775 Nanoparticle Synthesis

Hybrid magnetic polymer nanoparticles were synthesized by the "oil-in-water" microemulsion method followed by solvent evaporation based on the method previously described by us with modifications [40–42]. A total of 12 mg PLGA (RG 858 S, lactide/glycolide 85:15, MW 190–240 kDa, Sigma, Darmstadt, Germany), 100 µL of oleate-stabilized magnetite, and 250 µg IR775 (Sigma, Darmstadt, Germany) in 300 µL of dichloromethane were added to 3 mL of 3% aqueous PVA (Mowiol 4-88, Sigma, Darmstadt, Germany) supplemented with chitosan oligosaccharide lactate with a final concentration of 1 g/L (5 kDa, Sigma, Darmstadt, Germany). The mixture was treated with ultrasound for 1 min. After solvent evaporation, the particles were washed thrice with centrifugation for 5 min at 5000× g and finally resuspended in 10 mM HEPES (pH 7.0).

2.3. Electron Microscopy Analysis

The morphology of nanoparticles was studied with a MAIA3 Tescan (Tescan, Brno-Kohoutovice, Czech Republic) scanning electron microscope. The sample of nanoparticles at 10 µg/mL was applied on a silicon wafer on carbon film and air-dried, followed by SEM examination at an accelerating voltage of 7 kV. Magnetite was analyzed with transmitting electron microscopy as well at an accelerating voltage of 70 kV (JEOL JEM2100Plus transmitting electron microscope). The scanning electron microscopy images were processed in ImageJ as follows: the sizes of 250 nanoparticles were measured, and then the average particle size was calculated.

2.4. Spectroscopy

The absorbance spectra of nanoparticles were measured with the CLARIOstar microplate reader (BMG Labtech, Ortenberg, Germany) within the 300–1000 nm range for particles at 0.3 g/L in H_2O with subsequent subtraction of the absorbance spectrum of pure H_2O.

Fluorescent excitation and emission spectra were registered with an Infinite M1000 Pro microplate reader (Tecan, Grödig, Austria). The excitation spectrum was recorded at $\lambda em = 800$ nm within the 400–785 nm range, and the emission was recorded at $\lambda ex = 700$ nm within the 715–850 nm range. mPLGA/IR775 nanoparticles at 0.3 g/L in water were used for analysis. The obtained spectra were normalized. The dye loading efficiency was calculated by measuring the fluorescence intensity of mPLGA/IR775 nanoparticles at 0.3 g/L in a 50%/50% DMSO/H_2O mixture with $\lambda ex = 730$ nm and $\lambda em = 800$ nm. A calibration curve for free IR775 in the 50%/50% DMSO/H_2O mixture was used to determine dye loading content.

The generation of reactive oxygen species was measured as follows: nanoparticles at 1 g/L in H_2O were mixed with CM-H_2DCFDA to obtain a final concentration of 5 µM, irradiated with 808 nm laser for 5 min, and centrifuged for 5 min at $10,000\times g$; following this, the fluorescence of 100 µL of supernatant was measured with $\lambda ex = 500$ nm and $\lambda em = 525$ nm using a CLARIOstar microplate reader (BMG Labtech, Ortenberg, Germany), and the autofluorescence of blank wells was subtracted.

2.5. DLS Measurements

The hydrodynamic sizes and ζ-potentials of nanoparticles were analyzed using Zeta-Sizer Nano ZS (Malvern Instruments Ltd., Worcestershire, UK) in 10 mM HEPES (pH 7.0).

2.6. Chemical Conjugation

mPLGA/IR775 nanoparticles were conjugated to proteins using carbodiimide chemistry with EDC/sulfo-NHS as crosslinking agents. A total of 10 mg of mPLGA/IR775 were incubated with 5 mg EDC and 0.5 mg sulfo-NHS in 200 µL of 0.1 M MES buffer for 20 min at room temperature. Next, nanoparticles were purified from the excess of chemicals through centrifugation for 5 min at $5000\times g$. Following this, 200 µL of protein at 1 g/L in 0.1 M HEPES (pH 6.0) was added to the nanoparticles, and the suspension was sonicated for 10 s and incubated for at least 4 h at room temperature. Then, nanoparticles were purified from non-bound protein with triple centrifugation and resuspended in 10 mM HEPES (pH 7.0).

2.7. Cell Culture

EMT6/P, EA.hy926, and NIH/3T3 cells (Shemyakin-Ovchinnikov Institute RAS, Molecular Immunology Laboratory collection) were cultured in DMEM medium (HyClone, Logan, UT, USA) supplemented with 10% FBS (HyClone, Logan, UT, USA), penicillin/streptomycin (PanEko, Moscow, Russia), and 2 mM L-glutamine (PanEko, Moscow, Russia) at 37 °C and 5% CO_2.

2.8. Flow Cytometry

ROS generation was assessed with the flow cytometry test. A total of 0.2×10^6 EMT6/P cells in 200 µL PBS with 1% BSA were incubated with 0.2 g/L of mPLGA/IR775 nanoparticles and ROS sensor according to the manufacturer recommendations (Total Reactive Oxygen Species (ROS) Assay Kit 520 nm, Invitrogen, Thermo Fisher Scientific Inc., Waltham, MA, USA) for 30 min at 37 °C. Next, cells were washed from non-bound nanoparticles by centrifugation for 5 min at $100 \times g$ and irradiated with an 808 nm laser (600 mW) for 2 min. Thirty minutes later, the fluorescence of cells was analyzed with the Novocyte 3000 VYB flow cytometer (ACEA Biosciences, San Diego, CA, USA) in the BL1 channel (excitation laser 488 nm, emission filter 530/30 nm).

Particle binding efficiency was studied as follows. 0.2×10^6 cells in 200 µL PBS with 1% BSA were incubated with nanoparticle conjugates, then washed from non-bound nanoparticles by centrifugation for 5 min at $100 \times g$, and the fluorescence of cells was analyzed with the Accuri C6 (BD) flow cytometer in the FL4 channel (λex = 640 nm and λem = 675/25 nm).

2.9. Fluorescent Microscopy

EMT6/P cells were seeded on 96-well plates at 20×10^3 cells per well in 150 µL of full culture media and cultured overnight. Next, nanoparticles were added to wells to obtain a final concentration of 50 µg/mL, and plates were incubated for 30 min at 37 °C and 5% CO_2. Next, wells were washed thrice with full culture media and samples were analyzed using an epifluorescent Zeiss microscope at the following conditions: λex = 595–645 nm, λem = 670–725 nm.

2.10. Cell Toxicity Study

The cytotoxicity of nanoparticles was investigated using a resazurin-based cytotoxicity test. A total of 2×10^6 EMT6/P cells in 1 mL of full culture media were incubated with nanoparticles at different concentrations for 30 min at 37 °C with 5% CO_2. Then, the cells were washed from non-bound nanoparticles by centrifugation for 3 min at $100 \times g$. Next, cells were irradiated with an 808 nm laser (600 mW) for various time intervals. The cells were then diluted with full culture medium, and 5×10^3 cells in 150 µL of media were seeded into the wells of 96-well plates and cultured for 72 h at 37 °C with 5% CO_2. Then, the medium was removed from the wells, 100 µL of resazurin solution (13 mg/L in PBS) was added, and samples were incubated for 3 h at 37 °C and 5% CO_2. The fluorescence of wells was then measured using the CLARIOstar microplate reader (BMG Labtech, Ortenberg, Germany) at wavelengths of λex = 570 nm and λem = 600 nm. Data are presented as percentages from non-treated and non-irradiated cells.

2.11. Tumor Bearing Mice

Female BALB/c mice of 22–25 g weight were purchased from the Puschino Animal Facility (Shemyakin-Ovchinnikov Institute of Bioorganic Chemistry Russian Academy of Sciences, Pushchino branch of the Institute, Pushchino, Russia) and maintained at the Vivarium of the Shemyakin-Ovchinnikov Institute of Bioorganic Chemistry Russian Academy of Sciences (Moscow, Russia). All procedures were approved by the Institutional Animal Care and Use Committee (IACUC) of the Shemyakin-Ovchinnikov Institute of Bioorganic Chemistry Russian Academy of Sciences (Moscow, Russia) according to the IACUC protocol # 299 (1 January 2020–31 December 2022).

The animals were anesthetized with a mixture of Zoletil (Virbac, Carros, France) and Rometar (Bioveta, Ivanovice na Hané, Czech Republic) at a dose of 25/5 mg/kg.

To create tumor allografts, mice were injected with 4×10^6 EMT6/P cells in 100 µL of culture media into the right flank. The tumor volume was measured with a caliper and calculated using the following formula $V = a^2 \cdot b/2$, where a is the tumor width and b is the tumor length.

2.12. MPQ-Measurements

BALB/c mice with EMT6/P tumor were i.v. injected with 1 mg of nanoparticles through the retroorbital sinus injection with or without magnet application to the tumor site. Then, 24 h later, animals were euthanized with cervical dislocation, and organs were extracted, weighed, and fixed in the 4% formalin. The organ magnetic signal was measured using our original MPQ device [32,38].

2.13. Bioimaging Study

For bioimaging experiments, mice were anesthetized with tiletamine-HCl/zolazepam-HCl/xylazine-HCl and then placed into a chamber of LumoTrace FLUO bioimaging tomograph (Abisense, Sochi, Russia). The images were acquired with fluorescence excitation = 730 nm and 780LP nm emission filter and exposure of 1000 ms.

2.14. In Vivo PDT

For photodynamic therapy in vivo, mice with tumors of about 190 ± 67 mm^3 were selected and randomly divided into 3 groups (n = 6) and treated as follows: the first group served as the control non-treated cohort, and the second and third groups received the i.v. injections of 1 mg of nanoparticles with the magnet application to the tumor site. Then, 24 h later, mice from the third group only were irradiated with an 808 nm laser (600 mW) for 30 min. The dynamics of tumor growth were then measured every 2 days with a caliper.

3. Results

3.1. Design of the Experiment

The efficient PDT agent based on a polymer matrix was designed as follows (Figure 1): PLGA-based nanoparticles loaded with magnetite and a photodynamic sensitizer, IR775 dye, were synthesized (mPLGA/IR775). Magnetite in the composition of these nanoparticles makes it possible to quantitatively trace particle accumulation in the organism and realize magnetic-assisted drug delivery, while IR775 dye allows for the visualization of nanoparticle accumulation non-invasively in vivo, and, under external light irradiation, producing ROS, thus performing PDT.

Next, these nanoparticles were equipped with a spectrum of plant lectins using standard chemical conjugation. BSA-conjugated PLGA nanoparticles and unmodified PLGA nanoparticles were used as control samples. The efficiency of binding of these conjugates to cells was then screened by flow cytometry. Three types of cell lines were studied: cancer cells (mouse mammary breast cancer cells, EMT6/P), immortalized fibroblasts as a model of normal cells (NIH/3T3 cells), and immortalized endothelial cells (EA.hy926). Those particles were selected that have the maximum binding to cancer cells and endothelial cells while having a minimal effect on non-transformed fibroblast cells. Flow cytometry tests showed that such a leader is nanoparticles equipped with concanavalin A (mPLGA/IR775-ConA).

These nanoparticles were i.v. injected into mice with allograft tumors without and with magnetic delivery by applying a magnet to the site of the tumor. In vivo imaging tests showed a significant accumulation of nanoparticles in the tumor site, and exposure to the 808 nm laser led to complete remission in all mice from the light-treated group, thus realizing the concept of theranostics, namely, the diagnostics and the therapy using the same nanoformulation.

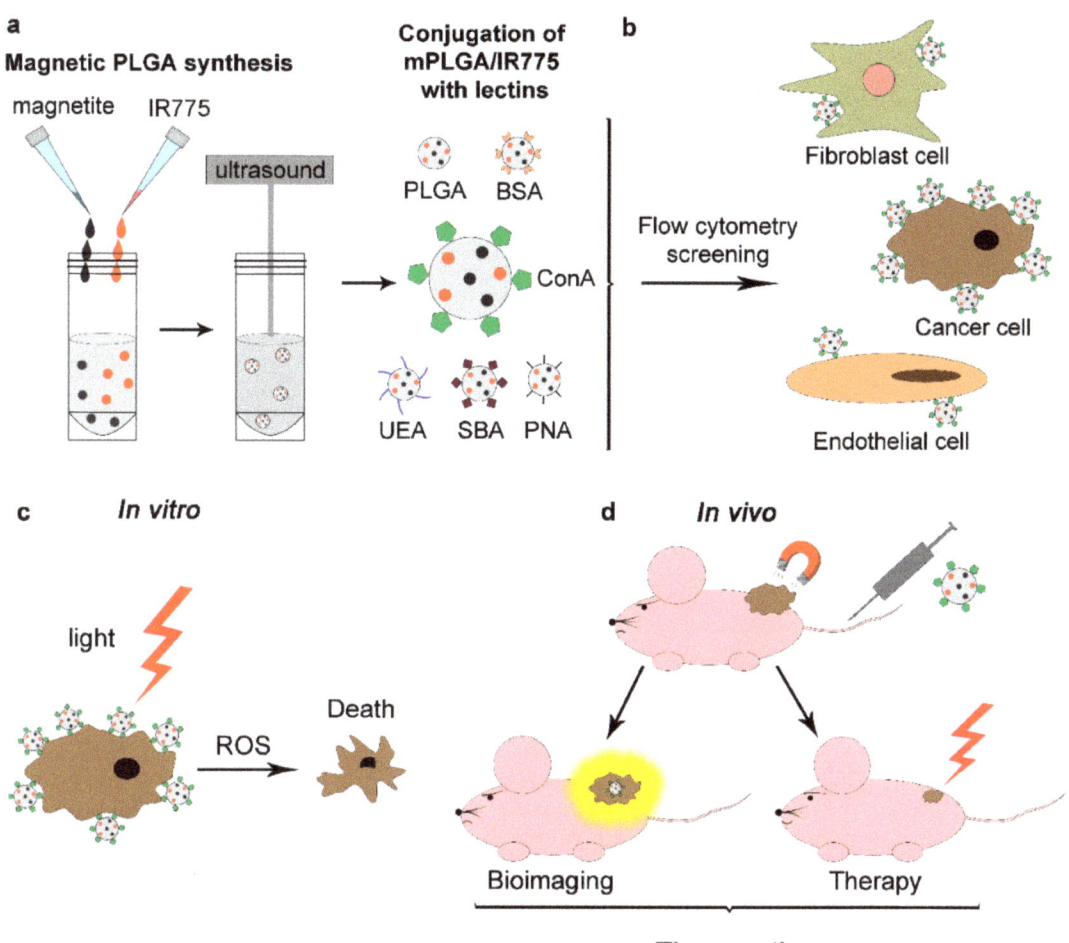

Figure 1. Lectin-conjugated magnetic PLGA for photodynamic therapy: scheme of the experiment. (**a**) Magnetite and photosensitizer IR775-loaded PLGA nanoparticles were synthesized via the "oil–water" microemulsion method. (**b**) As-synthesized mPLGA/IR775 nanoparticles were conjugated with different lectins and screened with flow cytometry for the most effective binding to cancer cells. (**c**) In vitro cell toxicity studies: cancer cells labeled with mPLGA/IR775-ConA were irradiated with an external IR light leading to cancer cell death. (**d**) In vivo diagnostics and therapy study: magnet-assisted delivery of mPLGA/IR775-ConA allowed for effective visualization and elimination of the tumor, thus realizing the theranostics concept.

3.2. Synthesis and Characterization of Magnetite-Loaded PLGA Nanoparticles

PLGA nanoparticles loaded with magnetite and photodynamic sensitizer IR775 dye were synthesized by the "oil–water" microemulsion method as schematically shown in Figure 1a. Prior to magnetic polymer nanoparticle synthesis, magnetic cores were synthesized and stabilized for the effective incorporation into the PLGA matrix.

Magnetite was synthesized according to the protocol in the Materials and Methods section using the standard co-precipitation technique with some modifications. The size of these magnetic cores determined by scanning electron microscopy was found to be 14.7 ± 5.5 nm, having a form close to spherical (Figure 2a,c).

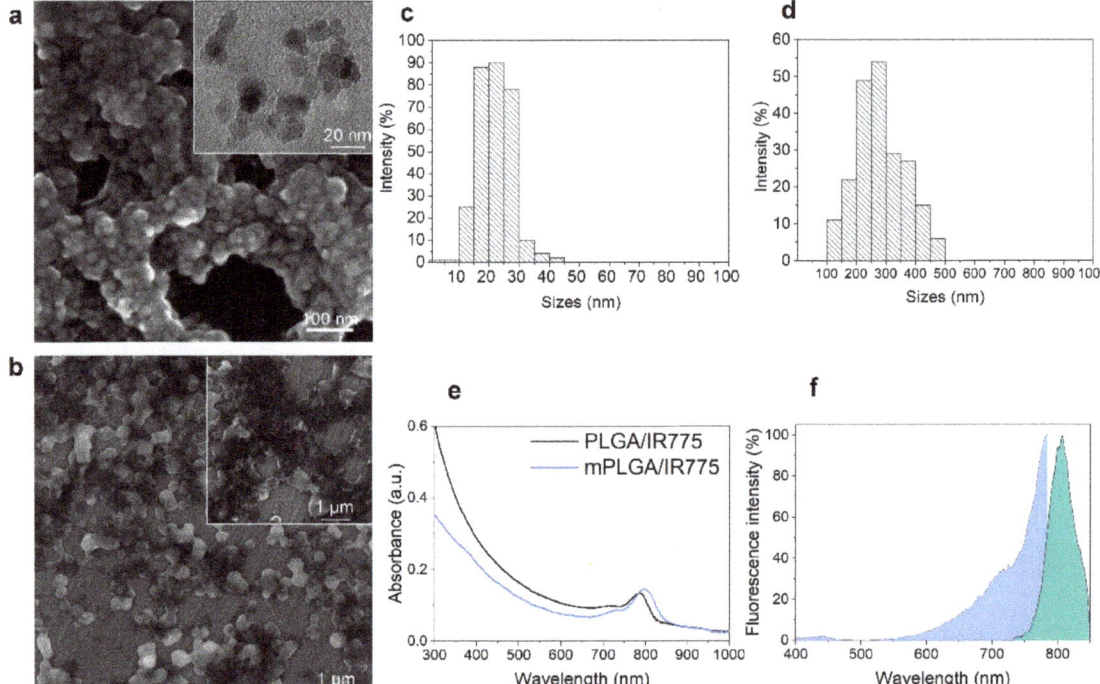

Figure 2. Synthesis and characterization of mPLGA/IR775 nanoparticles. (**a**) Scanning electron microscopy of magnetite synthesized with the coprecipitation technique. The inset is transmitting electron microscopy of magnetite. (**b**) Scanning electron microscopy of PLGA nanoparticle loaded with magnetite and IR775 photosensitizer. The SEM inset demonstrates the incorporation of magnetite: red arrows show white dots corresponding to magnetic cores. (**c**) Physical size distribution of magnetite obtained via SEM image processing. (**d**) Physical size distribution of PLGA nanoparticles loaded with magnetite and IR775 (mPLGA/IR775). (**e**) UV–VIS absorbance spectrum of mPLGA/IR775 and PLGA/IR775 nanoparticles. (**f**) Excitation and emission fluorescence spectra of mPLGA/IR775 nanoparticles: blue—fluorescence excitation (measured with λ_{em} = 800 nm), fluorescence emission (measured with λ_{ex} = 700 nm).

Next, particles were coated with sodium oleate/oleic acid. It was found that the particles coated with oleic acid were more stable than those coated with sodium oleate. Sodium hydroxide solution was added to the obtained magnetite to stabilize it, which resulted in the colloidal stability of the magnetite for at least one month.

The resultant oleic-acid-stabilized magnetic nanoparticles were used for PLGA microemulsion synthesis. As synthesized mPLGA/IR775 nanoparticles were analyzed with scanning electron microscopy, which confirmed the incorporation of magnetite. The analysis of nanoparticle sizes based on SEM image processing showed that particles were 281.2 ± 83.9 nm with spherical form (Figure 2b,d).

The absorbance spectra of mPLGA/IR775 as well as non-magnetic PLGA/IR775 nanoparticles are presented in Figure 2e: the pronounced peak near the 800 nm region corresponded to the incorporated IR775 dye in the nanoparticle structure, while the decrease in the spectrum intensity for mPLGA/IR775 nanoparticles in the UV region most likely occurred due to the increased light scattering by the magnetic cores within the nanoparticle.

The effective incorporation of IR775 to mPLGA/IR775 nanoparticles was confirmed by measuring the excitation and emission fluorescence spectra of as-synthesized nanoparticles

(Figure 2f). Data presented in Figure 2f confirm that the fluorescence of nanoparticles corresponded to the expected fluorescence of nanoparticles according to IR775 dye loading.

The measurement of IRR75 loading efficiency showed that the dye loading was equal to 16.8 ± 0.8 µg of IR775 per 1 mg of mPLGA/IR775, thus demonstrating 80.6% loading efficiency during the synthesis (250 µg of IR775 were used in synthesis per 12 mg of PLGA).

The effective reactive oxygen species (ROS) production of PLGA/IR775 and mPLGA/IR775 under light irradiation was evaluated using a CM-H_2DCFDA sensor pre-mixed with nanoparticles. The mixture was light-irradiated, particles were centrifuged, and the fluorescence of the supernatant was measured (Figure 3a). The data presented in Figure 3a confirm that particles produced ROS under light irradiation. Interestingly, the ROS production was 2.5 higher for nanoparticles loaded with magnetite and IR775 in comparison with particles loaded with IR775 only under light irradiation (Figure 3a).

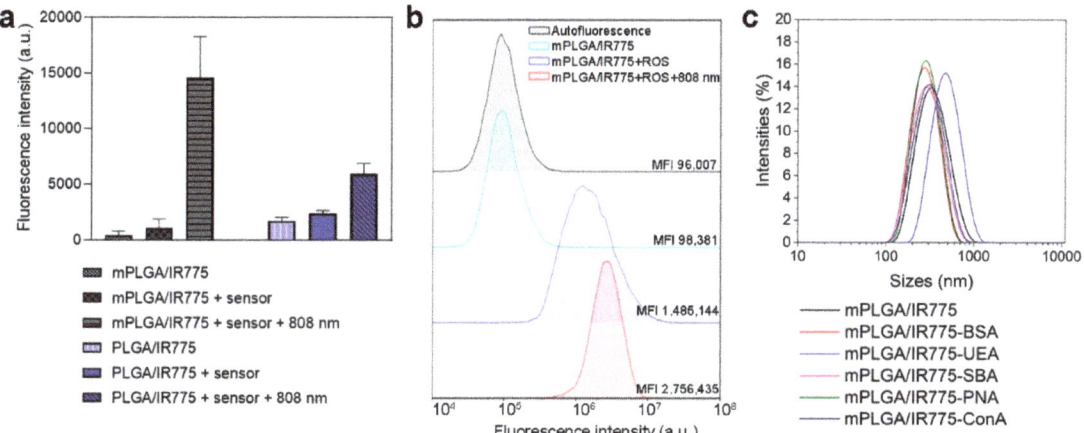

Figure 3. Characterization of mPLGA/IR775 nanoparticles as nanoagents for PDT. (**a**) ROS generation study of nanoparticles under the irradiation with an 808 nm laser: magnetic mPLGA/IR775 and non-magnetic PLGA/IR775 nanoparticles were incubated with ROS sensor and irradiated with 808 nm light and centrifuged, and the fluorescence of the supernatant was measured (λex = 500 nm, λem = 525 nm). (**b**) Flow cytometry study of ROS generation. Cells were incubated with mPLGA/IR775 nanoparticles and ROS sensor and irradiated with 808 nm laser for 10 min. Histograms represent cell populations in the fluorescent channel corresponding to the ROS sensor fluorescence (λex = 488 nm, λem = 525/20 nm). (**c**) mPLGA/IR775 were conjugated to different lectins as well as to BSA and characterized with the dynamic light scattering method, confirming the colloidal stability of nanoparticles after chemical conjugation.

Next, ROS generation was studied with flow cytometry tests. EMT6/P cells were incubated with mPLGA/IR775 nanoparticles and a total ROS sensor. Cells were washed from non-bound nanoparticles, then irradiated with an 808 nm laser and analyzed with flow cytometry in the fluorescent channel corresponding to ROS sensor fluorescence. The data presented in Figure 3b confirm that mPLGA/IR775 are effective PDT sensitizers since these particles generate ROS under light irradiation.

3.3. Chemical Conjugation of mPLGA/IR775 to Lectins

As-synthesized mPLGA/IR775 were conjugated to different plant lectins using standard carbodiimide chemistry with EDC/sulfo-NHS as cross-linking agents. Ulex Europaeus Agglutinin (UEA), soybean agglutinin (SBA), peanut agglutinin (PNA), and concanavalin A lectin from Canavalia ensiformis (Con A) were used for chemical conjugation. Particles conjugated with BSA as well as pristine non-conjugated nanoparticles served as control nanoparticles for the in vitro tests.

The stability of mPLGA/IR775–lectin conjugates was studied both visually and with the dynamic light scattering method.

Cumulant analysis showed the size of nanoparticles and their conjugated to be 314.3 ± 114.6 nm for mPLGA/IR775, 258.7 ± 80.67 nm for mPLGA/IR775-BSA, 442.3 ± 206.4 nm for mPLGA/IR775-UEA, 274.1 ± 94.63 nm for mPLGA/IR775-SBA, 276.0 ± 111.1 nm for mPLGA/IR775-PNA, and 279.4 ± 107 nm for mPLGA/IR775-ConA (Figure 3c), thus confirming the difference between conjugated and non-conjugated nanoparticles and proving their colloidal stability.

Electrophoretic light scattering measurements showed nanoparticle ζ-potentials to be equal to −20.2 ± 10.4 mV for mPLGA/IR775, −21.3 ± 4.84 mV for mPLGA/IR775-BSA, −19.9 ± 6.42 mV for mPLGA/IR775-UEA, −20.3 ± 6.91 mV for mPLGA/IR775-SBA, −19.2 ± 6.05 mV for mPLGA/IR775-PNA, and −17.7 ± 5.99 mV for mPLGA/IR775-ConA. Since the isoelectric points of conjugated lectins are in the slightly acidic region, pI 4.5–5.1 for UEA, pI 5.8–6.0 for SBA, pI 5.5–6.5 for PNA, and pI 4.5–5.5 for ConA, all of the studied lectins were charged neutrally or slightly negatively in buffers with pH 7.0. A possible problem would be that when a significant number of proteins are conjugated to the nanoparticle surface, the particles can aggregate due to a strong increase in surface charge. However, measurements showed only a slight increase in ζ-potentials due to the optimal concentration of proteins during conjugation, and the particles retained colloidal stability.

3.4. mPLGA/IR775–Lectin Interaction with Cells: Flow Cytometry Screening and In Vitro Cytotoxicity Tests

The obtained spectrum of synthesized nanoparticles modified with lectins was studied for interaction with different cell lines by flow cytometry. Thus, the binding of PLGA conjugates to the following cell lines was studied: (1) EMT6/P mouse breast cancer cells, (2) vascular endothelial cells EA.hy926, (3) non-cancerous cells—immortalized fibroblasts NIH/3T3 cells. The aim of this test was to find such conjugates that would more efficiently bind to cancer breast cells (EMT6/P) and endothelial cells (EA.hy926) to block tumor angiogenesis processes and at the same time have minimal binding to non-transformed cells—in this case, fibroblasts—for minimizing the negative effect on healthy tissues in the organism.

The described cells were incubated with the nanoparticle conjugates, washed from non-bound nanoparticles, and analyzed by flow cytometry in the fluorescence channel corresponding to the fluorescence of nanoparticles (Figure 4a). The binding efficiency was quantified using the *stain index* calculated as *stain index* = $(MFI_{pos} - MFI_{auto})/2SD_{auto}$, where MFI_{pos} and MFI_{auto} are medians of the fluorescence intensity of labeled and non-labeled cell populations, respectively, and SD_{auto} is the standard deviation of MFI_{auto}. The obtained calculations are shown in Figure 4b. The data presented indicate that it was the conjugates of nanoparticles with ConA that most effectively bound to mouse breast cancer cells EMT6/P and endothelial cells EA.hy926 while minimally affecting fibroblasts NIH/3T3. Thus, we suggested that mPLGA/IR775-ConA are leader nanoparticles for breast cancer PDT.

Next, we studied the binding of these nanoparticles to cells by fluorescence microscopy. The data presented in Figure 4c indicate a more efficient penetration of mPLGA/IR775-ConA nanoparticles into cells compared to control mPLGA/IR775-BSA nanoparticles. Since breast cancer cells overexpressed more mannose N-glycan units on the membrane proteins than that of normal cells, the binding of mPLGA/IR775-ConA occurred due to the interaction of ConA with mannose residues on several proteins types in cancer cells. N-glycosylation, the sequential addition of complex sugars to adhesion proteins, ion channels, and receptors, is one of the most frequent protein glycosylation processes, leading to the fact that ConA-based nanoparticles interact both with internalizing and non-internalizing proteins on the cell membrane, resulting in the nanoparticle accumulation in different cell compartments, resting on the cell membrane, localizing in endosomes and in the cytoplasm [43–45]. Maybe this feature is one of the most interesting aspects of lectin-modified nanoparticles' interaction with cancer cells gathering the attention of researchers, since this fact allows for the affecting several types of cell compartments simultaneously, which is very important on the road to the development of multifunctional cancer-fighting strategies, especially those based on PDT.

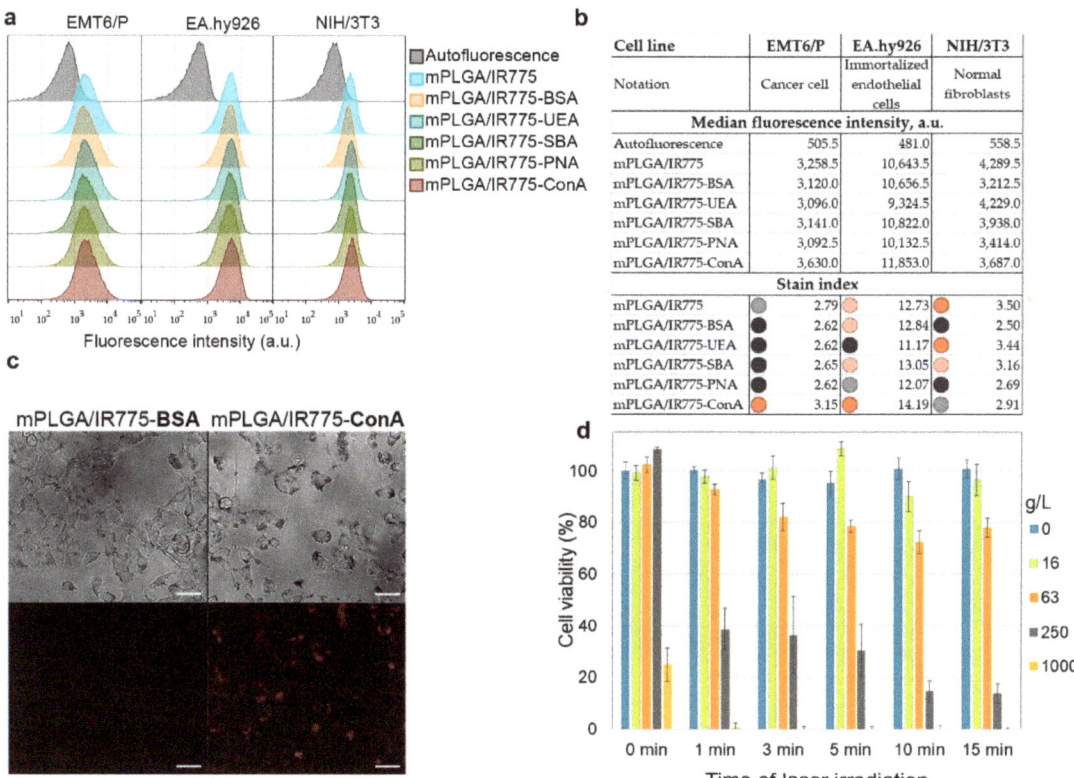

Figure 4. In vitro study of mPLGA/IR775 interaction with cells. (**a**) Flow cytometry assay on evaluation of the binding efficiency of lectin-modified nano-PLGA. EMT6/P cells were labeled with lectin-conjugated particles and analyzed with flow cytometry in the fluorescence channel corresponding to IR775 fluorescence. (**b**) Flow cytometry histograms are accompanied by the median fluorescence intensity data and stain index calculation for three cell lines. (**c**) Fluorescent microscopy of EMT6/P cell labeled with mPLGA/IR775-BSA and mPLGA/IR775-ConA nanoparticles. Scale bars, 50 µm. (**d**) Resazurin test on the evaluation of mPLGA/IR775-ConA nanoparticle cell toxicity under 808 nm laser light irradiation.

The effective interaction of nanoparticles with cells made it possible to achieve effective light-induced cytotoxicity. Thus, EMT6/P cells were incubated with nanoparticles, washed from non-bound ones, and irradiated with an 808 nm laser. Seventy-two hours later, a resazurin-based test was performed, which showed that, in a wide range of concentrations, nanoparticles are cytotoxic only when exposed to external radiation, and the cytotoxicity depends on the irradiation time (Figure 4d).

3.5. Diagnostic Properties of mPLGA/IR-775-ConA Nanoparticles: In Vivo and Ex Vivo Biodistribution Study

As-synthesized hybrid polymer magnetic nanoparticles of mPLGA/IR775-ConA were then i.v. injected into tumor-bearing mice in order to study their diagnostic and therapeutic capabilities. BALB/c mice with EMT6/P allograft tumors in the right flanks were used for this study. The magnetically guided targeted delivery of nanoparticles was realized, and 24 h after the nanoparticle injection, the accumulation of nanoparticles was studied both in vivo and ex vivo.

In vivo bioimaging was performed using the fluorescent properties of IR775 dye inside the nanoparticles using the LumoTrace Fluo (Abisense, Sochi, Russia) bioimaging device. Two targeting modes were compared, namely, magnetically guided delivery with the magnet applied near the tumor site and delivery without the applied magnet. Figure 5a presents fluorescent images of mice that were injected with nanoparticles without and with the use of an external magnetic field. These images indicate an increase in delivery efficiency when exposed to a magnetic field. In vivo images were accompanied by ex vivo photographs of the extracted organs, and the intensity of fluorescence of tumors in mice from the group with a magnet was to a certain degree higher.

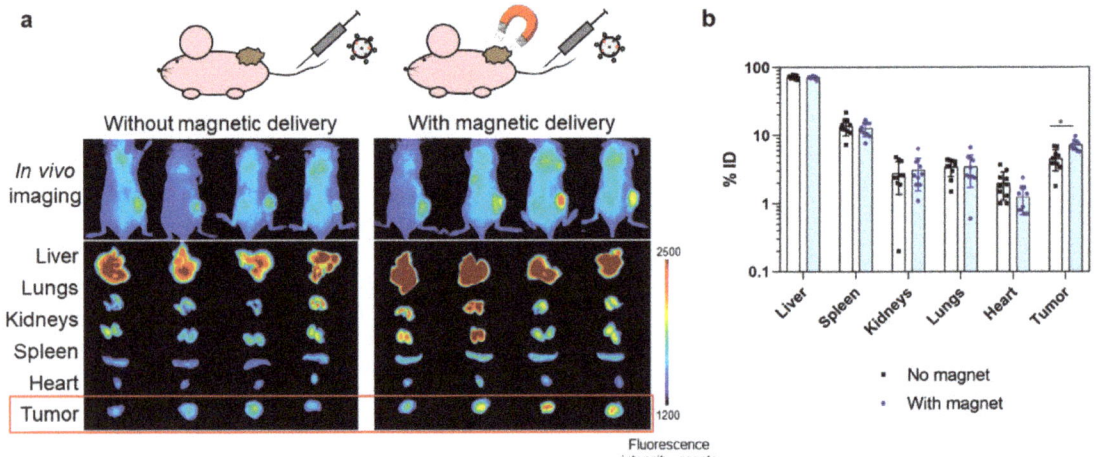

Figure 5. Diagnostic properties of mPLGA/IR-775-ConA nanoparticles: in vivo imaging and ex vivo evaluation of particle accumulation in tumors with and without magnet-assisted delivery. (**a**) Fluorescent images of mice 24 h after the i.v. injection of mPLGA/IR775-ConA with and without magnetic targeting. Data are accompanied by the ex vivo fluorescent images of mice organs (λex = 730 nm, λem = 780 nm). (**b**) Percent of injected dose accumulated in the main organs according to MPQ measurements. Data are presented on a logarithmic scale (n = 10 for each group). * $p < 0.001$.

Since fluorescent images do not provide information on the quantitative biodistribution of nanoparticles in mice organs and are made available only a qualitative picture of particle biodistribution, we studied the accumulation of nanoparticles in organs using our original MPQ method (magnetic particle quantification) [38] to assess the efficacy of magnetic delivery. MPQ is an efficient method for the quantitative measurement of the accumulation of magnetic nanoparticles in the organism. The method has a sensitivity limit of 0.33 ng nanoparticles and is devoid of background signals from liquids and tissues; each measurement takes a few seconds and does not require complex sample preparation or data processing. This is possible due to the way the device works: a magnetic coil generates an alternating electromagnetic field with two components, thereby affecting magnetic particles. Linear magnetic materials (dia- and paramagnetics) respond only at the frequencies of the applied field, whereas non-linear magnetic materials (e.g., superparamagnetic materials) respond at the combinatorial frequencies of the field [32,38,46].

Extracted organs of mice from two groups: with and without magnetic delivery (n = 10 in each group) were measured with the MPQ device in order to obtain particle distribution in organs. The data presented in Figure 5b demonstrate the difference in nanoparticle accumulation in tumors for these groups: 4.5 ± 1.5% vs. 7.3 ± 1.3% of the injected dose for groups without and with the applied magnetic field, respectively. Significant differences between the control group and experimental group determined with two-sided unequal variances t-tests (Welch's test) for two-group comparisons showed the p-value to be equal to 0.0004, thus proving the difference in nanoparticle accumulation in tumors.

Here, we showed that the application of a magnetic field increased the delivery of mPLGA/IR775-ConA nanoparticles into the tumor by 1.6 times. We should emphasize the problem that always arises on the way to the development and verification of human anti-cancer drugs using laboratory mice. Since mice are small rodents and their tumors are correspondingly small, it is quite difficult to predict how targeted nanoparticles will behave in terms of quantitative accumulation in the large human tumor, especially considering the fact that there are no successful cases of introducing targeted nanoparticles into the clinic and there is no way to obtain relevant data from the literature for at least one of the existing targeted nanoparticles. We can only try to model the processes in rodents that maximally reflect the processes of carcinogenesis in humans. For this purpose, the allograft tumor models were used in this study, namely, mouse cancer cells were injected in mice, while the tumors formed for a rather long time period (for almost 2 weeks) and were allowed to grow to a fairly large size, 190 ± 67 mm^3 (thus, corresponding to approximately 1% from total body weight), developing a normal vasculature, thereby minimizing the EPR effect of fast-growing tumors such as in B16 or CT26 models.

However, it is rather presumptuous to suggest that such efficiency and such a difference in the accumulation of nanoparticles with and without a magnetic delivery can be reproduced in human tumors. To shed light on this problem, it is necessary to perform a series of sequential experiments using larger mice, e.g., ICR (CD-1) instead of Balb/c, large rats, and other larger animals, and this is included in our plans for further research.

3.6. Therapeutic Properties of Lectin-Modified Nano-PLGA

The therapeutic properties of lectin-conjugated magnetic polymer nanoparticles were then studied. Mice with allograft EMT6/P tumors were randomly divided into three groups when the tumor sizes reached 190 ± 67 mm^3 (n = 6). Mice were treated as follows: (i) the first group served as the control non-treated cohort, (ii) the second group was treated with mPLGA/IR775-ConA nanoparticles, and (iii) the third group was treated with mPLGA/IR775-ConA nanoparticles followed by light irradiation. Mice received single i.v. injections of 1 mg of mPLGA/IR775-ConA nanoparticles with a magnet applied to the tumor site. Twenty-four hours later, the tumor sites of mice from the third group were irradiated with an 808 nm laser for 30 min (Figure 6b). The tumor growth dynamics that were monitored by caliper measurements every two days are presented in Figure 6d. Data presented in Figure 6a–d confirm the efficacy of mPLGA/IR775-ConA as tumor-targeting agents for light-PDT: tumor growth inhibition was found to be 100% in the light-treated group with complete tumor elimination.

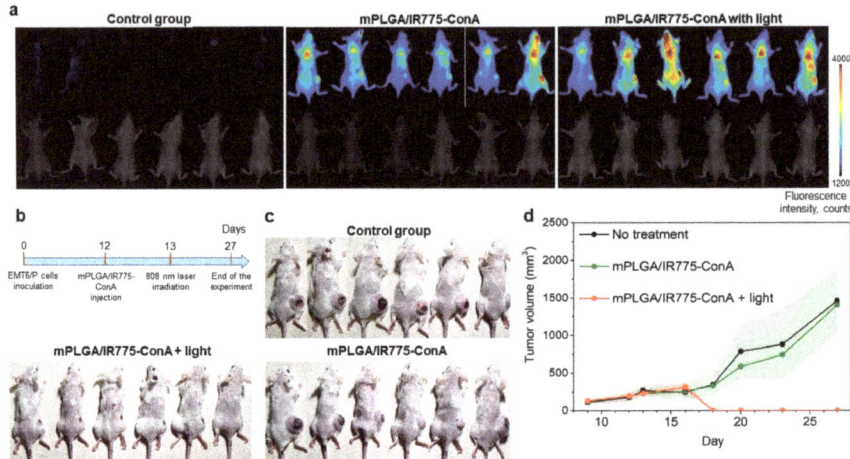

Figure 6. Therapeutic properties of lectin-modified nano-PLGA: mPLGA/IR775-ConA sensitizer-assisted photodynamic therapy of EMT6/P tumors. Mice were i.v. injected with 1 mg of mPLGA/IR775-ConA nanoparticles and 24 h later were irradiated with an 808 nm light (600 mW). (**a**) Evaluation of tumor targeting ability of mPLGA/IR775-ConA nanoparticles: fluorescent images of mice 24 h after the injection of nanoparticles (control group, group that received the nanoparticle injection only, and group that afterward was treated with light irradiation). (**b**) Scheme of the treatment. (**c**) Smartphone camera images at the end of the experiment (day 27): images of mice from all treated groups (n = 6). (**d**) Tumor growth dynamic of mice treated with mPLGA/IR775-ConA nanoparticles.

4. Discussion

The main goal of this work was the development of the most effective nanoagent for large solid tumor PDT. We synthesized polymer nanoparticles loaded with magnetite and the PDT agent, IR775 dye (mPLGA/IR775). The magnetite incorporation into PLGA particle structure allows for quantitative tracking of their accumulation in different organs and performing magnetic-assisted delivery, while IR775 makes fluorescent in vivo bioimaging possible, as well as light-induced PDT. To equip PLGA nanoparticles with targeting modality, the particles were conjugated with lectins of different origins, and the flow cytometry screening revealed that the most effective candidate for breast cancer cell labeling is ConA, a lectin from *Canavalia ensiformis*.

PDT is one of the most promising cancer treatment strategies, being based on the absorption of light by a photosensitizer and the conversion of oxygen into reactive oxygen species (ROS). Previously, various scientific groups have shown that PLGA nanoparticles are excellent nanoagents for PDT. Thus, different PLGA-based formulations loaded with photosensitizers were developed: Rose Bengal [42], indocyanine green [47], zinc phthalocyanine [48,49], and others, and their applications in vitro and in vivo were demonstrated for selective cancer cell destruction under the external light source. For example, Fadel et al. showed the efficacy of PLGA loaded with zinc(II) phthalocyanine (ZnPc) in vivo using Ehrlich ascite carcinoma cells under 880 nm light irradiation [48]. Zhang et al. demonstrated both the bioimaging capabilities of IR775-loaded PLGA using BALB/c mice with CT26 tumors and showed that the therapeutic modalities of particles led to the tumor growth inhibition of 87.28% [50]. The heptamethine cyanine derivative, IR775 dye (2-[2-[2-chloro-3-[2-(1,3-dihydro-1,3,3-trimethyl-2H-indol-2-ylidene)-ethylidene]-1-cyclohexen-1-yl]-ethenyl]-1,3,3-trimethyl-3H-indolium chloride), was shown to be one of the most effective PDT sensitizers, especially within the composition of different nanoparticles due to its hydrophobic nature [35–37].

To achieve targeted delivery, we chemically conjugated PLGA nanoparticles with concanavalin A. ConA is a lectin purified from jack bean (*Canavalia ensiformis*) that specifically binds D-mannose/D-glucose residues on the cell surface [51–53]. Several studies confirmed the ability of ConA-coated nanoparticles of various natures to preferentially bind to cancer cells. Thus, Chen et al. demonstrated that the binding capacity of ConA-conjugated silica–carbon hollow spheres was higher for liver cancer cells than for normal cells in vitro [54]. Martínez-Carmona et al. reported that ConA-conjugated mesoporous silica nanoparticles loaded with doxorubicin specifically bind and kill human osteosarcoma cells in contrast to healthy preosteoblast cells, MC3T3-E1 [55]. Chowdhury et al. showed ConA-conjugated quantum dots loaded with doxorubicin possess higher specific cytotoxicity against HeLa cancer cells compared to normal cells in vitro [53]. Khopade et al. reported ConA-coated multiple emulsion bearing 6-mercaptopurine [56] outperformed uncoated emulsion (ME) and free drug with IC50 0.7, 2.5, and 2.8 µM on murine leukemia cell line L-1210 in vitro. The same study demonstrated that the median survival time of mice treated with ConA ME was superior to that of ME and free drug [56]. In addition, it was shown that concanavalin A can modulate signaling pathways, inducing apoptosis in melanoma A375 cells and autophagy in glioblastoma and hepatoma cells, as well as inhibit proliferation of melanoma B16 cells and fibroblast 3T3 cells [30,57–59]. Hence, concanavalin A is a promising molecule that can not only recognize cancer cells but also lead to the suppression of their vital activity.

It should be also highlighted that ConA, like most lectins, is a homotetramer with one subunit of 26.5 kDa, and thus problems with steric hindrances that constantly arise during the conjugation of nanoparticles with antibodies or peptides are much less pronounced. Even if one of the homotetramer subunits is attached to the surface of the particle, the other three will still be active. In this way, most of the standard problems of chemical conjugation are removed, and the optimization of ligation protocols requires much less effort.

The combination of photodynamic properties, magnetite incorporation, and lectin modification within the single PLGA matrix allows for the obtaining of trifunctional nanoparticles for magnetically assisted targeted drug delivery. Such hybrid mPLGA/IR775-ConA nanoparticles were shown to (i) efficiently target the allograft tumors, (ii) perform magnetically assisted delivery (the efficacy of the delivery was 1.6 times greater under the application of the magnetic field), and (iii) induce tumor elimination under external light irradiation with tumor growth inhibition = 100% in all mice from the treated group.

5. Conclusions

This study is a step forward in the development of new-generation nanomedications for cancer PDT. Biocompatible PLGA nanoparticles are already widely used in clinical applications as well as some methods of PDT. We believe that the combination of lectin-assisted magnetic targeted delivery and all the advantages of PLGA carriers and PDT efficacy would result in a new smart nanoagent for oncotheranostics.

Author Contributions: Conceptualization, V.O.S.; methodology, V.L.K., E.N.K., A.S.S., E.D.T., O.A.K., A.B.M., A.M.I. and P.I.N.; software, V.L.K., E.N.K., A.S.S. and E.D.T.; validation, V.O.S.; formal analysis, V.O.S.; investigation, V.O.S.; resources, P.I.N. and V.O.S.; data curation, V.O.S.; writing—original draft preparation, V.O.S.; visualization, A.S.S.; supervision, V.O.S.; project administration, V.O.S.; funding acquisition, A.V.Z., P.I.N. and V.O.S. All authors have read and agreed to the published version of the manuscript.

Funding: Different aspects and parts of this multidisciplinary research was funded by the Russian Science Foundation grant No. 22-73-10141 (polymer nanoparticle synthesis) and grant No. 21-74-30016 (flow cytometry tests), the Ministry of Science and Higher Education of Russia, strategic academic leadership program "Priority 2030", agreement 075-02-2021-1316, 30.09.2021 (magnetic nanoparticle synthesis, characterization) and Sirius University, project NMB-RND-2120 (ROS generation study).

Institutional Review Board Statement: All animal procedures were approved by the Institutional Animal Care and Use Committee (IACUC) of the Shemyakin-Ovchinnikov Institute of Bioorganic Chemistry Russian Academy of Sciences (Moscow, Russia) according to the IACUC protocol # 299 (1 January 2020–31 December 2022).

Informed Consent Statement: Not applicable.

Data Availability Statement: Samples of nanoparticles are available from the authors.

Acknowledgments: We thank the Applied Genetics Resource Facility of MIPT (075-15-2021-684) for providing access to the scientific equipment (nanoparticle fluorescence and absorbance measurements).

Conflicts of Interest: The authors declare no conflict of interest.

References

1. Jain, V.; Kumar, H.; Anod, H.V.; Chand, P.; Gupta, N.V.; Dey, S.; Kesharwani, S.S. A review of nanotechnology-based approaches for breast cancer and triple-negative breast cancer. *J. Control. Release* **2020**, *326*, 628–647. [CrossRef] [PubMed]
2. Nikitin, M.P.; Zelepukin, I.V.; Shipunova, V.O.; Sokolov, I.L.; Deyev, S.M.; Nikitin, P.I. Enhancement of the blood-circulation time and performance of nanomedicines via the forced clearance of erythrocytes. *Nat. Biomed. Eng.* **2020**, *4*, 717–731. [CrossRef] [PubMed]
3. Sokolov, I.L.; Cherkasov, V.R.; Tregubov, A.A.; Buiucli, S.R.; Nikitin, M.P. Smart materials on the way to theranostic nanorobots: Molecular machines and nanomotors, advanced biosensors, and intelligent vehicles for drug delivery. *Biochim. Biophys. Acta Gen. Subj.* **2017**, *1861*, 1530–1544. [CrossRef] [PubMed]
4. Tregubov, A.A.; Nikitin, P.I.; Nikitin, M.P. Advanced Smart Nanomaterials with Integrated Logic-Gating and Biocomputing: Dawn of Theranostic Nanorobots. *Chem. Rev.* **2018**, *118*, 10294–10348. [CrossRef] [PubMed]
5. Zhong, H.; Chan, G.; Hu, Y.; Hu, H.; Ouyang, D. A Comprehensive Map of FDA-Approved Pharmaceutical Products. *Pharmaceutics* **2018**, *10*, 263. [CrossRef]
6. Gabizon, A.A. Pegylated liposomal doxorubicin: Metamorphosis of an old drug into a new form of chemotherapy. *Cancer Investig.* **2001**, *19*, 424–436. [CrossRef]
7. Danhier, F. To exploit the tumor microenvironment: Since the EPR effect fails in the clinic, what is the future of nanomedicine? *J. Control. Release* **2016**, *244*, 108–121. [CrossRef]
8. Dimov, I.P.; Tous, C.; Li, N.; Häfeli, U.O.; Martel, S.; Soulez, G. Future Advances in Diagnosis and Drug Delivery in Interventional Radiology Using MR Imaging-Steered Theranostic Iron Oxide Nanoparticles. *J. Vasc. Interv. Radiol.* **2021**, *32*, 1292–1295.e1. [CrossRef] [PubMed]
9. Nikitin, M.P.; Shipunova, V.O.; Deyev, S.M.; Nikitin, P.I. Biocomputing based on particle disassembly. *Nat. Nanotechnol.* **2014**, *9*, 716–722. [CrossRef]
10. Nikitin, M.P.; Vetoshko, P.M.; Brusentsov, N.A.; Nikitin, P.I. Highly sensitive room-temperature method of non-invasive in vivo detection of magnetic nanoparticles. *J. Magn. Magn. Mater.* **2009**, *321*, 1658–1661. [CrossRef]
11. Sharifianjazi, F.; Irani, M.; Esmaeilkhanian, A.; Bazli, L.; Asl, M.S.; Jang, H.W.; Kim, S.Y.; Ramakrishna, S.; Shokouhimehr, M.; Varma, R.S. Polymer incorporated magnetic nanoparticles: Applications for magnetoresponsive targeted drug delivery. *Mater. Sci. Eng. B* **2021**, *272*, 115358. [CrossRef]
12. Knežević, N.Ž.; Gadjanski, I.; Durand, J.-O. Magnetic nanoarchitectures for cancer sensing, imaging and therapy. *J. Mater. Chem. B* **2019**, *7*, 9–23. [CrossRef]
13. Alromi, D.A.; Madani, S.Y.; Seifalian, A. Emerging Application of Magnetic Nanoparticles for Diagnosis and Treatment of Cancer. *Polymers* **2021**, *13*, 4146. [CrossRef] [PubMed]
14. Shipunova, V.O.; Zelepukin, I.V.; Stremovskiy, O.A.; Nikitin, M.P.; Care, A.; Sunna, A.; Zvyagin, A.V.; Deyev, S.M. Versatile Platform for Nanoparticle Surface Bioengineering Based on SiO2-Binding Peptide and Proteinaceous Barnase*Barstar Interface. *ACS Appl. Mater. Interfaces* **2018**, *10*, 17437–17447. [CrossRef] [PubMed]
15. Shipunova, V.O.; Kolesnikova, O.A.; Kotelnikova, P.A.; Soloviev, V.D.; Popov, A.A.; Proshkina, G.M.; Nikitin, M.P.; Deyev, S.M. Comparative Evaluation of Engineered Polypeptide Scaffolds in HER2-Targeting Magnetic Nanocarrier Delivery. *ACS Omega* **2021**, *6*, 16000–16008. [CrossRef]
16. Kotelnikova, P.A.; Shipunova, V.O.; Aghayeva, U.F.; Stremovskiy, O.A.; Nikitin, M.P.; Novikov, I.A.; Schulga, A.A.; Deyev, S.M.; Petrov, R.V. Synthesis of Magnetic Nanoparticles Stabilized by Magnetite-Binding Protein for Targeted Delivery to Cancer Cells. *Dokl. Biochem. Biophys.* **2018**, *481*, 198–200. [CrossRef]
17. Novoselova, M.; Chernyshev, V.S.; Schulga, A.; Konovalova, E.V.; Chuprov-Netochin, R.N.; Abakumova, T.O.; German, S.; Shipunova, V.O.; Mokrousov, M.D.; Prikhozhdenko, E.; et al. Effect of Surface Modification of Multifunctional Nanocomposite Drug Delivery Carriers with DARPin on Their Biodistribution In Vitro and In Vivo. *ACS Appl. Bio Mater.* **2022**, *5*, 2976–2989. [CrossRef]
18. Shipunova, V.O.; Deyev, S.M. Artificial Scaffold Polypeptides As an Efficient Tool for the Targeted Delivery of Nanostructures In Vitro and In Vivo. *Acta Nat.* **2022**, *14*, 54–72. [CrossRef]
19. Proshkina, G.M.; Shramova, E.I.; Shilova, M.V.; Zelepukin, I.V.; Shipunova, V.O.; Ryabova, A.V.; Deyev, S.M.; Kotlyar, A.B. DARPin_9-29-Targeted Gold Nanorods Selectively Suppress HER2-Positive Tumor Growth in Mice. *Cancers* **2021**, *13*, 5235. [CrossRef]
20. Shipunova, V.O.; Shramova, E.I.; Schulga, A.A.; Shilova, M.V.; Deyev, S.M.; Proshkina, G.M. Delivery of Barnase to Cells in Liposomes Functionalized by Her2-Specific DARPin Module. *Russ. J. Bioorg. Chem.* **2020**, *46*, 1156–1161. [CrossRef]

21. Daniels, T.R.; Bernabeu, E.; Rodríguez, J.A.; Patel, S.; Kozman, M.; Chiappetta, D.A.; Holler, E.; Ljubimova, J.Y.; Helguera, G.; Penichet, M.L. The transferrin receptor and the targeted delivery of therapeutic agents against cancer. *Biochim. Biophys. Acta* **2012**, *1820*, 291–317. [CrossRef] [PubMed]
22. Kumari, S.; Kondapi, A.K. Lactoferrin nanoparticle mediated targeted delivery of 5-fluorouracil for enhanced therapeutic efficacy. *Int. J. Biol. Macromol.* **2017**, *95*, 232–237. [CrossRef] [PubMed]
23. Laakkonen, P.; Vuorinen, K. Homing peptides as targeted delivery vehicles. *Integr. Biol. (Camb)* **2010**, *2*, 326–337. [CrossRef] [PubMed]
24. Thomas, D.; Rathinavel, A.K.; Radhakrishnan, P. Altered glycosylation in cancer: A promising target for biomarkers and therapeutics. *Biochim. Biophys. Acta Rev. Cancer* **2021**, *1875*, 188464. [CrossRef]
25. Munkley, J. The glycosylation landscape of pancreatic cancer. *Oncol. Lett.* **2019**, *17*, 2569–2575. [CrossRef]
26. Pinho, S.S.; Reis, C.A. Glycosylation in cancer: Mechanisms and clinical implications. *Nat. Rev. Cancer* **2015**, *15*, 540–555. [CrossRef]
27. Costa, A.F.; Campos, D.; Reis, C.A.; Gomes, C. Targeting Glycosylation: A New Road for Cancer Drug Discovery. *Trends Cancer* **2020**, *6*, 757–766. [CrossRef]
28. Pinho, S.S.; Carvalho, S.; Marcos-Pinto, R.; Magalhães, A.; Oliveira, C.; Gu, J.; Dinis-Ribeiro, M.; Carneiro, F.; Seruca, R.; Reis, C.A. Gastric cancer: Adding glycosylation to the equation. *Trends Mol. Med.* **2013**, *19*, 664–676. [CrossRef]
29. Ferreira, J.A.; Magalhães, A.; Gomes, J.; Peixoto, A.; Gaiteiro, C.; Fernandes, E.; Santos, L.L.; Reis, C.A. Protein glycosylation in gastric and colorectal cancers: Toward cancer detection and targeted therapeutics. *Cancer Lett.* **2017**, *387*, 32–45. [CrossRef]
30. Munkley, J.; Elliott, D.J. Hallmarks of glycosylation in cancer. *Oncotarget* **2016**, *7*, 35478–35489. [CrossRef]
31. Shipunova, V.O.; Nikitin, M.P.; Zelepukin, I.V.; Nikitin, P.I.; Deyev, S.M.; Petrov, R.V. A comprehensive study of interactions between lectins and glycoproteins for the development of effective theranostic nanoagents. *Dokl. Biochem. Biophys.* **2015**, *464*, 315–318. [CrossRef] [PubMed]
32. Shipunova, V.O.; Nikitin, M.P.; Belova, M.M.; Deyev, S.M. Label-free methods of multiparametric surface plasmon resonance and MPQ-cytometry for quantitative real-time measurements of targeted magnetic nanoparticles complexation with living cancer cells. *Mater. Today Commun.* **2021**, *29*, 102978. [CrossRef]
33. Osorno, L.L.; Brandley, A.N.; Maldonado, D.E.; Yiantsos, A.; Mosley, R.J.; Byrne, M.E. Review of Contemporary Self-Assembled Systems for the Controlled Delivery of Therapeutics in Medicine. *Nanomaterials* **2021**, *11*, 278. [CrossRef] [PubMed]
34. Mo, Y.; Lim, L.-Y. Paclitaxel-loaded PLGA nanoparticles: Potentiation of anticancer activity by surface conjugation with wheat germ agglutinin. *J. Control. Release* **2005**, *108*, 244–262. [CrossRef] [PubMed]
35. Sun, Y.; Fang, K.; Hu, X.; Yang, J.; Jiang, Z.; Feng, L.; Li, R.; Rao, Y.; Shi, S.; Dong, C. NIR-light-controlled G-quadruplex hydrogel for synergistically enhancing photodynamic therapy via sustained delivery of metformin and catalase-like activity in breast cancer. *Mater. Today Bio* **2022**, *16*, 100375. [CrossRef]
36. Wu, W.; Bazan, G.C.; Liu, B. Conjugated-Polymer-Amplified Sensing, Imaging, and Therapy. *Chem* **2017**, *2*, 760–790. [CrossRef]
37. Zhou, Z.; Zhang, B.; Wang, S.; Zai, W.; Yuan, A.; Hu, Y.; Wu, J. Perfluorocarbon Nanoparticles Mediated Platelet Blocking Disrupt Vascular Barriers to Improve the Efficacy of Oxygen-Sensitive Antitumor Drugs. *Small* **2018**, *14*, e1801694. [CrossRef]
38. Shipunova, V.O.; Nikitin, M.P.; Nikitin, P.I.; Deyev, S.M. MPQ-cytometry: A magnetism-based method for quantification of nanoparticle-cell interactions. *Nanoscale* **2016**, *8*, 12764–12772. [CrossRef]
39. Shipunova, V.O.; Nikitin, M.P.; Lizunova, A.A.; Ermakova, M.A.; Deyev, S.M.; Petrov, R.V. Polyethyleneimine-coated magnetic nanoparticles for cell labeling and modification. *Dokl. Biochem. Biophys.* **2013**, *452*, 245–247. [CrossRef]
40. Shipunova, V.O.; Komedchikova, E.N.; Kotelnikova, P.A.; Zelepukin, I.V.; Schulga, A.A.; Proshkina, G.M.; Shramova, E.I.; Kutscher, H.L.; Telegin, G.B.; Kabashin, A.V.; et al. Dual Regioselective Targeting the Same Receptor in Nanoparticle-Mediated Combination Immuno/Chemotherapy for Enhanced Image-Guided Cancer Treatment. *ACS Nano* **2020**, *14*, 12781–12795. [CrossRef]
41. Shipunova, V.O.; Kovalenko, V.L.; Kotelnikova, P.A.; Sogomonyan, A.S.; Shilova, O.N.; Komedchikova, E.N.; Zvyagin, A.V.; Nikitin, M.P.; Deyev, S.M. Targeting Cancer Cell Tight Junctions Enhances PLGA-Based Photothermal Sensitizers' Performance In Vitro and In Vivo. *Pharmaceutics* **2021**, *14*, 43. [CrossRef] [PubMed]
42. Shipunova, V.O.; Sogomonyan, A.S.; Zelepukin, I.V.; Nikitin, M.P.; Deyev, S.M. PLGA Nanoparticles Decorated with Anti-HER2 Affibody for Targeted Delivery and Photoinduced Cell Death. *Molecules* **2021**, *26*, 3955. [CrossRef]
43. Ahangama Liyanage, L.; Harris, M.S.; Cook, G.A. In Vitro Glycosylation of Membrane Proteins Using N-Glycosyltransferase. *ACS Omega* **2021**, *6*, 12133–12142. [CrossRef] [PubMed]
44. Reily, C.; Stewart, T.J.; Renfrow, M.B.; Novak, J. Glycosylation in health and disease. *Nat. Rev. Nephrol.* **2019**, *15*, 346–366. [CrossRef] [PubMed]
45. Christiansen, M.N.; Chik, J.; Lee, L.; Anugraham, M.; Abrahams, J.L.; Packer, N.H. Cell surface protein glycosylation in cancer. *Proteomics* **2014**, *14*, 525–546. [CrossRef] [PubMed]
46. Zelepukin, I.V.; Yaremenko, A.V.; Shipunova, V.O.; Babenyshev, A.V.; Balalaeva, I.V.; Nikitin, P.I.; Deyev, S.M.; Nikitin, M.P. Nanoparticle-based drug delivery via RBC-hitchhiking for the inhibition of lung metastases growth. *Nanoscale* **2019**, *11*, 1636–1646. [CrossRef] [PubMed]
47. Miele, D.; Sorrenti, M.; Catenacci, L.; Minzioni, P.; Marrubini, G.; Amendola, V.; Maestri, M.; Giunchedi, P.; Bonferoni, M.C. Association of Indocyanine Green with Chitosan Oleate Coated PLGA Nanoparticles for Photodynamic Therapy. *Pharmaceutics* **2022**, *14*, 1740. [CrossRef]

48. Fadel, M.; Kassab, K.; Fadeel, D.A. Zinc phthalocyanine-loaded PLGA biodegradable nanoparticles for photodynamic therapy in tumor-bearing mice. *Lasers Med. Sci.* **2010**, *25*, 283–292. [CrossRef] [PubMed]
49. Ricci-Júnior, E.; Marchetti, J.M. Zinc(II) phthalocyanine loaded PLGA nanoparticles for photodynamic therapy use. *Int. J. Pharm.* **2006**, *310*, 187–195. [CrossRef]
50. Zhang, Z.; Wang, L.; Ding, Y.; Wu, J.; Hu, Y.; Yuan, A. Synergy of hypoxia relief and chromatin remodeling to overcome tumor radiation resistance. *Biomater. Sci.* **2020**, *8*, 4739–4749. [CrossRef]
51. Sumner, J.B.; Howell, S.F. Identification of Hemagglutinin of Jack Bean with Concanavalin A. *J. Bacteriol.* **1936**, *32*, 227–237. [CrossRef] [PubMed]
52. Bhattacharyya, L.; Brewer, C.F. Isoelectric focusing studies of concanavalin A and the lentil lectin. *J. Chromatogr.* **1990**, *502*, 131–142. [CrossRef] [PubMed]
53. Dutta Chowdhury, A.; Ganganboina, A.B.; Tsai, Y.-C.; Chiu, H.-C.; Doong, R.-A. Multifunctional GQDs-Concanavalin A@Fe3O4 nanocomposites for cancer cells detection and targeted drug delivery. *Anal. Chim. Acta* **2018**, *1027*, 109–120. [CrossRef] [PubMed]
54. Chen, Y.-C.; Chiu, W.-T.; Chen, J.-C.; Chang, C.-S.; Hui-Ching Wang, L.; Lin, H.-P.; Chang, H.-C. The photothermal effect of silica-carbon hollow sphere-concanavalin A on liver cancer cells. *J. Mater. Chem. B* **2015**, *3*, 2447–2454. [CrossRef] [PubMed]
55. Martínez-Carmona, M.; Lozano, D.; Colilla, M.; Vallet-Regí, M. Lectin-conjugated pH-responsive mesoporous silica nanoparticles for targeted bone cancer treatment. *Acta Biomater.* **2018**, *65*, 393–404. [CrossRef]
56. Khopade, A.J.; Nandakumar, K.S.; Jain, N.K. Lectin-functionalized multiple emulsions for improved cancer therapy. *J. Drug Target.* **1998**, *6*, 285–292. [CrossRef]
57. Liu, B.; Li, C.-y.; Bian, H.-j.; Min, M.-w.; Chen, L.-f.; Bao, J.-k. Antiproliferative activity and apoptosis-inducing mechanism of Concanavalin A on human melanoma A375 cells. *Arch. Biochem. Biophys.* **2009**, *482*, 1–6. [CrossRef]
58. Pratt, J.; Roy, R.; Annabi, B. Concanavalin-A-induced autophagy biomarkers requires membrane type-1 matrix metalloproteinase intracellular signaling in glioblastoma cells. *Glycobiology* **2012**, *22*, 1245–1255. [CrossRef]
59. Lei, H.-Y.; Chang, C.-P. Lectin of Concanavalin A as an anti-hepatoma therapeutic agent. *J. Biomed. Sci.* **2009**, *16*, 10. [CrossRef]

Disclaimer/Publisher's Note: The statements, opinions and data contained in all publications are solely those of the individual author(s) and contributor(s) and not of MDPI and/or the editor(s). MDPI and/or the editor(s) disclaim responsibility for any injury to people or property resulting from any ideas, methods, instructions or products referred to in the content.

Article

Laser Synthesized Core-Satellite Fe-Au Nanoparticles for Multimodal In Vivo Imaging and In Vitro Photothermal Therapy

Olga Yu. Griaznova [1,2,3,†], Iaroslav B. Belyaev [1,3,†], Anna S. Sogomonyan [1,3], Ivan V. Zelepukin [1,3,*], Gleb V. Tikhonowski [3], Anton A. Popov [3], Aleksei S. Komlev [4], Petr I. Nikitin [3,5], Dmitry A. Gorin [2], Andrei V. Kabashin [3,6] and Sergey M. Deyev [1,3,*]

[1] Shemyakin-Ovchinnikov Institute of Bioorganic Chemistry, Russian Academy of Sciences, Moscow 117997, Russia; olga.griaznova@skoltech.ru (O.Y.G.); yarbel9@mail.ru (I.B.B.); annasogomonyan2012@mail.ru (A.S.S.)
[2] Center for Photonic Science and Engineering, Skolkovo Institute of Science and Technology, 3 Nobel Str, Moscow 121205, Russia; d.gorin@skoltech.ru
[3] Institute for Physics and Engineering in Biomedicine (PhysBio), National Research Nuclear University MEPhI (Moscow Engineering Physics Institute), Moscow 115409, Russia; gvtikhonovskii@mephi.ru (G.V.T.); aapopov@mephi.ru (A.A.P.); nikitin@kapella.gpi.ru (P.I.N.); andrei.kabashin@univ-mrs.fr (A.V.K.)
[4] Faculty of Physics, M.V. Lomonosov Moscow State University, Moscow 119991, Russia; komlev.as16@physics.msu.ru
[5] Prokhorov General Physics Institute of the Russian Academy of Sciences, Moscow 119991, Russia
[6] Campus de Luminy—CNRS, LP3, Aix Marseille University, Case 917, 13288 Marseille, France
* Correspondence: zelepukin@phystech.edu (I.V.Z.); deyev@ibch.ru (S.M.D.)
† These authors contributed equally to this work.

Abstract: Hybrid multimodal nanoparticles, applicable simultaneously to the noninvasive imaging and therapeutic treatment, are highly demanded for clinical use. Here, Fe-Au core-satellite nanoparticles prepared by the method of pulsed laser ablation in liquids were evaluated as dual magnetic resonance imaging (MRI) and computed tomography (CT) contrast agents and as sensitizers for laser-induced hyperthermia of cancer cells. The biocompatibility of Fe-Au nanoparticles was improved by coating with polyacrylic acid, which provided excellent colloidal stability of nanoparticles with highly negative ζ-potential in water (-38 ± 7 mV) and retained hydrodynamic size (88 ± 20 nm) in a physiological environment. The ferromagnetic iron cores offered great contrast in MRI images with $r_2 = 11.8 \pm 0.8$ mM^{-1} s^{-1} (at 1 T), while Au satellites showed X-ray attenuation in CT. The intravenous injection of nanoparticles enabled clear tumor border visualization in mice. Plasmonic peak in the Fe-Au hybrids had a tail in the near-infrared region (NIR), allowing them to cause hyperthermia under 808 nm laser exposure. Under NIR irradiation Fe-Au particles provided 24.1 °C/W heating and an IC$_{50}$ value below 32 μg/mL for three different cancer cell lines. Taken together, these results show that laser synthesized Fe-Au core-satellite nanoparticles are excellent theranostic agents with multimodal imaging and photothermal capabilities.

Keywords: pulsed laser ablation in liquids; multimodal imaging; MRI; CT; photothermal therapy; iron-gold nanoparticles; pharmacokinetics

1. Introduction

Magnetic nanoparticles (NPs) consisting of iron, gadolinium, manganese, and other metals are widely studied for biomedical applications, e.g., drug delivery, magnetic resonance imaging (MRI), or visualization with magnetic spectral approaches [1–4]. However, independently of doping metals (Mn, Co, and Ni), most magnetic NPs are iron oxide based. After accumulation in an organism, NPs can degrade in lysosomal conditions in cells, releasing metal ions [5]. It can reduce contrasting properties of NPs and cause adverse toxic effects [6,7]. It was shown that T_1 MRI contrast agents based on chelates of gadolinium can release Gd^{3+} ions, which subsequently accumulate in brain, muscles, and other tissues, causing apoptosis, the competition of Gd^{3+} and Ca^{2+}, and other cytotoxic effects [8]. However, iron-based MRI contrast agents tend to have lower toxicity since there is a plethora of mechanisms of iron metabolism in the body. Moreover, a lysosomal-induced release of Fe^{3+} ions is used for the treatment of iron deficiency anemia [9].

Plasmonic NPs, such as gold, silver, TiN, ZrN, and HfN-based materials have unique optical properties such as increased extinction at resonant frequencies [10]. This property can be used for surface-enhanced Raman spectroscopy (SERS) sensing and visualization [11,12], as well as for local hyperthermia treatment of wound infection or tumors [13,14], which allows avoiding the usage of chemotherapy and, thus, lowers toxicological effects to a body. On the other hand, light absorbance causes an increase in temperature, subsequent nanoparticle volume expansion, and hence the generation of acoustic waves, which is a basis for photoacoustic visualization [15]. Among other plasmonic NPs, the non-reactivity of gold makes it low in toxicity, while a high atomic number renders the X-ray attenuation coefficient high enough for the in vivo visualization of NPs by computed tomography (CT) [16].

The combination of magnetic and plasmonic properties of iron and gold nanoparticles in one core-shell or alloy system renders possible not only multiple visualizations of pathologies by MRI and CT techniques for increasing diagnosis quality but also allows the performance of hyperthermia treatment of cancer cells. Recently, several chemically synthesized Fe-Au composites were investigated for multimodal in vivo imaging [17–19]. The chemical preparation of bimetallic nanoparticles can provide a wide variety of different crystal structures and compositions; however, their surface can be contaminated by absorbed reagents and surfactants, which potentially increase nanoparticle toxicity. To solve this problem, the pulsed laser ablation in liquids (PLAL) method was developed [20], which provides nanoparticles with contaminated-free surface and can generate particles with various composition and size. Different PLAL-fabricated particles, including silicon, gold, and TiN nanomaterials, were investigated for biomedical applications and showed low toxicity in vitro and in vivo [21–24]. A recent synthesis approach of Fe-Au core-satellite hybrid nanoparticles by the PLAL method was proposed by Popov A.A. et al. [25]. It provided unform decorations of nanosized iron cores with smaller gold nanoparticles. This type of Fe-Au nanocomposite assembly is preferred for in vivo applications due to the high biocompatibility of both components—iron oxides are biodegradable in physiological conditions via iron-recycling pathways, while gold nanoparticles are biodegradable in lysosomes during 2–6 months following biomineralization [26]. The resultant uncoated material has strong magnetization (12.6 emu/g) and broad plasmon peaks centered at 550 nm with a long tale in the near-infrared region [25]. These first data provide promising multimodal imaging and photothermal therapy relative to cancer cells using laser-synthesized Fe-Au hybrids, but the potential of this new material has yet to be confirmed in vitro and in vivo.

In this paper, the syntheses of bimetallic Fe-Au core-satellite nanoparticles by the PLAL method and the investigation of their biomedical applications in terms of their MRI, CT contrast properties, and photothermal therapeutic efficiency under irradiation by near-infrared red (NIR) light were reported. Polyacrylic acid coated Fe-Au particles showed low toxicity to cancer cells of different tissues in vitro. Under NIR irradiation, nanoparticles were heated and caused 100% cell death at concentrations of particles higher than 25 µg/mL. Moreover, Fe-Au particles acted as negative T_2 contrast agents in MRI and positive contrast material for CT imaging in vitro and in vivo. Our results make laser-synthesized Fe-Au core-satellite nanoparticles a perspective agent for multimodal imaging and therapy.

2. Materials and Methods

2.1. Materials

2.1.1. Chemicals

Sigma-Aldrich (St. Louis, MO, USA): sodium chloride ($\geq$99%), potassium ferrocyanide ($\geq$98.5%), hydrochloric acid (37%), nitric acid (70%), poly(acrylic acid sodium salt) (Mw ~5100) (>86%), 3-(4,5-Dimethyl-2-thiazolyl)-2,5-diphenyl-2H-tetrazoliumbromide (MTT) ($\geq$97.5%), and eosin (99%); Chimmed (Moscow, Russia): dimethyl sulfoxide (DMSO) (99.9%), crystal violet (98%), and formaldehyde (37%); Gibco (Waltham, MA, USA): DMEM medium; Capricorn (Ebsdorfergrund, Germany): fetal bovine serum; PanEko (Moscow, Russia): L-glutamine, penicillin, and streptomycin; Virbac (Carros, France): Zoletil; Bioveta (Komenského, Czech Republic): Rometar.

Targets for laser ablation: Au target (99.99%, GoodFellow, Delson, QC, Canada) and metallic Fe target (99.99%, GoodFellow, Delson, QC, Canada).

2.1.2. Cell Lines

Human tumor cell lines: lung carcinoma A549 (CCL-185™; ATCC), mammary ductal adenocarcinoma BT-474 (HTB-20™; ATCC), and ovarian adenocarcinoma SKOV3-1ip (collection of Laboratory of Molecular Immunology IBCh RAS); mouse mammary cell line EMT6/P (ECACC) and Chinese hamster ovary cell line CHO (Russian Cell Culture Collection) were used for in vitro studies. All cell lines were grown in colorless DMEM medium with 10% fetal bovine serum and 5000 U/mL and 5000 µg/mL of penicillin and streptomycin in culture flasks at 5% CO_2 and 37 °C.

2.2. Synthesis of Fe-Au Nanoparticles

The synthesis of Fe-Au core-satellite nanoparticles was performed by a two-step method of femtosecond (fs) pulsed laser ablation in liquids. A schematic diagram of the synthesis procedure is shown in Figure 1. At the first step of PLAL, an aqueous solution of Au NPs was synthesized. The target was placed vertically in a glass cuvette filled with 50 mL of 1mM NaCl aqueous solution. The thickness of liquid layer between the target surface and cuvette wall was 4 mm. Radiation from a Yb:KGW laser (1030 nm wavelength, 250 fs pulse duration, 30 µJ pulse energy, 100 kHz repetition rate; TETA 10 model, Avesta, Moscow, Russia) with 3 mm beam diameter was focused by a 100 mm F-theta lens on the surface of the target, through a side wall of the ablation vessel. To avoid ablation from one place and increase synthesis efficiency, the laser beam was moved over a 10 × 30 mm area on the surface of the target, with 4 m/s speed using a galvanometric scanner. The duration of the first step of laser ablation experiment was 40 min. Then, the prepared colloidal solution was centrifuged with 15,000× g for 15 min and the nanoparticles in the supernatant were taken for further steps. During the second step of PLAL a metallic Fe target was placed in the chamber filled with Au NP solution. The ablation of Fe target was performed in the same manner as for the first step. The duration of the second step was 15 min.

Non-reacted Au NPs were ejected by a step of magnetic separation. For this, a NdFeB cylindric (diameter—5 cm; length—3 cm) magnet was placed for 5 min next to the bottom of the cuvette with a colloidal solution of Fe-Au NPs. The magnetic fraction of NPs was attracted to the vessel's bottom, while nonmagnetic NPs remained dispersed in the solution. Then, the upper part of the solution containing free Au NPs and nonmagnetic Fe NPs was discarded, leaving 2 mL of solution at the bottom. The NPs were redispersed by an ultrasonication.

Obtained Fe-Au nanoparticles were stabilized by polyacrylic acid. Polymer solution measuring 1 mL (50 mM) was heated to 80 °C and added to 1 mL of Fe-Au nanoparticle water dispersion with a concentration of 1 g/L. Then, the nanoparticles were incubated for 15 min at 80 °C and washed 3 times with water from unbound polymer via centrifugation at 4000× g for 10 min.

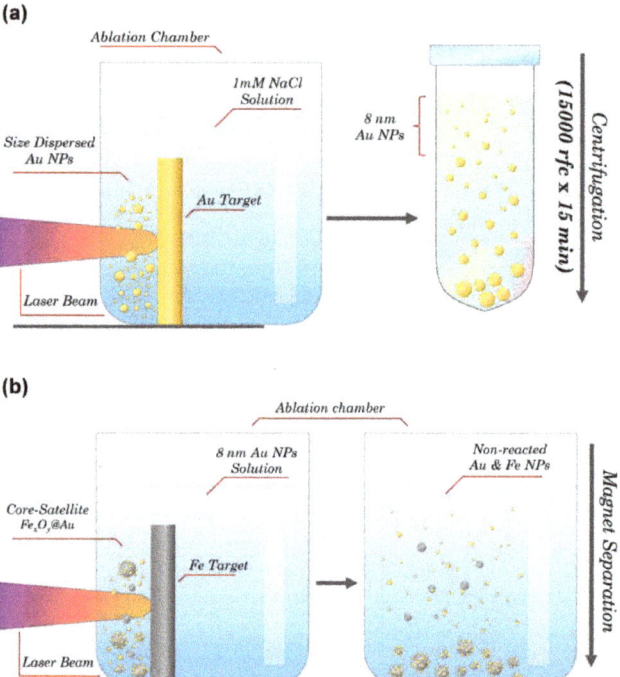

Figure 1. Schematic representation of two-step PLAL synthesis of core-satellite Au-Fe NPs. (**a**) Laser ablation of gold target with further separation step of 8 nm Au NPs fraction via centrifugation. (**b**) Laser ablation of iron target in presence of Au NPs with further separation step of magnetic Fe-Au NPs fraction via external magnetic field.

2.3. Nanoparticle Characterization

Hydrodynamic size and ζ-potential measurements were performed using a Malvern Zetasizer Nano ZS (Malvern Instruments, Malvern, UK). The experiments were performed in distilled water for hydrodynamic size measurement and in 10 mM NaCl for ζ-potential analysis. Number distribution was used for size analysis. Stability measurements were carried out in phosphate-buffered saline (PBS, 100 mM, pH 7.4) at 60 µg/mL concentration.

UV-Vis spectra were measured by an Infinite M1000 PRO (Tecan, Salzburg, Austria) microplate reader.

Surface morphology and the elemental composition of NPs were characterized by a scanning electron-microscope (SEM) MAIA3 (Tescan, Brno, Czech Republic) coupled with an energy dispersive X-ray spectroscope (EDS) X-act (Oxford Instruments, Abingdon, UK). Nanoparticle solution in water was dropped on a silicon wafer and dried under ambient conditions. Nanoparticle images were obtained at 20 kV accelerating voltage in secondary electron (SE) detection mode. The mean size of the Fe-Au core satellites was measured using ImageJ software; 210 nanoparticles were analyzed.

The concentrations of iron and gold in Fe-Au NPs were quantified via inductively coupled plasma mass-spectrometry using a NexION 2000 spectrometer (Perkin Elmer, Waltham, MA, USA). Particles were dissolved in concentrated aqua regia, 3:1 mixture of $HCl:HNO_3$, and diluted by water to 30% acid concentration for measurements. ^{57}Fe and ^{197}Au peaks were used for the analysis.

Magnetic hysteresis loop was measured using a VSM LakeShore 7407 Series magnetometer (Lake Shore Cryotronics, Westerville, OH, USA). The field dependence of magnetization was measured for NP water dispersion sealed in a quartz capillary at room temperature. The measurement was carried out up to 2 T field. Diamagnetic background from the quartz tube and holder were subtracted from measurement results.

2.4. Study of Photothermal Properties

To evaluate the photothermal properties of PAA-coated Fe-Au NPs, 1 mL of nanoparticle solution in water was irradiated in a square optical polystyrene cuvette either with 808 nm or 532 nm laser at room temperature. Temperature measurements were performed using a thermal imaging camera FLIR C3 (FLIR Systems, Wilsonville, OR, USA), and the temperature of the hottest spots in the irradiated region were used for analysis. The test on photostability was performed via repetitive cycles of 5 min heating under irradiation with an 808 nm laser (1.23 W) followed by a 5 min cooling period.

Photothermal conversion efficiency (η) of nanoparticles under near infrared irradiation (808 nm, 1.23 W, 10 min duration) was calculated via an approach developed in [27]:

$$\eta(\%) = \frac{hS\left(T_{max}^{NPs} - T^{amb}\right) - Q_s}{I\left(1 - 10^{-A^{808}}\right)}$$

where h is the heat transfer coefficient, S is the surface area for heat transfer, T_{max}^{NPs} is the maximum temperature reached in particle suspension under irradiation, T^{amb} is the ambient temperature, Q_s is the heat produced by water due to light absorption, I is the incident laser intensity, and A^{808} is the light absorbance of particles at λ = 808 nm. To derive hS, the following relation was used:

$$hS = \frac{m_s c_s}{\tau}$$

where m_s and c_s are mass and specific heat capacity of water, respectively, and τ is the time constant determined as the slope coefficient in cooling time vs. natural logarithm of the driving force ($\theta = \left(T - T^{amb}\right)/(T_{max}^{NPs} - T^{amb})$) dependence. The value of Q_s was determined by measuring temperature increments in a cuvette filled with distilled water under the same irradiation conditions.

2.5. In Vitro Studies

The 3-(4,5-dimethylthiazol-2-yl)-2,5-diphenyltetrazolium bromide (MTT) test was used to determine the cytotoxic effect of Fe-Au@PAA nanoparticles. For this, A549, BT474, SKOV3-1ip, EMT6/P, and CHO cells at a concentration of 6×10^4 cells in 600 µL of colorless DMEM medium were incubated with particles at concentrations of 3.12, 6.25, 12.5, 25, 50, and 100 µg/mL for 1 h at 5% CO_2 atmosphere and 37 °C. After that, cells were introduced into a 96-well plate at a concentration of 10^4 cells in 100 µL of medium per well. Cells were incubated for 48 h at 5% CO_2 and 37 °C. Next, the medium was removed, 100 µL of the MTT solution per well was added, and cells were incubated for 1 h at 37 °C. Then, MTT was removed and 100 µL of DMSO was added to each well. The measurement was carried out on the Infinite M1000 Pro spectrophotometer at a wavelength of 570 nm. Cell viability is shown as a percentage normalized relative to the control cells incubated without particles.

For the measurement of cytotoxicity under near-infrared irradiation, 6×10^4 cells of different lines, A549, EMT6/P, or CHO, were added to 1.5 mL tubes and incubated in 600 µL of colorless DMEM medium with particles at concentrations from 3.12 to 100 mg/L for 1 h at 5% CO_2 and temperature 37 °C. Next, tubes were covered with foil and samples were irradiated with NIR 808 nm laser (0.76 W) at room temperature while continuously shaking (285 rpm). Then, the irradiated cells were transferred into a 96-well plate at a concentration of 10^4 cells in 100 µL of medium per well. Cells were incubated for 48 h at 5% CO_2 atmosphere and 37 °C. Then, cell viability was measured by a MTT test, as described above. Cell viability is shown as a percentage normalized to control cells, incubated without particles and irradiation. The IC_{50} values were calculated using the OriginPro 2015 software (OriginLab, Northampton, MA, USA) with a dose–response function.

For setting clonogenic analysis, the same conditions were used as for the study of cytotoxicity by the MTT test. After the incubation of cells with nanoparticles and their irradiation, the cells were diluted 30 times, and 2×10^3 cells in 1 mL of DMEM medium were added to the each well of a 12-well plate and incubated for 8 days at 5% CO_2 and 37 °C. Next, the nutrient medium was removed, cells were washed with 600 μL of PBS, pH 7.4. Then, 600 μL of 70% ethanol was added to the wells to fix the cells, following incubation at room temperature for 15 min. Then, 70% ethanol was removed and 600 μL of 95% ethanol was added, repeating incubation conditions. After removal of 95% ethanol, the plates were washed with water. Then, the cells were stained by 600 μL of 1% crystal violet water solution for 30 min at room temperature. Then, the wells of the plates were washed 10 times with water.

2.6. Animals

All animal studies were approved by the Institutional Animal Care and Use Committee (IACUC) of Shemyakin–Ovchinnikov Institute of Bioorganic Chemistry (Moscow, Russia), protocol № 240 from 16 June 2020.

Female BALB/c mice of 18–22 g weight were obtained from Pushchino Animal Facility (Pushchino, Russia) and maintained in a vivarium at the Shemyakin-Ovchinnikov Institute of Bioorganic Chemistry. Before the experiments, mice were anesthetized by an intraperitoneal injection of 150 μL of Zoletil/Rometar solution in a concentrations 90 g/L and 0.16 g/L, respectively.

To perform a tumor model, EMT6/P mice mammary carcinoma cells were inoculated in animals. EMT6/P cells were grown in DMEM supplemented with 10% fetal bovine serum and 2 mM L-glutamine at 37 °C in 5% CO_2 atmosphere. The cells were harvested from the culture dish and transferred to PBS to obtain the concentration of 10^7 cells per mL. Then, 100 μL of EMT6/P cell suspension was administered subcutaneously into the flank region of BALB/c mice. Mice bearing tumors of 150–250 mm^3 volume were used for the experiments.

2.7. Pharmacokinetics Study

A magnetic particle quantification (MPQ) method [28] was used to evaluate Fe-Au nanoparticle biodistribution and blood circulation kinetics as described elsewhere [29]. Mice were anesthetized and their tail was placed into the magnetic coil of an MPQ reader and gently fixed with tape. Then, 1 mg of Fe-Au@PAA particles in 100 μL of PBS was injected into the retro-orbital sinus. The concentration of nanoparticles in blood was measured with MPQ in real-time in tail veins and arteries of mice. The 10–90% central part of kinetic curves was fitted with a monoexponential function to calculate the blood circulation time of magnetic NPs.

For the investigation of nanoparticle biodistribution, 2 h after particle administration, mice were euthanized by cervical dislocation, and the major organs (liver, spleen, lungs, kidneys, heart, femur bones, brain, muscle sample, and tumor) were harvested and placed into a measuring coil of a MPQ reader. To increase delivery to the tumor site, magnetic field-assisted delivery was used by placement of a 2 cm × 1 cm × 0.5 cm NdFeB magnet near the tumor. To consider differences in the actual injected dose among different animals, the particle quantities in organs were normalized by the sum of the magnetic signals from all studied organs. Each signal was then divided by the organ mass to obtain the signal per gram of tissue.

2.8. Magnetic Resonance Imaging

MRI analysis of mice and phantoms was carried out in an ICON 1T MRI system (Bruker, Billerica, MA, USA) using a mouse whole-body-volume radio frequency coil. For image acquisition, a two-dimensional gradient echo FLASH sequence was used with the following parameters: repetition time/echo time (TR/TE), 1000/5.16 ms; flip angle (FA), 60°; resolution, 200 µm; field of view (FOV), 85 × 30 mm; 1 signal average; 20 slices per scan; slice thickness, 1.5 mm. T_2 mapping was used for acquiring T_2 values of NP suspensions.

2.9. Computed Tomography

IVIS Spectrum CT (PerkinElmer, Waltham, MA, USA) small animal imaging system was used for computed tomography of phantoms and mice. Images were acquired with 50 kV tube at 1 mA current with 20 milliseconds of exposure time. A total of 720 projections spaced 0.5° apart were acquired, and the CT volume was reconstructed using Living Image software (PerkinElmer Inc., Waltham, MA, USA) with a FOV 8 cm × 8 cm × 2 cm.

2.10. Histological Analysis

Samples of the organs were fixed in 4% formaldehyde for 24 h, dehydrated in a series of alcohol solutions of increasing alcoholic concentration, and embedded in paraffin. Paraffin sections measuring 4 µm thick were stained with eosin and Perls Prussian Blue. Microscopy studies were carried out using Leica DM4000 B LED light microscope (Leica Microsystems, Wetzlar, Germany) equipped with a digital camera Leica DFC7000T.

2.11. Statistical Analysis

All experiments were performed at least in triplicates. All numerical data are presented as mean ± standard deviation. Significant differences were determined using 2-tailed Welch's t-test. The statistical difference was considered statistically significant when p value was < 0.05 (*), 0.01 (**), and 0.001 (***).

3. Results and Discussion

3.1. Nanoparticle Synthesis and Characterization

Highly pure Fe_xO_y@Au nanoparticles (Fe-Au NPs) were synthesized by a two-step femtosecond pulsed laser ablation in liquid method according to a previously reported procedure [25]. The resulting nanoparticles had core-satellite architecture, where spherical iron oxide NPs formed cores, while Au NPs formed satellites (Figure S1). The mass ratio of Fe:Au was 2.1 ± 0.3, according to analysis by inductively coupled mass spectrometry.

For in vivo applications of nanoparticles, they should possess colloidal stability in high ionic strength liquids. As-prepared Fe-Au nanoparticles had mean hydrodynamic size of (73 ± 40) nm in water, while in physiologically relevant environment of phosphate buffered saline (PBS, 100 mM, pH 7.4), a rapid aggregation to sub-micron range was observed with hydrodynamic size of (673 ± 200) nm (Figure 2a). To improve the poor colloidal stability of Fe-Au, NP surface was coated with a biocompatible organic polymer—polyacrylic acid (PAA). PAA is a polymer, which bears all qualities suited for biomedical applications, such as biocompatibility, nontoxic nature, and biodegradability [30]. The obtained PAA-coated Fe-Au core-satellite nanocomposites (Fe-Au@PAA) possessed slightly higher diameter in water (88 ± 40) nm, indicating the appearance of an additional polyelectrolyte kinetic layer. The PAA coating effectively prevented nanoparticle aggregation in PBS for at least 1 day, which was confirmed by the retainment of the mean particle size at the same value: (88 ± 20) nm (Figure 2b).

The formation of the polyelectrolyte layer of PAA on the surface of Fe-Au nanoparticles was proved by changes in ζ-potential from (+10 ± 5) mV before to (−38 ± 7) mV after the coating procedure (Figure 2c). Low positive ζ-potential of bare NPs indicated weak interparticle repulsion. According to theory [31], nanoparticles, which have ζ-potential less than 5 mV by modulus, undergo aggregation, while |ζ| > 30 mV provides a good colloidal stability of NP solution. Fe-Au@PAA NPs demonstrated a significantly higher ζ-potential magnitude with the negative charge, which is a distinct indication of PAA carboxylate anions presented on the particle's surface. A 1 mV shift in ζ-potential peak value was observed after 25 days of coated particles incubation at 4 °C (Figure S2), thus showing long-term stability of the organic layer during storage. Considering high chemical stability of both Au nanoparticles and oxidized iron in water, Fe-Au@PAA NPs could provide a particularly long half-life in aqueous colloid form, which is highly relevant for biomedical implementation.

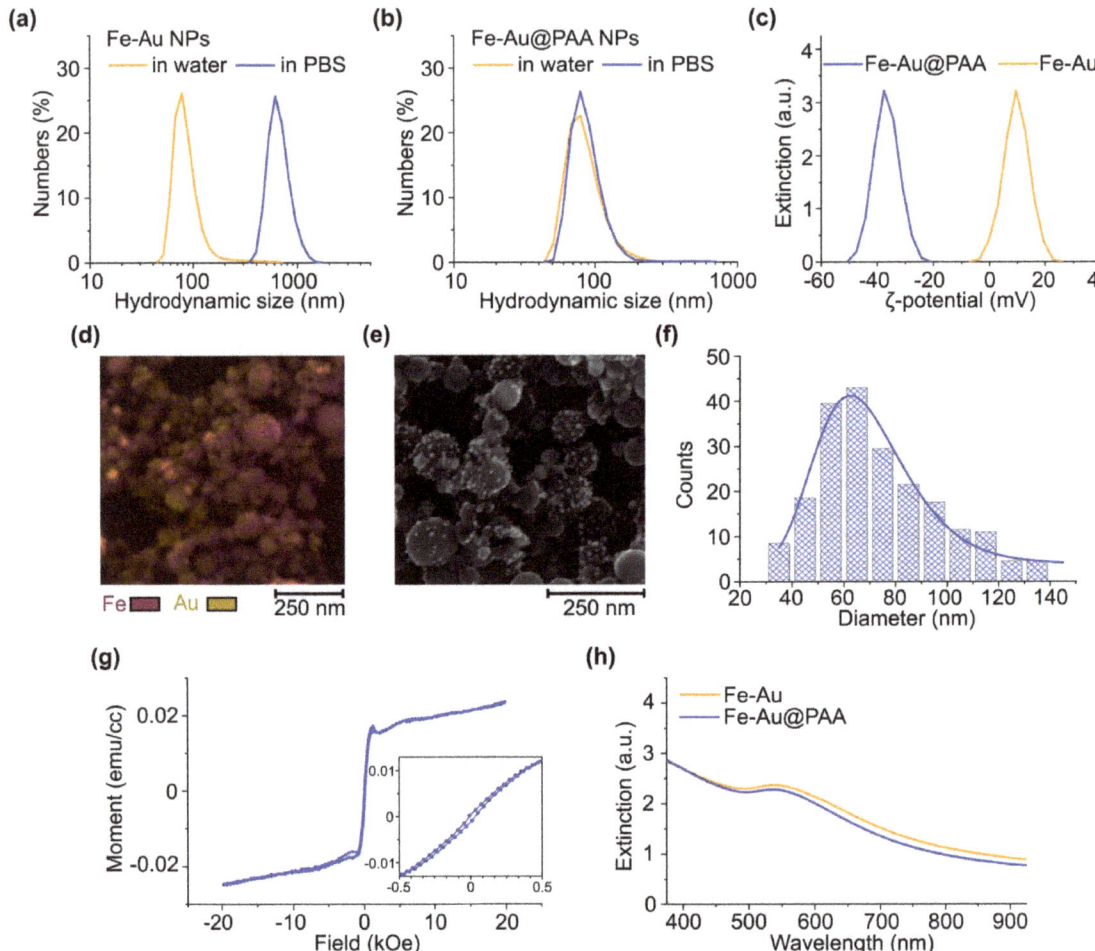

Figure 2. Fe-Au NP characterization before and after stabilization with a polymer coating. (**a,b**) Hydrodynamic size distribution of Fe-Au (**a**) or Fe-Au@PAA (**b**) NPs in water and PBS. (**c**) ζ-potential distributions of Fe-Au nanoparticles before and after the coating with PAA. (**d**) EDS mapping of Au and Fe abundance in nanoparticles. Scale bar—250 nm. (**e**) SEM image of Fe-Au@PAA NPs. Scale bar—250 nm. (**f**) Distribution of Fe-Au@PAA NP physical diameter. (**g**) Magnetization curves of Fe-Au NPs measured at 300 K. The inset is an enlarged view of the low-field region. (**h**) UV-Vis spectra of Fe-Au NPs before and after the PAA coating.

The preservation of core-satellite structure after the coating was confirmed by SEM imaging coupled with EDS mapping, demonstrating the sparse distribution of gold around spherical iron-based cores (Figure 2d,e). The mean physical diameter of cores in the coated nanoparticles was determined to be (74 ± 20 nm) (Figure 2f). The ratio of hydrodynamic and physical sizes indicated that Fe-Au@PAA NPs rather exist in solutions in a single form than in clusters.

Core-satellite nanoparticles possesses both magnetic and plasmonic properties. Magnetic hysteresis loop at 300 K temperature showed ferromagnetic behavior with coercivity ≈ 41 Oe (Figure 2g). Magnetization was measured for 1 mL of water dispersion of Fe-Au NPs with 1 mg/mL concentration. The volume magnetization saturation of Fe-Au NPs was M_S ≈ 0.022 emu/cc or 22 emu/g. This value is lower than for bare laser-synthesized iron oxide nanoparticles with M_S = 44.7 emu/g [25] for ultrasmall iron nanoparticles Fe-NPs (M_S = 117–163 emu/g) and other Fe_3O_4-Au (M_S = 86 emu/g) hybrids [32,33]. It should be noted that artifacts in magnetization curves at |H| > 0.5 kOe may stem from the aggregation of nanoparticles at high magnetic fields, which is challenging to avoid in liquids.

Plasmonic gold NPs contributed to the extinction spectra of iron oxides by appearance of absorbance peak in green region of the spectra with a broad tail in the infra-red part (Figure 2h), which is in an optical transparency window of biological tissue [25]. After the coating of Fe-Au NPs by PAA, the plasmon peak was shifted from 542 nm to 536 nm with decrease in its intensity, probably due to a slight release of loosely bound plasmon Au nanoparticles during the coating procedure.

3.2. Photothermal Properties of Fe-Au@PAA Nanoparticles

Nanoparticles demonstrating strong light absorption in the visible and near-infrared regions and then converting it into heat are promising candidates for photothermal therapy. Thus, owing to the strong extinction of Fe-Au at the plasmon resonance band, the photothermal properties of NPs under irradiation by a green light from a 532 nm laser source were evaluated firstly. Green lasers have been extensively used in surgery for photocoagulation [34] and tested for cancer therapy with nanoparticles [35]. Considering relatively low intensities of 532 nm lasers used in clinics (100–750 mW) [36] and high absorption of visible light by tissues, Fe-Au@PAA NPs were exposed to irradiation with 100 mW laser power. Under the irradiation, the steady-state temperatures increased from 26.5 to 31 °C with the increase in particle concentration from 25 to 400 µg/mL (Figure S3). This result suggests the weak opportunities of these particles to cause even mild hyperthermia under such irradiation conditions.

Nevertheless, the intense extinction in the near-infrared region associated with the light absorption by iron cores and clustered Au satellites makes Fe-Au NPs attractive for photothermal therapy with an 808 nm laser. This wavelength belongs to the first biological transparency window (650–950 nm), where low absorption and scattering of light by tissues provide highest light penetration depth [37]. Indeed, the aqueous suspensions of Fe-Au@PAA NPs under irradiation with a continuous 808 nm laser demonstrated excellent thermal response at concentrations 10–400 µg/mL (Figure 3a,b). For example, the heating of particles at 50 µg/mL from temperature of 37 °C to the conventional upper temperature limit of mild hyperthermia (42 °C) was reached just after 2.5 min of irradiation. The saturation in heating efficiency clearly appeared after 100 µg/mL NP concentration and reached 39 °C temperature increase at concentrations of 400 µg/mL. A linear dependence of heating efficiency on laser power was observed (Figure 3c). The high value of the slope coefficient (24.1 °C/W) suggests that particles can provide high photothermal efficiency under even low NIR light intensities that should be safe for healthy tissues.

Fe-Au NP suspension demonstrated a decent photothermal stability under on/off irradiation cycles, although particle sedimentation during the heating caused a slight decrease in peak temperature at the fifth cycle. The time constant of suspension cooling τ = 318 s was used for calculation of photothermal conversion efficiency coefficient (η) (Figure 3e,f). Fe-Au@PAA particles show η = 38%; thus, their efficiency exceeds those reported for gold nanoparticles (10.2%) [38], gold nanoshells (13%) [34], other iron-oxide nanoparticles with modified surface (13.1–35.7%) [39], and is comparable to another Fe-Au core-shell nanocomposites (42.6%) [17] evaluated under NIR irradiation.

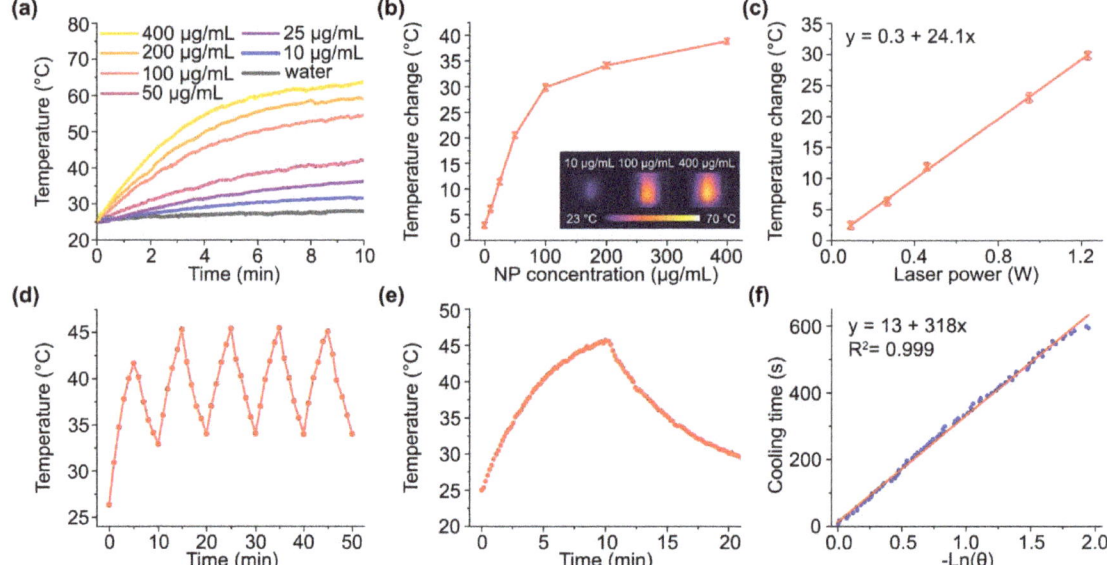

Figure 3. Photothermal properties of Fe-Au@PAA NPs under NIR-irradiation with 808 nm laser. (**a**) Temperature changes for aqueous solutions of NPs under irradiation at 1.23 W laser power. (**b**) Dependence of maximal temperature change on particle concentration. Insert shows thermal images of several particle samples. (**c**) Dependence of temperature change on laser power. (**d**) Photothermal stability cycling test with 5 min heating and 5 min cooling steps. Ⓔ Kinetics of the temperature changes illustrating 1 cycle of 10 min heating and 10 min cooling. (**f**) The dependence of cooling time on negative natural logarithm of driving force. Solution with 100 μg/mL (**c**) or 50 μg/mL (**d–f**) particle concentration was used for the studies.

3.3. In Vitro Studies of Fe-Au@PAA Biocompatibility and Toxicity after the Photothermal Treatment

The MTT assay [40] was used to evaluate the cytotoxic effects of Fe-Au@PAA nanoparticles on cell viability. This colorimetric test is based on the cellular reduction of tetrazolium dye (3-(4,5-dimethylthiazol-2-yl)-2,5-diphenyl-tetrazolium bromide), where living cells recover yellow colored tetrazolium dye to insoluble purple colored formazan in response to dehydrogenase activity. The MTT test measures cell viability in terms of reductive activity since the enzymatic conversion of tetrazolium occurs only in mitochondria and other organelles of living cells [41].

Five cell lines of different tissues and organisms were used in toxicity study: human (A549, SKOV3-1ip, BT-474), mouse (EMT6/P), and Chinese hamster (CHO) cell lines. To evaluate cytotoxic effects of Fe-Au@PAA, NPs were incubated with cells for 48 h as described in the Materials and Methods section.

The cytotoxicity of Fe-Au@PAA NPs at concentrations 3.12–50 μg/mL in BT-474 and SKOV3-1ip cells within tested time was negligible. A slight decline in cell viability was observed for 100 μg/mL concentration of nanoparticles with cell viability values being $67 \pm 9\%$ and $82 \pm 3\%$ for BT-474 and SKOV3-1ip cell lines, respectively (Figure 4a).

Nanoparticles possessed slight dose-dependent cytotoxic effects on A549 and CHO cells (Figure 1b–d green columns). However, cell viability for the 50 μg/mL dose was $66 \pm 3\%$ (CHO), $79 \pm 5\%$ (A549); for the 100 μg/mL dose, it was $21.6 \pm 0.6\%$ (CHO) and $73 \pm 3\%$ (A549). Note that, even at high concentrations of nanoparticles, cell viability was above 70%, which is consistent with the results obtained by other authors [17,42]. It confirms the biocompatibility of obtained Fe-Au@PAA nanoparticles for the studied cells.

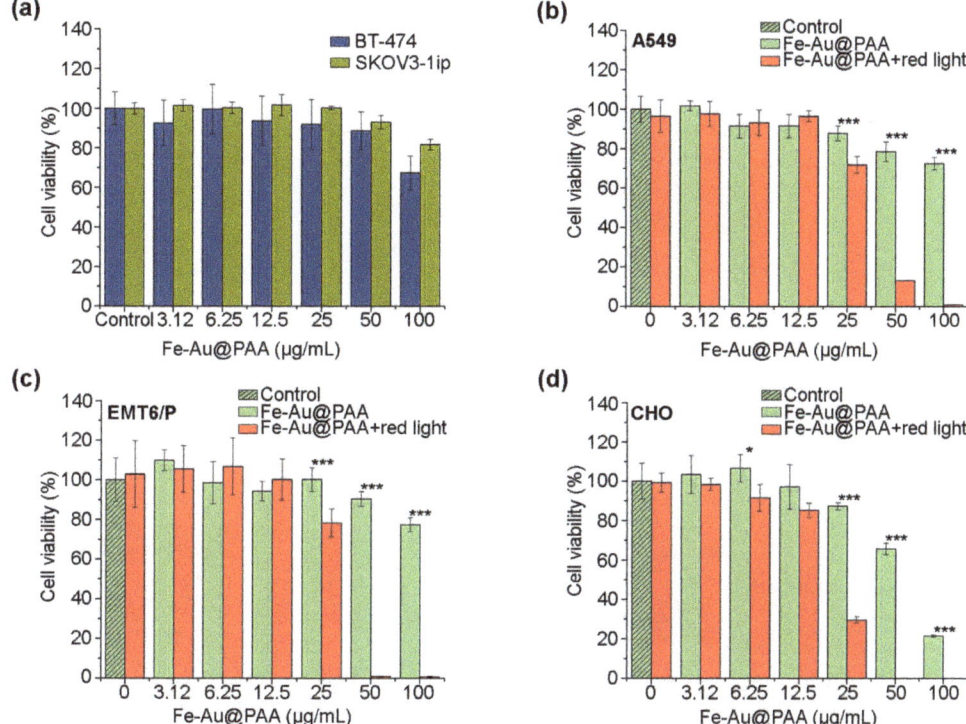

Figure 4. Analysis of the Fe-Au@PAA nanoparticle cytotoxicity using the MTT test. (**a**) Study of the toxic effect of particles upon exposure to cells BT-474 and SKOV3-1ip. (**b**–**d**) Investigation of the photothermic properties of particles. Cells were incubated with Fe-Au@PAA NPs at various concentrations and irradiated for 7 min with an 808 nm laser at 0.76 W power. Cell viability is shown as a percentage normalized to control cells incubated without particles and irradiation. The statistical differences were considered significant when the p value was < 0.05 (*), 0.001 (***), Welch's t-test.

The study of the effectiveness of photothermal treatment via Fe-Au@PAA nanoparticles was performed on three cell lines: A549, EMT6/P, and CHO (Figure 4b–d). To evaluate effects from laser-induced hyperthermia, cells were incubated with NPs at different concentrations for 1 h, following sample irradiation by 808 nm laser with a 0.76 W power for 7 min. These conditions were determined experimentally to not cause photothermal damage to cell lines without NPs (Figure 4b–d, see red columns at "0" NP dose).

Almost 100% of cell death was observed at concentrations 50 and 100 µg/mL for all cell lines. IC_{50} of Fe-Au@PAA NPs with irradiation for A549 cells was (32 ± 1) µg/mL; EMT6/P— (29 ± 3) µg/mL; and CHO— (21 ± 1) µg/mL, while the IC_{50} of NPs without exposure by laser could not be determined. Fe-Au@PAA NPs apparent had the most cytotoxic effects on the CHO cell line with and without laser exposure. The higher toxicity induced by photothermal treatment was caused by CHO higher sensitivity to heating, as was shown by other authors [43].

The MTT test reflects cellular activity, while clonogenic analysis is a method for determining the survival and proliferation of cells for a long period after their exposure to various agents [44,45]. Thus, a clonogenic assay was used to assess the long-term cytotoxic effects as it is a more sensitive test than MTT. The clonogenic test was carried out under the same experimental conditions. Cell lines: EMT6/P and CHO were incubated with Fe-Au@PAA NPs following irradiation by 808 nm laser. After dilution and incubation of cells for 8 days, the growing cell colonies were stained with crystal violet, and the results were evaluated visually. Cells without NP exposure and light irradiation were used as controls (Figure 5). The considerable inhibition of cell proliferation was observed at NP doses higher 12.5 µg/mL after the laser induced hyperthermia for both cell lines. At two-fold higher doses, no forming colonies were observed. There was no decrease in proliferative activity at lower concentrations of nanoparticles—3.12 and 6.25 µg/mL. Thus, the clonogenic assay showed that even smaller concentrations of Fe-Au nanoparticles can damage cancer cells under heating in comparison to the MTT test. In addition, while the MTT test of cell viability showed moderate cytotoxic effects of Fe-Au@PAA NPs to CHO cells with 21.6% enzyme activity at 100 µg/mL particle dose, clonogenic analysis did not confirm such dramatic toxicity (Figure 5). Thus, nanoparticles at high doses have biocompatibility in terms of cell survival but can affect enzyme activities of cells.

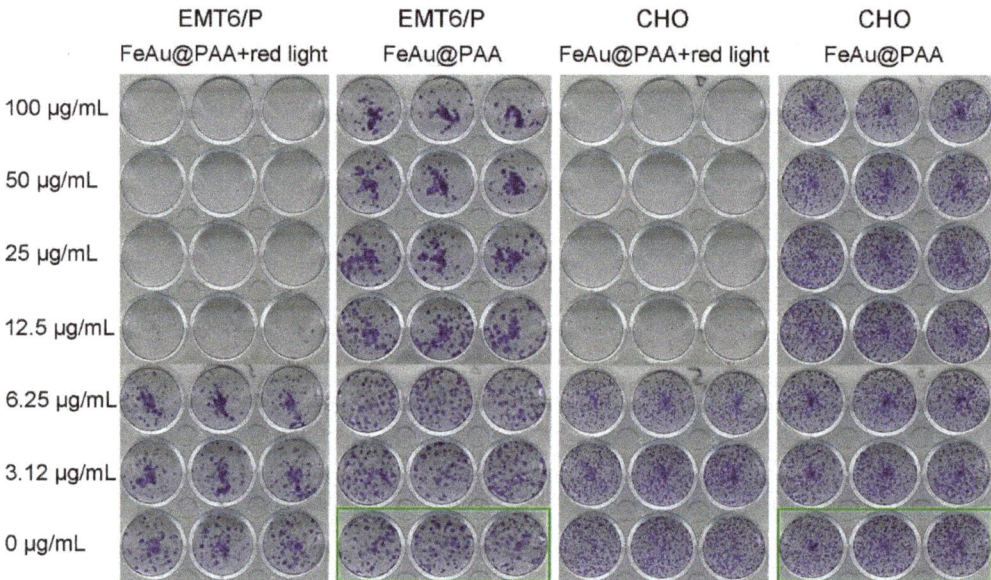

Figure 5. Colony formation assay of EMT6/P and CHO cells treated with Fe-Au@PAA NPs for 8 days with or without 808 nm laser irradiation. Control samples not treated with nanoparticles and irradiation are shown in green frames.

Despite the demonstrated promise of the photothermal properties and low in vitro toxicity of Fe-Au core-satellite nanoparticles, in vivo toxicity issues were left behind and warrant further studies. Natural and synthetic inorganic nanoparticles potentially can cause toxic effects for the organism. One of the common mechanisms is the formation of reactive oxygen species, which lead to increases in the level of lipid peroxidation, cause oxidative DNA damage, and result in carcinogenesis. However, the composition and physicochemical properties of NPs greatly alter toxicity caused by NP exposure. A large number of studies reported that iron oxide nanoparticles regardless of their physicochemical properties may cause cytotoxicity only at concentrations of 100 g/mL in specific organ sites, which is difficult to achieve in the organism [46]. In addition, many clinical studies of iron oxide-based NPs showed that such particles can be categorized as biocompatible. For example, different iron oxides are used for iron-deficient anemia treatment in humans with doses up to 7 mg/kg of body weight without notable toxicity [47]. The toxicity of gold nanoparticles highly depends on size, shape, and chemical groups on their surface. However, many anticancer drugs or diagnostic tools based on gold or gold composite NPs are being tested in clinical trials with several gold NP-based technologies recently approved [48]. In 2019, gold nanoshells were successfully investigated for the photothermal treatment of prostate cancer in humans in a 36 mg/kg dose with rare advert events reported during 3 month after the treatment [49]. As such, we believe that Fe-Au composites can also show low toxicity in vivo in doses required for photothermal therapy and imaging.

3.4. Pharmacokinetics of Fe-Au@PAA Nanoparticles

Due to the ferromagnetic nature of Fe-Au nanoparticles, it was possible to non-invasively measure their concentration in vivo via MPQ (magnetic particle quantification) method [28]. This spectral approach allows the quantitative detection of superparamagnetic and ferromagnetic nanoparticles without the interference of diamagnetic or paramagnetic materials that are naturally presented in the organism [29].

To measure kinetics of Fe-Au@PAA NP circulation in blood, mice were anesthetized, animal's tail was placed into the magnetic coil, and nanoparticles were administered into the retro-orbital sinus at a 40 mg/kg dose. After that, a rapid increase in magnetic signal was observed, indicating the entry of NPs to the bloodstream, following by the monoexponential decrease due to the nanoparticle's clearance (Figure 6a). After 20 min of circulation Fe-Au@PAA nanoparticles were almost completely eliminated from the bloodstream, with circulation half-life time $t_{1/2}$ = (4.4 ± 0.6) min. It is well known that polyacrylic acid coated NPs have small blood circulation time since this polymer did not protect nanoparticles from the adsorption of opsonins in serum [50].

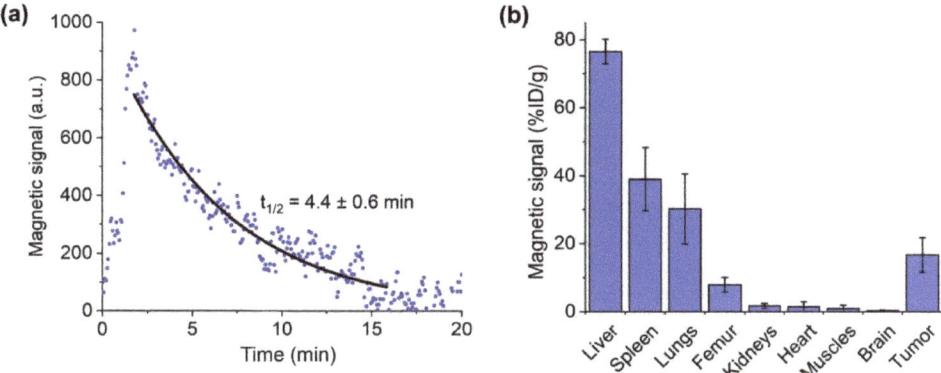

Figure 6. Pharmacokinetics of Fe-Au@PAA NPs. Blood circulation (**a**) and biodistribution (**b**) of nanoparticles.

To increase NP accumulation in tumor, magnetic field-assisted delivery was implemented by placement of 2 cm × 1 cm × 0.5 cm NdFeB magnet near the tumor. Biodistribution of NPs 30 min after injection was studied by analyzing magnetic signals in main organs, such as liver, spleen, lungs, femur bones, kidneys, heart, muscles, brain, and tumor (Figure 6b). Major part of Fe-Au@PAA NPs accumulated in the liver (77 ± 4% ID/g) and spleen (39 ± 9% ID/g), which are the main organs of mononuclear phagocyte system (MPS). The histological slices of liver and spleen stained by Perls Prussian blue and eosin indicated uptakes of Fe-Au@PAA NPs by Kupffer cells of liver and by the cells in the marginal zone of spleen (Figure S4). Fe-Au@PAA NP concentration in lungs was 30 ± 10% ID/g. The possible reason is increased particle hydrodynamic size by the formation of protein corona and the partial aggregation of nanoparticles following retention in capillaries. This behavior was supported by observing the aggregates of nanoparticles in histological slices of lung tissue (Figure S4). Fe-Au@PAA NPs were also accumulated in femur bones possibly due to the uptake by bone-marrow derived macrophages or osteoclasts (8 ± 2% ID/g). Less than 2% ID/g concentration of Fe-Au@PAA nanoparticles were observed in kidneys, heart, muscles, and brain.

The concentration of nanoparticles in tumor reached 17 ± 5% ID/g, which is 17-fold higher than in muscles due to magnetic field-assisted delivery and enhanced permeability and retention effects [51]. Tumor vasculature is leaky and possesses larger endothelial gaps and pores than the capillaries of healthy organs. Histology slices of tumor showed that a major part of nanoparticles accumulated at the borders of tumor. Recently, it was shown that the binding of NPs to the extracellular matrix can prevent their penetration deep into the tissue [52]. Nevertheless, most NPs, which were not trapped by cells of MPS, accumulated in tumor sites, providing a contrast in pathology relative to the surrounding tissues.

3.5. Magnetic Resonance Imaging

The hybrid Fe-Au@PAA nanoparticles possess all featured properties of theranostic agents, such as biocompatibility, photothermic therapeutic effect for cancer treatment, and dual-modal imaging for diagnostics. To evaluate Fe-Au@PAA NPs as contrast agents for magnetic resonance imaging, T_2-wheighted images of different concentrations (NP: 1–1000 mg/L; or Fe: 0.009–9 mM) of NPs in water were measured at 1T MR scanner. Fe-Au@PAA NPs decreased signals on T_2-weighted images and exhibited concentration-dependent negative contrast enhancement effect (Figure 7a). The inverse proton $1/T_2$ relaxation time had non-linear dependence on the concentration of Fe-Au@PAA NPs at studied concentrations. However, the curve could be fitted at lower concentrations with a linear function shown as a black line through the data with a correlation coefficient above 0.98 (Figure 7b). The relaxation rate (r_2) was determined to be (309 ± 20) s^{-1}/(g/L) or (11.8 ± 0.8) s^{-1}/mM of iron atoms. This value is lower than those reported in the literature for different structures of Fe$_3$O$_4$@Au nanoparticles [53,54] due to low crystallinity of laser synthesized Fe$_x$O$_y$ core of Fe-Au NPs.

For testing in vivo MR contrast properties, 1 mg of core-satellite Fe-Au@PAA NPs was injected into retro-orbital sinus of mice (Figure 7c), and magnetic guided-delivery to the tumor was performed. MR images were made 30 min after the injection to ensure full NP clearance from the blood. Before the injection of nanoparticles, there was no difference in signals of tumor and muscles, while after the injection NPs showed negative contrast in a border region of the tumor as well as in volume. However, nanoparticles were mostly accumulated in the liver and spleen, which is in agreement with the biodistribution analysis by the MPQ method. Nevertheless, the amount of NPs trapped in the tumor was enough for pathology visualization, which is beneficial for MRI-guided surgery or external therapy.

3.6. Computed Tomography

Gold nanoparticles are widely used as contrast agents for computed tomography since Au has high atomic number (Z = 79) and, thus, strong X-ray attenuation properties [55]. Due to the presence of Au NPs in a satellite shell, which provided ~32% of the particle mass, Fe-Au NPs should enhance X-ray attenuation with positive contrast at computed tomography. To illustrate the capability of Fe-Au@PAA NPs to enhance CT signals, firstly, the phantom images of aqueous solutions of NPs of different concentrations (0.5–20 g/L) were obtained (Figure 8a). The X-rays were generated with a tube at 50 kV voltage and 1 mA current. Results showed that Fe-Au@PAA NPs at concentrations lower than 1 mg/mL (0.026 mM of Au) had no difference in X-ray enhancement compared to distilled water. X-ray attenuation linearly depended on NP concentration with 0.97 Pearson's correlation coefficient (Figure 8b). The slope value was (57.7 ± 0.4) HU/(g/L) or (3.75 ± 0.02) HU/mM if recalculated to the molar concentration of Au atoms. The slope value of Fe-Au@PAA NPs was higher than that of Iohexol 2.7 HU/mM, a CT contrast agent commonly used in clinics [56].

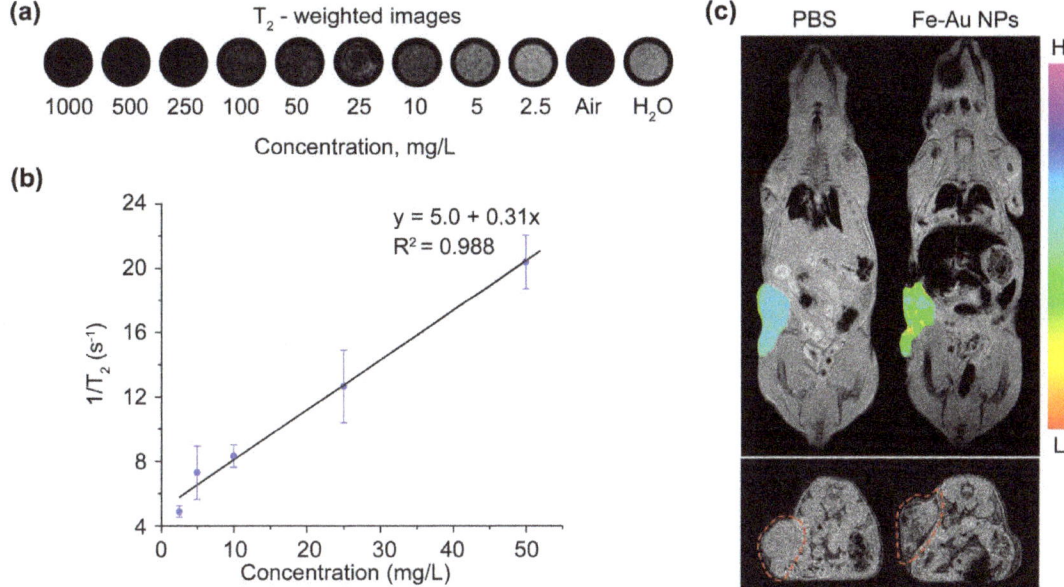

Figure 7. Application of Fe-Au@PAA NPs as MRI contrast agents. (**a**) T_2-weighted images of NP solutions at 1–1000 mg/L concentrations, distilled water, and air. (**b**) Inverse of the proton relaxation time T_2 as a function of Fe-Au@PAA NP concentration. (**c**) MR images of EMT6/P tumor bearing mouse after intravenous injection of PBS or Fe-Au@PAA NPs. In the coronal projections (top), a color image of tumor was combined with grey-level image, and scale represents signal intensity. In the axial projections (down), the tumor is bordered by a red dashed line.

To test Fe-Au@PAA as CT contrast agents, the in vivo X-ray attenuation properties of nanoparticles were measured before and after intravenous and intratumoral injections. After intravenous injections of 1 mg of Fe-Au@PAA NPs, no significant effects of contrast were achieved (Figure 8c). According to the literature data, sufficient signal enhancement in vivo can be provided after intravenous administrations only of 5–7 mg of pure gold NPs, which is approximately 15-times higher than the Au dose used in this study [57,58]. Intravenous injection in a rat tail of Fe_3O_4@Au composite nanoparticles in doses 28-times higher than that used in this work led to the contrast enhancement of the aorta and whole liver [59]. After an intratumoral injection of NPs in 40 μg dose, remarkably improved contrast enhancement was observed, which rendered it possible to visualize boundaries between tumor and normal adjacent tissues on axial projections.

Fe-Au@PAA NPs is a promising candidate for dual MRI and CT imaging. However, for further improvements of contrasting properties of both modalities, one needs to increase the accumulation of NPs at the targeted region. For this, different target proteins for specific tumor types could be anchored to Fe-Au@PAA NPs surface, e.g., antibody, affibody, or darpins [60]. Another approach is to modulate the accumulation of NPs by the prolongation of their blood circulation time by means of surface modification with stealth polymers or cell membrane coating [61,62] or by the blockade of mononuclear phagocyte system cells [63]. The implementation of both strategies at once might greatly increase the efficiency of nanoparticle trapping at a tumor site.

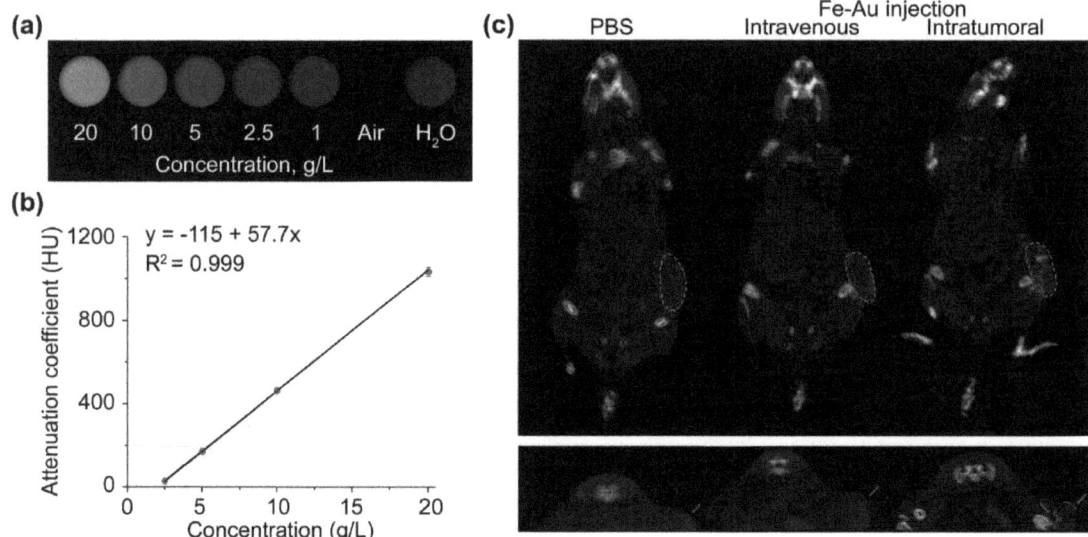

Figure 8. Application of Fe-Au@PAA NPs as CT contrast agents. (**a**) CT images of NP solutions at 1–20 g/L concentrations, distilled water, and air. (**b**) X-ray attenuation coefficient as a function of Fe-Au@PAA NP concentration. (**c**) CT images of EMT6/P tumor bearing mice before and after intravenous or intratumoral injections of Fe-Au@PAA NPs. The tumor boundary is indicated by a red dashed line in coronal projections (top) or with red arrows at axial projections (down).

4. Conclusions

This study presents the first assessment of biocompatibility, pharmacokinetics, and the application of laser-synthesized Fe-Au NPs for MRI/CT imaging and photothermal treatment. NPs were synthesized via laser ablation in liquids, which enabled the obtainment of materials with chemically pure surfaces. Coating by polyacrylic acid strongly improved the colloidal stability of Fe-Au NPs in a biologically relevant fluid, making them suitable for biomedical applications. Plasmon peaks in the visible region with a long near-infrared tail allowed heating Fe-Au core-satellites under 808 nm laser exposure. At 25 µg/mL concentration, Fe-Au@PAA NPs showed low cytotoxic effects in vitro, while under laser irradiation, they induced 100% cell death. Due to hybrid core-satellite architecture of Fe-Au@PAA, NPs had contrasting properties both in MRI and CT imaging modalities. In vivo injection of NPs allowed the visualization of tumor boundaries in T_2-weighted MR images and CT scans. Our results provided promise for the development of novel phototherapy and imaging agents, profiting from the superior properties of laser-synthesized bimetallic nanoparticles with core-satellite structure.

Supplementary Materials: The following are available online at https://www.mdpi.com/article/10.3390/pharmaceutics14050994/s1, Figure S1: SEM image of the Fe-Au nanoparticles obtained by the method of pulsed laser ablation in liquids. Scale bar: 500 nm, Figure S2: ζ-potential of PAA-coated Fe-Au NPs 25 days after their storage at 4 °C. Figure S3: Temperature curves of Fe-Au@PAA NPs (a) and dependence of maximal temperature change on the particle concentration (b). NPs were irradiated with 100 mW 532 nm laser source for 10 min. Figure S4: Histological evaluation of iron biodistribution. Tissues were stained with eosin and Perls Prussian blue. Trivalent iron in liver, spleen, lungs and tumor is colored black. Scale bar: 100 µm.

Author Contributions: Conceptualization, I.V.Z. and A.V.K.; formal analysis, O.Y.G., I.B.B., and A.S.S.; funding acquisition, S.M.D.; investigation, O.Y.G., I.B.B., A.S.S., I.V.Z., G.V.T., A.A.P., and A.S.K.; methodology, P.I.N., D.A.G., and A.V.K.; supervision, I.V.Z., A.V.K., and S.M.D.; writing—original draft, O.Y.G., I.B.B., and A.S.S.; writing—review and editing, I.V.Z., D.A.G., P.I.N., A.V.K., and S.M.D. All authors have read and agreed to the published version of the manuscript.

Funding: The work was supported by the Ministry of Science and Higher Education of the Russian Federation, agreement no. 075-15-2020-773.

Institutional Review Board Statement: This study was performed in strict accordance with the provisions of the European Convention for the Protection of Vertebrate Animals Used for Experimental and Other Scientific Purposes (ETS No. 123 of 18 March 1986) and was approved by the Institutional Animal Care and Use Committee of Shemyakin–Ovchinnikov Institute of Bioorganic Chemistry (Moscow, Russia) with the protocol № 240 from 16 June 2020.

Informed Consent Statement: Not applicable.

Data Availability Statement: All the data are presented within this article and its Supplementary Material. The raw datasets are available on request from the corresponding authors.

Conflicts of Interest: The authors have no conflicts of interest to declare.

References

1. Sizikov, A.A.; Kharlamova, M.V.; Nikitin, M.P.; Nikitin, P.I.; Kolychev, E.L. Nonviral Locally Injected Magnetic Vectors for in vivo Gene Delivery: A Review of Studies on Magnetofection. *Nanomaterials (Basel)* **2021**, *11*, 1078. [CrossRef] [PubMed]
2. Vallabani, N.V.S.; Singh, S. Recent advances and future prospects of iron oxide nanoparticles in biomedicine and diagnostics. *3 Biotech* **2018**, *8*, 279. [CrossRef] [PubMed]
3. Sargazi, S.; Hajinezhad, M.R.; Rahdar, A.; Mukhtar, M.; Karamzadeh-Jahromi, M.; Almasi-Kashi, M.; Alikhanzadeh-Arani, S.; Barani, M.; Baino, F. CoNi alloy nanoparticles for cancer theranostics: Synthesis, physical characterization, *in vitro* and in vivo studies. *Appl. Phys. A* **2021**, *127*. [CrossRef]
4. Arkaban, H.; Khajeh Ebrahimi, A.; Yarahmadi, A.; Zarrintaj, P.; Barani, M. Development of a multifunctional system based on CoFe2O4@polyacrylic acid NPs conjugated to folic acid and loaded with doxorubicin for cancer theranostics. *Nanotechnology* **2021**, *32*. [CrossRef]
5. van de Walle, A.; Kolosnjaj-Tabi, J.; Lalatonne, Y.; Wilhelm, C. Ever-Evolving Identity of Magnetic Nanoparticles within Human Cells: The Interplay of Endosomal Confinement, Degradation, Storage, and Neocrystallization. *Acc. Chem. Res.* **2020**, *53*, 2212–2224. [CrossRef]
6. Briley-Saebo, K.; Bjørnerud, A.; Grant, D.; Ahlstrom, H.; Berg, T.; Kindberg, G.M. Hepatic cellular distribution and degradation of iron oxide nanoparticles following single intravenous injection in rats: Implications for magnetic resonance imaging. *Cell Tissue Res.* **2004**, *316*, 315–323. [CrossRef]
7. Zelepukin, I.V.; Yaremenko, A.V.; Ivanov, I.N.; Yuryev, M.V.; Cherkasov, V.R.; Deyev, S.M.; Nikitin, P.I.; Nikitin, M.P. Long-Term Fate of Magnetic Particles in Mice: A Comprehensive Study. *ACS Nano* **2021**. [CrossRef]
8. Rogosnitzky, M.; Branch, S. Gadolinium-based contrast agent toxicity: A review of known and proposed mechanisms. *Biometals* **2016**, *29*, 365–376. [CrossRef]
9. Alphandéry, E. Biodistribution and targeting properties of iron oxide nanoparticles for treatments of cancer and iron anemia disease. *Nanotoxicology* **2019**, *13*, 573–596. [CrossRef]
10. Naik, G.V.; Shalaev, V.M.; Boltasseva, A. Alternative plasmonic materials: Beyond gold and silver. *Adv. Mater.* **2013**, *25*, 3264–3294. [CrossRef]
11. Ivanov, V.; Lizunova, A.; Rodionova, O.; Kostrov, A.; Kornyushin, D.; Aybush, A.; Golodyayeva, A.; Efimov, A.; Nadtochenko, V. Aerosol Dry Printing for SERS and Photoluminescence-Active Gold Nanostructures Preparation for Detection of Traces in Dye Mixtures. *Nanomaterials (Basel)* **2022**, *12*, 448. [CrossRef] [PubMed]
12. Amendola, V.; Scaramuzza, S.; Litti, L.; Meneghetti, M.; Zuccolotto, G.; Rosato, A.; Nicolato, E.; Marzola, P.; Fracasso, G.; Anselmi, C.; et al. Magneto-plasmonic Au-Fe alloy nanoparticles designed for multimodal SERS-MRI-CT imaging. *Small* **2014**, *10*, 2476–2486. [CrossRef]
13. Terentyuk, G.S.; Maslyakova, G.N.; Suleymanova, L.V.; Khlebtsov, N.G.; Khlebtsov, B.N.; Akchurin, G.G.; Maksimova, I.L.; Tuchin, V.V. Laser-induced tissue hyperthermia mediated by gold nanoparticles: Toward cancer phototherapy. *J. Biomed. Opt.* **2009**, *14*, 21016. [CrossRef]
14. Cao, M.; Chang, Z.; Tan, J.; Wang, X.; Zhang, P.; Lin, S.; Liu, J.; Li, A. Superoxide Radical-Mediated Self-Synthesized Au/MoO3-x Hybrids with Enhanced Peroxidase-like Activity and Photothermal Effect for Anti-MRSA Therapy. *ACS Appl. Mater. Interfaces* **2022**, *14*, 13025–13037. [CrossRef] [PubMed]
15. Mantri, Y.; Jokerst, J.V. Engineering Plasmonic Nanoparticles for Enhanced Photoacoustic Imaging. *ACS Nano* **2020**, *14*, 9408–9422. [CrossRef]

16. Xu, C.; Tung, G.A.; Sun, S. Size and Concentration Effect of Gold Nanoparticles on X-ray Attenuation As Measured on Computed Tomography. *Chem. Mater.* **2008**, *20*, 4167–4169. [CrossRef] [PubMed]
17. Caro, C.; Gámez, F.; Quaresma, P.; Páez-Muñoz, J.M.; Domínguez, A.; Pearson, J.R.; Pernía Leal, M.; Beltrán, A.M.; Fernandez-Afonso, Y.; de La Fuente, J.M.; et al. Fe3O4-Au Core-Shell Nanoparticles as a Multimodal Platform for in vivo Imaging and Focused Photothermal Therapy. *Pharmaceutics* **2021**, *13*, 416. [CrossRef] [PubMed]
18. Kang, N.; Xu, D.; Han, Y.; Lv, X.; Chen, Z.; Zhou, T.; Ren, L.; Zhou, X. Magnetic targeting core/shell Fe3O4/Au nanoparticles for magnetic resonance/photoacoustic dual-modal imaging. *Mater. Sci. Eng. C Mater. Biol. Appl.* **2019**, *98*, 545–549. [CrossRef]
19. Wang, W.; Hao, C.; Sun, M.; Xu, L.; Xu, C.; Kuang, H. Spiky Fe 3 O 4 @Au Supraparticles for Multimodal in vivo Imaging. *Adv. Funct. Mater.* **2018**, *28*, 1800310. [CrossRef]
20. Kabashin, A.V.; Meunier, M. Femtosecond laser ablation in aqueous solutions: A novel method to synthesize non-toxic metal colloids with controllable size. *J. Phys.: Conf. Ser.* **2007**, *59*, 354–359. [CrossRef]
21. Baati, T.; Al-Kattan, A.; Esteve, M.-A.; Njim, L.; Ryabchikov, Y.; Chaspoul, F.; Hammami, M.; Sentis, M.; Kabashin, A.V.; Braguer, D. Ultrapure laser-synthesized Si-based nanomaterials for biomedical applications: In vivo assessment of safety and biodistribution. *Sci. Rep.* **2016**, *6*, 25400. [CrossRef] [PubMed]
22. Bailly, A.-L.; Correard, F.; Popov, A.; Tselikov, G.; Chaspoul, F.; Appay, R.; Al-Kattan, A.; Kabashin, A.V.; Braguer, D.; Esteve, M.-A. In vivo evaluation of safety, biodistribution and pharmacokinetics of laser-synthesized gold nanoparticles. *Sci. Rep.* **2019**, *9*, 12890. [CrossRef]
23. Zelepukin, I.V.; Popov, A.A.; Shipunova, V.O.; Tikhonowski, G.V.; Mirkasymov, A.B.; Popova-Kuznetsova, E.A.; Klimentov, S.M.; Kabashin, A.V.; Deyev, S.M. Laser-synthesized TiN nanoparticles for biomedical applications: Evaluation of safety, biodistribution and pharmacokinetics. *Mater. Sci. Eng. C Mater. Biol. Appl.* **2021**, *120*, 111717. [CrossRef] [PubMed]
24. Zelepukin, I.V.; Mashkovich, E.A.; Lipey, N.A.; Popov, A.A.; Shipunova, V.O.; Griaznova, O.Y.; Deryabin, M.S.; Kurin, V.V.; Nikitin, P.I.; Kabashin, A.V.; et al. Direct photoacoustic measurement of silicon nanoparticle degradation promoted by a polymer coating. *Chem. Eng. J.* **2022**, *430*, 132860. [CrossRef]
25. Popov, A.A.; Swiatkowska-Warkocka, Z.; Marszalek, M.; Tselikov, G.; Zelepukin, I.V.; Al-Kattan, A.; Deyev, S.M.; Klimentov, S.M.; Itina, T.E.; Kabashin, A.V. Laser-Ablative Synthesis of Ultrapure Magneto-Plasmonic Core-Satellite Nanocomposites for Biomedical Applications. *Nanomaterials (Basel)* **2022**, *12*, 649. [CrossRef] [PubMed]
26. Balfourier, A.; Luciani, N.; Wang, G.; Lelong, G.; Ersen, O.; Khelfa, A.; Alloyeau, D.; Gazeau, F.; Carn, F. Unexpected intracellular biodegradation and recrystallization of gold nanoparticles. *Proc. Natl. Acad. Sci. USA* **2020**, *117*, 103–113. [CrossRef] [PubMed]
27. Roper, D.K.; Ahn, W.; Hoepfner, M. Microscale Heat Transfer Transduced by Surface Plasmon Resonant Gold Nanoparticles. *J. Phys. Chem. C Nanomater. Interfaces* **2007**, *111*, 3636–3641. [CrossRef]
28. Nikitin, P.I.; Vetoshko, P.M.; Ksenevich, T.I. New type of biosensor based on magnetic nanoparticle detection. *J. Magn. Magn. Mater.* **2007**, *311*, 445–449. [CrossRef]
29. Zelepukin, I.V.; Yaremenko, A.V.; Yuryev, M.V.; Mirkasymov, A.B.; Sokolov, I.L.; Deyev, S.M.; Nikitin, P.I.; Nikitin, M.P. Fast processes of nanoparticle blood clearance: Comprehensive study. *J. Control. Release* **2020**, *326*, 181–191. [CrossRef]
30. Pandey, S.P.; Shukla, T.; Dhote, V.K.; Mishra, D.K.; Maheshwari, R.; Tekade, R.K. Use of Polymers in Controlled Release of Active Agents. In *Basic Fundamentals of Drug Delivery*; an imprint of Elsevier; Tekade, R.K., Ed.; Academic Press: London, UK; San Diego, CA, USA, 2019; ISBN 9780128179093.
31. Kumar, A.; Dixit, C.K. Methods for characterization of nanoparticles. In *Advances in Nanomedicine for the Delivery of Therapeutic Nucleic Acids*; Nimesh, S., Chan, R., Gupta, N., Eds.; Woodhead Publishing: Duxford, UK, 2017; pp. 43–58. ISBN 9780081005576.
32. Efremova, M.V.; Naumenko, V.A.; Spasova, M.; Garanina, A.S.; Abakumov, M.A.; Blokhina, A.D.; Melnikov, P.A.; Prelovskaya, A.O.; Heidelmann, M.; Li, Z.-A.; et al. Magnetite-Gold nanohybrids as ideal all-in-one platforms for theranostics. *Sci. Rep.* **2018**, *8*, 11295. [CrossRef]
33. Kura, H.; Takahashi, M.; Ogawa, T. Synthesis of Monodisperse Iron Nanoparticles with a High Saturation Magnetization Using an Fe(CO) x —Oleylamine Reacted Precursor. *J. Phys. Chem. C Nanomater. Interfaces* **2010**, *114*, 5835–5838. [CrossRef]
34. Paliwal, S.R.; Kenwat, R.; Maiti, S.; Paliwal, R. Nanotheranostics for Cancer Therapy and Detection: State of the Art. *Curr. Pharm. Des.* **2020**, *26*, 5503–5517. [CrossRef] [PubMed]
35. Mendes, R.; Pedrosa, P.; Lima, J.C.; Fernandes, A.R.; Baptista, P.V. Photothermal enhancement of chemotherapy in breast cancer by visible irradiation of Gold Nanoparticles. *Sci. Rep.* **2017**, *7*, 10872. [CrossRef] [PubMed]
36. Xie, X.; Liu, Q.; Paulus, Y.M. Innovations in Retinal Laser Technology. *OPJ* **2018**, *8*, 173–186. [CrossRef]
37. Hemmer, E.; Benayas, A.; Légaré, F.; Vetrone, F. Exploiting the biological windows: Current perspectives on fluorescent bioprobes emitting above 1000 nm. *Nanoscale Horiz.* **2016**, *1*, 168–184. [CrossRef]
38. Zhang, P.; Wang, J.; Huang, H.; Yu, B.; Qiu, K.; Huang, J.; Wang, S.; Jiang, L.; Gasser, G.; Ji, L.; et al. Unexpected high photothemal conversion efficiency of gold nanospheres upon grafting with two-photon luminescent ruthenium(II) complexes: A way towards cancer therapy? *Biomaterials* **2015**, *63*, 102–114. [CrossRef]
39. Chang, T.-W.; Ko, H.; Huang, W.-S.; Chiu, Y.-C.; Yang, L.-X.; Chia, Z.-C.; Chin, Y.-C.; Chen, Y.-J.; Tsai, Y.-T.; Hsu, C.-W.; et al. Tannic acid-induced interfacial ligand-to-metal charge transfer and the phase transformation of Fe3O4 nanoparticles for the photothermal bacteria destruction. *Chem. Eng. J.* **2022**, *428*, 131237. [CrossRef]
40. Mosmann, T. Rapid colorimetric assay for cellular growth and survival: Application to proliferation and cytotoxicity assays. *J. Immunol. Methods* **1983**, *65*, 55–63. [CrossRef]

41. Stockert, J.C.; Blázquez-Castro, A.; Cañete, M.; Horobin, R.W.; Villanueva, A. MTT assay for cell viability: Intracellular localization of the formazan product is in lipid droplets. *Acta Histochem.* **2012**, *114*, 785–796. [CrossRef]
42. Christou, E.; Pearson, J.R.; Beltrán, A.M.; Fernández-Afonso, Y.; Gutiérrez, L.; de La Fuente, J.M.; Gámez, F.; García-Martín, M.L.; Caro, C. Iron-Gold Nanoflowers: A Promising Tool for Multimodal Imaging and Hyperthermia Therapy. *Pharmaceutics* **2022**, *14*, 636. [CrossRef]
43. Cameron, R.B.; Hou, D. Intraoperative hyperthermic chemotherapy perfusion for malignant pleural mesothelioma: An *in vitro* evaluation. *J. Thorac. Cardiovasc. Surg.* **2013**, *145*, 496–504. [CrossRef] [PubMed]
44. Buch, K.; Peters, T.; Nawroth, T.; Sänger, M.; Schmidberger, H.; Langguth, P. Determination of cell survival after irradiation via clonogenic assay versus multiple MTT Assay–a comparative study. *Radiat. Oncol.* **2012**, *7*, 1. [CrossRef] [PubMed]
45. Chen, D.; Pan, X.; Xie, F.; Lu, Y.; Zou, H.; Yin, C.; Zhang, Y.; Gao, J. Codelivery of doxorubicin and elacridar to target both liver cancer cells and stem cells by polylactide-co-glycolide/d-alpha-tocopherol polyethylene glycol 1000 succinate nanoparticles. *Int. J. Nanomedicine* **2018**, *13*, 6855–6870. [CrossRef] [PubMed]
46. Vakili-Ghartavol, R.; Momtazi-Borojeni, A.A.; Vakili-Ghartavol, Z.; Aiyelabegan, H.T.; Jaafari, M.R.; Rezayat, S.M.; Arbabi Bidgoli, S. Toxicity assessment of superparamagnetic iron oxide nanoparticles in different tissues. *Artif. Cells Nanomed. Biotechnol.* **2020**, *48*, 443–451. [CrossRef]
47. Anselmo, A.C.; Mitragotri, S. Nanoparticles in the clinic. *Bioeng. Transl. Med.* **2016**, *1*, 10–29. [CrossRef]
48. Singh, P.; Pandit, S.; Mokkapati, V.R.S.S.; Garg, A.; Ravikumar, V.; Mijakovic, I. Gold Nanoparticles in Diagnostics and Therapeutics for Human Cancer. *Int. J. Mol. Sci.* **2018**, *19*, 1979. [CrossRef]
49. Rastinehad, A.R.; Anastos, H.; Wajswol, E.; Winoker, J.S.; Sfakianos, J.P.; Doppalapudi, S.K.; Carrick, M.R.; Knauer, C.J.; Taouli, B.; Lewis, S.C.; et al. Gold nanoshell-localized photothermal ablation of prostate tumors in a clinical pilot device study. *Proc. Natl. Acad. Sci. USA* **2019**, *116*, 18590–18596. [CrossRef]
50. Aires, A.; Cabrera, D.; Alonso-Pardo, L.C.; Cortajarena, A.L.; Teran, F.J. Elucidation of the Physicochemical Properties Ruling the Colloidal Stability of Iron Oxide Nanoparticles under Physiological Conditions. *ChemNanoMat* **2017**, *3*, 183–189. [CrossRef]
51. Matsumura, Y.; Maeda, H. A new concept for macromolecular therapeutics in cancer chemotherapy: Mechanism of tumoritropic accumulation of proteins and the antitumor agent smancs. *Cancer Res.* **1986**, *46*, 6387–6392.
52. Lee, S.; Han, H.; Koo, H.; Na, J.H.; Yoon, H.Y.; Lee, K.E.; Lee, H.; Kim, H.; Kwon, I.C.; Kim, K. Extracellular matrix remodeling in vivo for enhancing tumor-targeting efficiency of nanoparticle drug carriers using the pulsed high intensity focused ultrasound. *J. Control. Release* **2017**, *263*, 68–78. [CrossRef]
53. Cai, H.; Li, K.; Shen, M.; Wen, S.; Luo, Y.; Peng, C.; Zhang, G.; Shi, X. Facile assembly of Fe3O4@Au nanocomposite particles for dual mode magnetic resonance and computed tomography imaging applications. *J. Mater. Chem.* **2012**, *22*, 15110. [CrossRef]
54. Zhou, T.; Wu, B.; Xing, D. Bio-modified Fe 3 O 4 core/Au shell nanoparticles for targeting and multimodal imaging of cancer cells. *J. Mater. Chem.* **2012**, *22*, 470–477. [CrossRef]
55. Storm, L.; Israel, H.I. Photon cross sections from 1 keV to 100 MeV for elements Z=1 to Z=100. *At. Data Nucl. Data Tables* **1970**, *7*, 565–681. [CrossRef]
56. Dou, Y.; Li, X.; Yang, W.; Guo, Y.; Wu, M.; Liu, Y.; Li, X.; Zhang, X.; Chang, J. PB@Au Core-Satellite Multifunctional Nanotheranostics for Magnetic Resonance and Computed Tomography Imaging in vivo and Synergetic Photothermal and Radiosensitive Therapy. *ACS Appl. Mater. Interfaces* **2017**, *9*, 1263–1272. [CrossRef] [PubMed]
57. Reuveni, T.; Motiei, M.; Romman, Z.; Popovtzer, A.; Popovtzer, R. Targeted gold nanoparticles enable molecular CT imaging of cancer: An in vivo study. *Int. J. Nanomedicine* **2011**, *6*, 2859–2864. [CrossRef] [PubMed]
58. Kim, D.; Park, S.; Lee, J.H.; Jeong, Y.Y.; Jon, S. Antibiofouling polymer-coated gold nanoparticles as a contrast agent for in vivo X-ray computed tomography imaging. *J. Am. Chem. Soc.* **2007**, *129*, 7661–7665. [CrossRef] [PubMed]
59. Li, J.; Zheng, L.; Cai, H.; Sun, W.; Shen, M.; Zhang, G.; Shi, X. Facile one-pot synthesis of Fe3O4@Au composite nanoparticles for dual-mode MR/CT imaging applications. *ACS Appl. Mater. Interfaces* **2013**, *5*, 10357–10366. [CrossRef] [PubMed]
60. Tolmachev, V.M.; Chernov, V.I.; Deyev, S.M. Targeted nuclear medicine. Seek and destroy. *Russian Chem. Rev.* **2022**, *91*. [CrossRef]
61. Zhen, X.; Cheng, P.; Pu, K. Recent Advances in Cell Membrane-Camouflaged Nanoparticles for Cancer Phototherapy. *Small* **2019**, *15*, e1804105. [CrossRef]
62. Li, S.-D.; Huang, L. Stealth nanoparticles: High density but sheddable PEG is a key for tumor targeting. *J. Control. Release* **2010**, *145*, 178–181. [CrossRef]
63. Mirkasymov, A.B.; Zelepukin, I.V.; Nikitin, P.I.; Nikitin, M.P.; Deyev, S.M. in vivo blockade of mononuclear phagocyte system with solid nanoparticles: Efficiency and affecting factors. *J. Control. Release* **2021**, *330*, 111–118. [CrossRef] [PubMed]

Article

Silica Coating of Ferromagnetic Iron Oxide Magnetic Nanoparticles Significantly Enhances Their Hyperthermia Performances for Efficiently Inducing Cancer Cells Death In Vitro

Cristian Iacoviță [1,†], Ionel Fizeșan [2,†], Stefan Nitica [1], Adrian Florea [3,*], Lucian Barbu-Tudoran [4,5], Roxana Dudric [6], Anca Pop [2], Nicoleta Vedeanu [1], Ovidiu Crisan [7], Romulus Tetean [6], Felicia Loghin [2] and Constantin Mihai Lucaciu [1,*]

1. Department of Pharmaceutical Physics-Biophysics, Faculty of Pharmacy, "Iuliu Hatieganu" University of Medicine and Pharmacy, 6 Pasteur St., 400349 Cluj-Napoca, Romania; cristian.iacovita@umfcluj.ro (C.I.); stefan_nitica@yahoo.com (S.N.); nicoletavedeanu@yahoo.com (N.V.)
2. Department of Toxicology, Faculty of Pharmacy, "Iuliu Hațieganu" University of Medicine and Pharmacy, 6A Pasteur St., 400349 Cluj-Napoca, Romania; ionel.fizesan@umfcluj.ro (I.F.); anca.pop@umfcluj.ro (A.P.); floghin@umfcluj.ro (F.L.)
3. Department of Cell and Molecular Biology, Faculty of Medicine, "Iuliu Hatieganu" University of Medicine and Pharmacy, 6 Pasteur St., 400349 Cluj-Napoca, Romania
4. Electron Microscopy Center "Prof. C. Craciun", Faculty of Biology & Geology, "Babes-Bolyai" University, 5-7 Clinicilor St., 400006 Cluj-Napoca, Romania; lucian.barbu@ubbcluj.ro
5. Electron Microscopy Integrated Laboratory, National Institute for Research and Development of Isotopic and Molecular Technologies, 67-103 Donath St., 400293 Cluj-Napoca, Romania
6. Faculty of Physics, "Babes Bolyai" University, Kogalniceanu 1, 400084 Cluj-Napoca, Romania; roxana.dudric@ubbcluj.ro (R.D.); romulus.tetean@phys.ubbcluj.ro (R.T.)
7. Department of Organic Chemistry, "Iuliu Hațieganu" University of Medicine and Pharmacy, 41 Victor Babes St., 400012 Cluj-Napoca, Romania; ocrisan@umfcluj.ro
* Correspondence: aflorea@umfcluj.ro (A.F.); clucaciu@umfcluj.ro (C.M.L.); Tel.: +40-752-265-123 (A.F.); +40-744-647-854 (C.M.L.)
† These authors contributed equally to this work.

Abstract: Increasing the biocompatibility, cellular uptake, and magnetic heating performance of ferromagnetic iron-oxide magnetic nanoparticles (F-MNPs) is clearly required to efficiently induce apoptosis of cancer cells by magnetic hyperthermia (MH). Thus, F-MNPs were coated with silica layers of different thicknesses via a reverse microemulsion method, and their morphological, structural, and magnetic properties were evaluated by multiple techniques. The presence of a SiO_2 layer significantly increased the colloidal stability of F-MNPs, which also enhanced their heating performance in water with almost 1000 W/g_{Fe} as compared to bare F-MNPs. The silica-coated F-MNPs exhibited biocompatibility of up to 250 µg/cm^2 as assessed by Alamar Blues and Neutral Red assays on two cancer cell lines and one normal cell line. The cancer cells were found to internalize a higher quantity of silica-coated F-MNPs, in large endosomes, dispersed in the cytoplasm or inside lysosomes, and hence were more sensitive to in vitro MH treatment compared to the normal ones. Cellular death of more than 50% of the malignant cells was reached starting at a dose of 31.25 µg/cm^2 and an amplitude of alternating magnetic field of 30 kA/m at 355 kHz.

Keywords: iron oxide magnetic nanoparticles; silica coating; magnetic hyperthermia; cancer cells; alamar blue; neutral red; A549; A35; BJ

Citation: Iacoviță, C.; Fizeșan, I.; Nitica, S.; Florea, A.; Barbu-Tudoran, L.; Dudric, R.; Pop, A.; Vedeanu, N.; Crisan, O.; Tetean, R.; et al. Silica Coating of Ferromagnetic Iron Oxide Magnetic Nanoparticles Significantly Enhances Their Hyperthermia Performances for Efficiently Inducing Cancer Cells Death In Vitro. *Pharmaceutics* **2021**, *13*, 2026. https://doi.org/10.3390/pharmaceutics13122026

Academic Editor: Hassan Bousbaa

Received: 28 October 2021
Accepted: 24 November 2021
Published: 27 November 2021

Publisher's Note: MDPI stays neutral with regard to jurisdictional claims in published maps and institutional affiliations.

Copyright: © 2021 by the authors. Licensee MDPI, Basel, Switzerland. This article is an open access article distributed under the terms and conditions of the Creative Commons Attribution (CC BY) license (https://creativecommons.org/licenses/by/4.0/).

1. Introduction

Magnetic nanoparticles (MNPs) subjected to the action of an externally applied alternating magnetic field (AMF) generate heat, leading to an exciting new technique in cancer treatment, named magnetic hyperthermia (MH) [1–4]. As attested by different clinical trials on prostate carcinoma [5,6], glioblastoma [7,8], and breast malignant tumors [9], the MH offers the possibility of heating the targeted tumoral tissue while preserving the healthy tissue. In clinical trials, the MH treatment has been carried out using superparamagnetic iron oxide nanoparticles (SPIONs), which were previously approved for clinical use by the US Food and Drug Administration [10]. Due to the limited efficiency of heat generation by SPIONs in the local tumor tissues for achieving a high performance in cancer therapy, very high dosages of SPIONs have been used. Moreover, to facilitate the complete elimination of the tumor, the MH treatment with SPIONs has been used in conjunction with chemotherapy and radiotherapy, which are prone to aggressive side effects. Therefore, for clinical safety, the MH treatment requires MNPs to display high heating efficiency to reduce the applied dose and minimize the risk of side effects.

MNPs dissipate the magnetic energy of interaction with alternating magnetic fields (AMF) in the environment through hysteresis losses [11]. The heating behavior of SPIONs can be simply described in the frame of the so-called Linear Response Theory [12]. In this model, the magnetization of the SPIONs depends linearly on the external AMF and the heat generated and its dependence on the MNPs' properties are explained in the terms of the Neel and Brown magnetic relaxations processes [11,12]. The cell internalization of SPIONs leads to an enhancement of intracellular clustering inside endosomes or lysosomes and inhibits their mobility in cells, restricting their physical rotation, and thus affecting the Brown relaxation process [13,14]. Overall, the specific absorption rate (SAR, the rate at which heat is dissipated by the MNPs) is dramatically reduced [15,16]. The replacement of SPIONs with larger MNPs in the ferromagnetic domain (F-MNPs) represents an alternative to improve the effectiveness of MH treatment in cancer therapy. The SAR values increase almost one order of magnitude in comparison with SPIONs, due to the increase in the size and to an increase in the dynamic hysteresis area resulting in enhanced hyperthermic efficacy [11,16–24]. Besides favorable MH performance, the F-MNPs have a significant drawback as magnetic dipole-dipole interactions among them are very strong and hinder the complete dispersion in solution. In this regard, several strategies of coating the surface of F-MNPs have been formulated aiming at improving their colloidal stability and to provide, at the same time, a high level of biocompatibility [25–28].

Silica coating represents a well-known technique for MNP surface modification concerning the main required features for their biomedical use [29,30]. For instance, the silica shell protects the magnetic core from corrosion and/or oxidation, maintaining its magnetic properties [31,32]; makes the MNPs hydrophilic and provides colloidal stability in biological solutions by avoiding interparticle interactions and agglomeration [33,34]; presents excellent biocompatibility, exhibiting no cell cytotoxicity at concentrations up to 500 µg/mL [35–39]; prevents iron release [40]; and generates high surface area available for further attachment of anticancer drugs or other biologically active moieties [41–45]. In terms of MH properties, individual MNPs coated with a silica shell, which protects them from aggregation when subjected to AMF, showed better heating performances in comparison with uncoated MNPs or with multiple MNPs encapsulated in a silica shell [35,37–39,41,46,47]. The excellent biocompatibility properties offered by the silica shell [35–40,48,49] facilitate the internalization of a large quantity of silica-coated MNPs inside cells which upon MH treatment effectively leads to the eradication of tumor cells in vitro [35,37–39,41,46,47]. The formation of a silica shell around MNPs has been mainly realized by the so-called Stöber process where tetraethyl orthosilicate (TEOS) is hydrolyzed in ethanol under the addition of aqueous ammonia. However, the Stöber method, widely used for silica coating of both either hydrophilic or hydrophobic SPIONs and F-MNPs, offers poor control over the silica shell thickness and uniformity and leads mainly to encapsulation of multiple MNPs [31,32,39,40,47,49–51]. Alternatively, a fine-tuning of silica shell thickness around single magnetic cores is offered by the water-in-oil micro-emulsion method. This method uses a non-ionic surfactant that facilitates ammonia-mediated TEOS hydrolysis on the surface of MNPs dispersed in an organic solvent [34–36,38]. To the best of our knowledge, the reverse microemulsion method provides the best results on SPIONs, as they exhibit superior colloidal dispersion as compared to F-MNPs. Additionally, there is no extensive study on the evolution of MH properties of F-MNPs upon their coating with a silica shell, or on their cytotoxicity and in vitro MH performances.

Based on the above description, the aim of this work was twofold: the optimization of the ferromagnetic polyhedral iron oxide MNPs (Fe_3O_4) coating with a homogeneous silica shell to increase their colloidal stability in water and their SAR, and, second, the evaluation of their cytocompatibility and their MH capabilities to induce cancer cell apoptosis in vitro. Employing the reverse microemulsion method, we were able to coat the polyhedral Fe_3O_4 with a silica shell of variable thickness. The structural, magnetic, and hyperthermia properties were studied as a function of the coating thickness and compared to the uncoated polyhedral Fe_3O_4. The silica-coated polyhedral Fe_3O_4 (sFe_3O_4) providing the best heating performance were further selected to evaluate their cytotoxicity, cellular uptake, and in vitro MH performance. The interaction of sFe_3O_4 with two types of cancer cell lines—human pulmonary cancer cells (A549) and human melanoma cancer cells (A375)—and one normal cell line, human foreskin fibroblasts (BJ), was considered. The cytotoxicity and intracellular MH were assessed using two complementary assays: the neutral red (NR)-uptake assay, which is related mainly to lysosomal activity of the cells, and the Alamar Blue (AB) assay, which provides more general information related to the whole-cell metabolism. The uptake of sFe_3O_4 MNPs by cells upon 24 h incubation was monitored by using scanning and transmission electron microscopy (SEM and TEM). These two experimental techniques together with the cytocompatibility assays were extensively used to evaluate the cellular viability and intracellular damages, respectively, upon exposure of cells loaded with different amounts of sFe_3O_4 MNPs to AMF of different amplitudes (up to 65 kA/m) at a constant frequency of 355 kHz.

2. Materials and Methods

2.1. Synthesis

All the reagents used for the synthesis of MNPs and their coating with a silica layer were of analytical grade and were used without any further purification. The following products were employed: iron (III) chloride hexahydrate ($FeCl_3 \cdot 6H_2O$) (Carl-Roth, Karlsruhe, Germany, $\geq$98%), polyethylene glycol 200 (PEG 200) (Roth, $\geq$99%), and sodium acetate trihydrate (NaAc) (Roth, $\geq$99.5%), ethanol (Chemical, Iași, Romania, Chemicals, 99.9 %), cyclohexane (Sigma Aldrich, Steinheim, Germany), Igepal CO-520 (Sigma Aldrich, St. Louis, MI, USA), ammonium hydroxide (25%) (Chemicals), tetraethyl orthosilicate (TEOS) (Sigma Aldrich, $\geq$99%), and APTES (Sigma Aldrich, 99%).

The synthesis of MNPs was performed following a polyol mediated synthetic route as presented previously by our group [52,53]. Upon multiple washing steps, the MNPs were redispersed in ethanol at a concentration of 4 mg_{MNPs}/mL. The MNPs were coated with silica via a reverse microemulsion method [35]. Briefly, 18 mL cyclohexane and 1.15 mL Igepal CO-520 were mixed for 30 min. Afterward, 4 mg of MNPs dispersed in 2 mL cyclohexane were added while stirring. After 5 min, 0.05 mL APTES and 0.05/0.1/0.2 mL TEOS were added, followed by 0.15 mL ammonium hydroxide (25%). The dispersions were stirred at room temperature for 24 h and the resulting $MNPs-SiO_2$ were precipitated by adding ethanol. The collected MNPs were washed in ethanol several times and finally redispersed in water.

2.2. Characterisation

Transmission electron microscopy (TEM) images of MNPs were obtained using a Hitachi HD2700 (Hitachi, Tokyo, Japan) microscope operating at 200 kV, coupled with an EDX (energy-dispersive X-ray) detector (Oxford Instruments, AZtec Software, version 3.3, Oxford, UK) used for elemental detection. Samples were prepared by depositing a drop of MNP dispersion on a carbon-coated copper grid and removing the excess water by filter paper after 2 min.

X-ray diffraction (XRD) measurements were performed using a Bruker D8 Advance diffractometer using Cu Kα radiation (Bruker AXS GmbH, Karlsruhe, Germany). The powder samples were realized by collecting the MNPs with a magnet and vacuum drying them overnight. The FullProf software (FullProf.2k, Version 7.00-May 2019-ILL JRC https://www.ill.eu/sites/fullprof/, Grenoble, France) was employed to detect the crystalline phases and to calculate the lattice parameters.

Fourier transform infrared spectroscopy (FTIR) spectra were recorded between 400 and 4000 cm^{-1} at 4 cm^{-1} resolution using a TENSOR II instrument (Bruker Optics Inc., Billerica, MA, USA) in attenuated total reflectance mode using the platinum attenuated total reflectance (ATR) accessory with a single reflection diamond ATR. A few microliters of the aqueous solution of MNPs were allowed to dry on the diamond crystal, and the average spectrum of 16 scans was recorded for each sample.

Hydrodynamic size and zeta potential measurements of the MNPs were determined using a Zetasizer Nano ZS90 (Malvern Instruments, Worcestershire, UK) in a 90° configuration. Particle solutions with concentration of 0.1 and 0.01 mg_{MNPs}/mL were used for measurements at room temperature. Each assay was made in triplicate.

The magnetic characterization of MNPs was carried out using a Cryogenic Limited (London, UK) vibrating sample magnetometer (VSM). The magnetization and hysteresis curves were measured from 0 to 10 T and −4 and 4 T, respectively, at both 4 and 300 K. The samples consisted of powder obtained in a similar way as for XRD examination.

The heating ability of MNPs was determined using a commercially available magnetic hyperthermia system, the Easy Heat 0224 from Ambrell (Scottsville, NY, USA) equipped with a fiber-optic thermometer. An AC magnetic field with fixed frequency (355 kHz) and variable amplitude (5–65 kA/m) was generated by an 8-turn coil that was made of a water-cooled copper tube. For measurements, 0.5 mL of MNPs suspended in water at different concentrations was placed in the center of the coil using a thermally isolated Teflon holder. Details about SAR calculation and iron concentration determination are provided in the Supplementary Materials.

2.3. Cell Lines

Two types of cancerous cancer cell lines (human pulmonary cancer cells—A549 and human melanoma cancer cells—A375) and one normal cell line (human foreskin fibroblasts—BJ) were used in the current study. All cells were maintained in Dulbecco's Modified Eagle Medium (DMEM, Gibco, Paisley, UK) supplemented with 10% Fetal Bovine Serum (FBS, Sigma Aldrich, Steinheim, Germany). Cells were cultured in flasks at 37 °C in a humidified incubator with 5% CO_2 supplementation while cellular media was changed every other day. At 70–80% confluence cells were subcultured or harvested for experiments.

2.4. In Vitro Cytocompatibility Assays

The cytocompatibility of NPs was evaluated on the cancerous and normal cell phenotypes by performing two complementary assays, namely Alamar Blue (AB) and Neutral Red (NR) assays. AB was used to measure the metabolic ability of viable cells to convert resazurin to resorufin, whereas the NR assay was used to measure the ATP content of exposed cells. In the case of both assays, 750,000 cells for A549 and A375 and 275,000 cells for BJ suspended in 2 mL of media were seeded in 6-well plates and left to attach for 24 h before MNP exposure. Cell concentration was selected to achieve a confluence of 70% before the exposure to the MNPs. Cells were exposed to 1000 µL of cell medium containing MNPs to reach a concentration of 500, 250, 125, 62.5, and 31.25 $\mu g_{MNPs}/cm^2$. After a 24 h incubation, the cells were washed three times with PBS (Gibco, Paisley, UK) and further incubated with the AB and NR dye, and dissolved in cellular media, for measuring the cellular viability. For the AB assay, cells were incubated for 3 h with 200 µM resazurin solution, and the fluorescence was measured at $\lambda_{excitation}$ = 530/25 nm; $\lambda_{emission}$ = 590/35 nm, using a Synergy 2 Multi-Mode Microplate Reader. For the NR assay, cells were incubated with a filtered neutral red dye (40 µg/mL) solution for 2 h, followed by a wash with PBS and the extraction of the accumulated cellular dye in a solution containing 50% ethanol, 49% water, and 1% glacial acetic acid. Supernatant fluorescence was measured at $\lambda_{excitation}$ = 530/25 nm;

$\lambda_{emission}$ = 620/40 nm, using a Synergy 2 Multi-Mode Microplate Reader. The experiments were conducted on at least three biological replicates and included a negative control (cells exposed to culture medium). The results were normalized to the value of negative control (100%).

2.5. Evaluation of Cellular Uptake

The uptake of silica-coated MNPs in cells was quantitatively determined by the thiocyanate assay and qualitatively evaluated by SEM and TEM image analysis. In the case of both assays, the same number of cells as above suspended in 2 mL of media were seeded in 6-well plates and left to attach for 24 h. Afterward, the cells were exposed to 1000 µL of cell medium containing MNPs to reach a concentration of 62.5, 31.25, 15.62, and 7.81 $\mu g_{MNPs}/cm^2$. Upon 24 h incubation, the cells were washed 3 times with PBS to remove attached MNPs from the cellular surface.

For quantitative measurement of the intracellular Fe^{3+} using the thiocyanate assay, cells were trypsinized and centrifuged at $4500 \times g$ for 5 min, then further processed as described in Section S6.

SEM analysis of all three cell lines grown on glass coverslips (previously carefully cleaned and disinfected) was performed. At the end of incubation with silica-coated Fe_3O_4, culture media containing suspended nanoparticles were removed from the wells, and cells were washed 3 times with PBS to remove unbound MNPs. Cells were pre-fixed with 2.7% glutaraldehyde (Agar Scientific, Stansted, UK) solution in 0.1 M phosphate buffer (pH = 7.4) for 2 h and next washed 4 times with the same buffer. Fixation was performed at 4 °C with 1.5% OsO_4 (Sigma-Aldrich, Steinheim, Germany) solution in 0.15 M phosphate buffer (pH = 7.4) for 1.5 h. Finally, an ethanol (VWR International, Fontenay-sous-Bois, France) series of increasing concentrations (50%-absolute—15 min each at room temperature) was used for dehydration. The coverslips fragments with the adherent cells were fastened on 10 mm/Ø9 mm aluminum stubs (Bio-Rad, Hercules, CA, USA), using carbon adhesive tabs (Electron Microscopy Sciences, Hatfield, PA, USA) and colloidal silver (Polaron Equipment, Watford, UK). Two samples were prepared for each group. They were next sputter-coated with gold in a Polaron E-5100 sputter coater (Polaron Equipment, Watford, UK). The examination was performed with a Jeol JSM 25-S (Jeol, Tokyo, Japan) at 30 kV, and relevant images were recorded with a Pixie 3000 system (Deben, Debenham, UK) and subjected both to a qualitative and quantitative evaluation. The general aspect of cells, presence, and the number of thick branches, and the number of filopodia were considered.

For TEM analysis, the cell suspensions were centrifuged for 5 min at 300 g and the pellets were further processed for TEM examination. Pre-fixation was performed for 2 h at 4 °C with 2.7% glutaraldehyde solution in 0.1 M phosphate buffer; the pellets were washed 4 times with the 0.1 M phosphate buffer at 4 °C; post-fixation was performed for 1.5 h with 1.5% OsO_4 solution at 4 °C; for dehydration, an acetone (International Laboratory, Cluj-Napoca, Romania) series of increasing concentrations (30%-absolute—15 min each at room temperature) was used, and embedding was performed in three steps with EMBED 812 (Electron Microscopy Sciences, Hatfield, PA, USA) epoxy resin (the last two overnight) at room temperature. After polymerization (72 h at 60 °C), ultrathin sections were cut with a Diatome diamond knife (DiATOME, Hatfield, PA, USA) on an LKB Ultrotome III Bromma 8800 ultramicrotome (LKB Produckter AB, Stockholm-Bromma, Stockholm, Sweden), collected on 3 mm 300 mesh Cu grids (Agar) (covered with a formvar film—Electron Microscopy Sciences, Hatfield, PA, USA) and contrasted with 13% uranyl acetate (Merck, Darmstadt, Germany) for 7 min at room temperature. Examination of sections was performed with a Jeol JEM 1010 at 80 kV, and relevant images were recorded with a Mega View G2 camera (Olympus Soft Imaging Solutions, Münster, Germany). On the TEM images, the presence and location of MNPs within the cells were analyzed, in addition to the ultrastructural changes generated by the simple presence of the MNPs in the three types of cells, and by the in vitro induced magnetic hyperthermia.

2.6. In Vitro Magnetic Hyperthermia

A549, A375, and BJ cells were seeded in 6-well plates and exposed for 24 h to 62.5, 31.25, 15.62, and 7.81 µg/cm^2 of MNPs. Cells were washed thoroughly with PBS, trypsinized, equally split into two aliquots of 1500 µL, and centrifuged for 10 min at 100× g to obtain a pellet. After the removal of 1300 µL of cell culture media, one aliquot was subjected to an AMF for 30 min, whereas the other aliquot maintained at 37 °C in a water bath served as a negative control. For the MH assays, three intensities of the AMF—30, 45, and 60 kA/m—were used, while the frequency was 355 kHz. Upon exposure to the AMF, cells were seeded in 96-well plates (6 technical replicates) and the viability was evaluated after 24 h using the AB and NR assays, described in Section 2.4. Cellular viability was expressed as the relative values between the viability of cells loaded with MNPs and exposed to AMF and negative control (cells exposed to MNPs but not to AMF). For the evaluation of cellular and intracellular damages induced by the MH treatment, biological samples were prepared for TEM and SEM analysis as described in Section 2.5. The experiments were performed with three biological replicates.

2.7. Statistics

The results are presented as mean values ± standard deviation (SD). Unless stated otherwise, the normally distributed data sets were analyzed using a one-way analysis of variance (ANOVA), and for quantitative assessment of cellular parameters Bonferroni multiple test comparison was also applied. Data analyses and graphical representation were performed in SigmaPlot 11.0 computer software (Systat Software Inc., Chicago, IL, USA), and GraphPad Prism 7.00 (GraphPad Software, San Diego, CA, USA). Results showing p-values less than 0.05 were considered statistically significant.

3. Results and Discussion

3.1. Structural Characterization of sFe$_3$O$_4$

The preparation of silica-coated magnetite nanoparticles (sFe$_3$O$_4$) consisted of two steps. Firstly, ferromagnetic polyhedral Fe$_3$O$_4$ with an average length of 42 nm (Figure 1a) were prepared using a modified polyol synthesis method as described by us previously [52]. The as-synthesized polyhedral Fe$_3$O$_4$ MNPs were then coated with silica by employing the reverse microemulsion method [34–36,38]. The amount of TEOS used in the reaction mixture was varied, while the other components were kept constant, the resulting samples being denoted as Fe$_3$O$_4$@SiO$_2$-1, Fe$_3$O$_4$@SiO$_2$-2, and Fe$_3$O$_4$@SiO$_2$-3. When using 50 µL of TEOS, the silica shell formed around Fe$_3$O$_4$ MNPs was very thin and therefore difficult to observe in TEM images (Figure 1b). By increasing the amount of TEOS at 100 µL in the reaction mixture, the presence of silica coating around Fe$_3$O$_4$ MNPs was easily seen in TEM images (black arrows in Figure 1c). The amorphous silica shell, surrounding the Fe$_3$O$_4$ exhibiting a dark contrast, is seen as a bright fringe. Thicker silica coating around Fe$_3$O$_4$ MNPs was achieved using 200 µL of TEOS in the silica coating process (black arrows in Figure 1d). The polyhedral morphology of Fe$_3$O$_4$ is similar in all three cases, indicating that it is not affected by the silica coating. Please note that the polyhedral Fe$_3$O$_4$ are ferromagnetic at room temperature; thus, once dispersed in different solvents, they are stabilized in aggregates of different dimensions as a function of the MNPs' concentration [24]. Therefore, as compared to SPIONs, where the silica shell surrounding individual MNPs has a uniform thickness, in our case, the similar silica coating process implies the coating of assemblies of Fe$_3$O$_4$ MNPs (Figures 1 and S1). For this reason, it is difficult to perform a precise quantification of the silica coating thickness. However, a closer look at both TEM (Figure 1) and STEM (Figure S1) images reveals that the contrast with sFe$_3$O$_4$ MNPs decreases and their profile became less and less visible, upon increasing the TEOS amount in the preparation method. These observations indicate that our approach, based on the variation of TEOS amount in the reverse microemulsion method contributed to the formation of sFe$_3$O$_4$ MNPs with variable silica coating thickness.

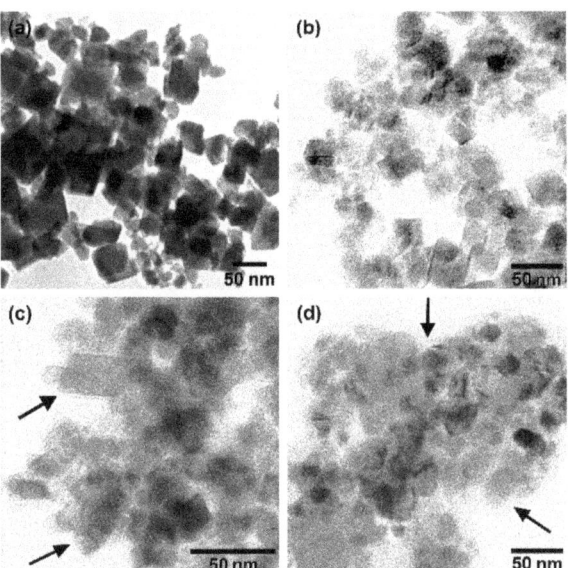

Figure 1. TEM images of (**a**) bare polyhedral Fe_3O_4 MNPs and sFe_3O_4 MNPs: (**b**) $Fe_3O_4@SiO_2$-1, (**c**) $Fe_3O_4@SiO_2$-2, and (**d**) $Fe_3O_4@SiO_2$-3. The black arrows indicate the silica shell.

The chemical composition and the distribution of chemical elements within the sFe_3O_4 MNPs were evaluated by energy-dispersive X-ray (EDX) spectroscopy and mapping. Figure 2 shows the EDX mapping of oxygen (blue), iron (orange), and silicon (red) of the as-produced bare and silica-coated Fe_3O_4 MNPs. Both iron and oxygen atoms were found to be homogeneously distributed within the total volume of polyhedral MNPs (Figure 2a) in agreement with the XRD patterns. Furthermore, in the case of $Fe_3O_4@SiO_2$-1 MNPs, the silicon atoms were mainly detected at the edges of polyhedral Fe_3O_4 MNPs and not mixed with iron atoms, which supports the formation of core/shell nanoparticles (Figure 2b). The red dots, representing the silicon atoms, appear more densely around polyhedral Fe_3O_4 MNPs together with oxygen atoms, indicating an increase in the silica shell (Figure 2c). For the last sample, $Fe_3O_4@SiO_2$-3 MNPs, the silicon atoms dominate the EDX map, whereas the iron atoms are faintly visible (Figure 2d), which reveals the formation of thicker silica shell around polyhedral Fe_3O_4 MNPs.

As demonstrated in our previous work [53], the high-temperature polyol synthesis method produces faceted magnetic nanoparticles containing mainly the magnetite phase. In this sense, the Rietveld refinement of the XRD pattern matched that of magnetite (PDF number: 88-0315) and was consistent with the cubic Fd-3m crystal structure for magnetite with no other phases being evident within the estimated uncertainty of the measurement (which is not possible for a sample containing predominantly maghemite). Moreover, the corresponding lattice parameter of polyhedral NPs (a = 0.8388 nm) was very close to that of pure magnetite NPs (a = 0.83952 nm) as compared to pure maghemite NPs (a = 0.8364 nm). The saturation magnetization had a high-value characteristic of magnetite. XRD analysis was further employed to determine whether the silica coating process affects the crystallinity and purity of Fe_3O_4 MNPs. For the three samples containing sFe_3O_4 MNPs (Figure S2b–d), the position and the relative intensities of all diffraction peaks correspond to those ascribed to Fe_3O_4 MNPs magnetite (Figure S2a). The identical XRD patterns found for all four samples indicate that the silica-coated process did not affect the crystallinity of the magnetite core. Instead, the related peaks of sFe_3O_4 MNPs progressively broaden and decrease in intensity as the amorphous silica shell thickness is increased (Figure S2). A slight decrease in the corresponding lattice parameter was also recorded for sFe_3O_4 MNPs, but the values are very close to that of bulk magnetite (a = 0.8375 nm), suggesting that the purity of the magnetic core is retained upon silica coating.

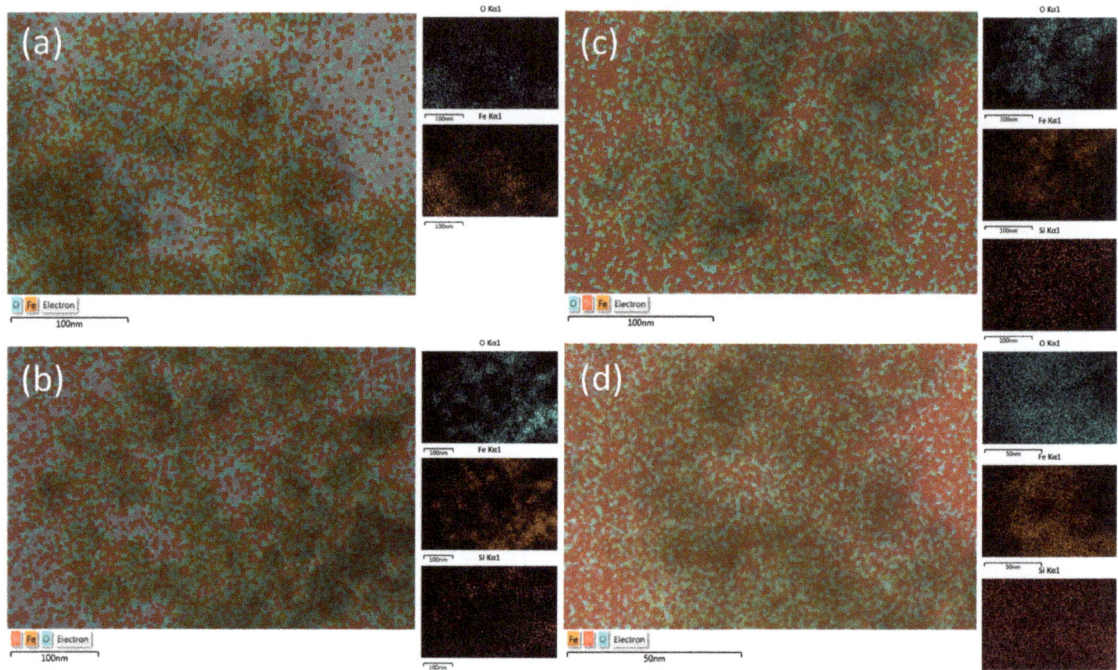

Figure 2. Elemental EDX mapping of (**a**) bare polyhedral Fe_3O_4 MNPs and sFe_3O_4 MNPs: (**b**) $Fe_3O_4@SiO_2$-1, (**c**) $Fe_3O_4@SiO_2$-2, and (**d**) $Fe_3O_4@SiO_2$-3. Oxygen (blue), iron (orange), and silicon (red) atoms.

3.2. FT-IR Spectroscopy and Colloidal Properties of sFe_3O_4

In addition to the direct observation of the silica layer around Fe_3O_4 MNPs in TEM images and EDX mapping, FT-IR spectroscopy confirmed the silica coating on Fe_3O_4 MNPs. The FT-IR spectrum of uncoated Fe_3O_4 MNPs, dominated by the Fe–O absorption band around 540 cm^{-1}, also reveals the characteristic absorption bands of PEG200 (in 750–1250 cm^{-1} region—purple rectangle) and those of the stretching vibrations of the carboxylate group of the Na-acetate (in 1300–1800 cm^{-1} region—green rectangle) (Figure 3a). Upon coating the polyhedral Fe_3O_4 MNPs with a silica shell, all vibration bands related to PEG200 and Na-acetate disappeared from the FT-IR spectra which exhibited the absorption band characteristics for SiO_2 (blue rectangle): Si–O–Si stretching vibrations between 1000 and 1250 cm^{-1} were the most prominent; the shoulder at 965 cm^{-1} is attributed to Si–OH vibrations, whereas the small absorption band at 475 cm^{-1} reflects the Si–O–Si bending [34,35,40,54]. As compared to the dominant absorption band representing the Fe–O bond, the intensity of the SiO_2 characteristics absorption bands increases when passing from $Fe_3O_4@SiO_2$-1 to $Fe_3O_4@SiO_2@$-2 and finally to $Fe_3O_4@SiO_2$-3 (Figure 3a). This supports the TEM and EDX analysis, according to which the increase in TEOS amount in the coating procedure leads to increasing silica shell thickness around polyhedral Fe_3O_4 MNPs. This is also reflected in a progressive shift of the Fe–O vibrational band towards higher wavenumbers (Figure 3b).

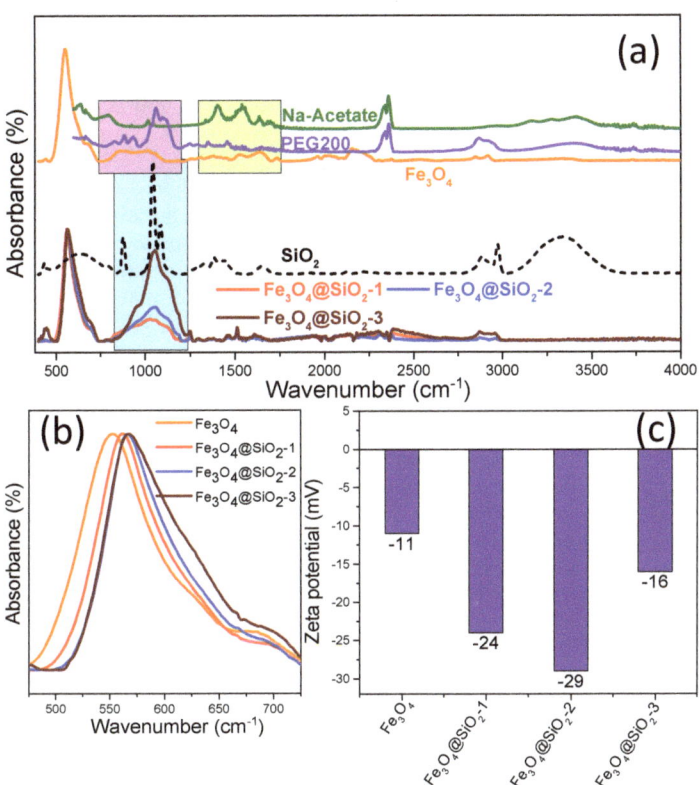

Figure 3. (a) FT-IR spectra of bare polyhedral Fe_3O_4 MNPs (orange), PEG200 (purple), Na-acetate (green), SiO_2 (black) and sFe_3O_4 MNPs: $Fe_3O_4@SiO_2$-1 (red), $Fe_3O_4@SiO_2$-2 (blue) and $Fe_3O_4@SiO_2$-3 (brown). The spectra are normalized to the highest absorption band and are vertically shifted for clarity. (b) Fe-O vibrational band of bare polyhedral Fe_3O_4 MNPs (orange), $Fe_3O_4@SiO_2$-1 (red), $Fe_3O_4@SiO_2$-2 (blue), and $Fe_3O_4@SiO_2$-3 (brown) MNPs. (c) Zeta potential values of Fe_3O_4, $Fe_3O_4@SiO_2$-1, $Fe_3O_4@SiO_2$-2, and $Fe_3O_4@SiO_2$-3 MNPs dispersed in water at 0.01 mg_{MNPs}/mL.

The polyhedral Fe_3O_4 MNPs presented a negative value of the zeta potential of −11 mV (Figure 3c), most probably due to the presence of acetate onto their surface (Figure 3a). The coating of polyhedral Fe_3O_4 MNPs with a silica shell rendered the sFe_3O_4 MNPs more negative: the zeta potentials of $Fe_3O_4@SiO_2$-1, $Fe_3O_4@SiO_2$-2, and $Fe_3O_4@SiO_2$-3 were −24, −29, and −16 mV, respectively (Figure 3c), in agreement with the results obtained in another study [40]. The hydrodynamic diameter of polyhedral Fe_3O_4 MNPs at a very low concentration of 0.01 mg_{MNPs}/mL was about 280 nm (Figure S3a), which is much greater than the average size resulting from TEM images [53]. Due to the ferromagnetic state of polyhedral Fe_3O_4 MNPs at room temperature [53], it is obvious that they form clusters in the colloidal suspension, as also revealed by TEM images. For the case of sFe_3O_4 MNPs, the hydrodynamic diameter increased as the silica shell increased: 325 nm for $Fe_3O_4@SiO_2$-1, 355 nm for $Fe_3O_4@SiO_2$-2, and 465 nm for $Fe_3O_4@SiO_2$-3 (Figure S3a). This highlights that the silica coating process involves clusters of MNPs to the detriment of individual MNPs.

The polydispersity index (PDI) for bare Fe_3O_4 MNPs is 0.27 (Table S1) which indicates a quasi-narrow size distribution of clusters. PDI decreases for $Fe_3O_4@SiO_2$-1 and $Fe_3O_4@SiO_2$-2 samples to 0.23 and 0.18, respectively. This suggests that the silica coating process involving up to 100 µL of TEOS reduces the size distribution of clusters. Instead, for the Fe_3O_4@SiO_2-3, the PDI increases to 0.34, which suggests that a higher amount of TEOS in the reaction mixture enables the formation of thicker silica later around clusters, increasing their size distribution.

DLS measurements performed at a concentration of 0.1 mg_{MNPs}/mL (ten times higher) revealed different colloidal behavior of the samples. For instance, the hydrodynamic diameter of both Fe_3O_4 and Fe_3O_4@SiO_2-3 MNPs increased considerably to 1650 and 830 nm, respectively (Figure S3b). Taking into account that these two samples presented the smallest zeta potential values, it can be considered that the negative charges on the MNPs surface do not produce sufficient repulsive force to balance the magnetically induced attractive force. The latter interaction dominates as the concentration of MNPs increases in the colloidal suspension, causing the formation of aggregates with larger sizes. On the contrary, the more negatively charged Fe_3O_4@SiO_2-1 and Fe_3O_4@SiO_2-2 samples were capable of producing higher repulsive forces that compete with the magnetic dipolar attractive forces, thus forming very stable dispersions in water. Hence, their hydrodynamic diameters change only slightly with increasing the concentration of MNPs (Figure S3b). All samples recorded a decrease in their PDI value with increasing the concentration in colloidal solutions. The samples Fe_3O_4@SiO_2-1 and Fe_3O_4@SiO_2-2 exhibit an insignificant decrease in PDI values to 0.22 and 0.17, respectively (Table S1). This tiny difference of 0.01 suggests that the sFe_3O_4 MNPs from these two samples are stable against aggregation. For the bare Fe_3O_4 MNPs and Fe_3O_4@SiO_2-3, the PDI values significantly decrease to 0.23 and 0.26, respectively. This observation together with the important increase in their hydrodynamic diameter indicates a spontaneous aggregation of silica-coated clusters in the case of these two samples with increasing their concentration in colloidal solutions.

3.3. Magnetic Characterization of sFe_3O_4

The magnetic properties of bare Fe_3O_4 MNPs and silica-coated MNPs were measured on powders through M-H curves and hysteresis loops at 4 and 300 K (Figure 4). The M-H curves recorded in a magnetic field as high as 10 T (8000 kA/m) reveal the value of the saturation magnetization (M_s) of the samples. As the magnetization is expressed in electromagnetic units on a per-gram basis, the values of M_s thus provide information about the silica thickness coating Fe_3O_4 MNPs. The bare Fe_3O_4 MNPs exhibited, at 4 K, a M_s of 90.6 emu/g, very close to that of the bulk magnetite (92 emu/g), suggesting the high quality in terms of crystallinity of the polyhedral Fe_3O_4 MNPs (Figure 4a). For the silica-coated Fe_3O_4 MNPs, the M-H curves (Figure 4a) showed a decrease in the M_s (Table 1) as expected due to the mass of non-magnetic silica coating. The M_s decreased proportionally with the amount of SiO_2 surrounding the magnetic core: 83.5 emu/g for Fe_3O_4@SiO_2-1, 77.1 emu/g for Fe_3O_4@SiO_2-2, and 63.5 emu/g for Fe_3O_4@SiO_2-3. The differences in the M_s of the sFe_3O_4 MNPs in respect to bare Fe_3O_4 MNPs indicate the following SiO_2 content within the silicated samples: 7.8% for Fe_3O_4@SiO_2-1, 14.9% emu/g for Fe_3O_4@SiO_2-2, and 29.9% for Fe_3O_4@SiO_2-3. As is typical for MNPs, the M_s decreases considerably, at room temperature, due to the existence of the magnetic dead layers and spin canting effects at the surface of MNPs (Figure 4b). In each case, the M_s value is reduced by about 10 emu/g (Table 1), which indicates that this is due to the magnetic core and not to the SiO_2 shell.

Table 1. Magnetic hysteresis parameters at 4 and 300 K for all four types of MNPs.

Sample	M_s (emu/g)		H_c (kA/m)	
	4 K	300 K	4 K	300 K
Fe_3O_4	90.6	80.7	32	10
Fe_3O_4-SiO_2-1	83.5	73.3	28	18
Fe_3O_4-SiO_2-2	77.1	67.5	26	16
Fe_3O_4-SiO_2-3	63.5	53.9	30	12

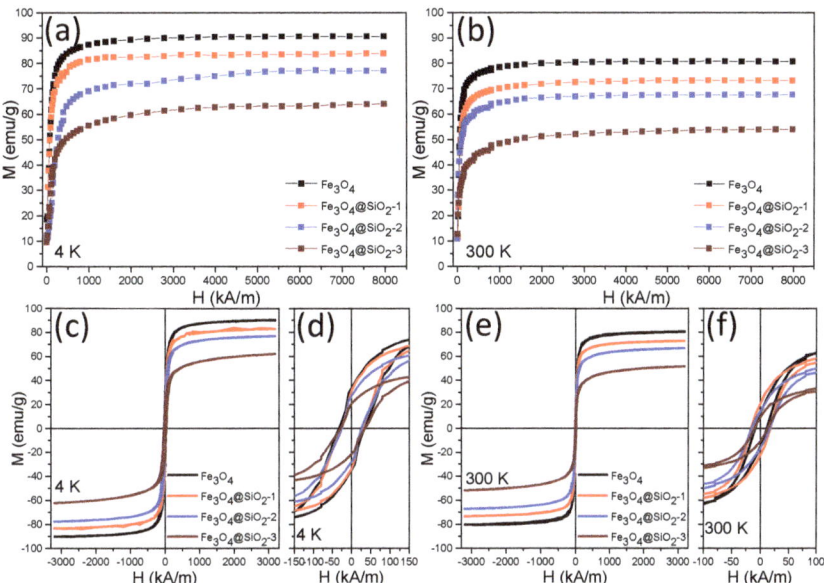

Figure 4. M(H) curves recorded at (**a**) 4 K and (**b**) 300 K, hysteresis loops recorded at (**c**) 4 K and (**e**) 300 K, and low-field regime of hysteresis loops at (**d**) 4 K and (**f**) 300 K for all four types of MNPs.

Hysteresis loops obtained at 4 K (Figure 4c,d) show that the coercive field (H_c) does not vary significantly over the four samples (Table 1). Concerning the bare Fe_3O_4 MNPs, which exhibit the highest H_c of 32 kA/m, a minute decrease in H_c is recorded for sFe_3O_4 MNPs: 28 kA/m for $Fe_3O_4@SiO_2$-1, 26 kA/m for $Fe_3O_4@SiO_2$-2, and 30 kA/m for $Fe_3O_4@SiO_2$-3 (Table 1). As shown in our previous paper [53], the polyhedral Fe_3O_4 MNPs preserve a soft ferromagnetic character at room temperature, which is also present on sFe_3O_4 MNPs (Figure 4e,f). The values of H_c decrease at room temperature for all four samples as indicated by the hysteresis loops (Figure 4e,f). The highest drop of H_c is recorded for bare $Fe_3O_4@SiO_2$ among all samples (Table 1). The ferromagnetic bare Fe_3O_4 MNPs form clusters, as shown by DLS measurements, whereas the strong dipole coupling interaction between clusters tends to assist magnetization reversal, thereby reducing the H_c value [55]. The $Fe_3O_4@SiO_2$-1 and $Fe_3O_4@SiO_2$-2 samples exhibit the highest H_c values at room temperature of 18 and 16 kA/m, respectively, followed by the $Fe_3O_4@SiO_2$-3 sample with a slightly lower H_c value of 12 kA/m (Table 1). Although the difference in H_c between silica-coated samples and bare Fe_3O_4 MNPs is not important, it can be related to the silica shell protecting the magnetic clusters. The increase in the H_c is a consequence of a decrease in the magnetic dipole coupling interactions between clusters due to the screening of silica shell coating [33].

3.4. Magnetic Hyperthermia Properties of sFe_3O_4

The magnetically induced heating capabilities of all four samples were measured in water (the heating curves are shown in Section S4). Specific absorption rate (SAR) values were obtained using Box–Lucas fitting [56] of the heating curves, following the procedure detailed in Section S5. The SAR values have been expressed as a function of the amplitude (H) of the applied alternating magnetic field ranging from 5 to 65 kA/m at a fixed frequency of 355 kHz. The iron concentration of MNP aqueous solutions was determined with a thiocyanate assay and used for data normalization (Section S6), expressing the SAR in watts per unit mass of iron (W/g_{Fe}). In the SAR determination, the contribution from pure water at each H was measured and subtracted as the background.

Figure 5 summarizes the SAR values obtained in water for all four samples presented in this study. Each panel displays the mean SAR values as a function of H for the case of four different iron concentrations, ranging from 0.8 to 0.1 mg$_{Fe}$/mL. It has to be mentioned that in the case of ferri/ferromagnetic MNPs, the main contribution to their hyperthermia capabilities is given by the energy losses quantified by the area of hysteresis loops. Additionally, the Brown relaxation—the reorientation of the MNPs itself in a fluid, resulting in their friction with the fluid—also contributes as a function of the medium viscosity. Hysteresis losses depend on the H and H$_c$ of the MNPs. For low values of H, the hysteresis areas are very small and, consequently, the SAR values are insignificant. For H values greater than H$_c$, the hysteresis area becomes larger as H increases, resulting in a steep increase in the SAR values up to a saturation value. Further increase in H will lead to a plateau of the SAR values. This typical evolution of SAR values with H (Figure 5) presents a sigmoidal shape that can be fitted (blue curves in Figure 5) phenomenologically with a simple logistic function (Section S7) as shown in our previous papers [24,53,57,58].

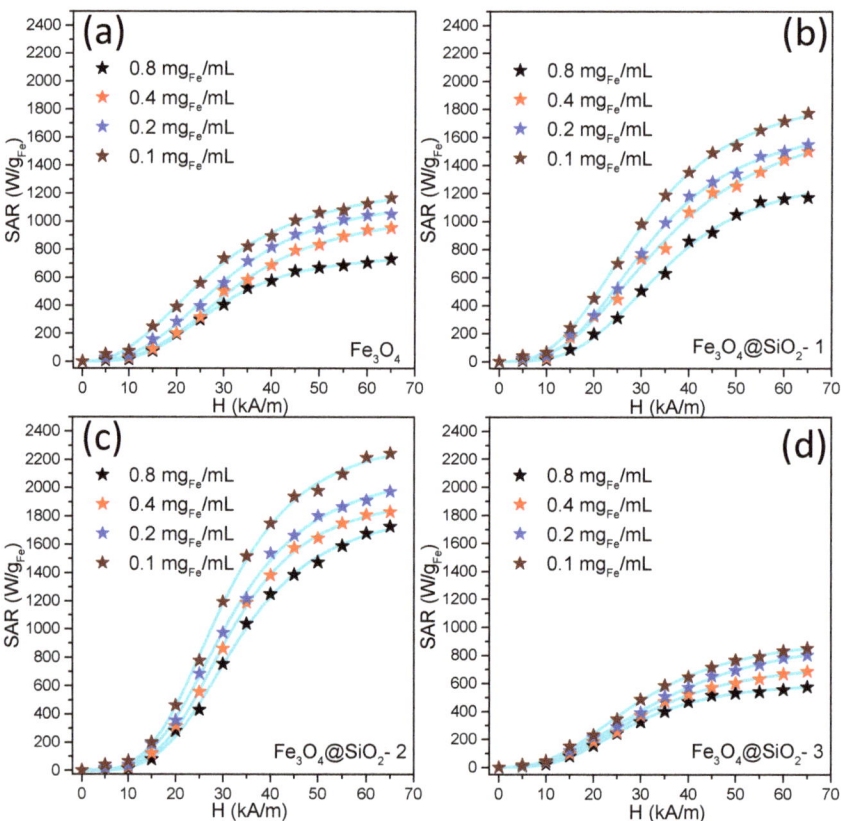

Figure 5. Specific absorption rate (SAR) dependence of H for (**a**) Fe$_3$O$_4$, (**b**) Fe$_3$O$_4$@SiO$_2$-1, (**c**) Fe$_3$O$_4$@SiO$_2$-2, and (**d**) Fe$_3$O$_4$@SiO$_2$-3 MNPs dispersed in water at iron concentration ranging from 0.8 to 0.1 mg$_{Fe}$/mL. Blue lines represent the fits with the logistic function.

The SAR evolution with H for bare Fe$_3$O$_4$ MNPs reveals a slight dependence on the concentration (c) of MNPs in aqueous solution (Figure 5a). For a c = 0.8 mg$_{Fe}$/mL the SAR values increase sigmoidally, reaching around 700 W/g$_{Fe}$ at the highest H values. With decreasing c by half, the SAR values do not change for H between 5 and 35 kA/m. Starting with H = 40 kA/m, higher SAR values are obtained, with a maximum of 950 W/g$_{Fe}$

at the highest H. The SAR values continue to increase in the entire range of H as the concentration is further decreased to 0.1 mg$_{Fe}$/mL (Figure 5a). In the literature, this behavior is associated with a decrease in the magnetic dipolar interactions between the MNPs as the concentration decreases [59,60]. A closer look at the panels in Figure S4a reveals the occurrence of kinks in the heating curves at each H, which are less pronounced as the c decreases. The kinks reduce the initial slopes of the heating curves leading to a decrease in SAR values. The kinks denote a spontaneous aggregation of MNPs clusters during the hyperthermia measurements.

For the case of Fe$_3$O$_4$@SiO$_2$-1 MNPs, the SAR values' dependence on the concentration followed the same trend observed for bare Fe$_3$O$_4$ MNPs (Figure 5b), which means as the c decreases the SAR values increase mainly starting at H = 15 kA/m. The differences became more pronounced as the H increased. The kinks are observed in the heating curves at 0.8 mg$_{Fe}$/mL, disappearing at lower concentrations (Figure S4b). This implies the absence of aggregation effect under AMF in the case of more diluted samples, resulting in the highest SAR values at the smallest concentration. The decrease in the heating curves slopes at 0.8 mg$_{Fe}$/mL, upon the formation of kinks, is less evident for Fe$_3$O$_4$@SiO$_2$-1 MNPs with respect to bare Fe$_3$O$_4$ MNPs, which suggests that the thin silica layer protects the clusters from making physical contact to form large aggregates responsible for decreasing SAR values. As compared to bare Fe$_3$O$_4$ MNPs, the SAR values of Fe$_3$O$_4$@SiO$_2$-1 MNPs are higher for each concentration, which can be explained by the better colloidal stability conferred by the thin silica layer. The difference in SAR values between the first two samples occurs starting with 30 kA/m, at each concentration. In the region of high H (50–60 kA/m), the difference increases from 450 to 600 W/g$_{Fe}$ as the concentration decreases from 0.4 to 0.1 mg$_{Fe}$/mL.

Among all the samples studied, the Fe$_3$O$_4$@SiO$_2$-2 MNPs display the highest SAR values over the H range between 20 and 65 kA/m (Figure 5c). Concerning the bare Fe$_3$O$_4$ MNPs, the difference between SAR values progressively increases with H (compare Figure 5a,c), attaining a difference around 1000 W/g$_{Fe}$ at the highest H (65 kA/m) for all four concentrations. Similar to the previous two samples, the SAR values of Fe$_3$O$_4$@SiO$_2$-2 MNPs strongly depend on the concentration: the maximum value of SAR increases from 1700 W/g$_{Fe}$ at 0.8 mg$_{Fe}$/mL up to 2200 W/g$_{Fe}$ for 0.1 mg$_{Fe}$/mL. The Fe$_3$O$_4$@SiO$_2$-2 MNPs exhibit a well-defined silica layer around them and the highest zeta potential (Figure 3c) as compared to Fe$_3$O$_4$@SiO$_2$-1 MNPs, which confer them excellent colloidal stability. This is evidenced by the absence of kinks in the heating curves for the entire concentration range (Figure S4c). Moreover, from Table S1, it can be observed that the exponent n—which indicates how steep is the dependence of SAR on H—presents the highest values among all samples for each concentration. This suggests that the heating behavior of these types of MNP is closer to an ideal Stoner Wohlfarth model [61]. In other words, the silica layer protects the clusters of MNPs from aggregation and reduces considerably the magnetic dipolar interactions among them, mainly at lower concentrations where the mean inter-clusters distance is high.

A thicker layer of silica around Fe$_3$O$_4$ MNPs, as in the case of Fe$_3$O$_4$@SiO$_2$-3 MNPs, offers a completely different picture of heating capabilities. As can be seen in Figure 5d, the SAR values decrease with respect to those recorded for bare Fe$_3$O$_4$ MNPs for each concentration. The reduction in SAR values is explained by the sedimentation of Fe$_3$O$_4$@SiO$_2$-3 MNPs at the bottom of the vial, as observed during measurements. The small value of the zeta potential (Figure 3c) leads to the aggregation of Fe$_3$O$_4$@SiO$_2$-3 MNPs, thus reducing their Brownian mobility and increasing their dipolar interactions.

We also performed SAR measurements with the MNPs dispersed in a solid matrix. Polyethylene glycol 8000 (PEG *K) with a melting point above 50 °C was used as a solvent. PEG 8K was heated to 90 °C and the MNPs were dispersed in the solved under sonication. The sample was allowed to cool at room temperature and afterward was submitted to the various AMF. As can be seen from Figure S6 panel a, the best SAR performances were recorded again for the sample Fe$_3$O$_4$@SiO$_2$-2 with a maximum SAR around 1200 W/g$_{Fe}$.

In addition, for this sample, we calculated the maximum drop in the SAR (from 1800 to 1200 W/g). This result indicates also that this sample has the largest mobility among all samples. Interestingly, when the SAR values were recorded as a function of the concentration (Figure S6 panel b) it appeared that the heating performances of these NPs do not depend on concentration. This result can be explained by the fact that the NPs are well dispersed in the matrix and after solidification, the solid matrix hinders the clusterization of the MNPs under the influence of the AMF field.

Among the four samples, the $Fe_3O_4@SiO_2$-2 MNPs, both dispersed in an aqueous solution or in a solid matrix, exhibited the best MH performance. In this regard, to verify their potential in MH applications, this class of MNPs was further tested in vitro on cancer and normal cell lines.

3.5. Cytocompatibility of $Fe_3O_4@SiO_2$-2 MNPs and Intracellular Modifications

The evaluation of in vitro cytocompatibility of $Fe_3O_4@SiO_2$-2 MNPs represents an essential step in their validation for MH therapy applications. Alamar Blue (AB) and Neutral Red (NR) assays were employed for cytocompatibility evaluation on two types of cancer cell lines (human pulmonary cancer cells—A549 and melanoma cancer cells—A375) and one normal cell line (human foreskin fibroblasts—BJ). Because the nanomaterials are prone to induce interference with the biochemical assays used to evaluate viability, leading to erroneous results [62], in a first step, the optical and biochemical interferences of the $Fe_3O_4@SiO_2$-2 MNPs with both assays were evaluated (Section S9). Similar to bare Fe_3O_4 MNPs [53], the emitted fluorescence gradually decreases as the concentration of $Fe_3O_4@SiO_2$-2 MNPs increases, for both assays (Figure S7). The presence of $Fe_3O_4@SiO_2$-2 MNPs quenches the emitted fluorescence, which can be optically avoided by the measurement of the supernatant fluorescence after removing the MNPs by centrifugation (Figure S7). On the contrary, no interference by reduction/oxidation or adsorption of the dyes was recorded for both assays (Figure S8).

The $Fe_3O_4@SiO_2$-2 displayed a low cytotoxicity profile towards the normal and cancerous cell phenotypes. Different viabilities were recorded, depending on the assay used, with the AB assay displaying a higher sensibility than the NR assay. In the case of the AB assay, a statistical decrease in viability was recorded starting from an exposure dose of 62.5 µg/cm^2 for A549 cells and 250 µg/cm^2 for A375 and BJ cells (Figure 6). However, considering the viability threshold of 80% for nanomaterials to be safe for biomedical applications [63], the $Fe_3O_4@SiO_2$-2 MNPs show a minor toxic behavior at the dose of 250 µg/cm^2 for A549 cells and 500 µg/cm^2 for A375 cells, and were nontoxic for BJ cells over the tested dose range (Figure 6). The results indicate that normal cells are more resilient to the toxicity elicited by $Fe_3O_4@SiO_2$-2, in agreement with our previously published results, where the toxicity of polyhedral iron oxide magnetic nanoparticles (IOMNPs) was slightly more pronounced on the A549 cell line than on normal human gingival fibroblast (HGF) cells [53]. In comparison with normal cells, cancerous cells show an increased ability to uptake MNPs, resulting in a higher cytotoxic susceptibility. The silica shell surrounding the MNPs slightly improves the biocompatibility as compared to citric acid-coated Fe_3O_4 MNPs. The viability of A549 cells decreases below 80% for a dose of 125 µg/cm^2 of $Fe_3O_4@SiO_2$-2 MNPs (Figure 6a), whereas in the case of citric acid-coated Fe_3O_4 this happens at a dose of 600 µg/mL (equivalent to 120 µg/cm^2) [53]. Conversely, in the case of NR assay, an overall increase in fluorescence was registered (Figure 6). As the NR assay is based on the ATP-dependent incorporation of the NR dye in the lysosomal compartment, we hypothesize that exposure to $Fe_3O_4@SiO_2$-2 increases the lysosomal compartment and thus the accumulation of the dye. These differences between formazan-related assays (AB, MTT, MTS) and the NR assay were previously reported in the literature by our group and others [53,62].

In the following, the internalization of $Fe_3O_4@SiO_2$-2 MNPs in different doses (15.62, 31.25, and 62.5 µg/cm^2) by all three types of cells is presented, aiming at revealing the internalization mechanism and possible intracellular modifications. Qualitative and some quantitative ultrastructural aspects are reported here for each experimental group of cells (for a detailed statistical analysis see Table 2).

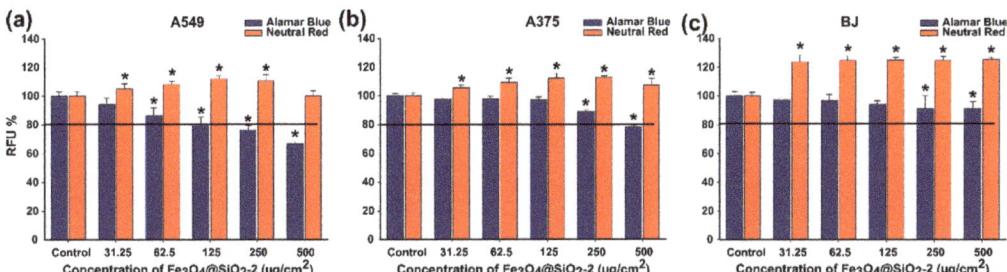

Figure 6. Cytocompatibility of $Fe_3O_4@SiO_2$-2 MNPs on (**a**) A549, (**b**) A365, and (**c**) BJ evaluated after 24 h exposure. Cellular viability was measured using two complementary assays, namely Alamar Blue and Neutral Red. The values are expressed as mean ± SD of six biological replicates. Data are expressed as relative values to the negative control. Asterisks (*) indicate significant differences compared to the negative control (ANOVA + Dunn's; $p < 0.05$). The horizontal straight line indicates the cell viability threshold of 80% to consider MNPs as biomedical safe.

First, untreated control cells from all three types of cells were visualized by both SEM and TEM. As shown and described in Section S10, the cells were adherent, presenting the characteristic shapes and internal structures (Figure S9).

The A549 cells generally preserved their adherence and the flat shape with a normal aspect and a with many filopodia still present upon incubation with $Fe_3O_4@SiO_2$-2 MNPs at doses of 15.62, 31.25, and 62.5 µg/cm^2 (Figure 7a,c,e). However, a tendency of retraction was obvious starting with a lower dose, resulting in reduced sizes of cells and prominence of their central region (Figure 7a), visualization of thicker branches (with statistical relevance when compared to control cells), and a progressive (significant) reduction in the number of filopodia, with the dose (also significant) when compared to control. The number of retracted cells increased with the exposed dose of $Fe_3O_4@SiO_2$-2 MNPs. Their shapes were almost spherical, with a low number of thin branches and almost no filopodia; those remaining filopodia were very short and thickened (Figure 7c,e). In the regions where the cells were fewer, the filopodia were even longer (Figure 7a). Cells during division were also observed (Figure 7c).

TEM examination revealed the presence of high amounts of $Fe_3O_4@SiO_2$-2 MNPs inside cells even starting with the dose of 15.62 µg/cm^2 (Figure 7b). A variation of the internalized amount of Fe3O4@SiO2-2 MNPs with increasing the dose was not evident in TEM investigations (Figure 7b,d,f). The $Fe_3O_4@SiO_2$-2 MNPs were internalized in large endosomes (electron-lucent polymorphous vesicles surrounded by a single membrane) and released within the unstructured fraction of cytoplasm (Figures 7b and S10). However, the number of free $Fe_3O_4@SiO_2$-2 MNPs dispersed in cytoplasm increased with increasing dose (Figure 7d,f). At a dose of 62.5 µg/cm^2, a few $Fe_3O_4@SiO_2$-2 MNPs were also found inside lysosomes (homogeneous or heterogeneous electron-dense organelles, surrounded by a single membrane, usually round, not shown—see Figure S10). The nucleus was not affected at all and the endoplasmic reticulum (which appeared as shorter or longer membranous channels, with or without ribosomes attached) displayed a slight enlargement at a dose of 31.25 µg/cm^2 (Figure 7d), and was difficult to detect at a higher dose (Figure 7f). The mitochondria (with double membrane—the inner one with internal foldings called cristae) preserved a round-oval shape and their cristae had a normal ultrastructural aspect (Figure 7b), becoming swollen, with electron-lucent matrix and altered cristae at the dose of 62.5 µg/cm^2 (inset Figure 7f). At this dose the cells were more affected by MNPs' interaction, displaying rarefied cytoplasm, many large vacuoles, and, in rare cases, even disrupted plasma membrane (not shown), resulting in cell death as indicated by AB and NR assay (Figure 6a).

Table 2. Quantitative assessment of several cellular parameters in the experimental groups of A549, A375, and BJ cells. Results are expressed as mean ± standard deviation (ANOVA + Bonferroni correction; statistical significance at $p < 0.05$).

Cell Type	A549			A375			BJ		
Experimental Conditions	Control	Internalization	In Vitro MH Treatment	Control	Internalization	In Vitro MH Treatment	Control	Internalization	In Vitro MH Treatment
sFe$_3$O$_4$ dose	0	15.62 µg/cm^2		0	15.62 µg/cm^2		0	15.62 µg/cm^2	
Cellular branches	0.76 ± 1.05 (n = 103)	2.22 ± 2.05 a (n = 100)	2.43 ± 1.82 a (n = 56)	4.11 ± 1.41 (n = 101)	2.78 ± 1.35 a (n = 128)	2.16 ± 1.41 a,b (n = 152)	3.93 ± 1.15 (n = 86)	3.71 ± 1.16 (n = 69)	3.32 ± 1.33 a (n = 87)
Filopodia	76.69 ± 13.5 (n = 83)	77.95 ± 15.7 (n = 38)	53.81 ± 22.75 a,b (n = 68)	66.49 ± 13.8 (n = 85)	35.77 ± 11.1 a (n = 86)	24.08 ± 11.2 a,b (n = 93)	96.32 ± 23.83 (n = 47)	81.43 ± 18.78 a,b (n = 61)	51.01 ± 32.76 a,b (n = 76)
Retracted cells (% of n)	7.91 (n = 215)	17.45 (n = 321)	33.33 (n = 312)	9.04 (n = 166)	30.40 (n = 273)	52.74 (n = 201)	6.12 (n = 147)	10.48 (n = 105)	34.74 (n = 95)
sFe$_3$O$_4$ dose	-	31.25 µg/cm^2		0	31.25 µg/cm^2		0	31.25 µg/cm^2	
Cellular branches	-	2.19 ± 1.62 a (n = 73)	1.42 ± 1.44 a,c,e (n = 107)	-	2.68 ± 1.23 a (n = 122)	1.81 ± 1.42 a,c (n = 163)	-	2.79 ± 1.53 a,b (n = 106)	2.04 ± 1.18 a,c,e (n = 105)
Filopodia	-	56.59 ± 19.73 a,b (n = 81)	16.45 ± 16.18 a,c,e (n = 87)	-	25.72 ± 8.74 a,b (n = 75)	11.25 ± 9.99 a,c,e (n = 72)	-	58.55 ± 28.95 a,b (n = 101)	46.89 ± 33.5 a (n = 97)
Retracted cells (% of n)	-	18.01 (n = 372)	82.52 (n = 452)	-	39.89 (n = 188)	63.75 (n = 309)	-	31.82 (n = 110)	50.26 (n = 189)
sFe$_3$O$_4$ dose	-	62.5 µg/cm^2		-	62.5 µg/cm^2		-	62.5 µg/cm^2	
Cellular branches	-	2.29 ± 1.44 a (n = 151)	0.07 ± 0.26 d,e,f (n = 72)	-	2.78 ± 1.35 a (n = 145)	0.31 ± 0.81 a,d,e,f (n = 65)	-	2.14 ± 1.26 a,b,c (n = 140)	1.77 ± 1.77 a,e (n = 107)
Filopodia	-	33.48 ± 16.58 a,b,c (n = 75)	0.63 ± 2.17 a,d,e,f (n = 72)	-	24.63 ± 10.1 a,b (n = 81)	1.83 ± 4.58 a,d,e,f (n = 63)	-	57.1 ± 34.63 a,b (n = 83)	44.75 ± 35.61 a (n = 101)
Retracted cells (% of n)	-	39.51 (n = 243)	100 (n = 72)	-	40.29 (n = 206)	100 (n = 76)	-	42.28 (n = 123)	56.67 (n = 120)

n—counted cells/experimental group; a—significant when compared to control; b—significant when compared to internalization 15.62 µg/cm^2; c—significant when compared to internalization 31.25 µg/cm^2; d—significant when compared to internalization 62.5 µg/cm^2; e—significant when compared to in vitro MH treatment 15.62 µg/cm^2; f—significant when compared to in vitro MH treatment 31.25 µg/cm^2.

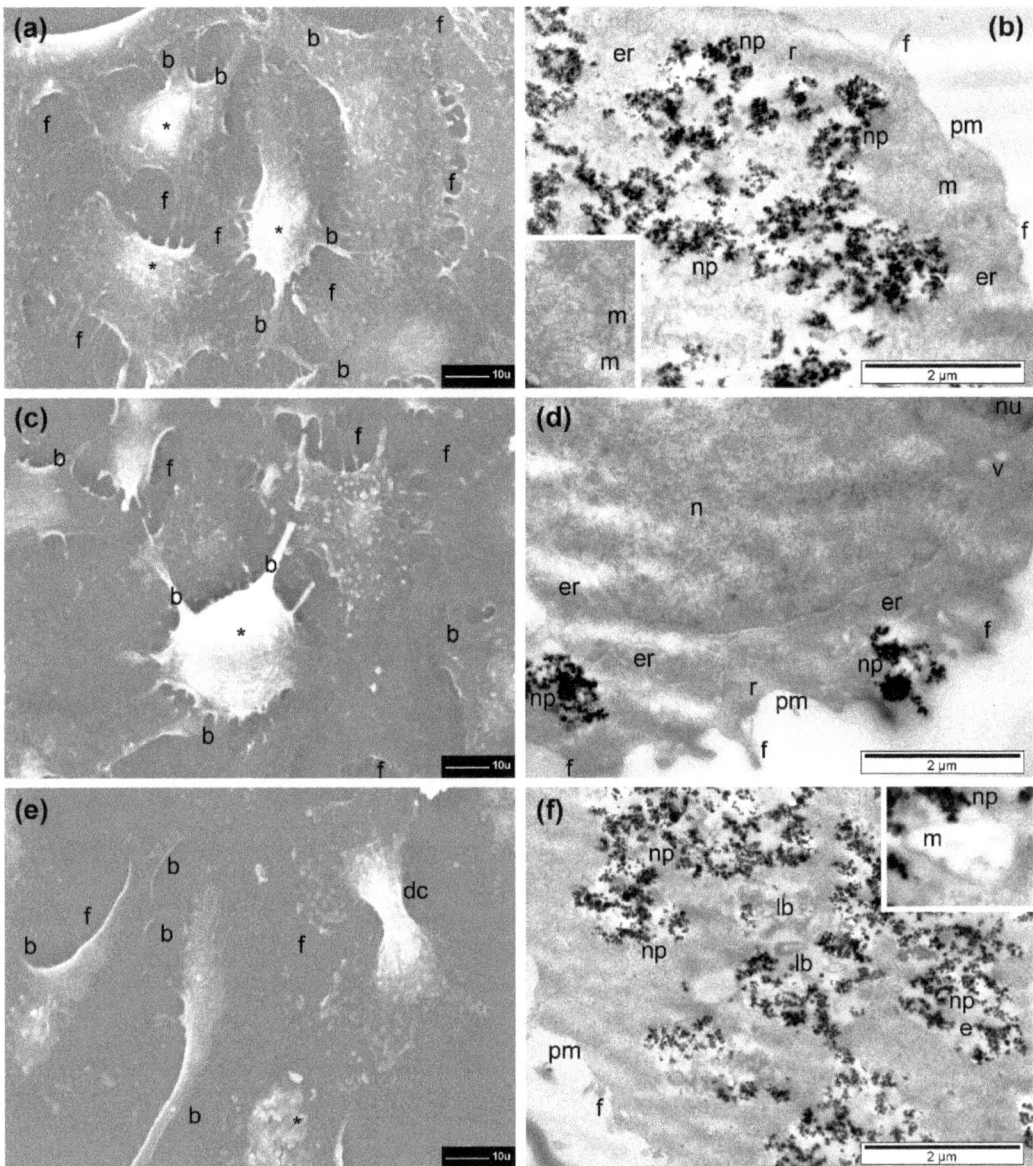

Figure 7. SEM (**left**) and TEM (**right**) images of A549 cells containing Fe_3O_4@SiO_2-2 MNPs after 24 h incubation time in a dose of (**a,b**) 15.62 µg/cm^2, (**c,d**) 31.25 µg/cm^2 and (**e,f**) 62.5 µg/cm^2. The significance of letters is: f = filopodia, b = branch, np = nanoparticles, er = endoplasmic reticulum, r = ribosome, m = mitochondria, pm = plasma membrane, n = nucleus, nu = nucleolus, v = vacuole, dc = divided cell and lb = lamellar bodies. The insets show (**b**, left) mitochondria and (**f**, right) mitochondria and free nanoparticles in cytoplasm. The asterisks indicate retracted cells. 10 u = 10 µm.

Most of the A375 cells displayed a retraction process due to the interaction with the MNPs, appearing to be smaller in size and more prominent, and were still adherent on the glass substrate (Figure S11a,c,e). Their branches were significantly fewer as compared to the control. Similar to the case of A549 cells, the percent of retracted A375 cells increased by increasing the exposure dose. As compared to those of the untreated control cells (Figure S9c), the filopodia of retracted cells were shorter and in reduced number (Figure S11a,c,e), along with a general progressive (significant) decrease in their mean number. Rare structures resembling lamellipodia were noticed at the periphery of some cells, also bearing filopodia, but they were likely only thinned regions of the cells presenting this unusual aspect during the cellular retraction (Figure S11a). Some of the filopodia had normal length establishing contacts between the cells separated during retraction (Figure S11e).

A high quantity of internalized $Fe_3O_4@SiO_2$-2 MNPs independent of the exposed dose was observed (Figure S11b,d,f). At a dose of 15.62 µg/cm^2, the $Fe_3O_4@SiO_2$-2 MNPs were seen inside endosomes, and dispersed, in direct contact with the aqueous medium of cytoplasm (Figures S10 and S11b). By increasing the dose, clusters of $Fe_3O_4@SiO_2$-2 MNPs of various sizes dispersed in cytoplasm and rare endosomes containing $Fe_3O_4@SiO_2$-2 MNPs were identified (Figure S11c,f). Neither relevant ultrastructural changes of the nucleus nor of the endoplasmic reticulum were found for the three doses. However, starting with a dose of 15.62 µg/cm^2, a mitochondrial reaction to the presence of $Fe_3O_4@SiO_2$-2 MNPs was identified, which consisted of swelling of some mitochondria, and also showed electron-lucent matrix and cristae alteration (Figure S11b). Mitochondria had an electron-lucent matrix and showed a tendency of cristae disorganization at the dose of 31.25 µg/cm^2 (Figure S11d), whereas at the higher dose the mitochondria were polymorphous, their matrix was rarefied, but cristae were still visible (Figure S11f). Two other important ultrastructural reactions were noted at a dose of 31.25 µg/cm^2: the presence of vacuoles in high numbers (and of various sizes), and the presence of lamellar bodies in a relatively high number (Figure S11d,f). It seems that the A375 cells are slightly more sensitive to the interaction with Fe3O4@SiO2-2 MNPs than A579 cells.

The BJ cells were adherent and generally preserved their size, and filopodia were present in high number at their surface. Between two and five branches of various sizes were still visible when incubated with the small dose of 15.62 µg/cm^2 of $Fe_3O_4@SiO_2$-2MNPs (Figure 8a), comparable to the control group. A retraction tendency of cells (more than 30%) was observed when increasing the dose to 31.25 µg/cm^2, displaying an elongated shape with significantly fewer visible branches and reduced number of filopodia (significant when compared to untreated control cells and to the previous dose) (Figure 8c). At the highest dose tested, most of the BJ cells showed a normal shape and aspect; however, many cells were visibly retracted. Normally developed filopodia were present in high numbers at the surface of the normal-looking cells, and in the less retracted regions of the other cells (Figure 8e).

Similar to both types of malignant cells, the $Fe_3O_4@SiO_2$-2 MNPs were internalized by the normal cells in high quantities: packed in large endosomes, dispersed freely in cytoplasm, or captured inside lysosomes (Figure 8b,d,f). No ultrastructural reactions were observed at the nuclear level (not shown), and in cytoplasm, a reduced number of autophagosomes (polymorphous and heterogeneous electron-dense vesicles surrounded by two or several membranes, belonging to the lysosomal system) was found, as compared to the untreated control cells (Figure 8b). By increasing the dose from 15.62 to 62.5 µg/cm^2, autophagosomes were present in high numbers and no notable ultrastructural changes were detected (Figure 8d). The only notable ultrastructural change found, at the higher dose, was that some mitochondria appeared to be swollen and with disrupted cristae (Figure 8f). It is quite evident that normal cells are less sensitive to the interaction with $Fe_3O_4@SiO_2$-2 MNPs than malignant cells, in agreement with the AB assay (Figure 6c). The overall increase in fluorescence, displayed by the NR assay, for the malignant cells (Figure 6a–c), can be explained by the increased number of autophagosomes, which enable the accumulation of a large amount of NR dye in lysosomes.

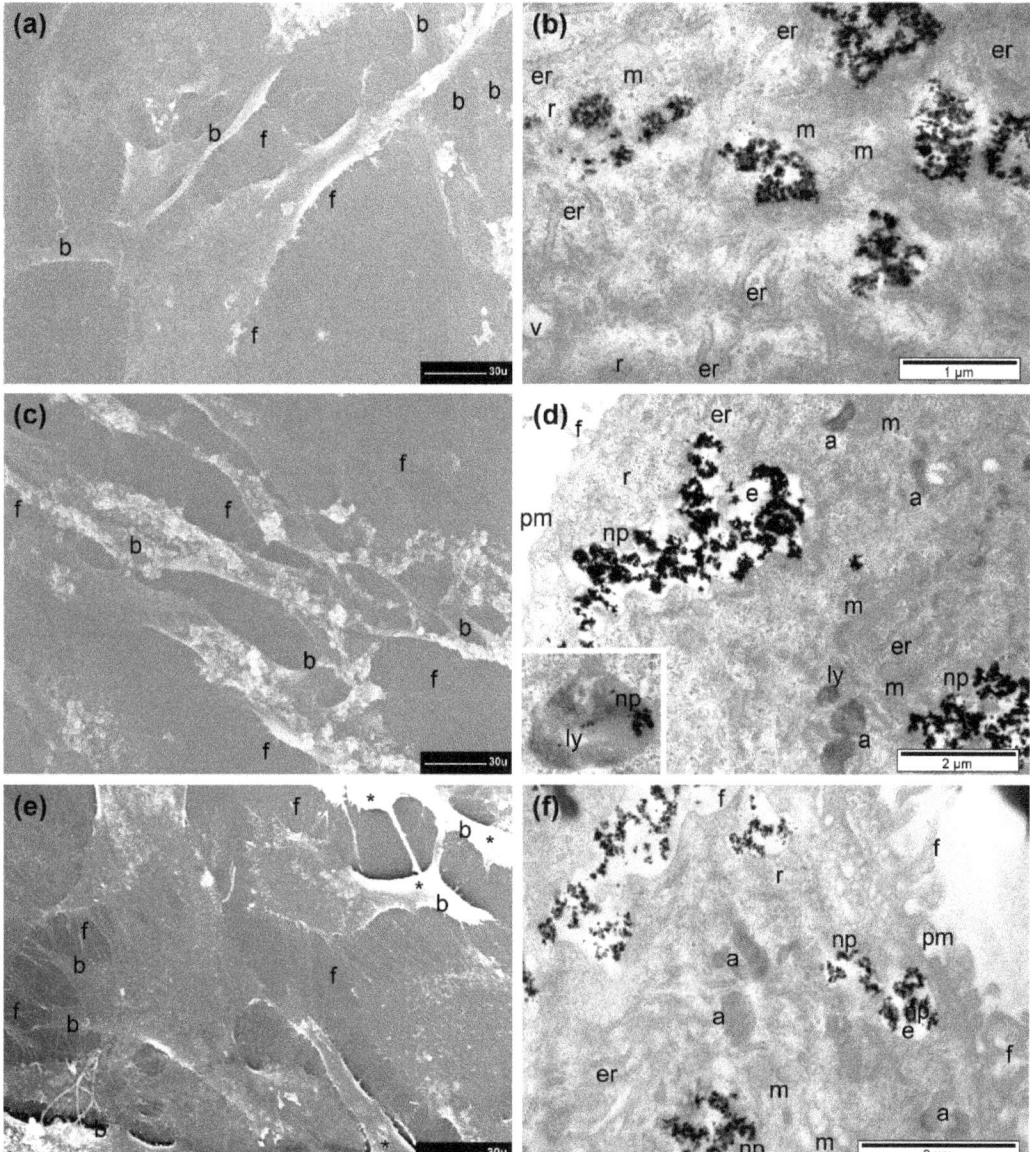

Figure 8. SEM (**left**) and TEM (**right**) images of BJ cells containing Fe_3O_4@SiO_2-2 MNPs after 24 h incubation time in a dose of (**a**,**b**) 15.62 µg/cm^2, (**c**,**d**) 31.25 µg/cm^2 and (**e**,**f**) 62.5 µg/cm^2. The significance of letters is: f = filopodia, b = branch, np = nanoparticles, er = endoplasmic reticulum, r = ribosome, m = mitochondria, v = vacuole, pm = plasma membrane, e = endosome, ly = lysosome and a = autophagosome. The inset show (**d**, left) a lysosome with nanoparticles. The asterisks indicate retracted cells. 30 u = 30 µm.

The cellular effects of the Fe_3O_4@SiO_2-2 MNPs' internalization in the three types of cells tested in our study were consistent with our results proving their cytocompatibility. We recorded moderate surface and internal ultrastructural alterations, both in the two lines of malignant cells (A549 and A375) and in the normal cells (BJ), mostly compatible with life. Even the percentage of the retracted cells was similar in the three lines (about 40%). It is worth noting here that we considered as retracted the cells ranging from those still adherent (without or with branches) and with their central zone more prominent (or more prominent branches), up to completely spherical cells, and all cells still viable despite their changes. Of course, our EM investigation revealed a low number of cells that responded more severely to the presence of the $Fe_3O_4@SiO_2$-2 MNPs. It is very likely that those cells, along with other dead/fragmented cells (probably lost during repeated centrifugations performed) covered a percentage of 5–15% of the cells, as also resulted from the cytocompatibility tests and literature data.

3.6. Quantitative Assessment of $Fe_3O_4@SiO_2$-2 MNPs Internalization

Different internalizations were observed between the cell lines, with the normal cell phenotype displaying the lowest internalization, independent of the dose used (Figure 9). As a general trend, for all three types of cells, the relative internalization increased with decreasing doses (Figure 9). For the lowest dose (7.81 µg/cm^2), the relative internalization for A549 cells was 95%—meaning that almost all amount of Fe_3O_4 MNPs was internalized —slightly decreasing to 82% for the highest dose (62.5 µg/cm^2). In the case of A375 cells, the relative internalization ranged between 85% and 70% by increasing the dose from 7.81 to 62.5 µg/cm^2. The normal BJ cells displayed the highest drop (40%) in the relative internalization, from 75% to 30% as the dose is increased from 7.81 to 62.5 µg/cm^2. The differences in the relative internalization are reflected in the total amount of internalized Fe^{3+}, which is the highest for the A549 cells, followed by A375 and BJ cells (Figure 9b). For the first two doses (7.81 and 15.62 µg/cm^2), the differences in the amount of Fe^{3+} internalized by the three types of cells are negligible (Figure 9b and Table S3). The differences are more pronounced as the dose increases (Figure 9b and Table S3). Statistical analysis (two-way ANOVA + Holm–Sidak) revealed that both the dose variable ($p < 0.001$) and the cellular type variable ($p < 0.001$) had a highly significant impact on the $Fe_3O_4@SiO_2$-2 cellular internalization. It is worth mentioning that, compared with citric acid-coated Fe_3O_4 MNPs, which exhibited a relative internalization between 37% and 21% for a dose ranging from 10 to 100 µg/cm^2 for A549 cells [53], the $Fe_3O_4@SiO_2$-2 MNPs display from 2.5 to 3.2 times higher relative internalization over the same dose range. This can be because the silica layer prevents the formation of micrometer-sized MNPs aggregates in cell culture media, thus leading to a homogeneous dispersion of sFe_3O_4 clusters over the entire well surface paved with cancer cells, which facilitate the internalization of a higher amount of MNPs inside cells.

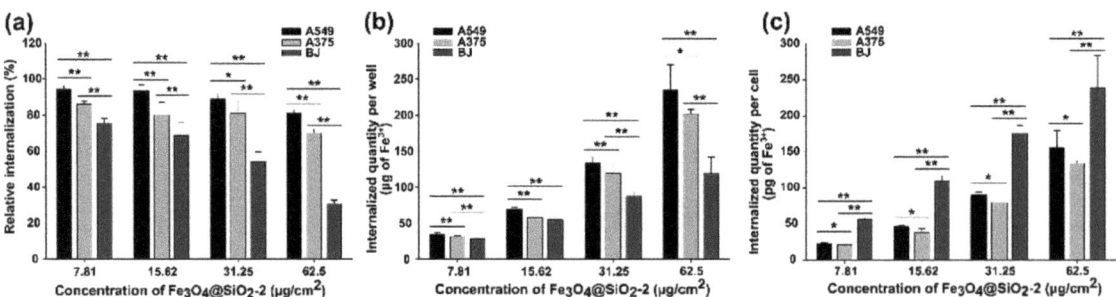

Figure 9. (a) The relative internalization (ratio between the internalized amount and the exposed amount) in A549, A375, and BJ; (b) the total amount of Fe^{+3} internalized per well; (c) the amount of Fe^{+3} internalized per cell evaluated after 24 h exposure to different concentrations of $Fe_3O_4@SiO_2$-2. The values are expressed as mean ± SD of at least three biological replicas. * $p < 0.05$ ** $p < 0.001$.

A common way of expressing the cellular internalization of Fe_3O_4 MNPs is to normalize the amount of internalized Fe^{3+} on the cell number. As BJ cells have a higher volume and surface than the cancerous cells, they are less numerous in a well plate and hence internalized a higher amount of Fe^{3+} per cell (Figure 9c). Over the dose range the BJ cells internalized between 50 to 240 pg_{Fe}/cell, whereas for the cancer cells the values were between 20 and 150 pg_{Fe}/cell (Figure 9c). Recently, Reczynska et al. [40] reported that A549 cells treated with SPIONs coated with a non-porous and mesoporous silica layer at a dose of 10 μg/mL (equivalent to 6.25 μg/cm^2) internalized approximately 20–25 pg_{Fe}/cell. Almost identical values (21, 22 pg_{Fe}/cell) were obtained in our study, at the lowest dose of 7.81 μg/cm^2 on both types of cancer cells. Similar results were also reported for zinc ferrite nanoparticles coated with silica, the Fe^{3+}/cell increasing from approximately 5 to 60 pg_{Fe}/cell during a 12 h incubation with a dose of 200 μg/mL (equivalent to 125 μg/cm^2) [38]. Overall, the results obtained suggest a passive transport process, as the amount of internalized Fe_3O_4 MNPs was almost linearly dependent on the exposure dose used, and are congruent with the results reported by Matsuda et al. [64], where the internalization of MNPs (≈10 nm) in three histological types of human mesothelioma cells increased linearly with the dose used.

3.7. In Vitro Magnetic Hyperthermia

The in vitro MH on the three types of cells without MNPs did not induce cell death as the temperature rise did not exceed 0.5–0.8 °C for H ranging from 30 to 60 kA/m at 355 kHz. On the contrary, all samples consisting of cells containing internalized MNPs exhibited a relevant increase in the temperature in the first 5 min followed by the formation of a plateau in the heating curves when the heat released by the internalized MNPs equalized the dissipated heat in the environment (Figure S12). The plateau defines a saturation temperature that strongly depends on the amount of internalized MNPs and on H (Figure S12). A correlation between saturation temperature (T_s) and the viability of cells can be established. For the lowest two doses (7.81 and 12.52 μg/cm^2), at H of 30 kA/m, no cell death was observed for all three types of cells as evaluated with both toxicity assays (Figure 10a–c). This is explained by the small T_s values between 40 and 41.4 °C obtained during MH treatment (Figure S13). At the higher doses of 31.25 and 62.5 μg/cm^2, the viability of both types of cancer cells, based mainly on the AB assays, was drastically reduced, ranging between 5 and 25%. The internalized MNPs were able to reach T_s from 43.5 to 46.4 °C (Figure S13) which suffices the temperature requirements of cancer thermal therapies. For almost the same T_s interval, the BJ cells showed a reduction in their viability by only 50–60% (AB assays in Figure 10c), proving that they are less sensitive to magnetic heating as compared to cancer cells.

Increasing the H at 45 kA/m resulted in increased T_s values over the entire dose range for all types of cells (Figure 10d–f). Nonetheless, for the lowest dose of 7.81 μg/cm^2, the cancerous and normal cells were less affected by the MH treatment as T_s was around 42 °C (Figure S13). A considerable loss of viability to approximately 20% was recorded starting with a dose of 15.62 μg/cm^2, mainly for cancerous cells (Figure 10d,e), which is in agreement with the T_s value of 44.6/44.7 °C reached during MH treatment (Figure S13). For the same dose, the normal cells exhibited a 50% reduction in their viability (Figure 10f). For the next two doses, the T_s increased to values between 48 and 50 °C, leading to cellular death of cancer cells as evidenced by both assays (Figure 10d,e). The normal cells were also profoundly affected by the MH treatment, exhibiting viabilities below 20% (Figure 10f).

The internalized MNPs at the lowest dose (7.81 μg/cm^2) in both types of cancer cells were able to sustain an increase in T_s up to 44 °C, when exposed to an H of 60 kA/m, which resulted in a decrease in viability around 40%, as indicated by the AB assay (Figures 10g,h and S11). A similar quantity of MNP internalized BJ cells did not substantially reduce the viability as the T_s increased to 42.2 °C (Figures 10i and S13). The viability dropped to near zero for the next three doses in the case of cancer cells (Figure 10g,h). The BJ cells were less sensitive at a dose of 15.62 μg/cm^2, as compared to cancer cells, however, their viability was close to zero for the higher two doses (Figure 10i).

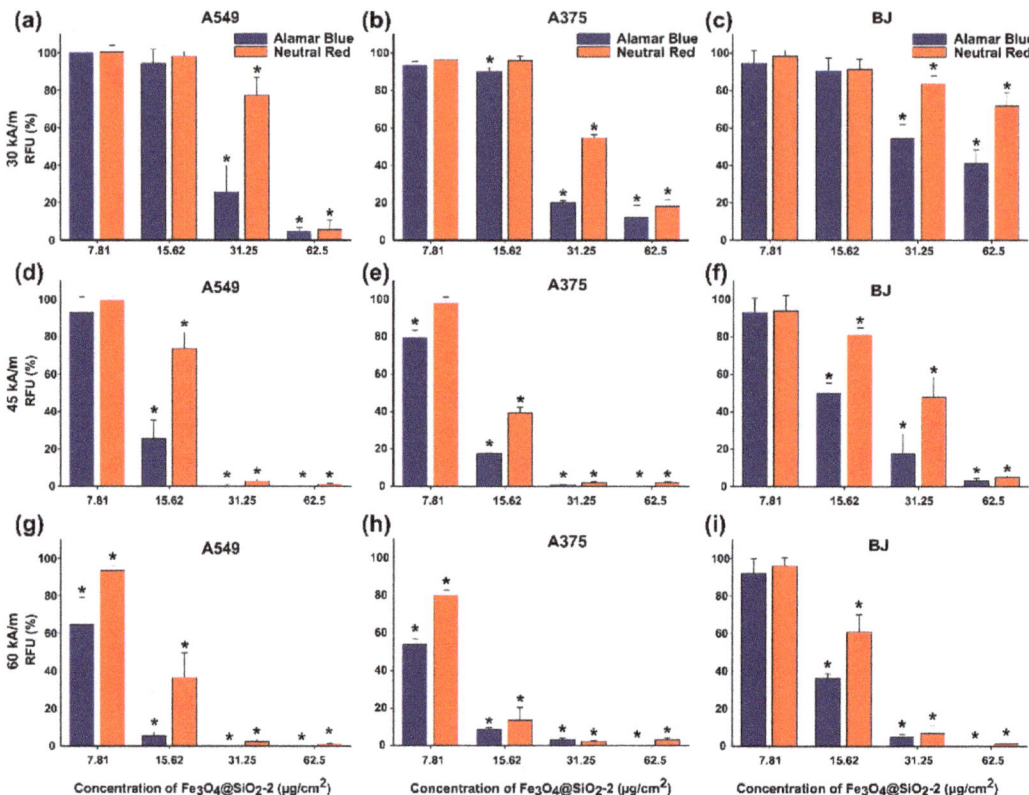

Figure 10. Cell viability (A549-(**a,d,g**), A375-(**b,e,h**), BJ-(**c,f,i**)) after exposure to MNPs and AMF of 30, 45, and 60 kA/m amplitudes. Data are expressed as relative values to the negative control, i.e., cells not exposed to AMF. Asterisks (*) indicate significant differences compared to the negative control (ANOVA + Dunn's; $p < 0.05$). The values are expressed as mean ± SD of at least three biological replicates.

For all three cell lines the viability data obtained using both assays could be fitted with a sigmoidal function:

$$C(T) = \frac{A}{1 + e^{\frac{T-T_0}{dT}}} \quad (1)$$

where A represents the viability of control cells (95–100%) and dT quantifies the temperature width for a given decrease in cell viability. T_0 represents the temperature at which the viability reaches a value of 50%. This equation was derived [65] from a two-state model of temperature-dependent cell damage as initially proposed by Feng et al. [66]. This simplified form was derived for comparing the magnetic hyperthermia heating with endogenous hyperthermia heating, under the condition of having the same time of exposure to high temperatures [65]. As in our case, in all experiments' cells were exposed for 30 min to MH; the experimental data were well fitted with the function depicted in Equation (1), as can be seen in Figure 11. The temperature T_0 represents the saturation temperature at which half of the cells were killed after they were exposed for 30 min to the MH treatment or, in other words, the temperature at which the cells were exposed for 30 min to MH receive 50% lethal dose (LD50%). As one can notice with both assays, the T_0 values are higher for BJ cells as compared to both cancer cell lines, but the difference is significantly higher in the case of the NR assay (Table S4). For this assay, T_0 is 47.3 °C for normal BJ cells, and

45.2 °C and 44.6 °C for A459 and A375 cells, respectively. The difference is much smaller for the AB assay (44.2 °C for normal and 43.6 °C for both cancer cell lines).

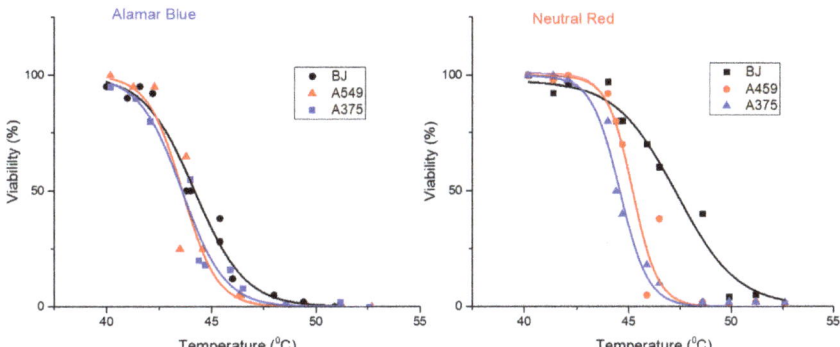

Figure 11. Viability using the Neutral Red (**left** panel) and Alamar Blue (**right** panel) assays as a function of saturation temperatures of cell cultures exposed to $Fe_3O_4@SiO_2$-2 MNPs and for 30 min to AMF (30, 45, and 60 kA/m, 355 kHz) for the three different cell lines (BJ—black squares, A459—red circles, A375—blue triangles).

These differences between the LD50% temperature values obtained using the two viability assays might be attributed to their different mechanisms of detection. The NR assay is based on the ATP-dependent lysosomal incorporation of the supravital dye and the measurement of the fluorescence of the incorporated dye. The apparent increase in the viability given by NR assay is most probably the result of an increased lysosomal compartment due to the IOMNPs' exposure and intra-lysosomal incorporation, as it was also proved by the electron microscopy study. Dissimilarities between viability assays that evaluate different mechanisms of toxicity were previously reported, reiterating the need for multiple viability assays when evaluating the cytocompatibility/cytotoxicity of nanomaterials [67,68]. Similar to the results obtained in this study, a slight decrease in the cellular viability, measured by the WST assay—an assay similar to Alamar Blue—and an increase in the NR-dye uptake upon exposure to IOMNPs were previously reported [62]. Thus, the increased uptake of NR dye may be correlated with the accumulation of IOMNPs in lysosomes. The higher sensitivity of either WST or Alamar Blue viability assay may be related to mitochondrial damage by IOMNPs, which results in a decreased conversion of formazan or resazurin by mitochondrial diaphorase. Moreover, the mitochondrial impairment of nanomaterials is presumed to be related to their redox-active surface that hinders the electron flow and the mitochondrial functionality [69].

Similar to HGF cells [53], the BJ normal cell line is less prone to MH treatment compared to the cancer cell lines. This may be a consequence of a lesser capability to internalize the $Fe_3O_4@SiO_2$-2 MNPs compared to cancer cells, mainly in the case of the higher doses (31.25 and 62.5 µg/cm^2). There is no significant difference between the cell viabilities exhibited by the two cancer cell types in the entire range of doses for the three values of H. This is supported by the small differences in saturation temperatures reached during MH, which are mainly due to the almost identical capacities of both types of cancer cells to internalize the $Fe_3O_4@SiO_2$-2 MNPs (Figure S12). At the lower doses (7.81 and 15.62 µg/cm^2), for which the total amount of internalized Fe^{3+} by the three types of cells does not vary significantly, the T_s recorded for cancer cells are higher than normal ones resulting in appreciable differences between cellular viabilities. This may be explained by the different mobility of internalized silica-coated MNPs on cancer versus normal cells. Inside cancer cells, the silica-coated MNPs can attain higher mobility when subjected to an AMF, as they are less numerous per cell (Figure 9c), leading to an increase in the saturation temperatures, and thus reduced viability.

It is also worth noting that for all cell types and all magnetic field strengths used there is a clear tendency of saturation of the T_s (Figure S12). This behavior may be explained by an increase in the concentration of the MNPs taken up by the cells within endosome/lysosome closed structures which can significantly reduce their mobility and thus their heating performance.

3.8. Cellular Modification upon MH Exposure of Cells

As in the control, and incubated but unexposed groups of cells, qualitative and some quantitative ultrastructural aspects are presented here, and a detailed statistical analysis is given in Table 2.

3.8.1. A549 Cells

The exposure of A549 cells, previously incubated with $Fe_3O_4@SiO_2$-2 MNPs at a dose of 15.62 μg/cm^2, to an AMF of 30 kA/m for 30 min, resulted in a higher number of retracted cells having significantly more visible branches compared with unexposed incubated cells. Rare contracted and prominent cells showed at their surface many small blebs formed at the level of the plasma membrane. Many other cells were normal, displaying numerous filopodia (Figure 12a); however, the overall mean number of filopodia/cell decreased significantly with regard to both the control group and the group of unexposed cells incubated with the same dose of MNPs (Table 2). Inside the cells, moderate ultrastructural changes were produced, such as mitochondrial alteration (swollen organelles, with electron-lucent cytoplasm and fewer cristae), cytoplasmic vacuolation, and the presence of the lamellar bodies as round or curved vesicles containing several concentric layers of membranes (Figure 12b). The number of normal A549 cells, incubated with the next dose and exposed to AMF of 30 kA/m for 30 min, decreased dramatically. Many other cells (more than 80% of the counted cells) showed accentuated retraction and rounding cells; more than a third of the identified cells were almost spherical, and also very small in diameter (about 10 μm or less). Surprisingly, these round cells had an almost completely smooth surface as seen in SEM, and only rare membrane blebs were observed. The other cells grown on the coverslip were either still flat and completely adherent to the substrate or partially retracted, showing 2–3 thick branches (with a significantly lower mean value than in the cells incubated with the same dose, and the cells incubated with the smaller dose and exposed to the magnetic field).

A similar (and also significant) response was found in this group of cells when we analyzed the number effect of hyperthermia on the number of filopodia. Only the flattest cells still showed filopodia and in small numbers (Figure 12c; Table 2). Cellular reactions were more extensive, consisting of irregular nuclear outline, and enlargement of the perinuclear space and of the endoplasmic reticulum lumen, associated with extensive vacuolation of cytoplasm. In most cells mitochondria were not identified—most likely these swollen organelles having no remaining cristae participated in the generation of the many vacuoles (Figure 12d). Rare normal A549 cells were observed in the case of the 62.5 μg/cm^2 dose. About a half of all identified cells (53%) were retracted (some still displaying a single visible branch and very few filopodia—Table 2) and rounded, covered with membrane blebs, whereas the other cells were in an advanced stage of fragmentation (Figure 12e). TEM examination confirmed the SEM results: in the smaller pellets remained suitable for analysis, the cells were either fragmented, with rarefied cytoplasm and disrupted plasma membrane on extensive regions (Figure 12f), or, in the case of still-intact cells, they were filled almost completely with numerous and large electron-dense vacuoles (inset of Figure 12f).

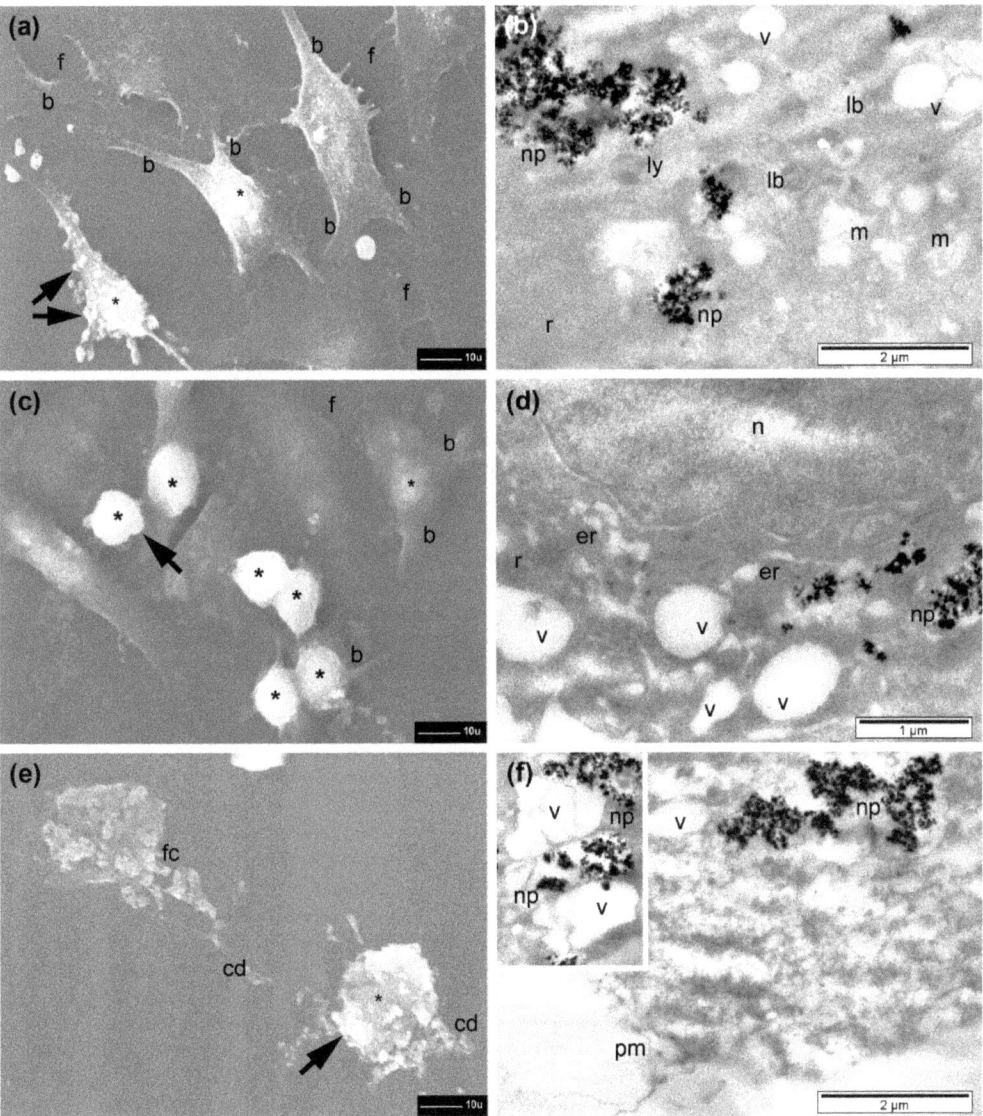

Figure 12. SEM (**left**) and TEM (**right**) images of A549 cells incubated with $Fe_3O_4@SiO_2$-2 MNPs for 24 h in a dose of (**a,b**) 15.62 μg/cm^2, (**c,d**) 31.25 μg/cm^2, and (**e,f**) 62.5 μg/cm^2 and exposed for 30 min to an AFM of 30kA/m, 355 kHz. The significance of letters is: f = filopodia, b = branch, np = nanoparticles, ly = lysosome, lb = lamellar bodies, v = vacuole, m = mitochondria, r = ribosome, n = nucleus, er = endoplasmic reticulum, pm = plasma membrane, fc = fragmented cells and cd = cellular debris. The inset (**f**, left) show large vacuoles and free nanoparticles in cytosol. The asterisks indicate retracted cells. The black arrows indicate membrane blebs. 10 u = 10 μm.

3.8.2. A375 Cells

In the case of A375 cells, a stronger response by retraction and plasma membrane blebbing at a dose of 15.62 μg/cm^2 was found, after MH treatment (Figure S14a,b). The most affected cells were rounded (losing their branches by retraction), and they lost almost completely their adherence to the glass surface. Such highly affected cells were more numerous compared to A549 cells (see Table 2), suggesting a slightly higher sensitivity of these cells to the AMF exposure, which was also indicated by AB assay (Figure 6a,b). Normal-looking cells were also found at SEM examination (Figure S14a). Statistical analysis showed a significantly lower mean number of branches and filopodia in the cells of this group, compared to the corresponding group of unexposed cells but incubated with the MNPs in the same dose (and of course, when compared to the control). The internal reactions of the affected cells were represented by rarefaction of the mitochondrial matrix; swelling of mitochondria and their cristae; cytoplasmic vacuolation; and accentuation of the nuclear polymorphism (Figure S14b). A higher number of cells (over 60%) were rounded and/or almost completely retracted at the dose of 31.25 µg/cm^2 (Figure S14c). Some cells were in an advanced stage of total fragmentation. A significantly low number of branches (compared to the cells incubated with the same dose), and very few filopodia (compared both to the cells incubated with the same dose and with the cells incubated with the lower dose and exposed to hyperthermia), were observed (average of about 11/cell), regardless of the general aspects of the cells. In the remaining viable cells, which were still adherent and more or less flat, the cytoplasm was filled with vacuoles of various sizes and shapes, some of which resulted from mitochondrial disorganization and swelling, whereas others may result from fragmentation and vesiculation of the endoplasmic reticulum (Figure S14d). Many lamellar bodies were also found. At the dose of 62.5 µg/cm^2, many of the A375 cells observed on the coverslips were spherical (27.6%), rarely with only one branch and very few filopodia (see Table 2), and the others were retracted and with various degrees of fragmentation. Cellular debris was also noted on the recorded images (Figure S14e). The A375 cells found on the sections at the TEM examination were deeply altered, with a disrupted plasma membrane and extremely rarefied cytoplasm, and sometimes still containing numerous vacuoles of very different sizes (Figure S14f).

3.8.3. BJ Cells

The BJ cells subjected to incubation with MNPs in the dose of 15.62 µg/cm^2 and to exposure to AMF also underwent an accentuated retraction and generation of membrane blebs, which were in higher numbers and much smaller compared to those observed in the malignant cells. Normal, but less branched cells (statistically not significant compared to the group of incubated cells with the same dose) were still present (Figure 13a). Over 30% of cells were retracted, and the mean number of filopodia significantly reduced when compared with the corresponding two groups (see Table 2). TEM examination showed a relatively normal ultrastructure, with the presence of numerous autophagosomes and normal profiles of the endoplasmic reticulum. Swollen mitochondria and the presence of rare lamellar bodies were noted (Figure 13b) compared to unexposed BJ cells. At the next dose (31.25 µg/cm^2), most of the BJ cells were adherent, but appeared as very long and thin cells (a significantly lower average number of branches/cell compared to the three corresponding groups—see Table S2), rather than triangular or flat (Figure 13c). More than half of the cells in this group were retracted. On the surface of their branches, the filopodia were also present but in a significantly lower number only when compared to the control group of BJ cells. Surface membrane blebbing was noted on almost all cells, whereas in some cells this blebbing was extremely advanced, leading to total fragmentation of the cells (Figure 13c). Examination using TEM showed mostly normal ultrastructure of the cells, but mild cellular vacuolation was also found in some cells (Figure 13d), whereas in rare regions only cellular debris was found on the sections (not shown). The BJ cells, loaded with $Fe_3O_4@SiO_2$-2 MNPs at a dose of 62.5 µg/cm^2, largely preserved their flat shapes (although some were extremely elongated and thick—significantly fewer than in the previous group, and in the control group, respectively) upon exposure to AMF. Filopodia were present in distinct regions, not covering the whole plasma membrane; the average number of these thin extensions was statistically significant only when compared to the control group. Some of the flat cells showed a slight tendency of retraction (Figure 13e), but overall, we recorded more than 55% of retracted cells. Moreover, 20% (24 out of 120) of the counted cells were fragmented. Regarding internal ultrastructural alterations, only rare mitochondrial deep alterations, and a certain degree of cytoplasmic rarefaction, are worth noting (Figure 13f). Basically, in the case of the BJ cell line, the effect of MH exposure had a much-reduced amplitude as compared with those recorded in the two lines of malignant cells, following the AB and NR assays (Figure 10a–c).

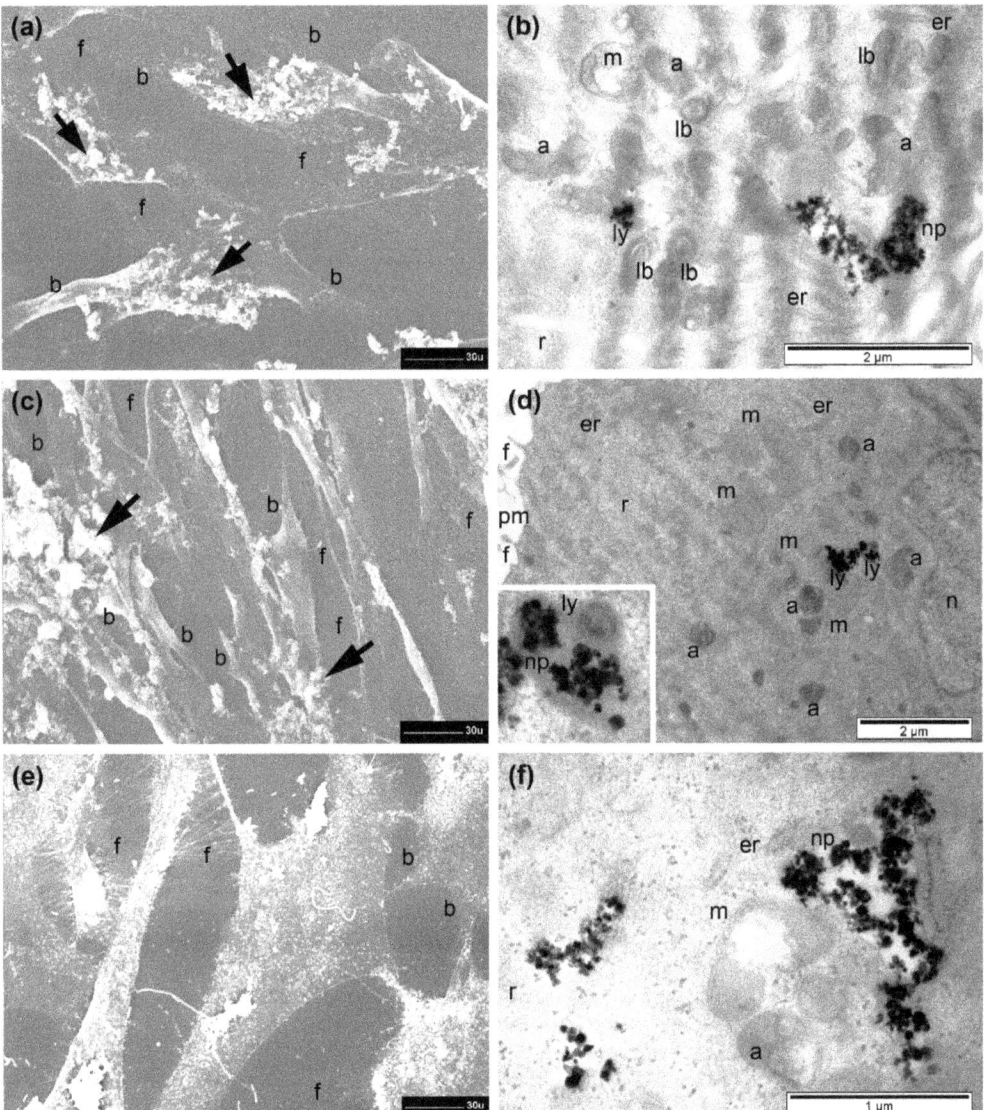

Figure 13. SEM and TEM images of BJ cells incubated with Fe_3O_4@SiO_2-2 MNPs for 24 h in a dose of (**a,b**) 15.62 µg/cm^2, (**c,d**) 31.25 µg/cm^2, and (**e,f**) 62.5 µg/cm^2 and exposed for 30 min to an AMF of 30 kA/m, 355 kHz. The significance of letters is: f = filopodia, b = branch, np = nanoparticles, ly = lysosome, lb = lamellar bodies, m = mitochondria, r = ribosome, er = endoplasmic reticulum, a = autophagosomes and pm = plasma membrane. The inset (**d**, left) show a heterogeneous lysosome loaded with nanoparticles. The black arrows indicate membrane blebs. 30 u = 30 µm.

The BJ cells also showed a more variable individual response to the AMF consecutive incubation with the three doses of MNPs, by comparison with the malignant cells. In the BJ group we found cells identical with those in the control group, in addition to others that were highly affected or dead. The reduced average number of branches and/or filopodia was due both to an unspecific reaction of all cells to the experimental conditions and a relatively high number of retracted cells that were more affected. The latter explanation may also apply to the two lines of malignant cells.

3.9. Intracellular SAR Values

The SAR values for the intracellular MH were calculated based on the Box–Lucas equation and the MNPs' cell uptake data. As can be seen in Figure 14, for all three cell lines there is a clear decrease in the SAR values as the intracellular concentrations of the MNPs increase. This type of dependence was also reported earlier in both normal and cancer cell lines [53].

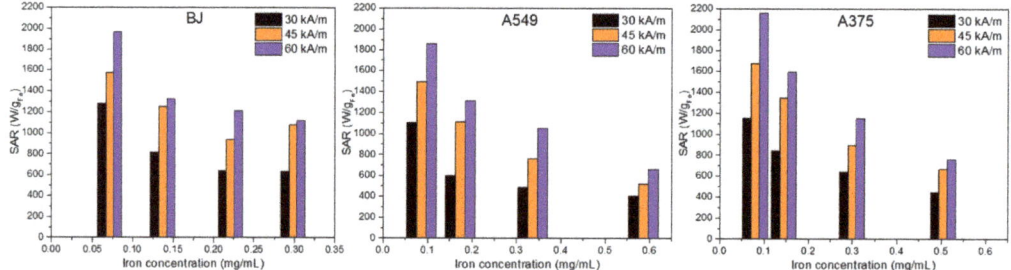

Figure 14. SAR values of the Fe_3O_4@SiO_2-2 MNPs calculated from the saturation temperatures reached during the intracellular MH and the quantity of MNPs inside the cells assessed by the thiocyanate method for BJ (**left** panel), A549 (**middle** panel), and A375 (**right** panel) for three values of the magnetic field intensity (30, 45, and 60 kA/m) at 355 kHz.

At the lowest value of the concentration (<0.1 mg/mL), the SAR values are close to those recorded in water. The decrease in the SAR is more pronounced in the case of the cancer cell lines as the quantity of internalized MNPs is almost double that of the BJ normal cancer cell line (Table S3). These results demonstrate that, at very low internalized concentrations, the MNPs preserve mobility comparable to that in water. The SAR decrease with the increase in the MNPs' concentration may be explained both by the MNPs' agglomeration in the endosome/lysosome-like structure with a subsequent decrease in their physical mobility (the only heat generation mechanism remaining the magnetization reversal), or the increase in their dipolar interactions due to their proximity.

This type of dependence of SAR on concentration explains the saturation of T_s with increasing MNPs' intracellular concentration presented in Figure S13.

4. Conclusions

In this study, ferromagnetic polyhedral Fe_3O_4 MNPs were successfully coated with silica shells of different thicknesses by varying the amount of TEOS within the reverse microemulsion technique. The colloidal stability of sFe_3O_4 MNPs increased with the silica amount up to a certain critical thickness and then decreased as the silica layer became thicker. The applied silica coating did not affect the intrinsic magnetic properties of the MNPs. The sFe_3O_4 MNPs with the highest colloidal stability presented enhanced MH performance in water, and the SAR values increased by almost 1000 W/g_{Fe} compared to bare Fe_3O_4 MNPs.

The malignant and normal cells internalized a high quantity of sFe_3O_4 MNPs inside large endosomes at a low dose (15.62 µg/cm^2); the sFe_3O_4 MNPs were dispersed in cytoplasm or accumulated inside lysosomes as the dose increased. Cytotoxicity studies using Alamar Blue and Neutral Red assays confirmed the SEM and TEM findings, revealing insignificant toxicity for normal cells over the entire dose range, whereas for A549 and A375 cell lines a drop in cellular viability to 80% was recorded starting with doses of 125 and 250 µg/cm^2, respectively.

Intracellular magnetic hyperthermia experiments revealed that the malignant cells were more sensitive to MH treatment compared to the normal ones. More than 50% of malignant cells, incubated at a dose of 31.25 µg/cm^2, underwent cellular death starting with an H of 30 kA/m (355 kHz). Increasing the H to 45 and 60 kA/m enabled the destruction of a large number of malignant cells at lower doses: 15.62 and 7.81 µg/cm^2, respectively. The affected cells were retracted and rounded, and covered with membrane blebs, displaying rarefied cytoplasm and disrupted plasma membrane on extensive regions, whereas many others were in an advanced stage of fragmentation. In the case of still-intact cells, upon MH treatment, they were filled almost completely with numerous and large vacuoles.

Our data demonstrate that controlled silica coating of ferromagnetic iron oxide nanoparticles significantly increases their hyperthermia performance, cellular uptake, and efficient destruction of cancer cells, making these MNPs excellent candidates for further in vivo studies.

Supplementary Materials: The following are available online at https://www.mdpi.com/article/10.3390/pharmaceutics13122026/s1, Figure S1: SEM images of silica-coated Fe_3O_4 MNPs, Figure S2: XRD patterns of MNPs, Figure S3: Hydrodynamic diameter of MNPs, Figure S4: Heating curves of silica-coated Fe_3O_4 MNPs dispersed in water at different concentrations, Figure S5: Calibration curve for iron concentration determination, Figure S6: SAR dependence on H for MH in PEG 8K, Figure S7: Optical interference of silica-coated Fe_3O_4 MNPs with Alamar Blue and Neutral Red assay, Figure S8: Biochemical interferences of silica-coated Fe_3O_4 MNPs with Alamar Blue and Neutral Red assays, Figure S9: SEM and TEM images of A549, A375, and BJ cells, Figure S10: Relevant TEM images of BJ and A375 cells incubated with silica-coated Fe_3O_4 MNPs, Figure S11: SEM and TEM images of A375 cells incubated with silica-coated Fe_3O_4 MNPs, Figure S12: Heating curves of cells incubated with silica-coated Fe_3O_4 MNPs, Figure S13: Saturation temperatures, Figure S14: SEM and TEM examination of A375 cells incubated with silica-coated Fe_3O_4 MNPs and exposed for 30 min to an AMF of 30 kA/m, 355 kHz. Table S1: PDI indexes as obtained from DLS data, Table S2: Fitting parameters of SAR evolution with H, Table S3: Amount of Fe^{3+} internalized in cells relevant for in vitro cytotoxicity assays and Fe^{3+} concentration of samples used in in vitro magnetic hyperthermia, Table S4: Fitting parameters of cytotoxicity data upon intracellular MH based on Equation (1).

Author Contributions: Conceptualization, C.I. and C.M.L.; methodology, C.I., I.F., A.F. and C.M.L.; software, C.I., I.F., A.F. and R.D.; validation, C.I., A.F., R.T., F.L. and C.M.L.; formal analysis, C.I., I.F., A.F., L.B.-T., R.D. and C.M.L.; investigation, C.I., I.F., S.N., A.F., L.B.-T., A.P., R.D., N.V. and O.C.; resources, C.I. and I.F. data curation, C.I., I.F., A.F. and C.M.L.; writing—original draft preparation, C.I., I.F., A.F. and C.M.L.; writing—review and editing, C.I., I.F., A.F. and C.M.L.; visualization, C.I., I.F., A.F. and C.M.L.; supervision, R.T., F.L. and C.M.L.; project administration, C.I. and I.F.; funding acquisition, C.I. and I.F. All authors have read and agreed to the published version of the manuscript.

Funding: This work was supported by two grants from the Romanian Ministry of Education and Research, CNCS—UEFISCDI, project numbers PN-III-P2-2.1-PED-2019-3283 and PN-III-P1-1.1-PD-2019-0804, within PNCDI III.

Institutional Review Board Statement: Not applicable.

Informed Consent Statement: Not applicable.

Data Availability Statement: All data available are reported in the article and in the supplementary materials.

Conflicts of Interest: The authors declare no conflict of interest. The funders had no role in the design of the study; in the collection, analyses, or interpretation of data; in the writing of the manuscript, or in the decision to publish the results.

References

1. Rubia-Rodríguez, I.; Santana-Otero, A.; Spassov, S.; Tombácz, E.; Johansson, C.; De La Presa, P.; Teran, F.; Morales, M.D.P.; Veintemillas-Verdaguer, S.; Thanh, N.; et al. Whither Magnetic Hyperthermia? A Tentative Roadmap. *Materials* **2021**, *14*, 706. [CrossRef] [PubMed]
2. Liu, X.; Zhang, Y.; Wang, Y.; Zhu, W.; Li, G.; Ma, X.; Chen, S.; Tiwari, S.; Shi, K.; Zhang, S.; et al. Comprehensive understanding of magnetic hyperthermia for improving antitumor therapeutic efficacy. *Theranostics* **2020**, *10*, 3793–3815. [CrossRef]
3. Etemadi, H.; Plieger, P.G. Magnetic Fluid Hyperthermia Based on Magnetic Nanoparticles: Physical Characteristics, Historical Perspective, Clinical Trials, Technological Challenges, and Recent Advances. *Adv. Ther.* **2020**, *3*, 2000061. [CrossRef]
4. Yu, X.; Ding, S.; Yang, R.; Wu, C.; Zhang, W. Research progress on magnetic nanoparticles for magnetic induction hyperthermia of malignant tumor. *Ceram. Int.* **2021**, *47*, 5909–5917. [CrossRef]
5. Johannsen, M.; Gneveckow, U.; Thiesen, B.; Taymoorian, K.; Cho, C.H.; Waldöfner, N.; Scholz, R.; Jordan, A.; Loening, S.A.; Wust, P. Thermotherapy of Prostate Cancer Using Magnetic Nanoparticles: Feasibility, Imaging, and Three-Dimensional Temperature Distribution. *Eur. Urol.* **2007**, *52*, 1653–1662. [CrossRef] [PubMed]
6. Johannsen, M.; Gneveckow, U.; Taymoorian, K.; Thiesen, B.; Waldöfner, N.; Scholz, R.; Jung, K.; Jordan, A.; Wust, P.; Loening, S.A. Morbidity and quality of life during thermotherapy using magnetic nanoparticles in locally recurrent prostate cancer: Results of a prospective phase I trial. *Int. J. Hyperth.* **2007**, *23*, 315–323. [CrossRef]
7. Maier-HauffFrank, K.; Ulrich, F.; Nestler, D.; Niehoff, H.; Wust, P.; Thiesen, B.; Orawa, H.; Budach, V.; Jordan, A. Efficacy and safety of intratumoral thermotherapy using magnetic iron-oxide nanoparticles combined with external beam radiotherapy on patients with recurrent glioblastoma multiforme. *J. Neuro-Oncol.* **2010**, *103*, 317–324. [CrossRef]
8. Grauer, O.; Jaber, M.; Hess, K.; Weckesser, M.; Schwindt, W.; Maring, S.; Wölfer, J.; Stummer, W. Combined intracavitary thermotherapy with iron oxide nanoparticles and radiotherapy as local treatment modality in recurrent glioblastoma patients. *J. Neuro-Oncol.* **2019**, *141*, 83–94. [CrossRef] [PubMed]
9. Kobayashi, T.; Kakimi, K.; Nakayama, E.; Jimbow, K. Antitumor immunity by magnetic nanoparticle-mediated hyperthermia. *Nanomedicine* **2014**, *9*, 1715–1726. [CrossRef]
10. Wilczewska, A.Z.; Niemirowicz, K.; Markiewicz, K.H.; Car, H. Nanoparticles as drug delivery systems. *Pharmacol. Rep.* **2012**, *64*, 1020–1037. [CrossRef]
11. Carrey, J.; Mehdaoui, B.; Respaud, M. Simple models for dynamic hysteresis loop calculations of magnetic single-domain nanoparticles: Application to magnetic hyperthermia optimization. *J. Appl. Phys.* **2011**, *109*, 083921. [CrossRef]
12. Rosensweig, R. Heating magnetic fluid with alternating magnetic field. *J. Magn. Magn. Mater.* **2002**, *252*, 370–374. [CrossRef]
13. Avolio, M.; Guerrini, A.; Brero, F.; Innocenti, C.; Sangregorio, C.; Cobianchi, M.; Mariani, M.; Orsini, F.; Arosio, P.; Lascialfari, A. In-gel study of the effect of magnetic nanoparticles immobilization on their heating efficiency for application in Magnetic Fluid Hyperthermia. *J. Magn. Magn. Mater.* **2019**, *471*, 504–512. [CrossRef]
14. Engelmann, U.M.; Seifert, J.; Mues, B.; Roitsch, S.; Ménager, C.; Schmidt, A.M.; Slabu, I. Heating efficiency of magnetic nanoparticles decreases with gradual immobilization in hydrogels. *J. Magn. Magn. Mater.* **2019**, *471*, 486–494. [CrossRef]
15. Cabrera, D.; Coene, A.; Leliaert, J.; Artés-Ibáñez, E.J.; Dupré, L.; Telling, N.D.; Teran, F.J. Dynamical Magnetic Response of Iron Oxide Nanoparticles Inside Live Cells. *ACS Nano* **2018**, *12*, 2741–2752. [CrossRef]
16. Nemati, Z.; Alonso, J.; Rodrigo, I.; Das, R.; Garaio, E.; García, J.Á.; Orue, I.; Phan, M.-H.; Srikanth, H. Improving the Heating Efficiency of Iron Oxide Nanoparticles by Tuning Their Shape and Size. *J. Phys. Chem. C* **2018**, *122*, 2367–2381. [CrossRef]
17. Mohapatra, J.; Zeng, F.; Elkins, K.; Xing, M.; Ghimire, M.; Yoon, S.; Mishra, S.R.; Liu, J.P. Size-dependent magnetic and inductive heating properties of Fe3O4 nanoparticles: Scaling laws across the superparamagnetic size. *Phys. Chem. Chem. Phys.* **2018**, *20*, 12879–12887. [CrossRef]
18. Tong, S.; Quinto, C.A.; Zhang, L.; Mohindra, P.; Bao, G. Size-Dependent Heating of Magnetic Iron Oxide Nanoparticles. *ACS Nano* **2017**, *11*, 6808–6816. [CrossRef]
19. Guardia, P.; Di Corato, R.; Lartigue, L.; Wilhelm, C.; Espinosa, A.; Garcia-Hernandez, M.; Gazeau, F.; Manna, L.; Pellegrino, T. Water-Soluble Iron Oxide Nanocubes with High Values of Specific Absorption Rate for Cancer Cell Hyperthermia Treatment. *ACS Nano* **2012**, *6*, 3080–3091. [CrossRef]
20. Bae, K.H.; Park, M.; Do, M.J.; Lee, N.; Ryu, J.H.; Kim, G.W.; Kim, C.; Park, T.G.; Hyeon, T. Chitosan Oligosaccharide-Stabilized Ferrimagnetic Iron Oxide Nanocubes for Magnetically Modulated Cancer Hyperthermia. *ACS Nano* **2012**, *6*, 5266–5273. [CrossRef] [PubMed]
21. Lv, Y.; Yang, Y.; Fang, J.; Zhang, H.; Peng, E.; Liu, X.; Xiao, W.; Ding, J. Size dependent magnetic hyperthermia of octahedral Fe3O4 nanoparticles. *RSC Adv.* **2015**, *5*, 76764–76771. [CrossRef]
22. Castellanos-Rubio, I.; Rodrigo, I.; Munshi, R.; Arriortua, O.; Garitaonandia, J.S.; Martinez-Amesti, A.; Plazaola, F.; Orue, I.; Pralle, A.; Insausti, M. Outstanding heat loss via nano-octahedra above 20 nm in size: From wustite-rich nanoparticles to magnetite single-crystals. *Nanoscale* **2019**, *11*, 16635–16649. [CrossRef] [PubMed]
23. Sugumaran, P.J.; Yang, Y.; Wang, Y.; Liu, X.; Ding, J. Influence of the Aspect Ratio of Iron Oxide Nanorods on Hysteresis-Loss-Mediated Magnetic Hyperthermia. *ACS Appl. Bio Mater.* **2021**, *4*, 4809–4820. [CrossRef]
24. Iacovita, C.; Florea, A.; Dudric, R.; Pall, E.; Moldovan, A.I.; Tetean, R.; Stiufiuc, R.; Lucaciu, C.M. Small versus Large Iron Oxide Magnetic Nanoparticles: Hyperthermia and Cell Uptake Properties. *Molecules* **2016**, *21*, 1357. [CrossRef] [PubMed]

25. Zhu, N.; Ji, H.; Yu, P.; Niu, J.; Farooq, M.U.; Akram, M.W.; Udego, I.O.; Li, H.; Niu, X. Surface Modification of Magnetic Iron Oxide Nanoparticles. *Nanomaterials* **2018**, *8*, 810. [CrossRef] [PubMed]
26. Tapeinos, C.; Tomatis, F.; Battaglini, M.; Larrañaga, A.; Marino, A.; Agirrezabal-Telleria, I.; Angelakeris, M.; Debellis, D.; Drago, F.; Brero, F.; et al. Cell Membrane-Coated Magnetic Nanocubes with a Homotypic Targeting Ability Increase Intracellular Temperature due to ROS Scavenging and Act as a Versatile Theranostic System for Glioblastoma Multiforme. *Adv. Healthc. Mater.* **2019**, *8*, e1900612. [CrossRef]
27. Castellanos-Rubio, I.; Rodrigo, I.; Olazagoitia-Garmendia, A.; Arriortua, O.; Gil de Muro, I.; Garitaonandia, J.S.; Bilbao, J.R.; Fdez-Gubieda, M.L.; Plazaola, F.; Orue, I.; et al. Highly Reproducible Hyperthermia Response in Water, Agar, and Cellular Environment by Discretely PEGylated Magnetite Nanoparticles. *ACS Appl. Mater. Interfaces* **2020**, *12*, 27917–27929. [CrossRef]
28. Zhang, L.; Xu, H.; Cheng, Z.; Wei, Y.; Sun, R.; Liang, Z.; Hu, Y.; Zhao, L.; Lian, X.; Li, X.; et al. Human Cancer Cell Membrane-Cloaked Fe3O4 Nanocubes for Homologous Targeting Improvement. *J. Phys. Chem. B* **2021**, *125*, 7417–7426. [CrossRef]
29. Guerrero-Martínez, A.; Pérez-Juste, J.; Liz-Marzán, L.M. Recent Progress on Silica Coating of Nanoparticles and Related Nanomaterials. *Adv. Mater.* **2010**, *22*, 1182–1195. [CrossRef]
30. Mamun, A.; Rumi, K.M.J.U.; Das, H.; Hoque, S.M. Synthesis, Properties and Applications of Silica-Coated Magnetite Nanoparticles: A Review. *Nano* **2021**, *16*, 2130005. [CrossRef]
31. Kolhatkar, A.G.; Nekrashevich, I.; Litvinov, D.; Willson, R.C.; Lee, T.R. Cubic Silica-Coated and Amine-Functionalized FeCo Nanoparticles with High Saturation Magnetization. *Chem. Mater.* **2013**, *25*, 1092–1097. [CrossRef]
32. Azadmanjiri, J.; Simon, G.P.; Suzuki, K.; Selomulya, C.; Cashion, J.D. Phase reduction of coated maghemite (γ-Fe2O3) nanoparticles under microwave-induced plasma heating for rapid heat treatment. *J. Mater. Chem.* **2012**, *22*, 617–625. [CrossRef]
33. Marcelo, G.; Perez, E.; Corrales, T.; Peinado, C. Stabilization in Water of Large Hydrophobic Uniform Magnetite Cubes by Silica Coating. *J. Phys. Chem. C* **2011**, *115*, 25247–25256. [CrossRef]
34. Ding, H.L.; Zhang, Y.X.; Wang, S.; Xu, J.M.; Xu, S.C.; Li, G.H. Fe3O4@SiO2 Core/Shell Nanoparticles: The Silica Coating Regulations with a Single Core for Different Core Sizes and Shell Thicknesses. *Chem. Mater.* **2012**, *24*, 4572–4580. [CrossRef]
35. He, S.; Zhang, H.; Liu, Y.; Sun, F.; Yu, X.; Li, X.; Zhang, L.; Wang, L.; Mao, K.; Wang, G.; et al. Maximizing Specific Loss Power for Magnetic Hyperthermia by Hard-Soft Mixed Ferrites. *Small* **2018**, *14*, e1800135. [CrossRef]
36. Mathieu, P.; Coppel, Y.; Respaud, M.; Nguyen, Q.T.; Boutry, S.; Laurent, S.; Stanicki, D.; Henoumont, C.; Novio, F.; Lorenzo, J.; et al. Silica Coated Iron/Iron Oxide Nanoparticles as a Nano-Platform for T2 Weighted Magnetic Resonance Imaging. *Molecule* **2019**, *24*, 4629. [CrossRef] [PubMed]
37. Gao, D.; Ring, H.L.; Sharma, A.; Namsrai, B.; Tran, N.; Finger, E.B.; Garwood, M.; Haynes, C.; Bischof, J.C. Preparation of Scalable Silica-Coated Iron Oxide Nanoparticles for Nanowarming. *Adv. Sci.* **2020**, *7*, 1901624. [CrossRef] [PubMed]
38. Wang, R.; Liu, J.; Liu, Y.; Zhong, R.; Yu, X.; Liu, Q.; Zhang, L.; Lv, C.; Mao, K.; Tang, P. The cell uptake properties and hyperthermia performance of $Zn_{0.5}Fe_{2.5}O_4/SiO_2$ nanoparticles as magnetic hyperthermia agents. *R. Soc. Open Sci.* **2020**, *7*, 191139. [CrossRef]
39. Kolosnjaj-Tabi, J.; Kralj, S.; Griseti, E.; Nemec, S.; Wilhelm, C.; Sangnier, A.P.; Bellard, E.; Fourquaux, I.; Golzio, M.; Rols, M.-P. Magnetic Silica-Coated Iron Oxide Nanochains as Photothermal Agents, Disrupting the Extracellular Matrix, and Eradicating Cancer Cells. *Cancers* **2019**, *11*, 2040. [CrossRef] [PubMed]
40. Reczyńska, K.; Marszałek, M.; Zarzycki, A.; Reczyński, W.; Kornaus, K.; Pamuła, E.; Chrzanowski, W. Superparamagnetic Iron Oxide Nanoparticles Modified with Silica Layers as Potential Agents for Lung Cancer Treatment. *Nanomaterials* **2020**, *10*, 1076. [CrossRef] [PubMed]
41. Horny, M.-C.; Gamby, J.; Dupuis, V.; Siaugue, J.-M. Magnetic Hyperthermia on γ-Fe2O3@SiO2 Core-Shell Nanoparticles for mi-RNA 122 Detection. *Nanomaterials* **2021**, *11*, 149. [CrossRef]
42. Dávila-Ibáñez, A.B.; Salgueirino, V.; Martinez-Zorzano, V.; Mariño-Fernández, R.; García-Lorenzo, A.; Maceira-Campos, M.; Úbeda, M.M.; Junquera, E.; Aicart, E.; Rivas, J.; et al. Magnetic Silica Nanoparticle Cellular Uptake and Cytotoxicity Regulated by Electrostatic Polyelectrolytes–DNA Loading at Their Surface. *ACS Nano* **2011**, *6*, 747–759. [CrossRef]
43. Pham, X.-H.; Hahm, E.; Kim, H.-M.; Son, B.S.; Jo, A.; An, J.; Thi, T.A.T.; Nguyen, D.Q.; Jun, B.-H. Silica-Coated Magnetic Iron Oxide Nanoparticles Grafted onto Graphene Oxide for Protein Isolation. *Nanomaterials* **2020**, *10*, 117. [CrossRef]
44. Královec, K.; Melounková, L.; Slováková, M.; Mannová, N.; Sedlák, M.; Bartáček, J.; Havelek, R. Disruption of Cell Adhesion and Cytoskeletal Networks by Thiol-Functionalized Silica-Coated Iron Oxide Nanoparticles. *Int. J. Mol. Sci.* **2020**, *21*, 9350. [CrossRef] [PubMed]
45. Abbas, M.; Torati, S.R.; Lee, C.S.; Rinaldi, C.; Kim, C. Fe3O4/SiO2 Core/Shell Nanocubes: Novel Coating Approach with Tunable Silica Thickness and Enhancement in Stability and Biocompatibility. *J. Nanomed. Nanotechnol.* **2014**, *5*, 1–8. [CrossRef]
46. Nemec, S.; Kralj, S.; Wilhelm, C.; Abou-Hassan, A.; Rols, M.-P.; Kolosnjaj-Tabi, J. Comparison of Iron Oxide Nanoparticles in Photothermia and Magnetic Hyperthermia: Effects of Clustering and Silica Encapsulation on Nanoparticles' Heating Yield. *Appl. Sci.* **2020**, *10*, 7322. [CrossRef]
47. Villanueva, A.; de la Presa, P.; Alonso, J.M.; Rueda, T.; Martínez, A.; Crespo, P.; Morales, M.D.P.; Gonzalez-Fernandez, M.A.; Valdés, J.; Rivero, G. Hyperthermia HeLa Cell Treatment with Silica-Coated Manganese Oxide Nanoparticles. *J. Phys. Chem. C* **2010**, *114*, 1976–1981. [CrossRef]
48. Starsich, F.H.L.; Sotiriou, G.A.; Wurnig, M.C.; Eberhardt, C.; Hirt, A.; Boss, A.; Pratsinis, S.E. Silica-Coated Nonstoichiometric Nano Zn-Ferrites for Magnetic Resonance Imaging and Hyperthermia Treatment. *Adv. Healthc. Mater.* **2016**, *5*, 2698–2706. [CrossRef]

49. Basu, P.; De, K.; Das, S.; Mandal, A.; Kumar, A.; Jana, T.K.; Chatterjee, K. Silica-Coated Metal Oxide Nanoparticles: Magnetic and Cytotoxicity Studies. *ChemistrySelect* **2018**, *3*, 7346–7353. [CrossRef]
50. Pinho, S.L.C.; Laurent, S.; Rocha, J.; Roch, A.; Delville, M.-H.; Mornet, S.; Carlos, L.D.; Elst, L.V.; Muller, R.N.; Geraldes, C.F.G.C. Relaxometric Studies of γ-Fe_2O_3@SiO_2 Core Shell Nanoparticles: When the Coating Matters. *J. Phys. Chem. C* **2012**, *116*, 2285–2291. [CrossRef]
51. Glaria, A.; Soulé, S.; Hallali, N.; Ojo, W.-S.; Mirjolet, M.; Fuks, G.; Cornejo, A.; Allouche, J.; Dupin, J.C.; Martinez, H.; et al. Silica coated iron nanoparticles: Synthesis, interface control, magnetic and hyperthermia properties. *RSC Adv.* **2018**, *8*, 32146–32156. [CrossRef]
52. Iacovita, C.; Stiufiuc, R.; Radu, T.; Florea, A.; Stiufiuc, G.; Dutu, A.G.; Mican, S.; Tetean, R.; Lucaciu, C.M. Polyethylene Glycol-Mediated Synthesis of Cubic Iron Oxide Nanoparticles with High Heating Power. *Nanoscale Res. Lett.* **2015**, *10*, 1–16. [CrossRef]
53. Iacovita, C.; Fizeșan, I.; Pop, A.; Scorus, L.; Dudric, R.; Stiufiuc, G.; Vedeanu, N.; Tetean, R.; Loghin, F.; Stiufiuc, R.; et al. In Vitro Intracellular Hyperthermia of Iron Oxide Magnetic Nanoparticles, Synthesized at High Temperature by a Polyol Process. *Pharmaceutics* **2020**, *12*, 424. [CrossRef]
54. Dippong, T.; Levei, E.; Cadar, O. Formation, Structure and Magnetic Properties of MFe_2O_4@SiO_2 (M = Co, Mn, Zn, Ni, Cu) Nanocomposites. *Materials* **2021**, *14*, 1139. [CrossRef] [PubMed]
55. Che, X.-D.; Bertram, H.N. Phenomenology of δM curves and magnetic interactions. *J. Magn. Magn. Mater.* **1992**, *116*, 121–127. [CrossRef]
56. Teran, F.J.; Casado, C.; Mikuszeit, N.; Salas, G.; Bollero, A.; Morales, M.D.P.; Camarero, J.J.; Miranda, R. Accurate determination of the specific absorption rate in superparamagnetic nanoparticles under non-adiabatic conditions. *Appl. Phys. Lett.* **2012**, *101*, 062413. [CrossRef]
57. Iacovita, C.; Florea, A.; Scorus, L.; Pall, E.; Dudric, R.; Moldovan, A.I.; Stiufiuc, R.; Tetean, R.; Lucaciu, C.M. Hyperthermia, Cytotoxicity, and Cellular Uptake Properties of Manganese and Zinc Ferrite Magnetic Nanoparticles Synthesized by a Polyol-Mediated Process. *Nanomaterials* **2019**, *9*, 1489. [CrossRef]
58. Iacovita, C.; Stiufiuc, G.F.; Dudric, R.; Vedeanu, N.; Tetean, R.; Stiufiuc, R.I.; Lucaciu, C.M. Saturation of Specific Absorption Rate for Soft and Hard Spinel Ferrite Nanoparticles Synthesized by Polyol Process. *Magnetochemistry* **2020**, *6*, 23. [CrossRef]
59. Branquinho, L.C.; Carrião, M.; Costa, A.S.; Zufelato, N.; Sousa, M.H.; Miotto, R.; Ivkov, R.; Bakuzis, A.F. Effect of magnetic dipolar interactions on nanoparticle heating efficiency: Implications for cancer hyperthermia. *Sci. Rep.* **2013**, *3*, srep02887. [CrossRef]
60. Martinez-Boubeta, C.; Simeonidis, K.; Makridis, A.; Angelakeris, M.; Iglesias, O.; Guardia, P.; Cabot, A.; Yedra, L.; Estradé, S.; Peiró, F.; et al. Learning from Nature to Improve the Heat Generation of Iron-Oxide Nanoparticles for Magnetic Hyperthermia Applications. *Sci. Rep.* **2013**, *3*, 1652. [CrossRef]
61. Stoner, E.C.; Wohlfarth, E.P. A mechanism of magnetic hysteresis in heterogeneous alloys. *Philos. Trans. R. Soc.* **1948**, *240*, 599–642. [CrossRef]
62. Ong, K.J.; MacCormack, T.; Clark, R.J.; Ede, J.D.; Ortega, V.A.; Felix, L.; Dang, M.K.M.; Ma, G.; Fenniri, H.; Veinot, J.G.C.; et al. Widespread Nanoparticle-Assay Interference: Implications for Nanotoxicity Testing. *PLoS ONE* **2014**, *9*, e90650. [CrossRef]
63. Gavard, J.; Hanini, A.; Schmitt, A.; Kacem, K.; Chau, F.; Ammar, S. Evaluation of iron oxide nanoparticle biocompatibility. *Int. J. Nanomed.* **2011**, *6*, 787–794. [CrossRef] [PubMed]
64. Matsuda, S.; Hitsuji, A.; Nakanishi, T.; Zhang, H.; Tanaka, A.; Matsuda, H.; Osaka, T. Induction of Cell Death in Mesothelioma Cells by Magnetite Nanoparticles. *ACS Biomater. Sci. Eng.* **2015**, *1*, 632–638. [CrossRef]
65. Sanz, B.; Calatayud, M.P.; Torres, T.E.; Fanarraga, M.; Ibarra, M.R.; Goya, G.F. Magnetic hyperthermia enhances cell toxicity with respect to exogenous heating. *Biomaterials* **2017**, *114*, 62–70. [CrossRef] [PubMed]
66. Feng, Y.; Oden, J.T.; Rylander, M.N. A Two-State Cell Damage Model Under Hyperthermic Conditions: Theory and In Vitro Experiments. *J. Biomech. Eng.* **2008**, *130*, 041016. [CrossRef] [PubMed]
67. Zanganeh, S.; Hutter, G.; Spitler, R.; Lenkov, O.; Mahmoudi, M.; Shaw, A.; Pajarinen, J.S.; Nejadnik, H.; Goodman, J.S.P.S.; Moseley, M.; et al. Iron oxide nanoparticles inhibit tumour growth by inducing pro-inflammatory macrophage polarization in tumour tissues. *Nat. Nanotechnol.* **2016**, *11*, 986–994. [CrossRef]
68. Dönmez Güngüneş, Ç.; Şeker, Ş.; Elçin, A.E.; Elçin, Y.M. A comparative study on the in vitro cytotoxic responses of two mammalian cell types to fullerenes, carbon nanotubes, and iron oxide nanoparticles. *Drug Chem. Toxicol.* **2017**, *40*, 215–227. [CrossRef] [PubMed]
69. Nedyalkova, M.; Donkova, B.; Romanova, J.; Tzvetkov, G.; Madurga, S.; Simeonov, V. Iron oxide nanoparticles–in vivo/in vitro biomedical applications and in silico studies. *Adv. Colloid Interface Sci.* **2017**, *249*, 192. [CrossRef]

Article

Biodistribution Profile of Magnetic Nanoparticles in Cirrhosis-Associated Hepatocarcinogenesis in Rats by AC Biosusceptometry

Guilherme A. Soares [1,*], Gabriele M. Pereira [1], Guilherme R. Romualdo [2,3], Gabriel G. A. Biasotti [1], Erick G. Stoppa [1], Andris F. Bakuzis [4], Oswaldo Baffa [5], Luis F. Barbisan [3] and Jose R. A. Miranda [1]

1. Department of Biophysics and Pharmacology, Institute of Biosciences, São Paulo State University—UNESP, Botucatu 18618-689, SP, Brazil
2. Department of Pathology, Botucatu Medical School, São Paulo State University (UNESP), Botucatu 18618-689, SP, Brazil
3. Department of Structural and Functional Biology, Institute of Biosciences, São Paulo State University—UNESP, Botucatu 18618-689, SP, Brazil
4. Institute of Physics, Federal University of Goiás, Goiânia 74690-900, GO, Brazil
5. Faculty of Philosophy, Sciences and Letters at Ribeirão Preto, University of São Paulo, Ribeirão Preto 14040-900, SP, Brazil
* Correspondence: guilherme.soares@unesp.br

Abstract: Since magnetic nanoparticles (MNPs) have been used as multifunctional probes to diagnose and treat liver diseases in recent years, this study aimed to assess how the condition of cirrhosis-associated hepatocarcinogenesis alters the biodistribution of hepatic MNPs. Using a real-time image acquisition approach, the distribution profile of MNPs after intravenous administration was monitored using an AC biosusceptometry (ACB) assay. We assessed the biodistribution profile based on the ACB images obtained through selected regions of interest (ROIs) in the heart and liver position according to the anatomical references previously selected. The signals obtained allowed for the quantification of pharmacokinetic parameters, indicating that the uptake of hepatic MNPs is compromised during liver cirrhosis, since scar tissue reduces blood flow through the liver and slows its processing function. Since liver monocytes/macrophages remained constant during the cirrhotic stage, the increased intrahepatic vascular resistance associated with impaired hepatic sinusoidal circulation was considered the potential reason for the change in the distribution of MNPs.

Keywords: AC biosuceptometry; magnetic nanoparticles; cirrhosis-associated rat hepatocarcinogenesis; nanotechnology

1. Introduction

The liver is a solid organ that is divided into two portions: (1) a parenchymal portion, which is composed of hepatocytes and biliary cells, and (2) a non-parenchymal portion, constituted by Kupffer cells (KCs), sinusoidal endothelial cells (LSECs), and resting hepatic stellate cells (HSCs) [1]. The KCs are resident macrophages that specialize in phagocytosis and cytokine release, acting as the liver's first immune defense [2]. KCs are associated with the LSECs that line the hepatic sinusoids. The HSCs are spread throughout the Disse space and are responsible for storing vitamin A and secreting limited amounts of extracellular matrix (ECM) proteins under physiological conditions [3].

The liver, under homeostasis, displays an extensive range of functions, such as the regulation of blood volume and immunity, drug detoxification, endocrine control of growth, lipid and cholesterol homeostasis, and the metabolism of nutrients. It also features a regenerative capacity through hepatocytes [4–6] Nonetheless, this organ may develop several chronic diseases, including non-neoplastic and neoplastic diseases. Hepatocellular carcinoma (HCC), the main primary hepatic malignancy, stands out for its current epidemiological burden, as it ranks fourth among the most common cancers and it is the sixth most common cause of cancer-related deaths worldwide [7]. HCC usually emerges in the context of hepatic fibrosis/cirrhosis (70–95% of cases) [8] and also features a poor prognosis, with a median survival time of 11 months and survival rates of 19 to 29% at 3 years after diagnosis [9,10]. Furthermore, a 53% to 60% growth in both the incidence of and mortality from HCC is estimated over the next 20 years [7]. Such epidemiological data elicit the need for new diagnostic, preventative, and therapeutic tools for this malignance.

Under the known risk conditions—i.e., chronic hepatitis B and C infections, non-alcoholic liver disease, and alcohol intake—HCC gradually emerges in the context of tumor-promoting inflammation/hepatocyte injury hallmarks, HSC activation, and HSCs and macrophage pro-inflammatory crosstalk, culminating on collagen production. Collagen progressively accumulates, leading to liver fibrosis, and the end stage of this process is called cirrhosis, which is characterized by a marked impairment of liver function and an increased risk for HCC development [11]. In order to investigate the different aspects of cirrhosis-associated hepatocarcinogenesis, experimental models, including chemically induced models, have been widely applied in translational research [11–14]. These models use chemical hepatotoxins that induce (pre)neoplastic lesions in a cirrhotic background, as in the diethylnitrosamine (DEN)-initiated and thioacetamide (TAA)-promoted model [15]. These murine models gather morphological and molecular similarities to the corresponding human disease, enabling translational research on hepatocarcinogenesis [11,13] and nanotechnology studies.

Despite liver fibrosis not having clear symptoms, its early detection is essential for preventing the further aggravation to other diseases such as cirrhosis and HCC and for providing beneficial future treatments [14,15]. Although a percutaneous liver biopsy is an invasive strategy and presents several drawbacks, such as sampling error and cost, this diagnostic procedure is usually always associated with non-invasive diagnostic methods and serum biochemistry [16–18]. In addition to non-invasive staging of hepatic fibrosis using magnetic resonance imaging (MRI) and computed tomography (CT), ultrasonography is a widely used accurate diagnostic imaging tool [19–22]. This diagnostic method is also inexpensive, supporting its practical use. Nevertheless, these imaging methods have drawbacks that make detecting fibrosis and cirrhosis at early stages difficult, and there are also drawbacks related to the experience level of the operator. Furthermore, these methods are not indicated for obese patients [18,23]. Despite a routine MRI examination presenting advantages such as its ability to reach deep tissue in the liver with a high spatial resolution, which allows for a complete characterization of liver disease processes, it has some limitations [14]. At the same time, the disadvantages of CTs are the need for ionizing radiation and the existence of respiratory motion artifacts.

Several conventional approaches have been employed to suppress hepatic inflammation/scar deposition using antifibrotic drugs to treat liver diseases. However, most of these conventional therapies are ineffective because the drug delivery is not specific, since specific hepatic cell types are responsible for hepatic inflammation/fibrosis [4,24,25]. In this way, the difficulty of delivering a sufficient dose of pharmacological agents to the liver is associated with the non-specificity of targeting cellular structures, indicating that treating liver diseases remains a challenge.

Nanomaterials have attracted attention in the development of nanotechnology due to the possibility of their use as multifunction probes for diagnoses and treatments in recent years [26–31]. Nanoparticles have great potential for several biomedical applications since they have interesting properties, such as a reduced size, shape manipulation, and the possibility of conjugation with other materials and molecules. A class of nanoparticles that has several advantages due to the intrinsic properties and biocompatibility of its members is magnetic nanoparticles (MNPs). Over the last few years, MNPs have been used in many theranostic applications [32–35], including diagnosing and treating hepatic diseases [32–34]. The magnetic nanoparticle-based diagnosis and treatment of liver diseases has shown great potential, since MNPs present advantages such as (i) easy functionalization and conjugation with molecules and surface markers, which allows for targeting drug delivery agents to specific cell-type agents [35]; (ii) their ability to act as magnetic vectors to specific liver locations, since they respond strongly to an external magnetic gradient [36]; and (iii) their ability to act as labeling and tracking agents in imaging modalities, thus enhancing non-invasive approaches for investigating liver fibrosis conditions [37,38]. Within the parameters that determine the blood clearance pharmacokinetics of MNPs, the hydrodynamic size of MNPs is one of the most critical parameters that affect their biodistribution kinetics and uptake by the mononuclear phagocyte system (MPS) [39,40]. The MPS, which comprises dendritic cells, blood monocytes, and resident-tissue macrophages in several organs, is a specialized and selective structure that takes up nanoparticles in general. Usually, it has been reported through consistent evidence that nanoparticles presenting with hydrodynamic sizes within 15–100 nm are captured mainly by the liver and the spleen [41–43]. In comparison, nanomaterials smaller than 10 nm are likely to be eliminated through renal clearance [44–46].

Once administered intravenously, MNPs are substantially captured and retained in the liver, depending on physical factors such as coating, dose, and size [47]. The presence of fenestrated vasculature (sinusoids) and many Kupffer cells supports the significant amount of MNPs in the liver. Literature reports have indicated that the liver takes up around 30–99% of the MNPs in a dose [48–51]. Therefore, the high abundance of MNPs in the liver after intravenous injection and their superparamagnetic properties increase the potential of these materials to be used as a contrast agent to enhance the signal-to-noise ratio, which makes magnetic imaging modalities feasible for diagnosing liver diseases [52–55].

Over the years, several methods have been used to detect MNPs in tissues. These methods are classified into direct and indirect methodologies. For in vivo studies, MRI and magnetic particle imaging (MPI) are techniques that detect and visualize particles by their inherent properties and can contribute to determining the pharmacokinetics and biodistribution of MNPs [56].

Despite MRI being a consolidated methodology for imaging, in general, it involves a high cost and also has drawbacks regarding the differentiation of the position of MNPs with a low signal [7].

MPI emerged as an alternative and promising technique for MNP detection. The technique is based on the nonlinear magnetic response of the IONPs to an applied AC magnetic field and presents no depth limitation when used to directly measure the MNP concentration. However, MPI presents limitations regarding the complexity and associated high cost, so it is not widely used [57]. Nanoparticles can be detected through their conjugation to contrast agents or radioactive markers by using imaging methodologies such as near-infrared (NIR) fluorescence, positron emission tomography (PET), and single-photon emission computed tomography (SPECT) [58–60].

Electron paramagnetic resonance (EPR) and a superconducting quantum interference device (SQUID) are magnetometry techniques that are able to carry out ex vivo assessments.

Within ex vivo methodologies, elemental analysis methodologies, such as inductively coupled plasma-atomic emission spectroscopy (ICP-AES) and Prussian blue analysis, show limitations in quantifying the exclusive iron from MNPs [61,62].

An alternate current biosusceptometry (ACB) system has been employed in biological applications involving MNPs because of its unique advantages, such as a low-cost versatility and a lack of specialized equipment required. The system also does not use ionization radiation and works in unshielded magnetic environments [63–67]. Recently, the system has been improved through a comprehensive mathematical and computational approach to quantitatively reconstruct 2D distributions of MNPs [68].

In a previous study, we undertook a pharmacological approach to understand hepatic MNP uptake through ACB imaging. However, the study was limited to quantifying the MNP distribution, potentially minimizing future pre-clinical applications.

We emphasize that the paper presented here represents a significant improvement to the ACB system, mainly regarding MNP quantification in real time using high-quality quantitative images through the inverse problem solution. In addition, this work describes the use of a new MC-ACB system with a higher temporal resolution due to the number and density of detector coils and an additional biodistribution analysis.

Therefore, we decided to implement the MC-ACB system associated with MNPs to investigate a chronic liver disease that significantly impacts morbidity and mortality worldwide.

2. Materials and Methods

2.1. Magnetic Nanoparticles

Solutions of iron (III) chloride hexahydrate ($FeCl_3$—purity 97–100%), manganese (II) chloride tetrahydrate ($MnCl_2\ 4H_2O$—purity 98–100%), iron nitrate ($Fe(NO_3)_3$—purity 98–100%), and methylamine (CH_3NH_2—purity 99.5%) were purchased from Sigma-Aldrich (St. Louis, MO, USA). Acetone (purity 99.6%) and sodium citrate ($Na_3C_6H_5O_7$—purity 99–100%) were purchased from Cromoline, Diadema, Brazil.

We used citrate-coated manganese ferrite nanoparticles (Cit-$MnFe_2O_4$) synthesized by co-precipitation as described before [69,70]. Dynamic light scattering (DLS, Zetasizer NanoS Malvern Instruments, Malvern, UK) measurements showed the hydrodynamic diameter of the particles and zeta potential. Through a Jeol transmission electron microscope, model JEM 2100 (Tokyo, Japan), operating at 200 kV, we obtained images of the core diameter distribution of the MNPs. The magnetization curve of the Cit-$MnFe_2O_4$ MNPs was obtained using an ADE Vibrating Sample Magnetometer (VSM), model EV9 (MicroSense, EastLowell, MA, USA). The Cit-$MnFe_2O_4$ composition was assessed using an energy-dispersive X-ray spectroscopy (EDS) detector (Jeol, JSM-6610).

The X-ray diffraction patterns of the MNP powders were analyzed using a Shimadzu 6000 diffractometer (Shimadzu Corporation, Kyoto, Japan) in order to study the structural parameters of the MNPs. To ensure the success of the coating and confirm the presence of the magnetic nanoparticles, Fourier-transform infrared (FTIR) analysis was carried out using Varian IR 640 equipment.

2.2. Alternate Current Biosusceptometry

The system was a magnetic material detector working as a double magnetic flux transformer, and was composed of 19 drive and pickup coils. Both pairs were arranged on a first-order gradiometer to provide a good signal-to-noise ratio while reducing environmental noise and leading to the cancelation of the common mode. When the magnetic sample was near the pickup coils, the magnetic flux balance was altered, inducing an electric current in the pickup coils proportional to the volume δv and magnetic susceptibility χ.

This signal was acquired using the same low-noise lock-in amplifier that recorded the excitation frequency components (10 kHz). After converting into a direct current signal (DC), the ACB signal was digitalized in real time using a National Instruments A/D board (20 Hz of the sampling rate).

The ACB signal intensity detected by the pickup coils depended on intrinsic coil parameters, such as the area of the detection coil, the number of turns, the magnetic flux change rate, and the amount of magnetic material. Detailed information can be found in [68].

We used two ACB setups for our measurements in this present study. The multichannel ACB system (MC-ACB) was employed to acquire the real-time biodistribution of MNPs simultaneously in blood circulation and the liver. Then, we utilized a suitable ACB sensor to quantify the final mass accumulated in each organ collected after the animal's death [43]. Figure 1 presents the two ACB setups used in this work.

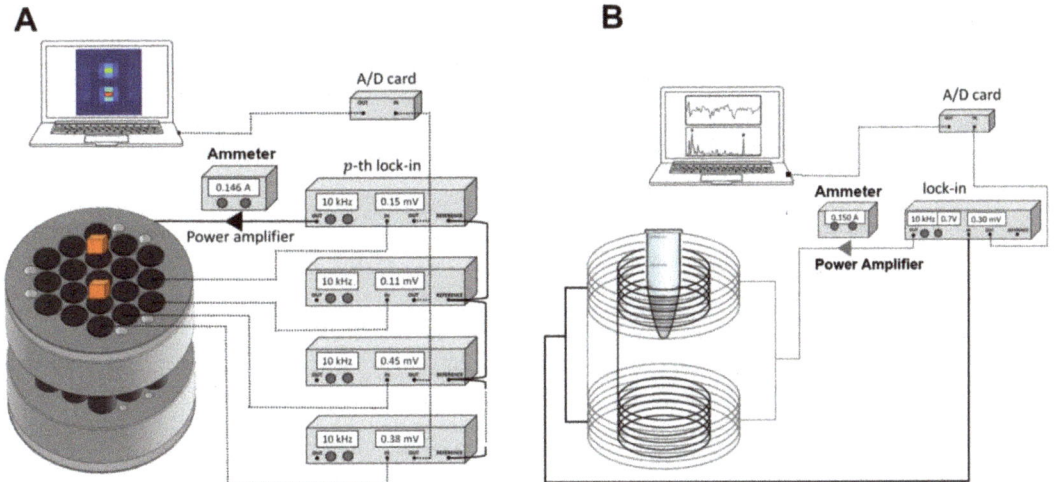

Figure 1. Schematic representation of both ABC setups utilized. (**A**) MC-ACB and (**B**) cavity ACB sensor.

2.3. Experimental Design of Cirrhosis-Associated Hepatocarcinogenesis Model

The current cirrhosis-associated hepatocarcinogenesis model was based on a previously published protocol [11]. In this rodent study, 36 males (*Rattus norvegicus albinus*, Wistar, weighing 250–300 g) provided by the UNESP animal facility (São Paulo State University) were divided into two groups. The animals were maintained under suitable conditions at 21 °C ± 1 °C with a 12 h/12 h light/dark cycle, constant air filtration, and ad libitum feeding. All animal experiments were previously approved and performed following the recommendations issued by the National Council for Control of Animal Experimentation (CONCEA) and were approved by the Ethics Committee on Animal Use of the São Paulo State University (IBB/UNESP) under protocol 7571041120.

The animals were randomly assigned to one of two groups, of which one was subjected to a NaCl 0.9% solution treatment (SAL-control group) and the other was subjected to the chemically-induced cirrhosis-associated hepatocarcinogenesis model (DEN/TAA group) [11].

The animals received a single intraperitoneal injection of diethylnitrosamine (DEN, 200 mg/kg in 0.9% saline solution) (Sigma-Aldrich, USA) to initiate liver carcinogenesis. After two weeks, we assigned the animals to three cycles of thioacetamide (TAA) (200 mg/kg in 0.9% saline solution) (Sigma-Aldrich, USA). During the fibrosis/cirrhosis induction, each TAA cycle was achieved after two intraperitoneal injections (twice a week), with an interval of one week without receiving treatment. The model of cirrhosis/hepatocarcinogenesis was carried out for eight weeks.

The animals were subjected to femoral vein cannulation surgery for the intravenous administration of MNPs (dose of 32 mg/kg) under anesthesia (99% urethane—1.5 mg/kg.) Then, the animals were positioned on the MC-ACB detection coils to carry out the magnetic in vivo biodistribution monitoring (further described in Section 2.4).

2.4. In Vivo Quantitative Imaging and Data Processing

We carried out quantitative MNP reconstruction by employing the MC-ACB, for which we had recently demonstrated the mathematical and computational approaches for improving the system's ability to acquire quantitative information [68]. We monitored the MNP biodistribution and recorded heart and liver signals using the MC-ACB system. We reconstructed the MNP biodistribution from the quantitative real-time signals, represented in sequential images (frames) at a sampling frequency of 20 Hz. Regions of interest (ROIs) were selected for the frames corresponding to the signals from the organs of interest (liver and heart). We quantified a series of pharmacokinetic parameters from the MNP distribution signals of both organs over time.

2.5. Histopathological and Immunohistochemical Analysis

Liver tissue samples were washed, fixed in formalin, embedded in paraffin, and sequentially sectioned into 5 μm sections. Slides were stained with hematoxylin and eosin (H&E) and Sirius red (collagen deposition) for microscopic analysis. Other sections obtained on silane-coated slides (Starfrost, Lowestoft, UK) were subjected to immunohistochemical reactions to evaluate the expression of placental glutathione-S-transferase (GSTP) (i.e., preneoplastic and neoplastic marker) and CD68 (monocyte, macrophage, and Kupffer cell marker) antigens, as previously described [11,12].

The placental GST-P (π isoform) is expressed in initiated hepatocytes, but not in normal or non-initiated hepatocytes, indicating their role in hepatocarcinogenesis [71]. CD68 is a glycosylated type I transmembrane glycoprotein that is considered an important cytochemical marker for macrophages, especially in the histochemical analysis of inflamed tissues [72,73].

We performed a morphometric analysis (lesions and nodules positive for GST-P and collagen content) using the Leica Q-win Software, version 3.1. The H&E- and picro-Sirius-stained sections were analyzed under a Leica DMLB 80 microscope connected to a Leica DC300FX camera. After image digitalization, we measured each experimental liver area per group under 20× magnification in five fields. The fibrosis degree analysis was performed using the criteria reported previously [74].

2.6. Pharmacokinetic Study

To determine the pharmacokinetic profile of the Cit-MnFe$_2$O$_4$, we determined classical pharmacokinetic parameters that are commonly used, such as $T_{1/2}$ (half-life) and the area under the curve (AUC). We adapted the concept of drug exposure to the MNP bioavailability, which was calculated from the area under the liver curves of the two experimental groups. Regarding the liver signals, we quantified the highest MNP level detected (C_{Max}) and the time to the highest MNP level (T_{Max}).

Analysis of the Ex Vivo Biodistribution of the Cit-MnFe$_2$O$_4$

After the in vivo measurement to collect the quantitative information, the rodents from both groups (SAL and DEN/TAA) were euthanized at 60 min by decapitation after the MNP injection. Subsequently, the organs of interest, such as the liver, spleen, heart, lungs, and kidneys, were collected by a laparoscopy procedure. In this experimental procedure, we also collected a blood sample. To quantify and certify the mass of MNPs accumulated in each organ, we randomly picked a sample (100 mg) of each organ and the blood, which had been previously lyophilized, homogenized, and stored in a volume-controlled flask. According to the previous protocol, the samples were positioned on the ACB sensor for signal detection to determine the mass of MNPs using a calibration curve that was previously constructed from an MNP batch (initially 28 mg/mL) diluted into fourteen vials with different concentrations while controlling the volume. This procedure allowed for the comparison of the measured ACB signals to established MNP concentrations [42].

2.7. Statistical Analyses

Data were expressed as means ± standard deviations. An unpaired Student's t-test was used to compare the control and the treated groups' pharmacokinetic parameters ($T_{1/2}$, AUC, and biodistribution quantifications). The incidence data from the histological analysis were analyzed using Fisher's exact test. The other data were compared using the Mann–Whitney test or Student's t-test, considering a significance level of $p < 0.05$. Analyses were performed using GraphPad Prism 6.0 software (GraphPad, San Diego, CA, USA).

3. Results

3.1. MNP Characterization

We synthesized manganese ferrite nanoparticles coated with citrate (Cit-MnFe$_2$O$_4$) through the co-precipitation method. These MNPs were applied due to their excellent field magnetic response to the ACB system. At a concentration of 28 mg/mL, the MNPs showed a superparamagnetic behavior. According to our TEM results, they presented a core diameter of 24 ± 4 nm. Once the organic molecule citrate was small (from 1.5 to 10 nm), it was assumed that the MNP core indicated by the TEM images was equal to the diameter of the Cit-MnFe$_2$O$_4$ MNPs. Through the DLS Zetasizer results, the MNPs presented with a hydrodynamic size of 65.6 ± 4 nm, a polydispersion index for the colloid sample of 0.25, and a zeta potential of −27.8 mV. We observed a negative zeta potential for the magnetic Cit-MnFe$_2$O$_4$ (−27.8 ± 1.7 mV), which resulted from their surface being coated with citrate ions due to the effect of citrate adsorption onto bare MNPs. This value is in agreement with literature reports [69,74].

The assessment of magnetic characterization for the powder (pure MNPs) and colloidal solution (magnetic fluid) using an ADE Vibrating Sample Magnetometer (VSM) indicated a magnetization saturation of 52.8 emu/g. The magnetization profile showed a quasi-static superparamagnetic behavior (no coercive field at DC conditions) (see Figure 2C of [69]). We also confirmed the presence of Mn and Fe in the MNPs through an EDS analysis. The XRD analysis showed the structural characterization of the Cit-MnFe$_2$O$_4$ MNPs. The XRD pattern of as-dried MnFe$_2$O$_4$ confirmed the ferrite phase's formation. The diffraction peaks matched the single crystalline MnFe2O4 (JCPDS card No. 074–2403). The XRD results for MnFe$_2$O$_4$ were comparable with the previously reported results [75]. It is worth pointing out that we did not detect any impurity phases in the ferrite group. See Figure S5, Supplementary Materials. We confirmed the presence of the magnetic core and citrate shell through a Fourier-transform infrared (FTIR) analysis, observing bands at 1581 and 1383 cm^{-1} for the Cit-MnFe$_2$O$_4$ MNPs (black curve), which were assigned to the citrate due to the C-O bonds of the carboxylic group present in the molecule [76]. The absorption peak within 500–600 cm^{-1} corresponded to the Fe–O vibration, which was related to the magnetic phase [77]. All results of the MNP characterization process are described in the Supplementary Materials.

3.2. Macroscopic Aspects of Animals Subjected to Cirrhosis Associated with Hepatocarcinogenesis

Macroscopically, the livers from animals of the SAL group (Figure 2A) presented typical features (regular and smooth surfaces). On the other hand, the livers from animals of the DEN/TAA group (Figure 2B) presented rough surfaces with numerous nodules. These findings indicate that the animals subjected to the DEN/TAA protocol presented well-defined features of cirrhosis. In addition, as expected, the DEN/TAA treatment increased the animals' absolute and relative liver weight (RLW) (Figure 2C,D, respectively).

3.3. Histopathological Analysis, Collagen Morphometry, and Immunostaining

The histopathological analysis revealed that 80% ($p = 0.049$) and 20% of animals in the DEN/TAA group developed adenomas and HCC, respectively (Figure 3A). Compared to the SAL (control) group, the livers from the DEN/TAA group presented with multiple preneoplastic lesions and nodules that were positive for GSTP (Figure 3B, $p = 0.0079$). Sirius red-stained DEN/TAA liver sections demonstrated extensive collagen deposition (Figure 3C, $p = 0.0079$) with bridging fibrosis and cirrhotic nodules, and most were positive for GST-P (fibrosis level 5, $p = 0.0079$). Figure 3D shows the data for the immunohistochemical analysis of the CD68 marker. The results indicate that no statistical difference was observed in the macrophage counts between the SAL and DEN/TAA groups ($p > 0.05$).

3.4. Dynamic ACB Monitoring

We acquired images that dynamically represented the biodistribution of MNPs in real time through MC-ACB monitoring. The images were acquired sequentially, allowing for a video representation of the circulation and accumulation processes. Figure 4 presents two frames showing moments of the MNP biodistribution for both experimental groups. In the first frame, at t = 820 s, an ROI was selected in the heart. The second frame, at t = 3600 s, corresponds to the liver region. These ROIs were applied to all the imaging frames, generating biodistribution curves. Previously, we positioned the animals on the MC-ACB system at the same projection to ensure the animal's anatomical references were kept during the biodistribution acquisition.

Consequently, Figure 4A shows the arrival of MNPs in the heart right after the MNP injection, while Figure 4B shows the final accumulation in the liver region. In order to demonstrate the difference between the MNP distributions, we quantified the average distribution of MNPs in the ROIs of the images. Figure 4C shows the MNPs in the bloodstream and liver. A high-intensity peak characterized the arrival of MNPs in the heart shortly after MNP injection. Then, the distribution of MNPs was represented by a rapid decay in the heart signal. Simultaneously, the liver captured and removed the MNPs from the bloodstream due to the high blood supply and many Kupffer cells in the hepatic tissue.

The liver signal can be associated with the uptake of macrophages and the accumulation of MNPs in the parenchyma. As depicted in Figure 4C (red curve), the liver showed a saturation tendency over time after a rapid intensity increase.

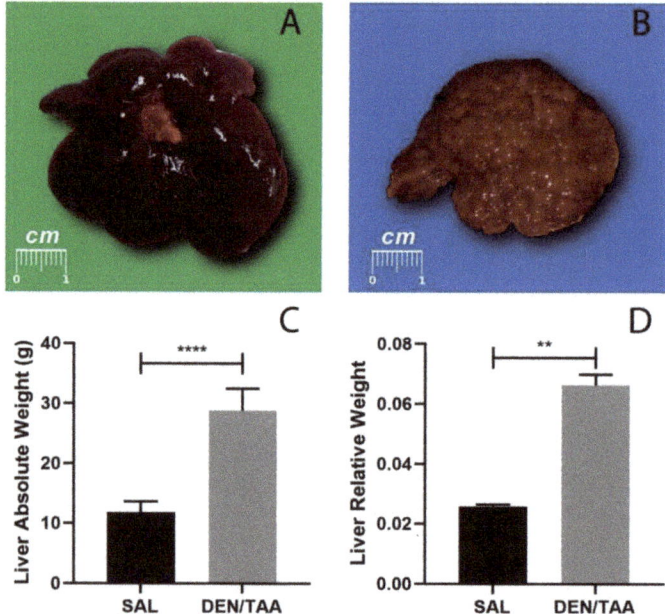

Figure 2. Representative images of macroscopic aspects of livers from (**A**) SAL-group animals and (**B**) DEN/TAA-group animals. Analyses of the absolute liver weight and relative liver weight are shown for (**C**) the SAL and (**D**) the DEN/TAA group. The relative liver weight (LW/BW) is expressed as the ratio between the liver weight (LW) and the body weight (BW). LW/BW ratio values are expressed as means ± sd; for the SAL and DEN/TAA groups, they were 0.25674 ± 0.000706 and 0.066253 ± 0.003554, respectively. (** $p < 0.05$) and (**** $p < 0.0001$).

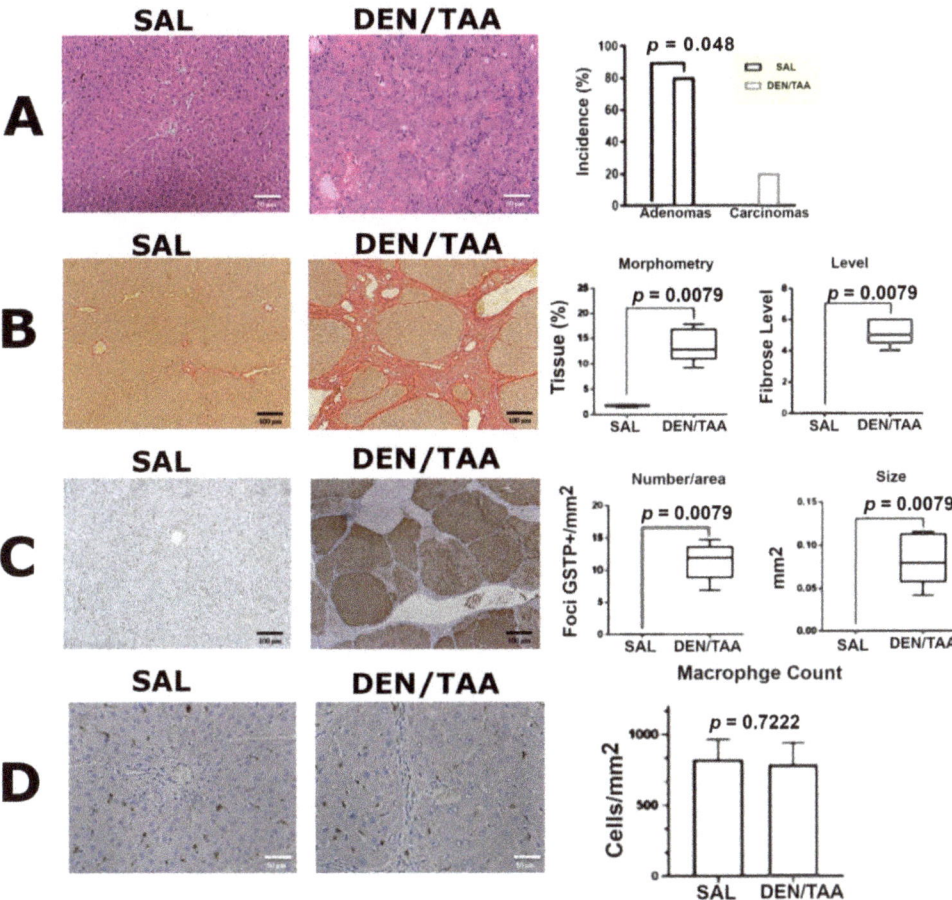

Figure 3. (**A**) Representative images of H&E-stained sections of SAL (left) and DEN/TAA (center) livers; HCC was characterized by profound cellular atypia and was composed of malignant hepatocytes arranged in acinar structures in the DEN/TAA group. Data on the incidence of adenomas and carcinomas are presented in terms of proportion (%) of affected animals and were analyzed by Fisher's exact test ($p < 0.05$) (right). (**B**) Collagen analysis shown with picro-Sirius red, showing sections from the SAL (left) and DEN/TAA (center) groups. Morphometry and degree of fibrosis data are presented in box plots and were analyzed using the Mann–Whitney test ($p < 0.05$) (right). (**C**) Immunohistochemistry sections from the SAL group (left) and the multiple GST-P+ nodules in the DEN/TAA group (center). The number and size of GST-P+ lesions are presented in box plots and were analyzed using the Mann–Whitney test ($p < 0.05$) (right). (**D**) Immunohistochemistry for the CD68 marker in sections from the SAL (left) and DEN/TAA (center) groups, showing macrophages. Macrophage count data are presented as means and standard deviations and were analyzed using a t-test ($p < 0.05$) (right).

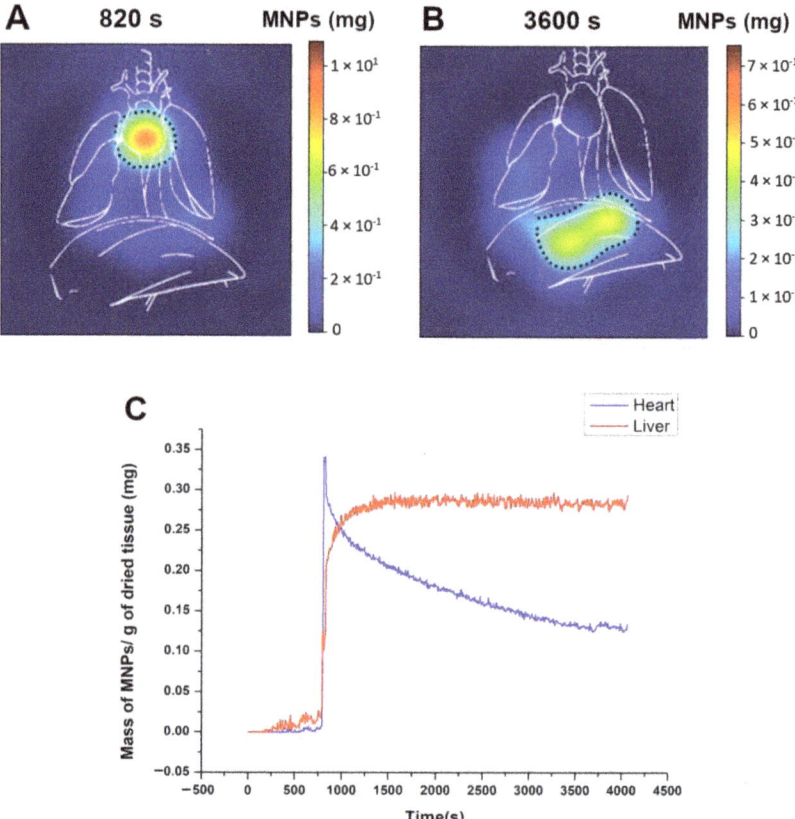

Figure 4. Representation of the biodistribution by frames after injection and the specific ROIs selected to access the pharmacokinetic parameters; (**A**) 820 s, showing the high and low concentrations of MNPs in the heart and the liver, respectively; (**B**) 3600 s, indicating the final biodistribution process, which was characterized by a solely higher intensity signal in the liver; and (**C**) the average intensity over time for ROI 1 (heart region) and ROI 2 (liver).

3.5. Pharmacokinetic Assessment and MNP Biodistribution

The kinetics of MNP accumulation were evaluated to prove the liver performance during MNP uptake. The kinetics of MNP accumulation were assessed by plotting graphs of the liver signals previously obtained from ROI imaging. To determine the liver's accumulation, we employed the classical concept of the AUC. We found a significant difference ($p < 0.0001$) between the rates of MNP deposition in the hepatic tissue of the animal groups. The pharmacokinetic assessment of hepatic curves also indicated significant differences ($p < 0.0001$) in C_{max} and T_{max} for the DEN/TAA group. The evaluation of liver signals revealed that the healthy livers reached C_{max} after 26.63 min (T_{max}).

On the other hand, the livers under a cirrhosis-induced process had an inversed profile, presenting with a lower peak concentration that was reached in a shorter time (T_{max} of 16.7 min) and remained constant over time. Through the plotted heart curves, we assessed the $T_{1/2}$ of the MNPs. The exponential decay analyses of the DEN/TAA group showed a half-life of 16.3 min compared to 28.3 min for the SAL group. All the pharmacokinetic parameters of the DEN/TAA group were found to be significantly different from the SAL group (Student *t*-test, $p < 0.05$). The pharmacokinetic parameters are summarized in Table 1.

Table 1. Pharmacokinetic parameters of Cit-MnFe$_2$O$_4$ MNPs after intravenous administration at a 32 mg/kg dose for the SAL and DEN/TAA groups. C_{max} = the highest MNP level detected; T_{max} = the time to the highest MNP level; AUC = the area under the curve; $T_{1/2}$ = half-life. **** $p < 0.0001$.

Pharmacokinetic Parameter	Evaluation (Mean ± SD)	
	SAL	DEN/TAA
C_{max} (mg MNP/dose injected)	0.4870 ± 0.01212	0.4150 ± 0.01621 ****
T_{max} (m)	26.63 ± 0.5145	16.71 ± 1.1 ****
$AUC_{0-60min}$	1472.6 ± 201	1198.5 ± 152 ****
$T_{1/2}$ (min)	19.6 ± 2.3	11.2 ± 3.1 ****

After acquiring the signals from the in vivo biodistribution measurements and the euthanasia of the animals, we started the protocol to analyze the ex vivo biodistribution from the collected organs as described in Analysis of the Ex Vivo Biodistribution of the Cit-MnFe$_2$O$_4$ section. From our system's ACB characterization, we found a limit of detection (LOD) of 12 µg and a limit of quantification (LOQ) of 40 µg for the MNP reference. The sensitivity was 0.9 (χ/mg (MNPs). Figure 5 presents the profile for the ex vivo MNP biodistribution. In general, we noticed a similar behavior of the MNP biodistribution between the experimental groups, where the spleen retained most of the particles, followed by the liver and lungs. However, we found significant differences between the control and treated groups for the same organs (spleen, liver, and lungs), indicating a higher accumulation for the SAL group. The ACB quantification revealed that the MNP accumulation in organs such as the heart and kidneys was minimal, with these organs presenting with very low values of MNP deposition. Both organs do not typically specialize in MNP retention and capture, which would explain the low signal. In addition, the two organs did not show significant differences between the groups, suggesting that besides the MNP properties, which would facilitate the splenic and hepatic uptake, the induced liver injury could not influence MNP deposition in these organs.

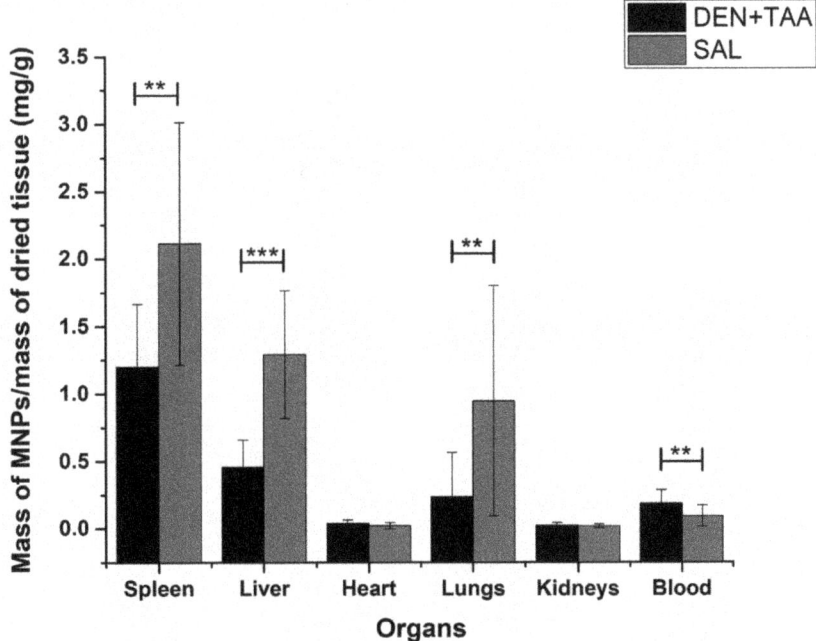

Figure 5. ACB data for MNP biodistribution in SAL and DEN/TAA; ** $p < 0.05$; *** $p < 0.01$.

The biodistribution analysis also indicated differences in the MNPs found in the blood samples. At the end of 60 min, the amounts of MNPs found in the SAL group's spleen, liver, and lungs were significantly higher than those in the DEN/TAA group.

4. Discussion

Besides its properties such as an excellent low-field magnetic response and a high magnetization saturation, Cit-MnFe$_2$O$_4$ presents with a negative zeta potential. According to the literature, nanomaterials with positive zeta potentials show an increased clearance [78–81]. It is generally known that positively charged nanoparticles have faster blood clearance, while neutral and negative particles exhibit a longer circulation time [82–85]. In addition, nanoparticles with a negative charge present with a lower Kupffer cell uptake in addition to their higher circulation time, which contributes to an increased tumor uptake [86,87]. Furthermore, a strongly negative potential allows the particles to be stable over a variety of pH levels and effectively prevents agglomeration due to steric and electrostatic forces from the citrate layer. The MNPs employed in this study were selected due to their ability to act exclusively as tracers for in vivo measurements, as the aim of this work was to assess liver cirrhosis under the MC-ACB system. Therefore, manganese ferrite nanoparticles are not functionalized with biotherapeutics and chemotherapeutics to work as cirrhosis treatment vehicles based on drug delivery systems and gene therapy.

Although there have been numerous alternatives to the treatment of chronic liver diseases such as cirrhosis and HCC, limitations such as non-specific targeting and adequate drug delivery concentrations have reduced the chances for the successful treatment of these illnesses. Therefore, new agents with improved therapeutic efficiencies have been investigated. However, new translational studies assessing the pathophysiological and pharmacokinetic profiles in liver cirrhosis are essential, as they will contribute to new perspectives in the approach and investigation of new drugs.

In the present study, we assessed the biodistribution profile of MNPs in a rat cirrhotic microenvironment associated with hepatocarcinogenesis, in which a complete assessment performed by the ACB system—which included real-time monitoring and the quantification of the accumulation of MNPs—was evaluated. It is noteworthy that the Cit-MnFe$_2$O$_4$ used here exhibited great potential for the diagnosis and study of hepatic diseases such as cirrhosis, as previously reported [33]. In addition, this magnetic particle system can be used as an efficient agent in magnetic hyperthermia due to its unique properties [88]. Thus, our real-time in vivo study was performed through image acquisition or distribution profiling of intravenously administered MNPs. The same system was employed to assess the biodistribution process in normal and injured livers in a DEN/TAA protocol.

The different ACB analyses allowed for a complete pharmacokinetic assessment, confirming that the spleen and the liver are the primary organs responsible for capturing MNPs after intravenous injection. The higher retention of MNPs in the spleen and liver can be attributed to the role of the MPS. According to several reports, most injected MNPs are cleared from the bloodstream by specialized cells, such as the resident macrophages of the liver (Kupffer cells) and spleen (red pulp macrophages) [89,90].

The in vivo results were confirmed by an ex vivo biodistribution analysis (spleen, liver, and lungs), indicating that in the DEN/TAA group, a lower uptake of MNPs occurred compared to that in healthy animals. The spleen plays a significant role in the clearance of MNPs from the bloodstream. Due to its close anatomic proximity to the liver, a communication axis between the liver and spleen ("liver–spleen axis") is commonly reported [91,92]. Furthermore, in the course of cirrhosis with portal hypertension, the spleen volume undergoes an increase in volume that is proportional to the degree of damage to the liver function [4,93]. However, the mechanisms underlying the splenic function under cirrhosis remain unknown. In a study that addressed liver cirrhosis, the authors found a decreased MNP accumulation in the liver. In contrast, the spleen under cirrhosis showed a higher uptake than in non-cirrhotic animals [94]. Despite the spleen's phagocytic activity increasing with splenomegaly [95], another study indicated that the hepatic uptake of nanocarriers is affected by liver disease, whereas the splenic uptake was partially affected [96]. In addition, another work also found that patients with liver cirrhosis presented with a decreased liver and spleen uptake of a superparamagnetic contrast agent for magnetic resonance imaging [97].

In this way, we believe that liver cirrhosis possibly induces a decreased uptake of the red zone macrophages in the spleen. It is worth pointing out that the red pulp is constituted by the macrophage population, which retains much of the administered nanoparticles.

Firstly, we also hypothesized that the DEN/TAA-induced protocol could have led to an abrupt depletion of the KCs, which is supported by several studies after toxic hepatic injury and infections [98,99]. However, our immunohistochemistry findings with the CD68 marker indicated no statistical differences in the counts of infiltrated macrophages and resident Kupffer cells between the SAL and DEN/TAA groups. It is worth pointing out that the depletion of KCs is not a mechanism related to all types of liver damage. For example, Kessler et al. [100] performed a DEN treatment to induce severe liver damage and did not find a significant KC loss in the short-term or long-term DEN models.

Concerning the accumulation by the liver, the IV administration of MNPs can reach the hepatocytes. The hepatocytes represent 70% of total liver cells and are separated from the sinusoids by the space of Disse [101]. As mentioned above, hepatic sinusoids are constituted by endothelial cells that have a fenestrated cytoplasm associated with a discontinuous basal lamina [101]. Through their fenestrae, the LSECs allow absorption and secretion to take place across the narrow space of Disse, creating a unique channel for blood–hepatocyte exchange across sinusoids [102]. In addition to being highly porous, hepatic sinusoids are characterized by the absence of an organized basal lamina in healthy conditions.

Nevertheless, hepatic disturbances and diseases such as fibrosis and cirrhosis induce a capillarization process in the LSECs, which leads to the loss of their fenestrated characteristics [103,104]. A continuous basal lamina characterizes the process of capillarization in hepatic sinusoids, thus avoiding the bidirectional traffic of molecules in the blood and the parenchyma, and vice versa [105,106]. Therefore, this could be considered the main reason for the change in the MNP biodistribution.

By comparing our data for MNP accumulation in the livers of the DEN/TAA group, the results suggest that the increases in intrahepatic vascular resistance, impaired hepatic sinusoidal circulation, collagen deposition, and portal hypertension, which are associated with cirrhosis [107], could influence the arrival of MNPs to the liver. When we analyzed the data for MNP accumulation in the DEN/TAA and SAL groups, the liver presented with a higher significant p-value, followed by the spleen and the lungs. We noticed a lower MNP accumulation in the DEN/TAA group, suggesting a decreased hepatic uptake and consequently more MNPs circulating, which was visualized through the statistical differences in the blood analyses (Figure 5). During the liver cirrhosis process, the hepatic tissue undergoes chronic damage and an inflammatory process, during which a repair process is initiated to regenerate damaged hepatocytes, resulting in scar formation. Therefore, we also assumed that besides the capillarization process, there is a loss of hepatocytes in the fibrosis state by connective tissue scars, which could affect MNP uptake, since an altered obstruction of blood circulation with portal hypertension can occur in this chronic disease [108].

Besides liver disorders, chronic liver disease can result in pulmonary complications. In this way, hepatopulmonary syndrome (HPS) is commonly associated with cirrhosis [109]. HPS is a pulmonary disorder characterized by arterial oxygen desaturation, pulmonary vascular vasodilation, and intrapulmonary shunts. The increased shunting associated with pulmonary vasodilatation is responsible for the imbalance between perfusion and ventilation, causing abnormalities in gas exchange [110,111]. This compromised lung profile could explain the lung biodistribution results for the DEN/TAA group, in which the MNPs did not reach the lung alveoli, and consequently did not remain in the tissue.

The pharmacokinetic assessment results confirmed that the DEN and TAA administration model caused damage that altered the liver architecture. Besides the biodistribution results, we noticed that the livers from the DEN/TAA group presented with a limited uptake efficiency of MNP, as observed by the C_{max}. We assumed that the altered basal lamina in hepatic sinusoids did not allow blood extravasation towards hepatocytes. In this way, we hypothesized that the new basal lamina in hepatic sinusoids affected the interaction between the blood and the hepatocytes in the DEN/TAA group. In contrast to the control group, the DEN/TAA protocol resulted in an altered blood flow in hepatocytes, which might have decreased liver uptake. Consequently, the non-extravasation towards hepatocytes resulted in early Kupffer cell saturation, according to our values for T_{max}. Despite the fact that hepatocytes are not specialized to retain MNPs, the possible deposition in the hepatocytes might be attributed to the diameter of the Cit-MnFe$_2$O$_4$ MNPs, which was smaller than the fenestrations of the sinusoids [112]. We noticed that the livers under cirrhosis took less time to exhibit a saturation profile, while the healthy livers continued the exchange between the hepatocytes and the hepatic sinusoidal blood. Therefore, we also assumed that a part of the injected MNPs penetrated the sinusoid and reached the hepatocytes in the SAL group. This condition would explain the significant differences in the observed pharmacokinetic profiles, such as the MNP accumulation.

Since our biodistribution results (Figure 5) indicated a lower hepatic uptake for the DEN/TAA group, it instantly led us to consider a longer circulation time of MNPs for this group. As depicted in Figure 2C,D, the DEN/TAA animals presented with an increased liver weight compared to the SAL group. However, to assess the biodistribution profile, we employed a protocol to calculate the mass of particles per gram of lyophilized tissue. The hepatic uptake of DEN/TAA would be higher due to its mass in a quantification using absolute values. It was evident by the non-normalized $T_{1/2}$ values that the livers under these conditions influenced the circulation time of MNPs.

Although the MC-ACB presented a high temporal resolution for acquiring the biodistribution of MNPs dynamically, but only for the liver and heart, we believe that an improvement mainly in the coil array might lead the system to detect the MNP biodistribution in other target organs.

This study reported an application of a new and improved MC-ACB system compared to the previous one [48], where the main progress was the acquisition of quantitative in vivo images of the MNP distribution in healthy and neoplastic animals.

In this context, we believe this methodology is adequate for investigating several organs and their functions, either in normal circumstances or while under dysfunction.

Nevertheless, nanotechnology-based magnetic systems are an alternative strategy to the conventional methods for the investigation of liver diseases. These systems can perform non-invasive imaging assessments to work towards an early diagnosis, which might contribute to the efficient delivery of therapeutics to the liver.

5. Conclusions

As highlighted by the presented findings, the pharmacokinetic profile of MNP distribution and accumulation was affected by pathophysiological factors induced by a cirrhosis state. Since the liver monocytes and macrophages remained stable, the differences found in the pharmacokinetic profile of cirrhotic animals strongly indicate that hepatic blood flow is most likely responsible for altering the distribution and accumulation profile of MNPs. Therefore, the feasibility of developing nanotechnology-based delivery platforms needs further investigation to address strategies to improve the interaction of therapeutic agents with injured hepatic tissue.

Through an in vivo and ex vivo information acquisition approach, the ACB system provided the ability to monitor and quantify the MNPs in healthy and cirrhosis conditions, providing the requirements necessary to assist in the diagnosis and therapy of hepatic disorders. By extrapolating the possibilities of evaluation to problems found in the clinical environment, the association of the ACB system with MNPs might offer a methodology with easy access, a low cost, and the absence of ionizing radiation to assess several biologic functions under disorders. Furthermore, it is expected that through instrumental improvements, the MC-ACB system will be enhanced to the level of relevant methodologies such as magnetic particle imaging and magnetorelaxometry.

Supplementary Materials: The following supporting information can be downloaded at: https://www.mdpi.com/article/10.3390/pharmaceutics14091907/s1, Figure S1: (A) Image of MNPs at 100 nm scale. (B) Image of MNPs at 50 nm scale. Figure S2: Hydrodynamic size obtained by the dynamic light scattering experiment. Figure S3: (A) Magnetization curve of MNPs in a fluid sample and a powder sample (B) acquired by the VSM experiment. Figure S4: EDS quantification of the MNPs composition quantification. (A) MNPs sample used, in which the numbers (1, 2, and 3) represent the studied region. (B) Representative example of EDS signal acquired and its quantification (region 2 of the MNPs image). Figure S5. X-ray diffractogram of Cit-MnFe$_2$O$_4$ MNPs. Figure S6: FTIR measurements of citrate and the Cit-MnFe$_2$O$_4$ MNPs.

Author Contributions: Conceptualization, G.A.S., G.R.R., L.F.B., and J.R.A.M.; methodology, L.F.B., J.R.A.M., and G.A.S.; software, G.G.A.B., G.R.R., and G.M.P.; validation, G.G.A.B., E.G.S., and G.R.R.; formal analysis, G.A.S., G.G.A.B., G.R.R., G.M.P., E.G.S., and J.R.A.M.; investigation, G.M.P. and G.A.S.; resources, G.A.S., G.M.P., and G.R.R.; data acquisition, G.A.S., G.M.P., E.G.S. and G.G.A.B.; writing—original draft preparation, G.A.S., O.B., E.G.S., G.M.P., and G.G.A.B.; writing—review and editing, G.A.S., A.F.B., G.R.R., L.F.B., and J.R.A.M.; supervision, O.B., A.F.B., J.R.A.M., G.R.R., and L.F.B.; project administration, O.B., A.F.B., and J.R.A.M.; funding acquisition, O.B., J.R.A.M., L.F.B., and A.F.B. All authors have read and agreed to the published version of the manuscript.

Funding: This work was supported by the Fundação de Amparo à Pesquisa do Estado de São Paulo (FAPESP), grants 2013/07699-0, 2021/06405-9, 2021/09829-4 and 2019/11277-0. It was also supported by the Conselho Nacional de Pesquisas e Desenvolvimento Tecnológico (CNPq), grants 312074/2018-9 and 311074/2018. The German Academic Exchange program DAAD in cooperation with Brazilian CAPES-PROBRAL (project ID 57446914, 88887.198747/2018-00, and 888 81.198748/2018-01) supported this work.

Institutional Review Board Statement: The study was conducted and approved by the São Paulo State University (UNESP) Committee for the Use and Care of Animals (protocol No. CEUA—IBB 7571041120).

Informed Consent Statement: Not applicable.

Data Availability Statement: Almost all data are presented within the manuscript (figures and tables). The raw data presented in this study are available upon request to the corresponding author.

Acknowledgments: We would like to thank Mirella Visnosvescki Fogaca for the support in the ex vivo measurements.

Conflicts of Interest: The authors declare no conflict of interest.

References

1. Malarkey, D.E.; Johnson, K.; Ryan, L.; Boorman, G.; Maronpot, R.R. New insights into functional aspects of liver morphology. *Toxicol. Pathol.* **2005**, *33*, 27–34. [CrossRef] [PubMed]
2. Binatti, E.; Gerussi, A.; Barisani, D.; Invernizzi, P. The Role of Macrophages in Liver Fibrosis: New Therapeutic Opportunities. *Int. J. Mol. Sci.* **2022**, *23*, 6649. [CrossRef] [PubMed]
3. Nianan, L.; Jiangbin, L.; Yu, W.; Jianguo, L.; Rui, D. Hepatic Stellate Cell: A Double-Edged Sword in the Liver. *Physiol. Res.* **2021**, *70*, 821.
4. Ezhilarasan, D. Advantages and challenges in nanomedicines for chronic liver diseases: A hepatologist's perspectives. *Eur. J. Pharmacol.* **2021**, *893*, 173832. [CrossRef]
5. Desmet, V.J.; Roskams, T. Cirrhosis reversal: A duel between dogma and myth. *J. Hepatol.* **2004**, *40*, 860–867. [CrossRef]
6. Hoekstra, L.T.; de Graaf, W.; Nibourg, G.A.; Heger, M.; Bennink, R.J.; Stieger, B.; van Gulik, T.M. Physiological and biochemical basis of clinical liver function tests: A review. *Ann. Surg.* **2013**, *257*, 27–36. [CrossRef]
7. Sung, H.; Ferlay, J.; Siegel, R.L.; Laversanne, M.; Soerjomataram, I.; Jemal, A.; Bray, F. Global cancer statistics 2020: GLOBOCAN estimates of incidence and mortality worldwide for 36 cancers in 185 countries. *CA A Cancer J. Clin.* **2021**, *71*, 209–249. [CrossRef]
8. Yang, J.D.; Kim, W.R.; Coelho, R.; Mettler, T.A.; Benson, J.T.; Sanderson, S.O.; Therneau, T.M.; Kim, B.; Roberts, L.R. Cirrhosis is present in most patients with hepatitis B and hepatocellular carcinoma. *Clin. Gastroenterol. Hepatol.* **2011**, *9*, 64–70. [CrossRef]

9. Greten, T.; Papendorf, F.; Bleck, J.; Kirchhoff, T.; Wohlberedt, T.; Kubicka, S.; Klempnauer, J.; Galanski, M.; Manns, M. Survival rate in patients with hepatocellular carcinoma: A retrospective analysis of 389 patients. *Br. J. Cancer* **2005**, *92*, 1862–1868. [CrossRef]
10. Op den Winkel, M.; Nagel, D.; Sappl, J.; op den Winkel, P.; Lamerz, R.; Zech, C.J.; Straub, G.; Nickel, T.; Rentsch, M.; Stieber, P. Prognosis of patients with hepatocellular carcinoma. Validation and ranking of established staging-systems in a large western HCC-cohort. *PLoS ONE* **2012**, *7*, e4506.
11. Romualdo, G.R.; Grassi, T.F.; Goto, R.L.; Tablas, M.B.; Bidinotto, L.T.; Fernandes, A.A.H.; Cogliati, B.; Barbisan, L.F. An integrative analysis of chemically-induced cirrhosis-associated hepatocarcinogenesis: Histological, biochemical and molecular features. *Toxicol. Lett.* **2017**, *281*, 84–94. [CrossRef]
12. Romualdo, G.R.; Prata, G.B.; da Silva, T.C.; Fernandes, A.A.H.; Moreno, F.S.; Cogliati, B.; Barbisan, L.F. Fibrosis-associated hepatocarcinogenesis revisited: Establishing standard medium-term chemically-induced male and female models. *PLoS ONE* **2018**, *13*, e0203879. [CrossRef]
13. Romualdo, G.R.; Leroy, K.; Costa, C.J.S.; Prata, G.B.; Vanderborght, B.; Da Silva, T.C.; Barbisan, L.F.; Andraus, W.; Devisscher, L.; Câmara, N.O.S. In vivo and in vitro models of hepatocellular carcinoma: Current strategies for translational modeling. *Cancers* **2021**, *13*, 5583. [CrossRef]
14. Salarian, M.; Turaga, R.C.; Xue, S.; Nezafati, M.; Hekmatyar, K.; Qiao, J.; Zhang, Y.; Tan, S.; Ibhagui, O.Y.; Hai, Y.; et al. Early detection and staging of chronic liver diseases with a protein MRI contrast agent. *Nat. Commun.* **2019**, *10*, 4777. [CrossRef]
15. Li, C.; Li, R.; Zhang, W. Progress in non-invasive detection of liver fibrosis. *Cancer Biol. Med.* **2018**, *15*, 124–136. [CrossRef]
16. Surendran, S.P.; Thomas, R.G.; Moon, M.J.; Jeong, Y.Y. Nanoparticles for the treatment of liver fibrosis. *Int. J. Nanomed.* **2017**, *12*, 6997. [CrossRef]
17. Sumida, Y.; Nakajima, A.; Itoh, Y. Limitations of liver biopsy and non-invasive diagnostic tests for the diagnosis of nonalcoholic fatty liver disease/nonalcoholic steatohepatitis. *World J. Gastroenterol.* **2014**, *20*, 475–485. [CrossRef]
18. Cheng, Z.; Lv, Y.; Pang, S.; Bai, R.; Wang, M.; Lin, S.; Xu, T.; Spalding, D.; Habib, N.; Xu, R. Kallistatin, a new and reliable biomarker for the diagnosis of liver cirrhosis. *Acta Pharm. Sin. B* **2015**, *5*, 194–200. [CrossRef]
19. Lin, Y.-S. Ultrasound Evaluation of Liver Fibrosis. *J. Med. Ultrasound* **2017**, *25*, 127–129. [CrossRef]
20. Petitclerc, L.; Gilbert, G.; Nguyen, B.N.; Tang, A. Liver Fibrosis Quantification by Magnetic Resonance Imaging. *Top Magn. Reson. Imaging* **2017**, *26*, 229–241. [CrossRef]
21. Huber, A.; Ebner, L.; Heverhagen, J.T.; Christe, A. State-of-the-art imaging of liver fibrosis and cirrhosis: A comprehensive review of current applications and future perspectives. *Eur. J. Radiol. Open* **2015**, *2*, 90–100. [CrossRef] [PubMed]
22. Salarian, M.; Ibhagui, O.Y.; Yang, J.J. Molecular imaging of extracellular matrix proteins with targeted probes using magnetic resonance imaging. *Wiley Interdiscip. Rev. Nanomed. Nanobiotechnol.* **2020**, *12*, e1622. [CrossRef] [PubMed]
23. Reddy, L.H.; Couvreur, P. Nanotechnology for therapy and imaging of liver diseases. *J. Hepatol.* **2011**, *55*, 1461–1466. [CrossRef] [PubMed]
24. Bai, X.; Su, G.; Zhai, S. Recent advances in nanomedicine for the diagnosis and therapy of liver fibrosis. *Nanomaterials* **2020**, *10*, 1945. [CrossRef]
25. Jin, Y.; Wang, H.; Yi, K.; Lv, S.; Hu, H.; Li, M.; Tao, Y. Applications of Nanobiomaterials in the Therapy and Imaging of Acute Liver Failure. *Nano-Micro Lett.* **2020**, *13*, 25. [CrossRef]
26. Liu, Q.; Song, L.; Chen, S.; Gao, J.; Zhao, P.; Du, J. A superparamagnetic polymersome with extremely high T2 relaxivity for MRI and cancer-targeted drug delivery. *Biomaterials* **2017**, *114*, 23–33. [CrossRef]
27. Patra, J.K.; Das, G.; Fraceto, L.F.; Campos, E.V.R.; Rodriguez-Torres, M.d.P.; Acosta-Torres, L.S.; Diaz-Torres, L.A.; Grillo, R.; Swamy, M.K.; Sharma, S.; et al. Nano based drug delivery systems: Recent developments and future prospects. *J. Nanobiotechnol.* **2018**, *16*, 71. [CrossRef]
28. Yao, Y.; Zhou, Y.; Liu, L.; Xu, Y.; Chen, Q.; Wang, Y.; Wu, S.; Deng, Y.; Zhang, J.; Shao, A. Nanoparticle-Based Drug Delivery in Cancer Therapy and Its Role in Overcoming Drug Resistance. *Front. Mol. Biosci.* **2020**, *7*, 193. [CrossRef]
29. Yu, Z.; Gao, L.; Chen, K.; Zhang, W.; Zhang, Q.; Li, Q.; Hu, K. Nanoparticles: A New Approach to Upgrade Cancer Diagnosis and Treatment. *Nanoscale Res. Lett.* **2021**, *16*, 88. [CrossRef] [PubMed]
30. Talaei, S.; Mellatyar, H.; Pilehvar-Soltanahmadi, Y.; Asadi, A.; Akbarzadeh, A.; Zarghami, N. 17-Allylamino-17-demethoxygeldanamycin loaded PCL/PEG nanofibrous scaffold for effective growth inhibition of T47D breast cancer cells. *J. Drug Deliv. Sci. Technol.* **2019**, *49*, 162–168. [CrossRef]
31. Ahlawat, J.; Hooda, R.; Sharma, M.; Kalra, V.; Rana, J.; Batra, B. Nanoparticles in Biomedical Applications. In *Green Nanoparticles*; Springer: Cham, Switzerland, 2020; pp. 227–250.
32. Li, Y.; Shang, W.; Liang, X.; Zeng, C.; Liu, M.; Wang, S.; Li, H.; Tian, J. The diagnosis of hepatic fibrosis by magnetic resonance and near-infrared imaging using dual-modality nanoparticles. *RSC Adv.* **2018**, *8*, 6699–6708. [CrossRef] [PubMed]
33. Saraswathy, A.; Nazeer, S.S.; Nimi, N.; Santhakumar, H.; Suma, P.R.; Jibin, K.; Victor, M.; Fernandez, F.B.; Arumugam, S.; Shenoy, S.J.; et al. Asialoglycoprotein receptor targeted optical and magnetic resonance imaging and therapy of liver fibrosis using pullulan stabilized multi-functional iron oxide nanoprobe. *Sci. Rep.* **2021**, *11*, 18324. [CrossRef] [PubMed]
34. Nagórniewicz, B.; Mardhian, D.F.; Booijink, R.; Storm, G.; Prakash, J.; Bansal, R. Engineered Relaxin as theranostic nanomedicine to diagnose and ameliorate liver cirrhosis. *Nanomed. Nanotechnol. Biol. Med.* **2019**, *17*, 106–118. [CrossRef]
35. Ezhilarasan, D.; Lakshmi, T.; Raut, B. Novel Nano-Based Drug Delivery Systems Targeting Hepatic Stellate Cells in the Fibrotic Liver. *J. Nanomater.* **2021**, *2021*, 4674046. [CrossRef]

36. Eslaminejad, T.; Noureddin Nematollahi-Mahani, S.; Ansari, M. Glioblastoma targeted gene therapy based on pEGFP/p53-loaded superparamagnetic iron oxide nanoparticles. *Curr. Gene Ther.* **2017**, *17*, 59–69. [CrossRef]
37. Li, F.; Yan, H.; Wang, J.; Li, C.; Wu, J.; Wu, S.; Rao, S.; Gao, X.; Jin, Q. Non-invasively differentiating extent of liver fibrosis by visualizing hepatic integrin αvβ3 expression with an MRI modality in mice. *Biomaterials* **2016**, *102*, 162–174. [CrossRef]
38. Ungureanu, B.S.; Teodorescu, C.M.; Saftoiu, A. Magnetic Nanoparticles for Hepatocellular Carcinoma Diagnosis and Therapy. *J. Gastrointest. Liver Dis. JGLD* **2016**, *25*, 375–383. [CrossRef]
39. Hume, D.A. Differentiation and heterogeneity in the mononuclear phagocyte system. *Mucosal Immunol.* **2008**, *1*, 432–441. [CrossRef]
40. Alexis, F.; Pridgen, E.; Molnar, L.K.; Farokhzad, O.C. Factors affecting the clearance and biodistribution of polymeric nanoparticles. *Mol. Pharm.* **2008**, *5*, 505–515. [CrossRef]
41. Colino, C.I.; Lanao, J.M.; Gutierrez-Millan, C. Targeting of hepatic macrophages by therapeutic nanoparticles. *Front. Immunol.* **2020**, *11*, 218. [CrossRef]
42. Próspero, A.G.; Soares, G.A.; Moretto, G.M.; Quini, C.C.; Bakuzis, A.F.; de Arruda Miranda, J.R. Dynamic cerebral perfusion parameters and magnetic nanoparticle accumulation assessed by AC biosusceptometry. *Biomed. Eng./Biomed. Tech.* **2020**, *65*, 343–351. [CrossRef]
43. Próspero, A.G.; Quini, C.C.; Bakuzis, A.F.; Fidelis-de-Oliveira, P.; Moretto, G.M.; Mello, F.P.; Calabresi, M.F.; Matos, R.V.; Zandoná, E.A.; Zufelato, N. Real-time in vivo monitoring of magnetic nanoparticles in the bloodstream by AC biosusceptometry. *J. Nanobiotechnol.* **2017**, *15*, 22. [CrossRef] [PubMed]
44. Zhang, H.; Li, L.; Liu, X.L.; Jiao, J.; Ng, C.-T.; Yi, J.B.; Luo, Y.E.; Bay, B.-H.; Zhao, L.Y.; Peng, M.L.; et al. Ultrasmall Ferrite Nanoparticles Synthesized via Dynamic Simultaneous Thermal Decomposition for High-Performance and Multifunctional T1 Magnetic Resonance Imaging Contrast Agent. *ACS Nano* **2017**, *11*, 3614–3631. [CrossRef]
45. Park, J.Y.; Daksha, P.; Lee, G.H.; Woo, S.; Chang, Y. Highly water-dispersible PEG surface modified ultra small superparamagnetic iron oxide nanoparticles useful for target-specific biomedical applications. *Nanotechnology* **2008**, *19*, 365603. [CrossRef]
46. Kumar, A.; Pandey, A.K.; Singh, S.S.; Shanker, R.; Dhawan, A. Cellular uptake and mutagenic potential of metal oxide nanoparticles in bacterial cells. *Chemosphere* **2011**, *83*, 1124–1132. [CrossRef] [PubMed]
47. Alphandéry, E. Biodistribution and targeting properties of iron oxide nanoparticles for treatments of cancer and iron anemia disease. *Nanotoxicology* **2019**, *13*, 573–596. [CrossRef]
48. Soares, G.; Próspero, A.; Calabresi, M.; Rodrigues, D.; Simoes, L.; Quini, C.; Matos, R.; Pinto, L.; Sousa, A.; Bakuzis, A. Multichannel AC Biosusceptometry system to map biodistribution and assess the pharmacokinetic profile of magnetic nanoparticles by imaging. *IEEE Trans. Nanobiosci.* **2019**, *18*, 456–462. [CrossRef]
49. Zhang, Y.-N.; Poon, W.; Tavares, A.J.; McGilvray, I.D.; Chan, W.C. Nanoparticle–liver interactions: Cellular uptake and hepatobiliary elimination. *J. Control. Release* **2016**, *240*, 332–348. [CrossRef]
50. Cole, A.J.; David, A.E.; Wang, J.; Galbán, C.J.; Yang, V.C. Magnetic brain tumor targeting and biodistribution of long-circulating PEG-modified, cross-linked starch-coated iron oxide nanoparticles. *Biomaterials* **2011**, *32*, 6291–6301. [CrossRef]
51. Duguet, E.; Vasseur, S.; Mornet, S.; Devoisselle, J.M. Magnetic nanoparticles and their applications in medicine. *Nanomedicine* **2006**, *1*, 157–168. [CrossRef]
52. Maurea, S.; Mainenti, P.P.; Tambasco, A.; Imbriaco, M.; Mollica, C.; Laccetti, E.; Camera, L.; Liuzzi, R.; Salvatore, M. Diagnostic accuracy of MR imaging to identify and characterize focal liver lesions: Comparison between gadolinium and superparamagnetic iron oxide contrast media. *Quant. Imaging Med. Surg.* **2014**, *4*, 181.
53. Lurie, Y.; Webb, M.; Cytter-Kuint, R.; Shteingart, S.; Lederkremer, G.Z. Non-invasive diagnosis of liver fibrosis and cirrhosis. *World J. Gastroenterol.* **2015**, *21*, 11567–11583. [CrossRef]
54. Faria, S.C.; Ganesan, K.; Mwangi, I.; Shiehmorteza, M.; Viamonte, B.; Mazhar, S.; Peterson, M.; Kono, Y.; Santillan, C.; Casola, G. MR imaging of liver fibrosis: Current state of the art. *Radiographics* **2009**, *29*, 1615–1635. [CrossRef] [PubMed]
55. Petitclerc, L.; Sebastiani, G.; Gilbert, G.; Cloutier, G.; Tang, A. Liver fibrosis: Review of current imaging and MRI quantification techniques. *J. Magn. Reson. Imaging* **2017**, *45*, 1276–1295. [CrossRef] [PubMed]
56. Alphandéry, E.J.R.a. Iron oxide nanoparticles as multimodal imaging tools. *RSC Adv.* **2019**, *9*, 40577–40587. [CrossRef] [PubMed]
57. Zheng, B.; Vazin, T.; Goodwill, P.W.; Conway, A.; Verma, A.; Ulku Saritas, E.; Schaffer, D.; Conolly, S.M.J.S.r. Magnetic particle imaging tracks the long-term fate of in vivo neural cell implants with high image contrast. *Sci. Rep.* **2015**, *5*, 14055. [CrossRef] [PubMed]
58. Madru, R.; Kjellman, P.; Olsson, F.; Wingårdh, K.; Ingvar, C.; Ståhlberg, F.; Olsrud, J.; Lätt, J.; Fredriksson, S.; Knutsson, L.; et al. 99mTc-labeled superparamagnetic iron oxide nanoparticles for multimodality SPECT/MRI of sentinel lymph nodes. *J. Nucl. Med.* **2012**, *53*, 459–463. [CrossRef] [PubMed]
59. Forte, E.; Fiorenza, D.; Torino, E.; Costagliola di Polidoro, A.; Cavaliere, C.; Netti, P.A.; Salvatore, M.; Aiello, M. Radiolabeled PET/MRI Nanoparticles for Tumor Imaging. *J. Clin. Med.* **2019**, *9*, 89. [CrossRef]
60. Shen, S.; Wang, S.; Zheng, R.; Zhu, X.; Jiang, X.; Fu, D.; Yang, W. Magnetic nanoparticle clusters for photothermal therapy with near-infrared irradiation. *Biomaterials* **2015**, *39*, 67–74. [CrossRef]
61. Seested, T.; Appa, R.S.; Christensen, E.I.; Ioannou, Y.A.; Krogh, T.N.; Karpf, D.M.; Nielsen, H.M. In vivo clearance and metabolism of recombinant activated factor VII (rFVIIa) and its complexes with plasma protease inhibitors in the liver. *Thromb. Res.* **2011**, *127*, 356–362. [CrossRef]

62. Levy, M.; Luciani, N.; Alloyeau, D.; Elgrabli, D.; Deveaux, V.; Pechoux, C.; Chat, S.; Wang, G.; Vats, N.; Gendron, F.; et al. Long term in vivo biotransformation of iron oxide nanoparticles. *Biomaterials* **2011**, *32*, 3988–3999. [CrossRef]
63. Soares, G.A.; Pires, D.W.; Pinto, L.A.; Rodrigues, G.S.; Prospero, A.G.; Biasotti, G.G.A.; Bittencourt, G.N.; Stoppa, E.G.; Corá, L.A.; Oliveira, R.B.; et al. The Influence of Omeprazole on the Dissolution Processes of pH-Dependent Magnetic Tablets Assessed by Pharmacomagnetography. *Pharmaceutics* **2021**, *13*, 1274. [CrossRef]
64. Soares, G.A.; Faria, J.V.C.; Pinto, L.A.; Prospero, A.G.; Pereira, G.M.; Stoppa, E.G. Long-Term Clearance and Biodistribution of Magnetic Nanoparticles Assessed by AC Biosusceptometry. *Materials* **2022**, *15*, 2121. [CrossRef]
65. Prospero, A.G.; Buranello, L.P.; Fernandes, C.A.; Dos Santos, L.D.; Soares, G.; Rossini, B.C.; Zufelato, N.; Bakuzis, A.F.; de Mattos Fontes, M.R.; de Arruda Miranda, J.R. Corona protein impacts on alternating current biosusceptometry signal and circulation times of differently coated $MnFe_2O_4$ nanoparticles. *Nanomedicine* **2021**, *16*, 2189–2206. [CrossRef]
66. Prospero, A.G.; Fidelis-de-Oliveira, P.; Soares, G.A.; Miranda, M.F.; Pinto, L.A.; Dos Santos, D.C.; Silva, V.D.S.; Zufelato, N.; Bakuzis, A.F.; Miranda, J.R. AC biosusceptometry and magnetic nanoparticles to assess doxorubicin-induced kidney injury in rats. *Nanomedicine* **2020**, *15*, 511–525. [CrossRef]
67. Quini, C.C.; Prospero, A.G.; Calabresi, M.F.F.; Moretto, G.M.; Zufelato, N.; Krishnan, S.; Pina, D.R.; Oliveira, R.B.; Baffa, O.; Bakuzis, A.F.; et al. Real-time liver uptake and biodistribution of magnetic nanoparticles determined by AC biosusceptometry. *Nanomedicine* **2017**, *13*, 1519–1529. [CrossRef]
68. Biasotti, G.G.d.A.; Próspero, A.G.; Alvarez, M.D.T.; Liebl, M.; Pinto, L.A.; Soares, G.A.; Bakuzis, A.F.; Baffa, O.; Wiekhorst, F.; Miranda, J.R.d.A. 2D Quantitative Imaging of Magnetic Nanoparticles by an AC Biosusceptometry Based Scanning Approach and Inverse Problem. *Sensors* **2021**, *21*, 7063.
69. Branquinho, L.C.; Carriao, M.S.; Costa, A.S.; Zufelato, N.; Sousa, M.H.; Miotto, R.; Ivkov, R.; Bakuzis, A.F. Effect of magnetic dipolar interactions on nanoparticle heating efficiency: Implications for cancer hyperthermia. *Sci. Rep.* **2013**, *3*, 2887. [CrossRef]
70. Nunes, A.D.; Gomes-Silva, L.A.; Zufelato, N.; Próspero, A.G.; Quini, C.C.; Matos, R.V.; Miranda, J.R.; Bakuzis, A.F.; Castro, C.H. Albumin coating prevents cardiac effect of the magnetic nanoparticles. *IEEE Trans. Nanobiosci.* **2019**, *18*, 640–650. [CrossRef]
71. Tatematsu, M.; Tsuda, H.; Shirai, T.; Masui, T.; Ito, N.J.T.P. Placental glutathione S-transferase (GST-P) as a new marker for hepatocarcinogenesis: In vivo short-term screening for hepatocarcinogens. *Toxicol. Pathol.* **1987**, *15*, 60–68. [CrossRef]
72. Chistiakov, D.A.; Killingsworth, M.C.; Myasoedova, V.A.; Orekhov, A.N.; Bobryshev, Y.V. CD68/macrosialin: Not just a histochemical marker. *Lab. Investig.* **2017**, *97*, 4–13. [CrossRef]
73. Ishak, K. Histological grading and staging of chronic hepatitis. *J. Hepatol.* **1995**, *22*, 696–699. [CrossRef]
74. Sousa-Junior, A.A.; Mendanha, S.A.; Carrião, M.S.; Capistrano, G.; Próspero, A.G.; Soares, G.A.; Cintra, E.R.; Santos, S.F.O.; Zufelato, N.; Alonso, A.; et al. Predictive Model for Delivery Efficiency: Erythrocyte Membrane-Camouflaged Magnetofluorescent Nanocarriers Study. *Mol. Pharm.* **2020**, *17*, 837–851. [CrossRef]
75. Islam, K.; Haque, M.; Kumar, A.; Hoq, A.; Hyder, F.; Hoque, S.M. Manganese ferrite nanoparticles ($MnFe_2O_4$): Size dependence for hyperthermia and negative/positive contrast enhancement in MRI. *Nanomaterials* **2020**, *10*, 2297. [CrossRef]
76. Jardim, K.V.; Palomec-Garfias, A.F.; Andrade, B.Y.G.; Chaker, J.A.; Báo, S.N.; Márquez-Beltrán, C.; Moya, S.E.; Parize, A.L.; Sousa, M.H. Novel magneto-responsive nanoplatforms based on $MnFe_2O_4$ nanoparticles layer-by-layer functionalized with chitosan and sodium alginate for magnetic controlled release of curcumin. *Mater. Sci. Eng. C* **2018**, *92*, 184–195. [CrossRef]
77. Darwish, M.S.; Stibor, I. Pentenoic acid-stabilized magnetic nanoparticles for nanomedicine applications. *J. Dispers. Sci. Technol.* **2016**, *37*, 1793–1798. [CrossRef]
78. Chertok, B.; David, A.E.; Yang, V.C. Polyethyleneimine-modified iron oxide nanoparticles for brain tumor drug delivery using magnetic targeting and intra-carotid administration. *Biomaterials* **2010**, *31*, 6317–6324. [CrossRef]
79. Sakulkhu, U.; Mahmoudi, M.; Maurizi, L.; Salaklang, J.; Hofmann, H. Protein corona composition of superparamagnetic iron oxide nanoparticles with various physico-chemical properties and coatings. *Sci. Rep.* **2014**, *4*, 5020. [CrossRef]
80. Feng, Q.; Liu, Y.; Huang, J.; Chen, K.; Huang, J.; Xiao, K. Uptake, distribution, clearance, and toxicity of iron oxide nanoparticles with different sizes and coatings. *Sci. Rep.* **2018**, *8*, 2082. [CrossRef]
81. Gupta, A.K.; Naregalkar, R.R.; Vaidya, V.D.; Gupta, M. Recent advances on surface engineering of magnetic iron oxide nanoparticles and their biomedical applications. *Future Med.* **2007**, *2*, 23–29. [CrossRef]
82. Han, S.S.; Li, Z.Y.; Zhu, J.Y.; Han, K.; Zeng, Z.Y.; Hong, W.; Li, W.X.; Jia, H.Z.; Liu, Y.; Zhuo, R.X.J.S. Dual-pH sensitive charge-reversal polypeptide micelles for tumor-triggered targeting uptake and nuclear drug delivery. *Small* **2015**, *11*, 2543–2554. [CrossRef] [PubMed]
83. Kenzaoui, B.H.; Vilà, M.R.; Miquel, J.M.; Cengelli, F.; Juillerat-Jeanneret, L. Evaluation of uptake and transport of cationic and anionic ultrasmall iron oxide nanoparticles by human colon cells. *Int. J. Nanomed.* **2012**, *7*, 1275.
84. Petri-Fink, A.; Chastellain, M.; Juillerat-Jeanneret, L.; Ferrari, A.; Hofmann, H. Development of functionalized superparamagnetic iron oxide nanoparticles for interaction with human cancer cells. *Biomaterials* **2005**, *26*, 2685–2694. [CrossRef] [PubMed]
85. Xiao, K.; Li, Y.; Luo, J.; Lee, J.S.; Xiao, W.; Gonik, A.M.; Agarwal, R.G.; Lam, K.S. The effect of surface charge on in vivo biodistribution of PEG-oligocholic acid based micellar nanoparticles. *Biomaterials* **2011**, *32*, 3435–3446. [CrossRef]
86. Imam, S.Z.; Lantz-McPeak, S.M.; Cuevas, E.; Rosas-Hernandez, H.; Liachenko, S.; Zhang, Y.; Sarkar, S.; Ramu, J.; Robinson, B.L.; Jones, Y.; et al. Iron oxide nanoparticles induce dopaminergic damage: In vitro pathways and in vivo imaging reveals mechanism of neuronal damage. *Mol. Neurobiol.* **2015**, *52*, 913–926. [CrossRef]

87. Elbialy, N.S.; Aboushoushah, S.F.; Alshammari, W.W. Long-term biodistribution and toxicity of curcumin capped iron oxide nanoparticles after single-dose administration in mice. *Life Sci.* **2019**, *230*, 76–83. [CrossRef]
88. Rodrigues, H.F.; Mello, F.M.; Branquinho, L.C.; Zufelato, N.; Silveira-Lacerda, E.P.; Bakuzis, A.F. Real-time infrared thermography detection of magnetic nanoparticle hyperthermia in a murine model under a non-uniform field configuration. *Int. J. Hyperth.* **2013**, *29*, 752–767. [CrossRef]
89. Wilhelm, S.; Tavares, A.J.; Dai, Q.; Ohta, S.; Audet, J.; Dvorak, H.F.; Chan, W.C.J.N.r.m. Analysis of nanoparticle delivery to tumours. *Nat. Rev. Mater.* **2016**, *1*, 16014. [CrossRef]
90. Tavares, A.J.; Poon, W.; Zhang, Y.-N.; Dai, Q.; Besla, R.; Ding, D.; Ouyang, B.; Li, A.; Chen, J.; Zhang, Y.N.; et al. Effect of removing Kupffer cells on nanoparticle tumor delivery. *Proc. Natl. Acad. Sci. USA* **2017**, *114*, E10871–E10880. [CrossRef]
91. Li, L.; Wei, W.; Li, Z.; Chen, H.; Li, Y.; Jiang, W.; Chen, W.; Kong, G.; Yang, J.; Li, Z. The Spleen Promotes the Secretion of CCL2 and Supports an M1 Dominant Phenotype in Hepatic Macrophages During Liver Fibrosis. *Cell. Physiol. Biochem.* **2018**, *51*, 557–574. [CrossRef]
92. Li, L.; Duan, M.; Chen, W.; Jiang, A.; Li, X.; Yang, J.; Li, Z. The spleen in liver cirrhosis: Revisiting an old enemy with novel targets. *J. Transl. Med.* **2017**, *15*, 111. [CrossRef]
93. Murotomi, K.; Arai, S.; Uchida, S.; Endo, S.; Mitsuzumi, H.; Tabei, Y.; Yoshida, Y.; Nakajima, Y. Involvement of splenic iron accumulation in the development of nonalcoholic steatohepatitis in Tsumura Suzuki Obese Diabetes mice. *Sci. Rep.* **2016**, *6*, 22476. [CrossRef]
94. Wei, Y.; Zhao, M.; Yang, F.; Mao, Y.; Xie, H.; Zhou, Q. Iron overload by Superparamagnetic Iron Oxide Nanoparticles is a High Risk Factor in Cirrhosis by a Systems Toxicology Assessment. *Sci. Rep.* **2016**, *6*, 29110. [CrossRef]
95. Han, X.; Lv, Y.; Li, Y.; Deng, J.; Qiu, Q.; Liu, N.; Zhao, S.; Liao, C. Distribution characteristics of cells in splenomegaly due to hepatitis B-related cirrhotic portal hypertension and their clinical importance. *J. Int. Med. Res.* **2018**, *46*, 2633–2640. [CrossRef]
96. Ergen, C.; Niemietz, P.M.; Heymann, F.; Baues, M.; Gremse, F.; Pola, R.; van Bloois, L.; Storm, G.; Kiessling, F.; Trautwein, C.; et al. Liver fibrosis affects the targeting properties of drug delivery systems to macrophage subsets in vivo. *Biomaterials* **2019**, *206*, 49–60. [CrossRef]
97. Hundt, W.; Petsch, R.; Helmberger, T.; Reiser, M. Signal changes in liver and spleen after Endorem administration in patients with and without liver cirrhosis. *Eur. Radiol.* **2000**, *10*, 409–416. [CrossRef]
98. Borst, K.; Frenz, T.; Spanier, J.; Tegtmeyer, P.-K.; Chhatbar, C.; Skerra, J.; Ghita, L.; Namineni, S.; Lienenklaus, S.; Köster, M. Type I interferon receptor signaling delays Kupffer cell replenishment during acute fulminant viral hepatitis. *J. Hepatol.* **2018**, *68*, 682–690. [CrossRef]
99. Borst, K.; Graalmann, T.; Kalinke, U. Reply to: "Lack of Kupffer cell depletion in diethylnitrosamine-induced hepatic inflammation". *J. Hepatol.* **2019**, *70*, 815–816. [CrossRef]
100. Kessler, S.M.; Hoppstädter, J.; Hosseini, K.; Laggai, S.; Haybaeck, J.; Kiemer, A.K. Lack of Kupffer cell depletion in diethylnitrosamine-induced hepatic inflammation. *J. Hepatol.* **2019**, *70*, 813–815. [CrossRef]
101. Sanz-García, C.; Fernández-Iglesias, A.; Gracia-Sancho, J.; Arráez-Aybar, L.A.; Nevzorova, Y.A.; Cubero, F.J. The Space of Disse: The Liver Hub in Health and Disease. *Livers* **2021**, *1*, 3–26. [CrossRef]
102. Ni, Y.; Li, J.-M.; Liu, M.-K.; Zhang, T.-T.; Wang, D.-P.; Zhou, W.-H.; Hu, L.-Z.; Lv, W.-L. Pathological process of liver sinusoidal endothelial cells in liver diseases. *World J. Gastroenterol.* **2017**, *23*, 7666–7677. [CrossRef] [PubMed]
103. Cheng, Q.-N.; Yang, X.; Wu, J.-F.; Ai, W.-B.; Ni, Y.-R. Interaction of non-parenchymal hepatocytes in the process of hepatic fibrosis. *Mol. Med. Rep.* **2021**, *23*, 364. [CrossRef] [PubMed]
104. Lafoz, E.; Ruart, M. The Endothelium as a Driver of Liver Fibrosis and Regeneration. *Cells* **2020**, *9*, 929. [CrossRef] [PubMed]
105. Sun, T.; Kang, Y.; Liu, J.; Zhang, Y.; Ou, L.; Liu, X.; Lai, R.; Shao, L. Nanomaterials and hepatic disease: Toxicokinetics, disease types, intrinsic mechanisms, liver susceptibility, and influencing factors. *J. Nanobiotechnol.* **2021**, *19*, 108. [CrossRef]
106. De Rudder, M.; Dili, A.; Stärkel, P.; Leclercq, I.A. Critical Role of LSEC in Post-Hepatectomy Liver Regeneration and Failure. *Int. J. Mol. Sci* **2021**, *22*, 8053. [CrossRef]
107. Iwakiri, Y.; Trebicka, J. Portal hypertension in cirrhosis: Pathophysiological mechanisms and therapy. *JHEP Rep. Innov. Hepatol.* **2021**, *3*, 100316. [CrossRef]
108. Hall, A.; Cotoi, C.; Luong, T.V.; Watkins, J.; Bhathal, P.; Quaglia, A. Collagen and elastic fibres in acute and chronic liver injury. *Sci. Rep.* **2021**, *11*, 14569. [CrossRef]
109. Soulaidopoulos, S.; Cholongitas, E.; Giannakoulas, G.; Vlachou, M.; Goulis, I. Review article: Update on current and emergent data on hepatopulmonary syndrome. *World J. Gastroenterol.* **2018**, *24*, 1285–1298. [CrossRef]
110. Cheng, T.-Y.; Lee, W.-S.; Huang, H.-C.; Lee, F.-Y.; Chang, C.-C.; Lin, H.-C.; Lee, S.-D. The effects of pioglitazone in cirrhotic rats with hepatopulmonary syndrome. *J. Chin. Med. Assoc.* **2017**, *80*, 683–689. [CrossRef]
111. Nuzzo, A.; Dautry, R.; Francoz, C.; Logeart, D.; Mégarbane, B. Hepatopulmonary syndrome-attributed extreme hypoxemia and polycythemia revealing liver cirrhosis. *Am. J. Emerg. Med.* **2019**, *37*, 175.e171–175.e172. [CrossRef]
112. Jiang, L.-Q.; Wang, T.-Y.; Wang, Y.; Wang, Z.-Y.; Bai, Y.-T. Co-disposition of chitosan nanoparticles by multi types of hepatic cells and their subsequent biological elimination: The mechanism and kinetic studies at the cellular and animal levels. *Int. J. Nanomed.* **2019**, *14*, 6035–6060. [CrossRef]

Article

The Surface Amine Group of Ultrasmall Magnetic Iron Oxide Nanoparticles Produce Analgesia in the Spinal Cord and Decrease Long-Term Potentiation

Guan-Ling Lu [1], Ya-Chi Lin [2], Ping-Ching Wu [3,4,5,6,*] and Yen-Chin Liu [1,2,7,*]

[1] Department of Anesthesiology, School of Post-Baccalaureate, College of Medicine, Kaohsiung Medical University, Kaohsiung 807, Taiwan; r93443013@ntu.edu.tw
[2] Department of Anesthesiology, National Cheng Kung University Hospital, College of Medicine, National Cheng Kung University, Tainan 701, Taiwan; tonyangiekiki@hotmail.com
[3] Department of Biomedical Engineering, National Cheng Kung University, Tainan 701, Taiwan
[4] Institute of Oral Medicine and Department of Stomatology, National Cheng Kung University Hospital, College of Medicine, National Cheng Kung University, Tainan 701, Taiwan
[5] Center of Applied Nanomedicine, National Cheng Kung University, Tainan 701, Taiwan
[6] Medical Device Innovation Center, Taiwan Innovation Center of Medical Devices and Technology, National Cheng Kung University Hospital, National Cheng Kung University, Tainan 701, Taiwan
[7] Department of Anesthesiology, Kaohsiung Medical University Hospital, Kaohsiung 807, Taiwan
* Correspondence: wbcxyz@gmail.com (P.-C.W.); anesliu@kmu.edu.tw (Y.-C.L.); Tel.: +886-6-2757575 (ext. 63436) (P.-C.W.); +886-7-3121101 (ext. 7035) (Y.-C.L.)

Citation: Lu, G.-L.; Lin, Y.-C.; Wu, P.-C.; Liu, Y.-C. The Surface Amine Group of Ultrasmall Magnetic Iron Oxide Nanoparticles Produce Analgesia in the Spinal Cord and Decrease Long-Term Potentiation. *Pharmaceutics* 2022, 14, 366. https://doi.org/10.3390/pharmaceutics14020366

Academic Editors: Bogdan Parakhonskiy, Francesca Garello, Miriam Filippi and Yulia I. Svenskaya

Received: 15 December 2021
Accepted: 3 February 2022
Published: 6 February 2022

Publisher's Note: MDPI stays neutral with regard to jurisdictional claims in published maps and institutional affiliations.

Copyright: © 2022 by the authors. Licensee MDPI, Basel, Switzerland. This article is an open access article distributed under the terms and conditions of the Creative Commons Attribution (CC BY) license (https://creativecommons.org/licenses/by/4.0/).

Abstract: Our previous studies have revealed the ultrasmall superparamagnetic iron oxide in the amine group USPIO-101 has an analgesic effect on inflammatory pain. Here, we further investigated its effect on the spinal cord and brain via electrophysiological and molecular methods. We used a mouse inflammatory pain model, induced by complete Freund's adjuvant (CFA), and measured pain thresholds via von Frey methods. We also investigated the effects of USPIO-101 via an extracellular electrophysiological recording at the spinal dorsal horn synapses and hippocampal Schaffer collateral-CA1 synapses, respectively. The mRNA expression of pro-inflammatory cytokines was detected by quantitative real-time polymerase chain reaction (RT-qPCR). Our results showed intrathecal USPIO-101 produces similar analgesic behavior in mice with chronic inflammatory pain via intrathecal or intraplantar administration. The potentiated low-frequency stimulation-induced spinal cord long-term potentiation (LTP) at the spinal cord superficial dorsal horn synapses could decrease via USPIO-101 in mice with chronic inflammatory pain. However, the mRNA expression of cyclooxygenase-2 was enhanced with lipopolysaccharide (LPS) stimulation in microglial cells, and we also found USPIO-101 at 30 μg/mL could decrease the magnitude of hippocampal LTP. These findings revealed that intrathecal USPIO-101 presented an analgesia effect at the spinal cord level, but had neurotoxicity risk at higher doses.

Keywords: ultrasmall magnetic iron oxide nanoparticles; inflammatory pain; analgesia; pro-inflammatory cytokines; neurotoxicity; long-term potentiation

1. Introduction

The therapeutic application of iron oxide nanoparticles has been developing over the years. Currently, the commercialized product of iron oxide nanoparticles are used for the treatment of cancer and iron-deficiency anemia [1,2]; however, a lesser known use for iron oxide nanoparticles is pain management. The ongoing nanoparticle-based therapeutics in pain management [3] have several advantages for chronic pain relief, for example, controlled release, prolonged circulation time, and limited side effects [4].

Our previous study revealed a form of amine-terminated (-NH$_2$) ultrasmall superparamagnetic iron oxide (USPIO) called USPIO-101, which has an analgesic effect on inflammatory pain [5]. This analgesic response probably happens by attenuating inflammatory cell infiltration and reducing reactive oxygen species (ROS) production in the paw [5]. However, this was the first report to demonstrate that USPIO itself could have analgesic ability not in conjugation with other pain-relieving drugs.

Spinal cord synaptic plasticity is an *in vitro*, cellular, molecular model for pain [6]. Long-term potentiation (LTP) at the superficial dorsal horn synapses could also represent the nociceptive nerve signal. On the contrary, the decreased LTP level could stand for the analgesic effect of the treated compound or anti-allodynia signal [7]. However, less is known about the analgesic effect of USPIO-101 in the spinal cord via electrophysiological evidence. Here, we use *in vitro* extracellular recording to measure the effect of USPIO-101 on the spinal LTP to demonstrate the analgesic effect of USPIO-101 on the spinal cord.

Before developing the application of USPIO-101, one crucial indication was to lower cell cytotoxicity. In addition, USPIO-101 in a small size could probably interfere with normal body function via crossing the blood–brain barrier [8]. Therefore, the cytotoxicity of USPIO-101 is unclear. However, some reports have revealed the cytotoxicity of other types of USPIO. For example, the acute intravenous (iv.) injection of USPIO caused thrombosis, cardiac oxidative stress, and DNA damage in mice [9]. Furthermore, USPIO also triggered interleukin (IL)-6-related acute-phase inflammation [10] with a mechanism of endoplasmic reticulum (ER)-mitochondria Ca^{2+} crosstalk, which was mediated by cyclooxygenase-2 (COX-2) [11] in hepatocytes. Finally, superparamagnetic iron oxide (SPIO) administration, either locally or systemically, gave an acute inflammatory response [12]. These reports inspired us to consider that USPIO-101 could probably have neurotoxicity in neuronal cells. Thus, we investigated the toxicity of USPIO-101 in neuron-like or microglial cells via measuring the ROS production or mRNA expression of pro-inflammatory cytokines in the present study.

Synaptic plasticity is fundamental to many neurobiological functions, including memory and pain [13]. Moreover, the hippocampus's long-lasting potentiated synaptic field potentials are a proposed cellular mechanism for memory [14]. Here, we examined the toxicity effect of USPIO-101 on hippocampal LTP, which represented a higher level of neurobiological functions via *in vitro* extracellular recording at Schaffer collateral/CA1 synapses.

In this study, we revealed further analgesic evidence for using USPIO-101 at the spinal cord and the possible neurotoxicity that should be concerned for future application.

2. Materials and Methods

2.1. Drugs and Administration

The amine-terminated (-NH_2) iron oxide nanoparticles (Fe_3O_4 NPs) were commercially purchased from TANBead (USPIO-101, Taiwan Advanced Nanotech Inc., Taoyuan, Taiwan), and the stock concentration was 10 mg/mL. For behavioral tests, the intrathecal or intraplantar injection, the stock solution was used in a volume of 10 μL. For *in vitro* electrophysiological study, 1000X dilution was used for perfusion.

2.2. The Particle Size, Zeta Potential, and Surface Group Measurement of Iron Oxide Nanoparticles

The particle size distribution and zeta potential were measured by dynamic light scattering (DLS) (Beckman Coulter DelsaTM Nano instrument, CA, USA) with deionized water (ddH_2O) as the solvent. Fourier transform infrared (FTIR) spectra analyzed the surface group of the iron oxide nanoparticles via Nicolet FTIR spectrometers (Thermo Scientific, MA, USA) in the range 500–4000 cm^{-1} using a resolution of 1 cm^{-1} and 10 scans. In advance of testing, the particles were placed in an oven (60 °C) overnight to remove water and then ground with potassium bromide (KBr) powder to increase the absorption of infrared light and eventually pressed to obtain self-supporting discs.

2.3. Animal

Male ICR mice were used in all the experiments: for spinal cord slices electrophysiological recording, we used 4~6-week-old mice; others were 6~8-week-old mice. All animals were purchased from Bio-LASCO Inc. (Taipei, Taiwan). Mice were housed 4~5 per cage under 12 h light/dark controlled (AM/PM 7:00) with free access to food and water. The animals used for electrophysiological recording were housed in the NHRI laboratory animal center and approved by the NHRI laboratory animal center. The animals used for the behavioral test were housed and performed in National Cheng Kung University (NCKU) Laboratory Animal Center and approved by NCKU Medical College Animal Care Guidelines.

2.4. CFA Inflammatory Pain Model and Behavior Tests

CFA (complete Freund's adjuvant; Sigma-Aldrich, Saint Louis, MO, USA) or saline 10 μL were injected into the plantar surface of the left hind paw to induce an inflammatory pain model [5]. Mice were placed in individual test boxes for the mechanical pain sensitivity test. Mice were habituated for at least two days in the testing environment daily for one hour. Before the examination, at least one hour of habituation was necessary. A series of von Frey hairs with logarithmically increasing stiffness (0.02–2.56 g, Stoelting, Wood Dale, IL, USA), perpendicular to the plantar surface of the left hind paw was applied for 1 s, until it buckled. We marked a positive response if the animal exhibited any nocifensive behaviors after removing the filament, including quick paw withdrawal, licking, or shaking the paw. The first filament was chosen to be close to the 50% withdrawal threshold. If there was no response, the next filament was a higher force; if there was a response, the next filament was a lower force. This continued until at least four readings were obtained after the first change of direction [15]. The analysis of the 50% paw withdrawal threshold was determined using the Dixon up–down method and calculated using the formula: 50% threshold (g) = $10^{(X+kd)}/10^4$, where X = the value (in log units) of the final von Frey filament, k = tabular value for the response pattern (see Appendix 1 in [16]) and d = the average increment (in log units) between von Frey filaments [16].

2.5. Electrophysiological Recordings

2.5.1. Spinal Cord Slice

Transverse spinal cord slices (350 μm) were dissected as described previously with modifications [17,18]. Mice were sacrificed with overdose isoflurane and transcardial perfusion with cold artificial cerebral spinal fluid (aCSF) immediately; then, the spinal cord was removed from the spinal column. After dissection of the spinal cord pia-arachnoid membrane in cold aCSF, the spinal cord's lumbosacral enlargement (L1–S3) was maintained. Then, we collected spinal cord slices from the L4~L6 region with a vibratome (DTK1000, Dosaka) and equilibrated slices at room temperature for at least one hour before recording. The aCSF consisted of (mM): NaCl 117, KCl 4.5, $CaCl_2$ 2.5, $MgCl_2$ 1.2, NaH_2PO_4 1.2, $NaHCO_3$ 25 and glucose 11, and was oxygenated with 95% O_2/5% CO_2 (pH 7.4).

We recorded the field excitatory postsynaptic potentials (fEPSPs) at the spinal cord superficial dorsal horn synapses of the mouse spinal cord slices with a continuously perfused oxygenated aCSF at 1~2 mL/min. Glass pipettes (resistance, 5~8 MΩ) were filled with aCSF. Then, according to Terman's report [19], we determined the position of the stimulating electrode and the recording glass pipette. First, the stimulating electrode was attached to the spinal cord slice's dorsal root remnant; second, the recording glass pipette was placed on the superficial dorsal horn area of the spinal cord slice. Signal acquisition was measured by Multiclamp 700B amplifier (Molecular Devices) and sampled by pCLAMP 10.2 and an analog-to-digital converter (Digidata 1322A), filtered at 2~5 kHz, digitized at 10 kHz, and stored for off-line analysis.

The stimulation signals were sequentially evoked (Grass S88) for thirty seconds, once, with a 0.5-ms pulse. Low-frequency stimulation (2 Hz, 120 s) was applied to induce long-term potentiation (LTP). The baseline of fEPSP was obtained 10 min from the beginning, and the slope of fEPSPs was normalized by the calculated average slope of 20 fEPSPs. The magnitude of LTP was the average slope of the last 20 fEPSPs recorded after high-frequency stimulation for 30 to 40 min. Each fEPSP was collected and analyzed by Clampfit software. Every mouse used 1~3 spinal cord slices for recording.

2.5.2. Hippocampal Slice

Coronal hippocampal slices (400 μm) were dissected as described previously with modifications [20]. Mice were sacrificed with overdose isoflurane and decapitated immediately, then the brain was transferred to cold aCSF. After dissection, slices were equilibrated at room temperature for two hours before recording.

The recorded fEPSPs were evoked on the Schaffer collateral fiber path and detected in the apical dendritic field (the stratum radiatum) in the CA1 region in each hippocampal slice. Basal stimulation was given at 0.03 Hz by constant current pulses (0.2 ms). LTP was induced by theta-burst stimulation (TBS), which contained three trains of five bursts separated by 300 ms, with each burst consisting of ten pulses at 100 Hz. The baseline of fEPSP was obtained before TBS and maintained for at least 10 min. The magnitude of LTP was calculated by the average slope of 20 fEPSPs recorded after TBS 30 to 40 min. Every mouse used 1~3 brain slices for recording. Signal acquisition and analysis were similar to spinal cord slices.

2.6. ROS Levels

The human neuroblastoma SH-SY5Y cells, and mouse SM826 microglia cell line, were cultured in DMEM growth medium (Gibco) supplemented with 10% fetal bovine serum (FBS, BI) and 0.1% penicillin/streptomycin (Gibco). Cells were incubated at 37°C in an atmosphere containing 5% CO_2.

An OxiSelect intracellular ROS assay kit (Cell Biolabs, San Diego, CA, USA) was used to measure the ROS levels in the SH-SY5Y and SM826 cells, respectively. Cells were seeded into 96-well plates (4×10^4 cells/well) and incubated for 16 h at 37°C. The SH-SY5Y cells were incubated with 2,7-Dichlorodihydrofluorescein diacetate (DCFH-DA) 0.05 mM/serum-free medium for 60 min at 37°C. DCFH-DA medium was removed and treated with USPIO-101 (10 or 30 μg/mL), H_2O_2 1 mM for 30 min, or H_2O_2 0.2 mM for 24 h stimulation. The SM826 cells were treated with USPIO-101 (10 or 30 μg/mL), or LPS (1 μg/mL) for 24 h, and then incubated with DCFH-DA (0.05 mM, 60 min)/serum-free medium. All cells had lysis buffer added 5 min before reading the fluorescence and were analyzed by a fluorometric microplate reader (SpectraMax M2) at 480 nm/530 nm.

2.7. Assay of mRNA Expression

Total RNA was extracted via GENEzol™ TriRNA Pure Kit (Geneaid, New Taipei City, Taiwan) following the manufacturer's instructions. Total RNA (500 ng) was utilized for the reverse-transcription polymerase chain reaction (RT-PCR) by Thermo Scientific™ RevertAid RT Reverse Transcription Kit (Thermo Fisher Scientific, Waltham, MA, USA). The RNA and cDNA products were stored at -80°C before the following experimental procedure.

The mRNA expression levels were determined by real-time quantitative polymerase chain reaction (RT-qPCR). The Applied Biosystems™ StepOnePlus™ Real-Time PCR System and StepOne™ Software v2.3 (Thermo Fisher Scientific, Taiwan) were used. The reagent was Thermo Scientific™ Luminaris Color HiGreen qPCR Master Mix (2X) high ROX and Yellow Sample Buffer (40X) (Thermo Fisher Scientific, Waltham, MA, US). The RT-qPCR conditions were initial denaturation, 95 °C for 15 s; annealing, 60 °C for 30 s; extension, 72 °C for 30 s; 40 cycles.

The primers were synthesis from MISSION BIOTECH CO., LTD., Taiwan. The relative mRNA expression was determined by the $2^{-\Delta\Delta Ct}$ method using GAPDH (glyceraldehyde-3-phosphate dehydrogenase) as a normalization control.

Forward and reverse primer sets for each cDNA were used as follows: 5′-ATCTCATAC CAGGAGAAAGTCAACCT-3′ and 5′-TGGGCTCATACCAGGGTTTG-3′ (for TNF-α); 5′-GCTGCCAAAGAAGGACACGACA-3′ and 5′-GGCAGGCTATTGCTCATCACAG -3′ (for NF-kB); 5′-GGCCATGGAGTGGACTTAAA-3′ and 5′-CACCTCTCCACCAATGACCT-3′ (for COX-2); 5′-TGTGTCCGTCGTGGATCTGA-3′ and 5′-GATGCCTGCTTCACCACCTT-3′ (for GAPDH).

2.8. Statistical Analysis

All results are expressed as the mean ± SEM (standard error of mean). Electrophysiological results, ROS level, or mRNA expression were analyzed by one-way analysis of variance (ANOVA) followed by Newman–Keuls multiple comparisons test for post-hoc analyses. Behavioral results were analyzed with repeated-measure two-way ANOVA followed by Tukey tests for post-hoc analyses. The criterion for statistical significance was $p < 0.05$ when compared with each group.

3. Results

3.1. The Particle Size, Zeta Potential, and Surface Group Analysis of USPIO-101

Before the following experiments, the commercialized USPIO-101 measured its particle size, zeta potential, and surface group (Figure 1). The hydrodynamic diameter of USPIO-101 was 63.3 ± 2.3 nm; polydispersity index was 0.43 ± 0.58; zeta potential was 36.8 ± 0.6 mV (triple measurements, mean ± standard error, Figure S1). Next, we measured the surface group of USPIO-101 via FTIR. The spectra of FTIR represented that the –NH$_2$ group expressed in the surface of USPIO-101 at the wavelength of 3300~3500 cm^{-1} or 1560~1640 cm^{-1}. In addition, we also measured the positive control of the surface group –COOH at the wavelength of 1550~1610 cm^{-1} (Figure 1C). This result revealed that USPIO-101 majorly expressed –NH$_2$ surface group.

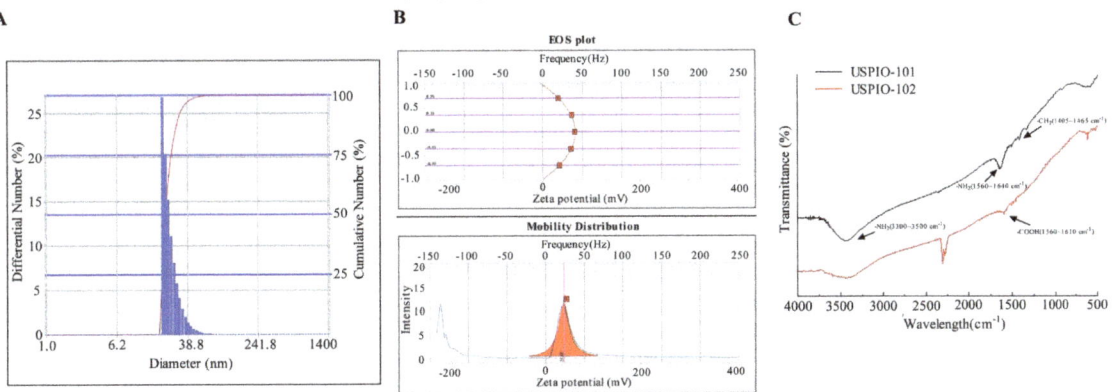

Figure 1. The dynamic light scattering (DLS), zeta potential analysis, and Fourier transform infrared spectroscopy (FTIR) spectra of ultrasmall magnetic iron oxide nanoparticles. (**A**) number distribution of USPIO-101, as measured by DLS. The concentration of USPIO-101 was 1 mg/mL. (**B**) Zeta potential analysis of USPIO-101. (**C**) Fourier transform infrared (FTIR) spectra are presented for USPIO-101 (black) and USPIO-102 (red), respectively. The -NH$_2$ and -CH$_2$ group signals are expressed in USPIO-101 at the wavelength of 3300~3500 cm^{-1}, 1560~1640 cm^{-1}, or 1405~1465 cm^{-1} (arrow). The -COOH group signal is expressed in USPIO-102 at the wavelength of 1550~1610 cm^{-1} (arrow).

3.2. USPIO-101 Alleviated the Allodynia Behavior via Intrathecal or Intraplantar Injection in Mice with Chronic Inflammatory Pain

The analgesic effect of USPIO-101 was measured in mice with chronic inflammatory pain via different administration routes, intrathecal or intraplantar injection. After paw injection of CFA for four days, the mice showed decreased paw withdrawal thresholds.

Both intrathecal (Figure 2A, PBS: 0.28 ± 0.02, USPIO-101: 2.1 ± 0.16 at 1.5 h, $p < 0.05$, two-way ANOVA and Bonferroni's multiple comparisons) and intraplantar (Figure 2B, PBS: 0.29 ± 0.0, USPIO-101: 2.0 ± 0.15 at 1.5 h, $p < 0.05$, two-way ANOVA and Bonferroni's multiple comparisons) injection of USPIO-101 (10 mg/mL, 10 µL) attenuated paw withdrawal thresholds.

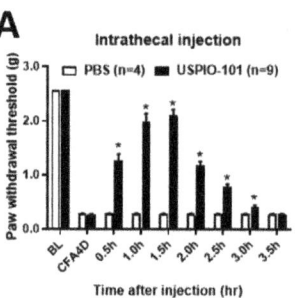 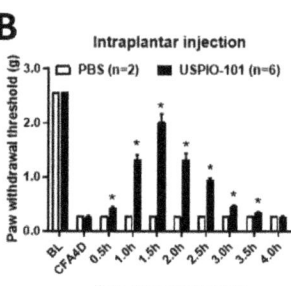

Figure 2. USPIO-101 attenuated chronic inflammatory pain via intrathecal or intraplantar injection. Mechanical pain sensitivity was measured using von Frey microfilaments. USPIO-101 attenuated the analgesia behavior in both (**A**) intrathecal and (**B**) intraplantar injection (10 mg/mL, 10 µL) after CFA paw injection for four days. The paw withdrawal thresholds were measured every 30 min until there was no difference between the three groups. Data were analyzed by two-way ANOVA and post-hoc with Tukey's test. *: $p < 0.05$ vs. PBS sham group.

3.3. USPIO-101 Decreased the Spinal Cord LTP at Spinal Cord Superficial Dorsal Horn Synapses in Mice with Chronic Inflammatory Pain and Naïve Mice

The potentiated spinal cord LTP at the spinal cord superficial dorsal horn synapses could stand for hyperalgesia [21]. The mice with chronic inflammatory pain showed the potentiated spinal cord LTP at the spinal cord superficial dorsal horn synapses was significantly higher than the control group ($p < 0.05$, t-test vs. saline-treated ipsilateral) (Figure 3A,C).

The concentration effect of USPIO-101 was tested on the LFS-evoked LTP at the superficial spinal dorsal horn in the spinal cord slices of mice with chronic inflammatory pain. USPIO-101 (10 or 30 µg/mL) was applied 7.5 min before LFS induction. The magnitude of LTP was significantly decreased in the treatments of 10 or 30 µg/mL ($p < 0.01$ or $p < 0.05$, one-way ANOVA vs. control), as shown in Figure 3D~3F. However, there was no concentration-dependent effect between 10 or 30 µg/mL ($p > 0.05$, t-test, 10 µg/mL vs. 30 µg/mL).

In the other part, we also tested the analgesic effect of USPIO-101 in naïve mice with the basal transmission or LFS-evoked LTP at the superficial spinal dorsal horn slices. No difference was observed in the basal transmission of naïve mice spinal cord slices when 10 µg/mL USPIO-101 was applied for 15 min (Figure 4A,B). However, USPIO-101 significantly inhibited the LTP in naïve mice spinal cord slices (Figure 4C,E). For USPIO-101 (10 µg/mL) applied 7.5 min before LFS induction, the magnitude of LTP significantly decreased in the treatment of USPIO-101 when compared with the control group ($p < 0.05$, Figure 4E).

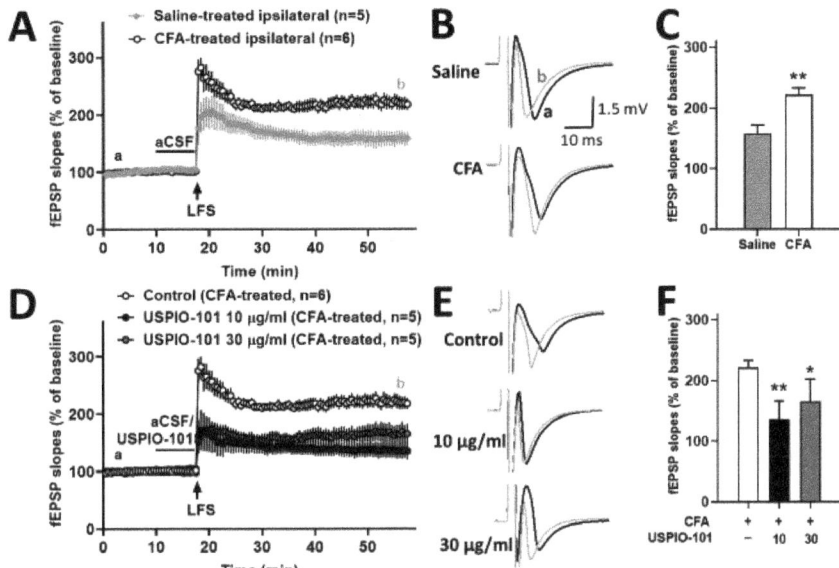

Figure 3. The effect of USPIO-101 on spinal cord LTP in CFA paw-injected mice. (**A**) Time courses of the slope of fEPSPs recorded before and after low-frequency stimulation (LFS, arrow) in spinal cord slices from CFA- or saline-treated mice. (**C**) The bar graph represents the magnitudes of potentiation, which averaged 20 fEPSPs recorded 30~40 min after LFS. **: $p < 0.01$ (unpaired t-tests) vs. CFA-treated ipsilateral group. (**D**) Time courses of the slope of fEPSPs recorded before and after LFS (arrow) in spinal cord slices from CFA-treated mice with USPIO-101 (10 or 30 μg/mL), respectively. USPIO-101 (10 or 30 μg/mL) applied for 7.5 min before LFS stimulation, respectively. (**F**) The bar graph represents the magnitudes of potentiation, which averaged 20 fEPSPs recorded 30~40 min after LFS at a concentration of 10 or 30 μg/mL. * or **: $p < 0.05$ or $p < 0.01$ vs. control group (one-way ANOVA). (**B,E**) Twenty recorded fEPSPs at time points a and b were averaged in each group. The slope of each fEPSP was expressed as % of the baseline fEPSP slope, which was the average of 20 fEPSPs at the beginning of 10 min recording. n indicates the number of slices recorded.

These results revealed that USPIO-101 had an analgesic effect on inflammatory pain in the spinal cord through electrophysiological evidence.

3.4. Effects of USPIO-101 on Intracellular ROS Levels

The iron oxide nanoparticle penetrates the cell and produces ROS [22], and the elevation of ROS induces neurotoxicity [23]. The ROS production ability of USPIO-101 was measured in neuron-like or microglial cells (SH-SY5Y or SM826 cells). Hydrogen peroxide (H_2O_2) was used as a positive control in SH-SY5Y cells for short (1 mM, 30 min) or long (0.2 mM, 24 h) stimulation, and compared to two USPIO-101 concentrations (10 or 30 μg/mL) as shown in Figure 5A,B. The intracellular ROS levels were only significantly increased in the group of H_2O_2 ($p < 0.05$, one-way ANOVA, vs. control, Figure 5A,B). In SM826 cells, USPIO-101 (10 or 30 μg/mL) did not induce significant elevation of ROS when compared with the control ($p < 0.05$, one-way ANOVA, vs. control, Figure 5C). The ROS level of co-treatment of LPS and USPIO-101 was close to the group of LPS alone (positive control, $p > 0.05$, one-way ANOVA, vs. LPS, Figure 5C) and significantly higher than the control ($p < 0.05$, one-way ANOVA, vs. control, Figure 5C).

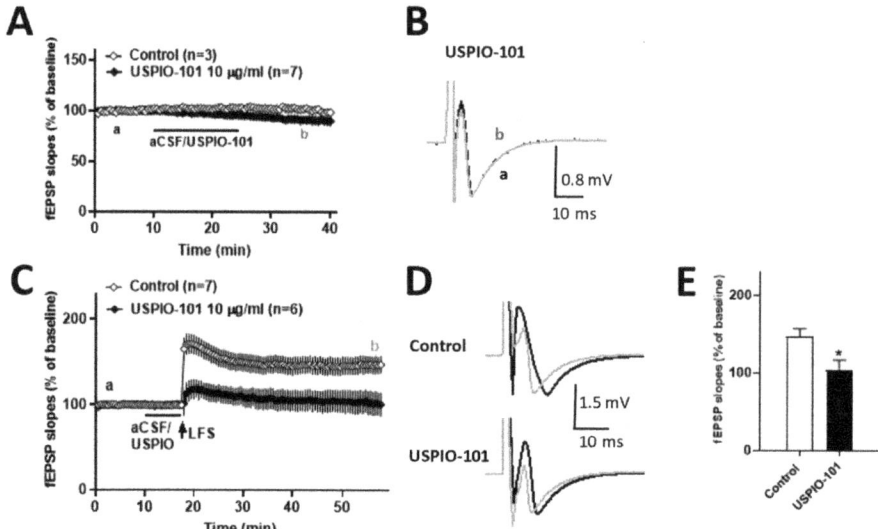

Figure 4. The effect of USPIO-101 on basal transmission (**A**) and spinal cord LTP (**B**,**C**) at the spinal cord superficial dorsal horn synapses, respectively. (**A**) After baseline recording for 10 min, USPIO-101 (10 μg/mL) was applied for another 15 min, then washed for 15 min. No difference was observed in the basal transmission after application of USPIO-101 to the spinal cord superficial dorsal horn synapses. (**C**) Time courses of the slope of fEPSPs recorded before and after LFS (arrow) in spinal cord slices. Drugs: USPIO-101 (10 μg/mL) was applied for 7.5 min before LFS stimulation. (**E**) The bar graph represents the magnitudes of potentiation, which averaged 20 fEPSPs recorded 30~40 min after LFS. *: $p < 0.05$ vs. control group (one-way ANOVA). (**B**,**D**) The traces shown in the graph are averaged 20 recordings of fEPSPs at times a and b in each group. The expression and analysis of the baseline fEPSP slope are the same as Figure 2. n indicates the number of slices recorded.

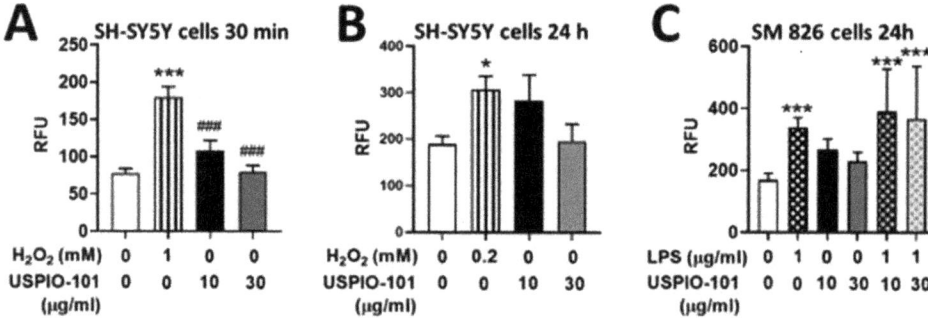

Figure 5. Intracellular reactive oxygen species level after treatment with USPIO-101 in SH-SY5Y or SM826 cells. (**A**,**B**) Hydrogen peroxide (H_2O_2) as positive control and USPIO-101 (10 or 30 μg/mL) were used in SH-SY5Y cells for 30 min (**A**, H_2O_2 1 mM) and 24 h (**B**, H_2O_2 0.2 mM), respectively. (**C**) Lipopolysaccharide (LPS) 1 μg/mL as positive control and USPIO-101 (10 or 30 μg/mL) were used in SM826 cells for 24 h, respectively. RFU means the relative fluorescence unit in 3×10^4 cells/well in 96-well plates. Data shown are the mean ± SEM of three independent experiments performed in triplicate. * or ***: $p < 0.05$ or $p < 0.001$ vs. control group. ###: $p < 0.001$ vs. the H_2O_2 group (one-way ANOVA).

These results suggested that USPIO-101 did not elicit ROS toxicity in neuron-like or microglial cells.

3.5. Effects of USPIO-101 on mRNA Expression with LPS Stimulation

To examine the effect of USPIO-101 on the pro-inflammatory cytokines' mRNA expression with LPS stimulation, the RT-qPCR was used for measuring mRNA extracted from microglia cells. USPIO-101 (10 μg/mL) was pretreated for 1 h before LPS (1 μg/mL) stimulation, and the cells were collected after LPS stimulation for 30, 60, or 120 min. LPS treatment for 120 min significantly upregulated the levels of transcripts encoding the pro-inflammatory cytokines TNF-α, NF-κB, and COX-2 ($p < 0.05$ vs. control, one-way ANOVA, Figure 6B–D). USPIO-101 significantly enhanced the expression of NF-κB and COX-2 after LPS stimulation for 120 min ($p < 0.05$ vs. LPS, one-way ANOVA, Figure 6C,D). USPIO-101 significantly enhanced the mRNA expression of COX-2 in the group of LPS + USPIO-101 after LPS 30-, 60-, or 120-min treatment ($p < 0.05$ vs. LPS, one-way ANOVA, Figure 6D). USPIO-101 treatment alone, for 3 h, did not affect the mRNA expression of these pro-inflammatory cytokines (Figure 6). These results suggest that USPIO-101 enhances the mRNA expression of pro-inflammatory cytokines, especially COX-2, when co-treated with LPS stimulation in the microglia cells.

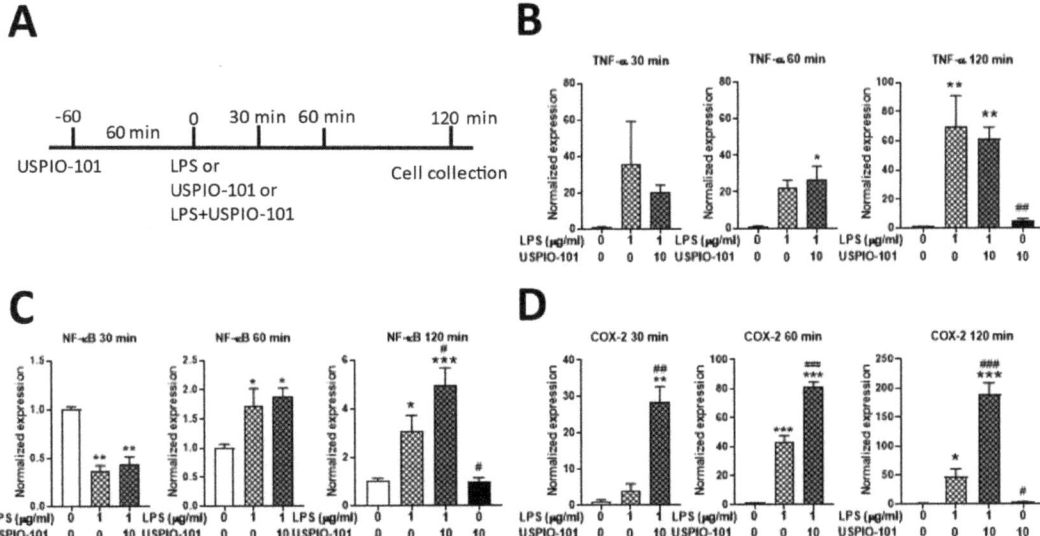

Figure 6. USPIO-101 potentiates the LPS-induced NF-kB and COX-2 mRNA expression in SM826 microglial cells. (**A**) The time scale for treating USPIO-101 and LPS stimulation. (**B–D**) The levels of mRNAs encoding the pro-inflammatory cytokines TNF-α (**B**), NF-κB (**C**), and COX-2 (**D**) were analyzed by qRT-PCR and normalized by the expression of GAPDH (glyceraldehyde-3-phosphate dehydrogenase) at three time points after LPS stimulation: 30, 60, or 120 min. Data are presented as mean ± SEM ($n = 4$). * or ** or ***: $p < 0.05$, or $p < 0.01$, or $p < 0.001$ vs. control; # or ## or ###: $p < 0.05$, or $p < 0.01$, or $p < 0.001$ vs. LPS.

3.6. USPIO-101 Impaired the Hippocampal LTP at the Schaffer Collateral-CA1 Synapses

The effect of USPIO-101 on the hippocampal LTP is still unknown. Here, we investigated whether the hippocampal LTP at the Schaffer collateral-CA1 synapses was affected by USPIO-101 in naïve mice.

The concentration effect of USPIO-101 was elucidated in the hippocampal slice, and USPIO-101 (10 or 30 μg/mL) was applied 7.5 min before TBS induction. The magnitude of

LTP was not affected at 10 µg/mL, but significantly decreased in the treatment of 30 µg/mL ($p < 0.01$, one-way ANOVA vs. control, Figure 7). These results revealed that USPIO-101 could impair hippocampal LTP at a higher concentration, and suggested the neurotoxicity possibility of USPIO-101.

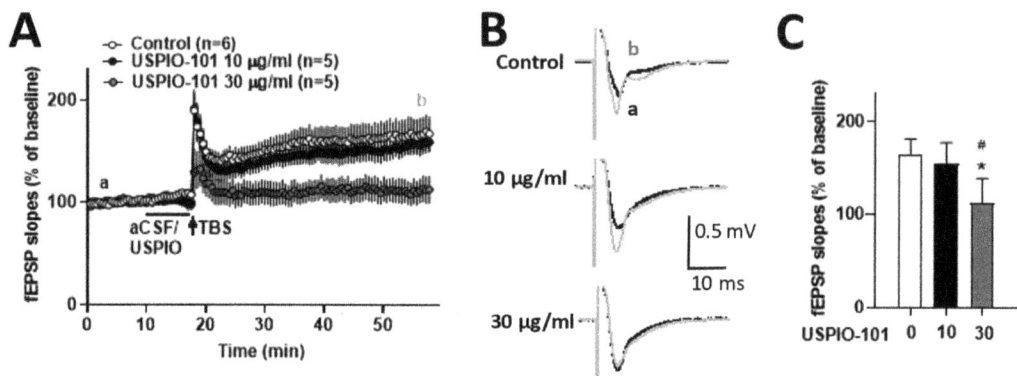

Figure 7. The effect of USPIO-101 on hippocampal LTP. (**A**) Time courses of the slope of fEPSPs recorded before and after theta-burst stimulation (TBS, arrow) in hippocampal slices. Drugs: USPIO-101 (10 or 30 µg/mL) was applied for 7.5 min before TBS stimulation. (**B**) The traces shown in the graph are the average of 20 recorded fEPSPs at times a and b in each group. (**C**) The bar graph represents the magnitudes of potentiation, with the average of 20 fEPSPs recorded 30~40 min after TBS at a concentration of 10 or 30 µg/mL. *: $p < 0.05$ vs. control group. #: $p < 0.05$ vs. USPIO-101 group (one-way ANOVA). n indicates the number of slices recorded. The expression and analysis of the baseline fEPSP slope are the same as Figure 3.

4. Discussions

Our results showed the dual effect of USPIO-101: one effect was the alleviation of inflammatory pain at the spinal cord; the other was the risk of neurotoxicity.

4.1. The Analgesic Effect of USPIO-101

Our results showed USPIO-101 (10 mg/mL, 10 µL) increased the mechanical paw withdrawal thresholds through intrathecal or intraplantar injection. Comparing intrathecal to intraplantar injection, intrathecal injection had the higher paw withdrawal thresholds at time 0.5 h, which suggested intrathecal injection was more potent than intraplantar injection at the onset time.

The in vitro electrophysiological study showed that USPIO-101 (10 or 30 µg/mL) partially decreased the potentiated spinal LTP at the spinal superficial dorsal horn synapses in mice with chronic inflammatory pain. The USPIO-101-treated spinal LTP level in CFA-treated mice was almost back to the level of saline-treated spinal LTP (Figure 3D). However, the LFS-induced potentiated spinal LTP in naïve mice could be entirely abolished by USPIO-101 (10 µg/mL) (Figure 4C). In the disease (inflammatory pain) model, the potentiated spinal LTP was more complicated than the naïve state. Even at higher concentrations, the potentiated spinal LTP would be no further decreased by USPIO-101 (30 µg/mL). The reasons why USPIO-101 only partially reduced the LFS-induced LTP in mice with chronic inflammatory pain are still unknown. However, this maintained spinal LTP in CFA-treated mice was probably why USPIO-101 only attenuated the inflammatory pain for a short duration (less than 3.5 h, Figure 2) in the behavioral tests.

The activity-dependent effect was another character of USPIO-101 revealed from our data. USPIO-101 did not affect the basal transmission but inhibited the LFS-evoked spinal LTP (Figure 4). The LFS-evoked LTP was associated with the injury or inflammatory situation at the spinal cord superficial dorsal horn synapses [24]. These results demonstrate the analgesic effect of USPIO-101 on inflammatory pain, which has been reported in our previous study [5,25] and our present study (Figure 2).

4.2. Cell Toxicity of USPIO-101

The iron oxide nanoparticles were less toxic than other ion nanoparticles, especially USPIO [26]. However, more evidence revealed that USPIO could induce cytotoxicity due to the size, shape, surface charge, or coating of the nanoparticles [27]. The physicochemical character of USPIO-101 showed that USPIO-101 probably had some aggregation with the self or other ions, but kept normal stability. Meanwhile, although USPIO-101 has proven analgesia ability, the cytotoxicity of USPIO-101 is less known.

Our data showed that USPIO-101 (10 or 30 µg/mL) did not induce the significant elevation of ROS in SH-SY5Y or SM826 cells for 30 min or 24 h, compared with the positive control (Figure 5). However, after the treatment of USPIO-101 for 24 h, the production of ROS was still high at the concentration of 10 µg/mL in a trend (SH-SY5Y or SM 826 cell: $p = 0.07$ or $p = 0.014$, control vs. USPIO-101 10 µg/mL 24 h, unpaired t-test). This suggested that USPIO-101 could induce ROS production with a chronic but not acute effect in neuron-like or microglial cells.

Our mRNA data showed that COX-2 was significantly up-regulated after LPS stimulation for 30~120 min in microglial cells (Figure 6D). COX-2 was an enzyme involved in synthesizing prostaglandins (PGs), and the induction of COX-2 enhanced nociception via increasing PG release [28]. This was controversial to our analgesic results. However, the spinal cord and brain microglia had a different response to inflammatory stimulation for unknown reasons [29]. The SM826 cell was derived from the brains of mice [30], which probably could not represent the actual situation in the spinal cord. Our data suggested that USPIO-101 could likely induce the elevation of COX-2 in brain microglial cells, but not the spinal cord.

The cytotoxicity of USPIO has been reported [9–12], but less is known about COX-2. Only one study has revealed that COX-2 is elevated after SPIO treatment in the liver [11]. We still do not know why USPIO-101 induces COX-2 so quickly during LPS stimulation in the microglial cells, because phagocytic cells, such as microglia cells, are not as sensitive to positive surface charge nanoparticles as they are to negative surface charges [31].

4.3. Hippocampal LTP Was Impaired by USPIO-101 at a Higher Concentration

We measured the effect of USPIO-101 on the hippocampal LTP to predict if USPIO-101 has neurotoxicity in the hippocampus. Our data demonstrated USPIO-101 could impair hippocampal LTP at a higher concentration (Figure 7). Another study revealed that USPIO could induce neurotoxicity in the hippocampus via in vivo study. A direct single injection of USPIO (1 µg/µL, size: 30 nm) into the mouse hippocampus for 7 or 14 days could impair the animals' spatial memory in the Morris water maze test [32]. Other controversial data showed no toxicity response to intranasally instilled Fe_3O_4 (1 mg/mL, size: 30 nm) nanoparticles in the brain [33]. One possible explanation of neurotoxicity in the hippocampus was that hippocampal neurons were more sensitive to SH-SY5Y cells when applied with exogenous iron, showing higher cell death [34].

Our data showed USPIO-101 could impair hippocampal LTP, which suggests that USPIO-101 probably has toxicity in hippocampal neurons or can antagonize some ionic receptors which are essential for early LTP induction, such as the N-methyl-D-aspartate (NMDA) receptor [35], the α-amino-3-hydroxy-5-methyl-4-isoxazolepropionic acid (AMPA) receptor [36], or other targets. More evidence is needed to elucidate this.

4.4. The Effect of Surface Group

Our used TANBead® USPIO-101 was a conventional product designed to conjugate with target-specific molecules through the amide-bond formation with carbodiimide-activated carboxylic acid groups. We investigated another carboxyl group product from the same company, TANBead® USPIO-102, on the analgesic effect of hippocampal neurotoxicity.

However, we observed that USPIO-102 had a less analgesic response and no neurotoxicity in the hippocampal slice, compared with USPIO-101 (Figures S2 and S3). The properties of USPIO-101 and USPIO-102 are almost the same, including size (6~10 nm), solvent (water), and stock concentration (10 mg/mL). The only different parts are the surface, amine group (USPIO-101), or carboxyl group (USPIO-102), as shown in Figure 1C. The surface group for Fe_3O_4 nanoparticles is critical, because naked Fe_3O_4 has a high surface energy, leading to aggregation and oxidation [37]. In addition, both amine and carboxyl groups are hydrophilic groups, which strongly attract water solubility, good biological compatibility, and biodegradability [38]. However, the physiological function of the different surfaces is still unclear.

5. Conclusions

Our results revealed the dual effect of USPIO-101: it could relieve inflammatory pain at the spinal cord, but also induce neurotoxicity in the central brain. These localized administration routes (e.g., intrathecal or intraplantar administration) of USPIO-101 did not elicit a toxicity response during our experiments; however, if the USPIO-101 leaks to the brain, it would probably impair hippocampal LTP at higher concentrations.

Supplementary Materials: The following supporting information can be downloaded at: https://www.mdpi.com/article/10.3390/pharmaceutics14020366/s1, Figure S1: the dynamic light scattering (DLS) and zeta potential analysis of USPIO-101. Figure S2: USPIO-101 and USPIO-102 attenuate chronic inflammatory pain via intrathecal or intraplantar injection. Figure S3: The effect of USPIO-101 or USPIO-102 on hippocampal LTP, respectively.

Author Contributions: Conceptualization, Y.-C.L. (Yen-Chin Liu); data collection, G.-L.L., P.-C.W. and Y.-C.L. (Ya-Chi Lin); data analysis, G.-L.L. and P.-C.W.; manuscript writing G.-L.L., P.-C.W. and Y.-C.L. (Yen-Chin Liu). All authors have read and agreed to the published version of the manuscript.

Funding: The authors thank the support provided by the Ministry of Science and Technology in Taiwan under MOST 108-2314-B-037-112-MY3, 110-2221-E-006-013-MY3, and Academia Sinica Healthy Longevity Grand Challenge Competition (AS-HLGC-110-07). This work was financially supported by the Center of Applied Nanomedicine, the Medical Device Innovation Center (MDIC) National Cheng Kung University, Taiwan.

Institutional Review Board Statement: Not applicable.

Informed Consent Statement: Not applicable.

Data Availability Statement: Not applicable.

Acknowledgments: Gratefully appreciate Hwei-Hsien Chen's lab (National Health Research Institutes) for supporting the electrophysiological and RNA analysis studies. The used cell lines in this study were kindly provided by Chia-Hsiang, Chen lab (Chang Gung University, human neuroblastoma SH-SY5Y cells) and Feng-Shiun, Shie lab (National Health Research Institutes, mouse SM826 microglia cells). Ren-Hau Tsai kindly served the DLS measurement.

Conflicts of Interest: The authors declare no conflict of interest.

References

1. Alphandery, E. Iron oxide nanoparticles for therapeutic applications. *Drug Discov. Today* **2020**, *25*, 141–149. [CrossRef] [PubMed]
2. Dadfar, S.M.; Roemhild, K.; Drude, N.I.; von Stillfried, S.; Knuchel, R.; Kiessling, F.; Lammers, T. Iron oxide nanoparticles: Diagnostic, therapeutic and theranostic applications. *Adv. Drug Deliv. Rev.* **2019**, *138*, 302–325. [CrossRef] [PubMed]
3. Andreu, V.; Arruebo, M. Current progress and challenges of nanoparticle-based therapeutics in pain management. *J. Control. Release Off. J. Control. Release Soc.* **2018**, *269*, 189–213. [CrossRef] [PubMed]
4. Chen, J.; Jin, T.; Zhang, H. Nanotechnology in Chronic Pain Relief. *Front. Bioeng. Biotechnol.* **2020**, *8*, 682. [CrossRef]
5. Wu, P.C.; Hsiao, H.T.; Lin, Y.C.; Shieh, D.B.; Liu, Y.C. The analgesia efficiency of ultrasmall magnetic iron oxide nanoparticles in mice chronic inflammatory pain model. *Nanomed. Nanotechnol. Biol. Med.* **2017**, *13*, 1975–1981. [CrossRef]
6. Woolf, C.J. Evidence for a central component of post-injury pain hypersensitivity. *Nature* **1983**, *306*, 686–688. [CrossRef]
7. Ruscheweyh, R.; Wilder-Smith, O.; Drdla, R.; Liu, X.G.; Sandkuhler, J. Long-term potentiation in spinal nociceptive pathways as a novel target for pain therapy. *Mol. Pain* **2011**, *7*, 20. [CrossRef]

8. Patil, R.M.; Thorat, N.D.; Shete, P.B.; Bedge, P.A.; Gavde, S.; Joshi, M.G.; Tofail, S.A.M.; Bohara, R.A. Comprehensive cytotoxicity studies of superparamagnetic iron oxide nanoparticles. *Biochem. Biophys. Rep.* **2018**, *13*, 63–72. [CrossRef]
9. Nemmar, A.; Beegam, S.; Yuvaraju, P.; Yasin, J.; Tariq, S.; Attoub, S.; Ali, B.H. Ultrasmall superparamagnetic iron oxide nanoparticles acutely promote thrombosis and cardiac oxidative stress and DNA damage in mice. *Part. Fibre Toxicol.* **2016**, *13*, 22. [CrossRef]
10. He, C.; Jiang, S.; Yao, H.; Zhang, L.; Yang, C.; Zhan, D.; Lin, G.; Zeng, Y.; Xia, Y.; Lin, Z.; et al. Endoplasmic reticulum stress mediates inflammatory response triggered by ultra-small superparamagnetic iron oxide nanoparticles in hepatocytes. *Nanotoxicology* **2018**, *12*, 1198–1214. [CrossRef]
11. Che, L.; Yao, H.; Yang, C.L.; Guo, N.J.; Huang, J.; Wu, Z.L.; Zhang, L.Y.; Chen, Y.Y.; Liu, G.; Lin, Z.N.; et al. Cyclooxygenase-2 modulates ER-mitochondria crosstalk to mediate superparamagnetic iron oxide nanoparticles induced hepatotoxicity: An in vitro and in vivo study. *Nanotoxicology* **2020**, *14*, 162–180. [CrossRef]
12. Vermeij, E.A.; Koenders, M.I.; Bennink, M.B.; Crowe, L.A.; Maurizi, L.; Vallee, J.P.; Hofmann, H.; van den Berg, W.B.; van Lent, P.L.; van de Loo, F.A. The in-vivo use of superparamagnetic iron oxide nanoparticles to detect inflammation elicits a cytokine response but does not aggravate experimental arthritis. *PLoS ONE* **2015**, *10*, e0126687. [CrossRef] [PubMed]
13. Ji, R.R.; Kohno, T.; Moore, K.A.; Woolf, C.J. Central sensitization and LTP: Do pain and memory share similar mechanisms? *Trends Neurosci.* **2003**, *26*, 696–705. [CrossRef]
14. Bliss, T.V.; Lomo, T. Long-lasting potentiation of synaptic transmission in the dentate area of the anaesthetized rabbit following stimulation of the perforant path. *J. Physiol.* **1973**, *232*, 331–356. [CrossRef] [PubMed]
15. Deuis, J.R.; Dvorakova, L.S.; Vetter, I. Methods Used to Evaluate Pain Behaviors in Rodents. *Front. Mol. Neurosci.* **2017**, *10*, 284. [CrossRef]
16. Chaplan, S.R.; Bach, F.W.; Pogrel, J.W.; Chung, J.M.; Yaksh, T.L. Quantitative assessment of tactile allodynia in the rat paw. *J. Neurosci. Methods* **1994**, *53*, 55–63. [CrossRef]
17. Zhu, M.; Zhang, D.; Peng, S.; Liu, N.; Wu, J.; Kuang, H.; Liu, T. Preparation of Acute Spinal Cord Slices for Whole-cell Patch-clamp Recording in Substantia Gelatinosa Neurons. *J. Vis. Exp. JoVE* **2019**. [CrossRef] [PubMed]
18. Lai, C.Y.; Ho, Y.C.; Hsieh, M.C.; Wang, H.H.; Cheng, J.K.; Chau, Y.P.; Peng, H.Y. Spinal Fbxo3-Dependent Fbxl2 Ubiquitination of Active Zone Protein RIM1alpha Mediates Neuropathic Allodynia through CaV2.2 Activation. *J. Neurosci. Off. J. Soc. Neurosci.* **2016**, *36*, 9722–9738. [CrossRef] [PubMed]
19. Terman, G.W.; Eastman, C.L.; Chavkin, C. Mu opiates inhibit long-term potentiation induction in the spinal cord slice. *J. Neurophysiol.* **2001**, *85*, 485–494. [CrossRef]
20. Lu, G.L.; Lee, C.H.; Chiou, L.C. Orexin A induces bidirectional modulation of synaptic plasticity: Inhibiting long-term potentiation and preventing depotentiation. *Neuropharmacology* **2016**, *107*, 168–180. [CrossRef]
21. Sandkuhler, J.; Gruber-Schoffnegger, D. Hyperalgesia by synaptic long-term potentiation (LTP): An update. *Curr. Opin. Pharmacol.* **2012**, *12*, 18–27. [CrossRef] [PubMed]
22. Abdal Dayem, A.; Hossain, M.K.; Lee, S.B.; Kim, K.; Saha, S.K.; Yang, G.M.; Choi, H.Y.; Cho, S.G. The Role of Reactive Oxygen Species (ROS) in the Biological Activities of Metallic Nanoparticles. *Int. J. Mol. Sci.* **2017**, *18*, 120. [CrossRef] [PubMed]
23. Ge, D.; Du, Q.; Ran, B.; Liu, X.; Wang, X.; Ma, X.; Cheng, F.; Sun, B. The neurotoxicity induced by engineered nanomaterials. *Int. J. Nanomed.* **2019**, *14*, 4167–4186. [CrossRef]
24. Ikeda, H.; Stark, J.; Fischer, H.; Wagner, M.; Drdla, R.; Jager, T.; Sandkuhler, J. Synaptic amplifier of inflammatory pain in the spinal dorsal horn. *Science* **2006**, *312*, 1659–1662. [CrossRef]
25. Wu, P.C.; Shieh, D.B.; Hsiao, H.T.; Wang, J.C.; Lin, Y.C.; Liu, Y.C. Magnetic field distribution modulation of intrathecal delivered ketorolac iron-oxide nanoparticle conjugates produce excellent analgesia for chronic inflammatory pain. *J. Nanobiotechnology* **2018**, *16*, 49. [CrossRef] [PubMed]
26. Singh, N.; Jenkins, G.J.; Asadi, R.; Doak, S.H. Potential toxicity of superparamagnetic iron oxide nanoparticles (SPION). *Nano Rev.* **2010**, *1*, 5358. [CrossRef] [PubMed]
27. Valdiglesias, V.; Fernandez-Bertolez, N.; Kilic, G.; Costa, C.; Costa, S.; Fraga, S.; Bessa, M.J.; Pasaro, E.; Teixeira, J.P.; Laffon, B. Are iron oxide nanoparticles safe? Current knowledge and future perspectives. *J. Trace Elem. Med. Biol. Organ Soc. Miner. Trace Elem.* **2016**, *38*, 53–63. [CrossRef]
28. Vanegas, H.; Schaible, H.G. Prostaglandins and cyclooxygenases [correction of cyclogenases] in the spinal cord. *Prog. Neurobiol.* **2001**, *64*, 327–363. [CrossRef]
29. Xuan, F.L.; Chithanathan, K.; Lillevali, K.; Yuan, X.; Tian, L. Differences of Microglia in the Brain and the Spinal Cord. *Front. Cell. Neurosci.* **2019**, *13*, 504. [CrossRef]
30. Lee, Y.H.; Lin, C.H.; Hsu, P.C.; Sun, Y.Y.; Huang, Y.J.; Zhuo, J.H.; Wang, C.Y.; Gan, Y.L.; Hung, C.C.; Kuan, C.Y.; et al. Aryl hydrocarbon receptor mediates both proinflammatory and anti-inflammatory effects in lipopolysaccharide-activated microglia. *Glia* **2015**, *63*, 1138–1154. [CrossRef]
31. Frohlich, E. The role of surface charge in cellular uptake and cytotoxicity of medical nanoparticles. *Int. J. Nanomed.* **2012**, *7*, 5577–5591. [CrossRef] [PubMed]
32. Liu, Y.; Li, J.; Xu, K.; Gu, J.; Huang, L.; Zhang, L.; Liu, N.; Kong, J.; Xing, M.; Zhang, L.; et al. Characterization of superparamagnetic iron oxide nanoparticle-induced apoptosis in PC12 cells and mouse hippocampus and striatum. *Toxicol. Lett.* **2018**, *292*, 151–161. [CrossRef] [PubMed]

33. Wu, J.; Ding, T.; Sun, J. Neurotoxic potential of iron oxide nanoparticles in the rat brain striatum and hippocampus. *Neurotoxicology* **2013**, *34*, 243–253. [CrossRef] [PubMed]
34. Aguirre, P.; Mena, N.; Tapia, V.; Arredondo, M.; Nunez, M.T. Iron homeostasis in neuronal cells: A role for IREG1. *BMC Neurosci.* **2005**, *6*, 3. [CrossRef]
35. Incontro, S.; Diaz-Alonso, J.; Iafrati, J.; Vieira, M.; Asensio, C.S.; Sohal, V.S.; Roche, K.W.; Bender, K.J.; Nicoll, R.A. The CaMKII/NMDA receptor complex controls hippocampal synaptic transmission by kinase-dependent and independent mechanisms. *Nat. Commun.* **2018**, *9*, 2069. [CrossRef]
36. Penn, A.C.; Zhang, C.L.; Georges, F.; Royer, L.; Breillat, C.; Hosy, E.; Petersen, J.D.; Humeau, Y.; Choquet, D. Hippocampal LTP and contextual learning require surface diffusion of AMPA receptors. *Nature* **2017**, *549*, 384–388. [CrossRef]
37. Shen, L.; Li, B.; Qiao, Y. Fe(3)O(4) Nanoparticles in Targeted Drug/Gene Delivery Systems. *Materials* **2018**, *11*, 324. [CrossRef]
38. Wu, W.; He, Q.; Jiang, C. Magnetic iron oxide nanoparticles: Synthesis and surface functionalization strategies. *Nanoscale Res. Lett.* **2008**, *3*, 397–415. [CrossRef]

Article

Magnetoresponsive Functionalized Nanocomposite Aggregation Kinetics and Chain Formation at the Targeted Site during Magnetic Targeting

Sandor I. Bernad [1,2,*], Vlad Socoliuc [1,*], Daniela Susan-Resiga [1,3], Izabell Crăciunescu [4], Rodica Turcu [4], Etelka Tombácz [5], Ladislau Vékás [1], Maria C. Ioncica [2] and Elena S. Bernad [6]

1. Romanian Academy-Timisoara Branch, Centre for Fundamental and Advanced Technical Research, Mihai Viteazul Str. 24, 300223 Timisoara, Romania
2. Research Center for Engineering of Systems with Complex Fluids, Politehnica University Timisoara, Mihai Viteazul Str. 1, 300222 Timisoara, Romania
3. Faculty of Physics, West University of Timisoara, Vasile Parvan Str. 1, 300222 Timisoara, Romania
4. National Institute for Research and Development of Isotopic and Molecular Technologies (INCDTIM), Donat Str. 67-103, 400293 Cluj-Napoca, Romania
5. Soós Ernő Water Technology Research and Development Center, University of Pannonia, Zrínyi M. Str. 18, 8800 Nagykanizsa, Hungary
6. Department of Obstetrics and Gynecology, University of Medicine and Pharmacy "Victor Babes" Timisoara, P-ta Eftimie Murgu 2, 300041 Timisoara, Romania
* Correspondence: sandor.bernad@upt.ro (S.I.B.); vsocoliuc@gmail.com (V.S.)

Abstract: Drug therapy for vascular disease has been promoted to inhibit angiogenesis in atherosclerotic plaques and prevent restenosis following surgical intervention. This paper investigates the arterial depositions and distribution of PEG-functionalized magnetic nanocomposite clusters (PEG_MNCs) following local delivery in a stented artery model in a uniform magnetic field produced by a regionally positioned external permanent magnet; also, the PEG_MNCs aggregation or chain formation in and around the implanted stent. The central concept is to employ one external permanent magnet system, which produces enough magnetic field to magnetize and guide the magnetic nanoclusters in the stented artery region. At room temperature (25 °C), optical microscopy of the suspension model's aggregation process was carried out in the external magnetic field. According to the optical microscopy pictures, the PEG_MNC particles form long linear aggregates due to dipolar magnetic interactions when there is an external magnetic field. During magnetic particle targeting, 20 mL of the model suspensions are injected (at a constant flow rate of 39.6 mL/min for the period of 30 s) by the syringe pump in the mean flow (flow velocity is Um = 0.25 m/s, corresponding to the Reynolds number of Re = 232) into the stented artery model. The PEG_MNC clusters are attracted by the magnetic forces (generated by the permanent external magnet) and captured around the stent struts and the bottom artery wall before and inside the implanted stent. The colloidal interaction among the MNC clusters was investigated by calculating the electrostatic repulsion, van der Waals and magnetic dipole-dipole energies. The current work offers essential details about PEG_MNCs aggregation and chain structure development in the presence of an external magnetic field and the process underlying this structure formation.

Keywords: magnetoresponsive nanocomposite; functional coating; particle targeting; particle aggregation; stent targeting; nanomedicine

1. Introduction

Although PCI (percutaneous coronary intervention) is the most widely used treatment for those with atherosclerosis [1,2], it is invasive and has a high risk of restenosis—up to 16 percent for DES (drug-eluting stent) and 16–44 percent for BMS (bare metal stent) [2,3]. Furthermore, patients with advanced atherosclerosis commonly have complex plaques, which are more prone to rupture and result in coronary thrombosis, myocardial infarction, or other potentially deadly clinical events if not caught in time. It has been proven that nanoparticles are more beneficial for diagnosing atherosclerosis when targeted ligands are added [3,4].

Drug therapy for vascular disease has been promoted to inhibit angiogenesis in atherosclerotic plaques and prevent restenosis following surgical intervention [5–8]. However, when used systemically, anti-angiogenic or cytostatic medications frequently display unfavourable side effects. To enable the local application of the compounds, the chemicals were linked to balloons or stents [9,10] used in percutaneous transluminal angioplasty. This technique, in particular the use of drug-eluting stents laden with anti-proliferative agents, has resulted in restenosis rates below 10% [11,12].

To lower the risks of stent-related vascular injury or end-organ damage, our team has developed a technique for endothelial cell repair with direct targeting of the vascular wall under flow conditions.

Magnetic iron oxide nanoparticles and their nanocomposites, created with various functional coatings, are the most promising magnetic carriers for use in medicine [13–17], in particular in magnetically controlled tissue engineering [18] and cardiac regenerative medicine [19]. This is because essential requirements are met by them, including simple and non-toxic cellular uptake, superparamagnetic behaviour, the significant field-induced magnetic moments of multicore particles, good response to moderate magnetic fields, inherent ability to cross biological barriers, protection of the drug from rapid degradation in the biological environment, and a sizable surface area for conjugating targeting ligands. These features, together with characteristic sizes up to approximately 100 nm of the magnetic core of functionalized magnetic nanoparticle clusters dispersed in a biocompatible (usually aqueous) carrier, distinguish a category of ferrofluids as bio-ferrofluids [20,21]. These external stimuli nanosystems synthesized and experimented with in this paper were typical bio-ferrofluids and were designed to exploit the advantages offered by the high magnetic moment multicore particles (magnetic carriers) towards remotely controlled therapeutic performances in a stented artery.

Polyethylene glycols (PEGs) with the built-in EPR (enhanced permeability and retention) effect (increased permeability and retention) protected nanocomposite particle against unwanted protein corona formation, allowing them to enter cells and be employed in medicine. Site-specific drug delivery is the main challenge when delivering drug-carrying particles into the bloodstream [14,15].

Since it is stable, biocompatible, and hydrophilic, the polymer polyethylene glycol (PEG) has been the subject of extensive research for its potential in drug administration [22–24]. Furthermore, PEG reduces the immune system's response to nanoparticles [25–27]. As a result, fewer nanoparticles are removed from the circulation, enhancing the concentration of nanoparticles at the target location, and lengthening the circulation duration [28–30].

Magnetic carriers can only partially reach their target areas before being recognized and expelled from the body by the mononuclear phagocytic system (MPS2) because of their propensity for aggregation and the presence of magnetic dipole-dipole interactions or van der Waals forces [31].

When an artery's stents are targeted with magnetic carriers and then coated with an antirestenotic substance, in-stent restenosis is inhibited at drug dosages far lower than DESs. The following goals guided the present research:

1. To investigate the arterial depositions and distribution of PEG_MNCs following local delivery in a stented artery model in a uniform magnetic field produced by a regionally positioned permanent magnet.
2. To create a novel concept using permanent magnet systems to guide and target the functionalized nanocomposite around the stent to test the approach's effectiveness for treating coronary heart disease (CHD).
3. PEG_MNCs aggregation or chain formation in and around the implanted stent (in the presence of the external magnetic field).

2. Materials and Methods

The competition between the magnetic force and the drag force exerted by the moving fluid, as well as the magnetic field gradient produced by the permanent external magnet and the ferromagnetic stent magnetic field is what causes the capture and deposition of the injected magnetic clusters in the flow stream (Figure 1).

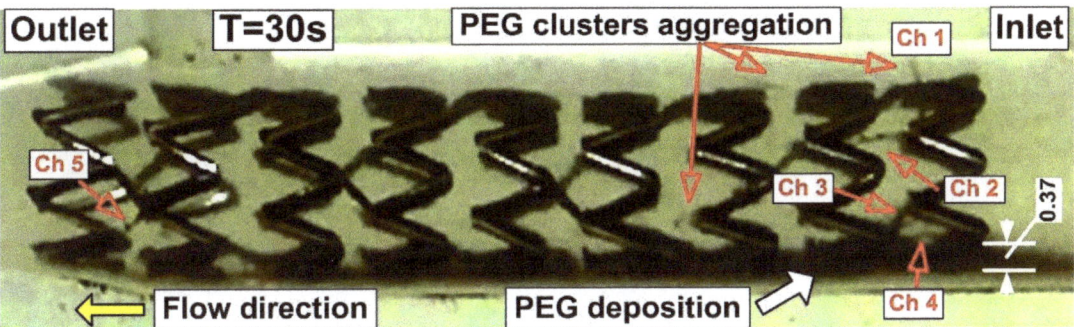

Figure 1. Flow-oriented large-scale magnetically induced aggregation of the PEG_MNC in the targeted region (red arrows) at the end of the injection period of 30 s. Magnetic cluster depositions correspond to the permanent magnet positions of 15 mm from the stent's bottom wall. Ch 1 ÷ Ch 5–chain-like magnetic particles structure, generated in a different part of the stent geometry in the presence of the external magnetic field. Particle depositions on the bottom wall of the artery model in the stent inlet section (0.37 mm) at the end of the injection period of 30 s.

We calculated that the employed permanent magnet provided a consistent magnetic field of about 0.18 T and 0.08 T at the different locations from the magnet surface (10 mm and 20 mm), respectively, using magnetic field measurements (according to Figures 2 and 3). Therefore, the magnetic force could be produced because the magnetic field stayed constant.

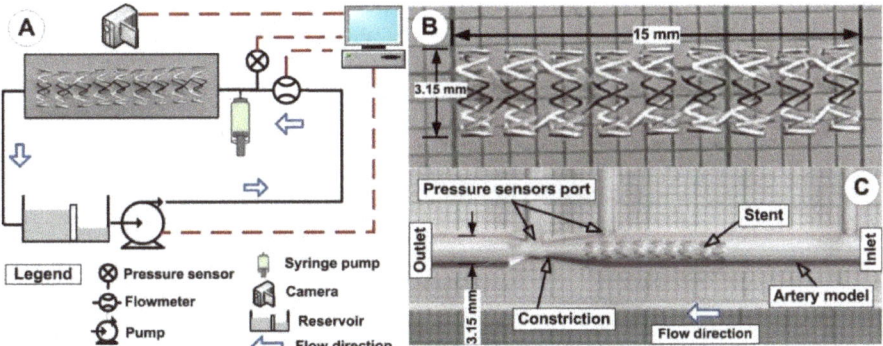

Figure 2. Implanted stent, particle targeting experimental setup. (**A**) The flowmeter, pressure sensors, suspension injection mechanism with a syringe pump, test section with magnetic stent model, reservoir, centrifugal pump, camera, and PC were all shown in the block diagram of the main recirculating flow loop. A suspension is injected using a syringe pump before the stent model enters part. (**B**) A general perspective of the stent geometry used and the distinctive dimension of the expanded stent. (**C**) Aerial view of the model artery showing the position of the stent.

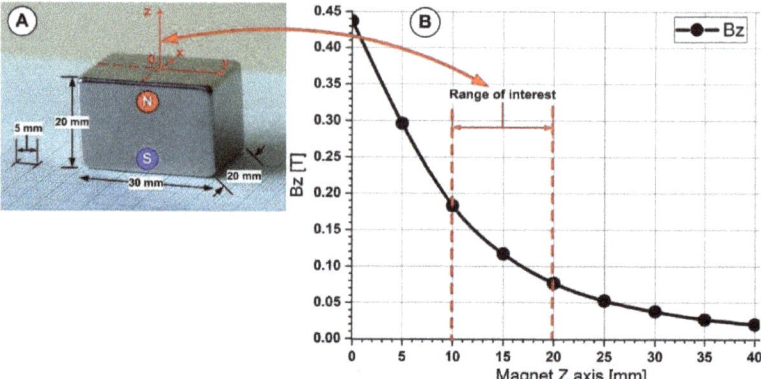

Figure 3. A magnetic field was generated by the NbFeB50 permanent magnet used in the experimental investigation. (**A**). The magnet's dimension and axis association. A used permanent magnet has polarization along the Z-axis. (**B**) The figure depicts the magnetic field measured at various Z-axis positions with the F.W. Bell Gaussmeter, model 5080.

During magnetic particle targeting, 20 mL of the model suspensions are injected (at a constant flow rate of 39.6 mL/min for the period of 30 s) by the syringe pump in the mean flow (flow velocity is Um = 0.25 m/s, corresponding to the Reynolds number of Re = 232) into the stented artery model. The PEG_MNC clusters are attracted by the magnetic forces (generated by the permanent external magnet) and captured around the stent struts and the bottom artery wall before and inside the implanted stent (Figure 1). The evolution of the particle build-up is investigated using an image analysis technique.

All experimental measures are conducted in an air-conditioned chamber. The fluid is assumed not to be recirculated and to pass through the test section just once in the direction of flow when measurements are taken in an open circuit. This prevents the working fluid from heating up or changing its rheological properties throughout the measurements, with an injection time of 30 s. Furthermore, studies were conducted with a DC magnetic field. Direct current (DC) magnetic fields, in contrast to high-frequency AC (alternating current) magnetic fields, do not transfer thermal energy to magnetic nanoparticles or clusters injected in the test section.

2.1. Experimental Setup

The test portions are made of acrylic glass and are precisely shaped, with an interior diameter of 3.15 mm (Figure 2C). Our earlier publication [32] provided a thorough explanation of the stent targeting methodology, and our previous work [33] provided a complete description of the setup's overall concept (Figure 2A). All flow tests were carried out in typical physiologic settings [34]. The main benefits of the acrylic glass model were its excellent transparency and ability to examine and evaluate the hemodynamic properties and the particle targeting process in the arterial stent segment.

The experimental setup demonstrated the removal of PEG_MNC from the flow flux and the deposition of the nanocomposite around the implanted stent's geometry.

As previously mentioned in our articles [30] and articles [35,36], the implanted stent (Figure 2B) was made from magnetic 2205 duplex stainless steel (2205 SS).

It is significant to note that the Resolute Integrity Zotarolimus-Eluting Coronary Stent System (Minneapolis, MN, USA) and the stent used in the experimental setup have nearly identical geometrical characteristics (internal diameter of 3.15 mm, length of 15 mm, and strut thickness of 0.09 mm) (inner diameter of 3 mm, length of 15 mm, strut thickness of 0.09 mm).

2.2. Magnetic Field Generation

The central concept is to employ one external permanent magnet system, which produces enough magnetic field to magnetize and guide the MNP's in the stented artery region, based on our prior findings [30,32,33].

We employed a Neodymium 50 type magnet (NdFeB50) with a maximum energy product (BxH) of 50 MGOe to create the magnetic field. According to our earlier research [30], the magnetic field produced by this neodymium magnet ranges from 0.44 T to 0.02 T or at a magnet distance of 0 mm (from the magnet surface) to 40 mm. We investigated the magnetic particle targeting in our studies for magnetic fields generated between 0.18 T and 0.08 T corresponding to the magnet position to the implanted stent between 10 mm and 20 mm (along with the magnet z axis—Figure 3).

2.3. Blood Analogue Fluid, Preparation, and Rheological Properties

In our studies, the carrier fluid (CF) was glycerol-water solutions, which were made by combining estimated weights of glycerol and distilled water and having a density (1060 kg/m^3) that is identical to that of blood [30]. This CF made it easier for the experimental investigations to accurately reproduce the rheological behaviour of the fluid flow at the spot of the implanted stent. The PEG_MNC dispersed in distilled water, with a 0.1 percent mass concentration, was combined with a carrier fluid (CF) to create the model suspension of magnetic carriers utilized in tests.

The agreement between the model suspension's viscosity curve at T = 25 °C, the blood sample (obtained from a 38-year-old female volunteer in good health), and the results presented in the literature [37] is shown in Figure 4. During these investigations, we did not use any blood samples. Instead, we used one example from the existing databases in our institution. The data from the databases were used in a comparative sense with data from the literature and prepared blood mimicking fluid.

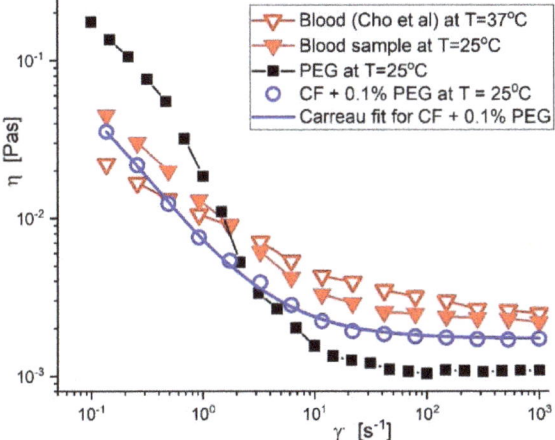

Figure 4. Viscosity curves for blood (values from literature and healthy volunteer), PEG—coated functionalized nanocomposite (PEG_MNC), and model suspension fluid (carrier fluid + 0.1% PEG_MNC).

The Carreau model with four pertinent parameters can be used to define the rheological values for the model suspensions, as shown in Table 1. (Equation (1)).

$$\eta(\dot{\gamma}) = \eta_\infty + (\eta_0 - \eta_\infty)\left[1 + (C\dot{\gamma})^2\right]^{-p} \quad (1)$$

in which C [s] represents the Carreau constant (the value of the slope of the viscosity curve in the log-log scale at high values of the shear rate $\dot{\gamma}$, p [-] is the Carreau exponent, η_0 is the viscosity at infinitely low shear rates, and η_∞ [Pas]- viscosity at infinitely high shear rates.

Table 1. Rheological characterization of the model suspensions.

Fluid	T [°C]	B [mT]	η_∞ [Pas]	η_0 [Pas]	C [s]	p [-]	r^2
CF + 0.1% PEG_MC	25	0	0.00154	0.095	18.08	0.664	0.999

Where: T [°C] is the carrier fluid temperature, B [T] is magnetic field intensity, η_∞ [Pas] is the viscosity at infinite shear rates, η_0 [Pas] is zero shear viscosity, C [s] is the characteristic time constant, and p [-] flows behaviour index. The r^2 values for all fits are close to unity, indicating an excellent fit (r^2 is the coefficient of determination used to evaluate the quality of the Carreau fits).

3. Results

3.1. Synthesis and Characterization of the PEG-Functionalized Magnetic Nanocomposite Clusters (PEG_MNC)

Our earlier work [22,23,30] goes to great length about the synthesis of PEG-coated magnetite nanoparticles, the magnetic and colloidal characteristics of the core-shell, their biocompatibility, and their potential for use in biomedicine. Briefly, we prepared the PEG-coated MNCs by the oil-in-water miniemulsion. As presented in our previous work [30], PEG (1.795 g), which functions as a surfactant, was added to an aqueous solution with a toluene-based ferrofluid (0.5 wt% Fe3O4). PEG molecules arranged themselves with the polar end in the water phase and the nonpolar end in the oil phase, forming micelles when they were present. The newly formed droplets included toluene-dispersed magnetic nanoparticles. First, the two-phase mixture was homogenized using an ultrasonic finger U.P. 400S for 2 min to produce a stable miniemulsion. The second step involved evaporating the toluene organic phase in an oil bath at 100 °C while being magnetically stirred at 500 rpm. After being cleaned of any excess reactants with the methanol-water solution (50 mL), the magnetic clusters were redispersed in distilled water. The PEG-coated MNCs were dispersed in an aqueous carrier (blood analogue fluid). Analog was used as a carrier fluid (CF) for blood and was made from glycerol-water solutions by combining distilled water and glycerol at predetermined weights.

A Hitachi HD2700 microscope was used for the electron microscopy (TEM) investigation into the morphology of the magnetic clusters. The magnetic nanoparticles highly packed into spherical clusters (Figure 5) with diameters of 40 to 120 nm are depicted in the representative TEM images of the PEG_MNC in Figure 1. The average diameter of the PEG-coated clusters (size range 40–120 nm), according to TEM image analysis of micrographs is 62 ± 17 nm (Figure 6).

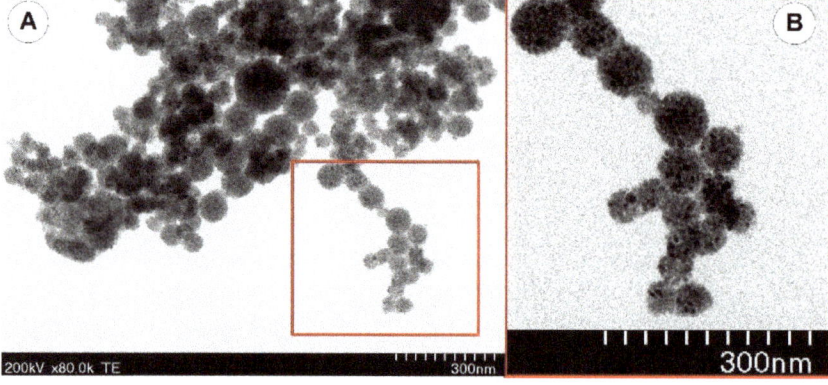

Figure 5. TEM image of the magnetic nanoclusters coated with PEG (**A**). Detail of the spherical clusters (**B**).

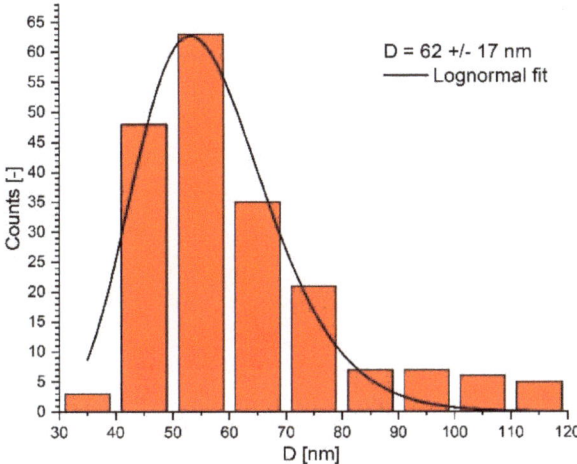

Figure 6. TEM size distribution and lognormal fit of the PEG−coated. The TEM size distribution was obtained by analysis of the TEM micrograph shown in Figure 1.

The PEG_MNC diameter distribution was obtained using ImageJ [https://imagej.nih.gov/ij/ (accessed on 3 August 2022)]. A number of 195 PEG_MNCs from three TEM images were measured. Figure 6 is presented the magnetic nanoparticles' diameter distribution. The diameter distribution shows positive skewness, with a 62 nm mean diameter. Due to the clustering during the TEM sample preparation evaporation process, the question remains whether the MNCs are soldered within the clusters or not.

A vibrating sample magnetometer (VSM 880-ADE Technologies, Massachusetts, MA, USA) with a field range of 0 kA/m to 1000 kA/m was used to measure the magnetic characteristics of the PEG-coated clusters. Figure 7 shows the PEG_MNCs' (A) and its aqueous dispersion (B) magnetization curves at room temperature.

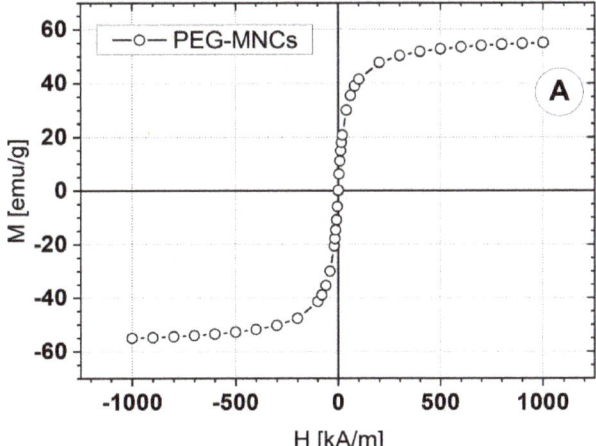

Figure 7. *Cont.*

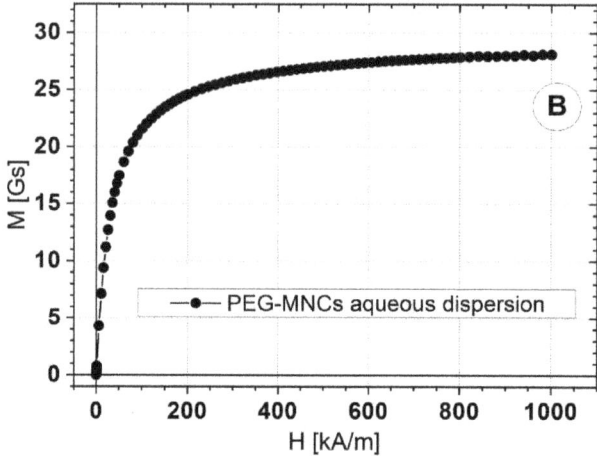

Figure 7. Magnetization curves for (**A**) a sample of dry PEG-MNCs, and (**B**) PEG_MNCs aqueous dispersion at room temperature (25 °C).

Both samples show superparamagnetic behaviour. The PEG − MNCs have zero remanent magnetization and 55 emu/g saturation magnetization. The PEG − MNC aqueous dispersion has 28 Gs saturation magnetization.

3.2. Rheological Properties of the PEG_MNC Aqueous Dispersion

An *Anton Paar* Physica MCR 300, Graz, Austria rheometer was used. The magnetorheological cell (plate-plate geometry PP20/MRD/TI-SN18581), has a diameter of 2R = 20 mm, and a gap fixed at h = 0.2 mm, was used for the rheological measurements in this article. The sample layer between the plates is subjected to a perpendicular magnetic field in this cell. The magnetic flux density of the applied magnetic field is determined using a Hall probe inserted under the bottom plate of the MR cell.

Numerous cardiovascular illnesses have mechanical blood artery wall behaviour and blood flow characteristics contributing to their onset and progression [31]. Blood exhibits non-Newtonian fluid behaviour in the small/capillary arteries while acting similarly to a Newtonian fluid in the major arteries [38]. Massive viscosity fluctuations in these blood arteries indicate the pseudoplastic nature of flow at low shear rates ($\dot{\gamma} < 100 \text{s}^{-1}$) [39].

Viscosity Curves of PEG_MNC Aqueous Dispersions

Viscosity curves at different values (B = 0, 42, 52, 117, 183 mT) of applied magnetic induction were measured in the range 0.1–1000s^{-1} of shear rates, at temperature value, T = 25 °C (Figure 8).

A shear-thinning behaviour of the PEG_MNC suspension is observed, both in the absence and in the presence of the magnetic field. The shear −thinning behaviour character is accentuated in the magnetic field's presence when the magnetic interactions' intensification induces agglomerations of magnetite clusters. As the shear rate increases, these cluster agglomerations are progressively destroyed, and as a result, the suspension's viscosity decreases.

The relative increase in viscosity induced by the magnetic field (magneto viscous effect-MV) is already significant when applying weak fields, and this phenomenon tends to saturate as the magnetic field intensifies [40,41].

The $\eta = f(\dot{\gamma})$ data were correlated with the Carreau model [42], Equation (1).

The values obtained for the fit parameters are listed in Table 2. As expected, in the presence of the magnetic field, the values of the fit parameters η_0, C, and p increase with the intensification of the field.

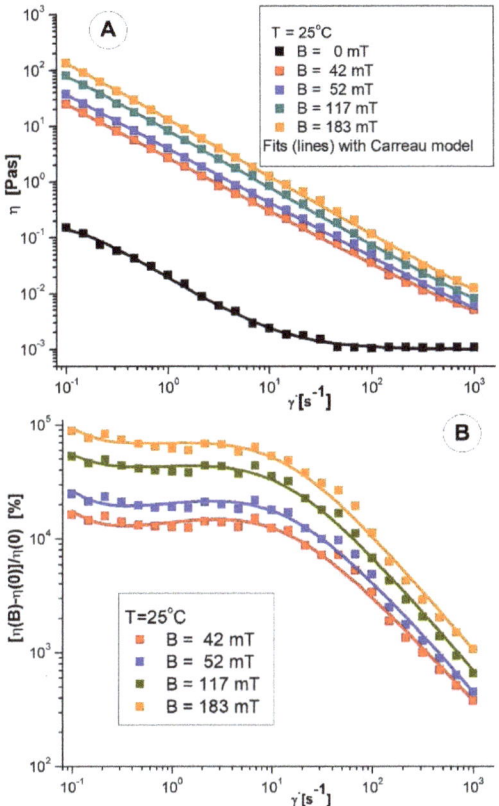

Figure 8. (**A**) The model suspension viscosity curves at T = 25 °C in the presence and absence of the magnetic field are shown; (**B**) MV effect as a function of shear rate at different values of the magnetic flux density, at T = 25 °C.

Table 2. The values of the fit parameters were obtained by fitting the viscosity curves with the Carreau model for different magnetic flux density values.

T [°C]	B [mT]	η_∞ [Pas]	η_0 [Pas]	C [s]	p [-]	r^2
	0	0.001	0.182	7.645	0.559	0.986
	42	0.001	86.6	12.20	0.478	0.996
25 °C	52	0.001	98.00	12.31	0.486	0.994
	117	0.001	155.35	17.00	0.513	0.996
	183	0.001	322.20	21.26	0.517	0.991

Where: T [°C] is the carrier fluid temperature, B [T] is magnetic flux density, η_∞ [Pas] is the viscosity at infinite shear rates, η_0 [Pas] is zero shear viscosity, C [s] is the characteristic time constant, and p [-] flows behaviour index. The r^2 values for all fits are close to unity, indicating an excellent fit (r^2 is the coefficient of determination used to evaluate the quality of the Carreau fits).

The influence of shear rate on the MV effect at different magnetic flux density values at T = 25 °C is represented in Figure 8B. It is observed that the effect of MV on the range of low shear rates ($\dot{\gamma} = (0.1 - 10)\ \text{s}^{-1}$)) is almost independent of the shear rate but significantly decreases at high speeds as cluster agglomerations are destroyed. The MNC agglomerates are destroyed for higher shear rates, and the MV effect decreases. The observed magnetoviscous behaviour is a consequence of the multicore nature of the magnetic component resulting in a high induced magnetic moment of particles, favouring their structuring in a magnetic field. The PEG-coated magnetic nanoclusters of 62 nm mean size suspended in a glycerol-water carrier (mass concentration 50 mg/mL) show a very high MV effect of about 10^4%. Commercial bio-ferrofluids with multi-core magnetic particles of hydrodynamic diameters of 50, 100 and 200 nm were investigated in [40]. They manifested a significant, up to 1500x increase, in the effective viscosity in the magnetic field already at a significantly lower mass concentration of multicore magnetic particles of about 25 mg/mL. At the same time, in the case of highly concentrated aqueous ferrofluids with predominantly single core (less than 10 nm) magnetite nanoparticles, the magnetoviscous effect is very small and manifests only up to a 50% increase in viscosity [41]. The careful design of the size and, implicitly, of the magnetic moment of nanoclusters is essential to meet the optimal biomedical size window (50–150 nm) requirements and the need for moderate magnetic fields to control the movement of particles.

Magnetorheology evidences the possibility of remote control by already a moderate magnetic field of the motion, interaction and collection of PEG_MNC particles in the required regions of the stent, a condition of efficient magnetic targeting.

3.3. PEG_MNC Sedimentation

The sedimentation profile of the PEG_MNC nanoclusters was also investigated during this experiment. For 240 min, the sedimentation profile was captured as a function of time. During the first 10 min, the PEG-coated nanocluster's sedimentation was visible, and after that, a steady state was attained (Figure 9). In other words, the PEG_MNC suspension's dispersion stability was sufficient for our experiment (suspension injection time in the targeted region range between 20 to 120 s).

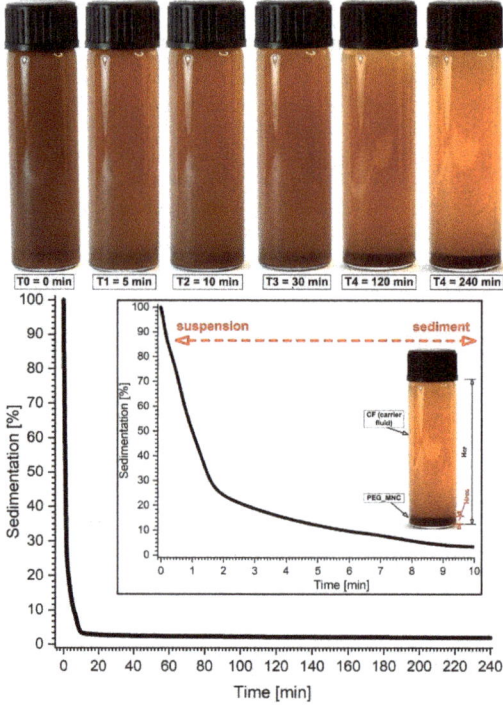

Figure 9. Sedimentation profile as a function of time for PEG_MNC suspension. Detail regarding sedimentation profile in the first 10 min. CF–carrier fluid; PEG_MNC-PEG-coated magnetic nanoclusters.

3.4. PEG_MNC Delivery at the Targeted Site

Flow-mediated particle transport achieves the PEG_MNC delivery at the targeted site in the present experiment.

The evolution of the functionalized magnetoresponsive clusters build-up is acquired at a frame rate of 30 frames/s. The MP's deposition and the magnetically induced chain length are investigated at the end of the injections period (after 30 s) (Figure 10D). These parameters are estimated as a function of the initial PEG_MNC quantity dispersed in the model suspensions and the injections period.

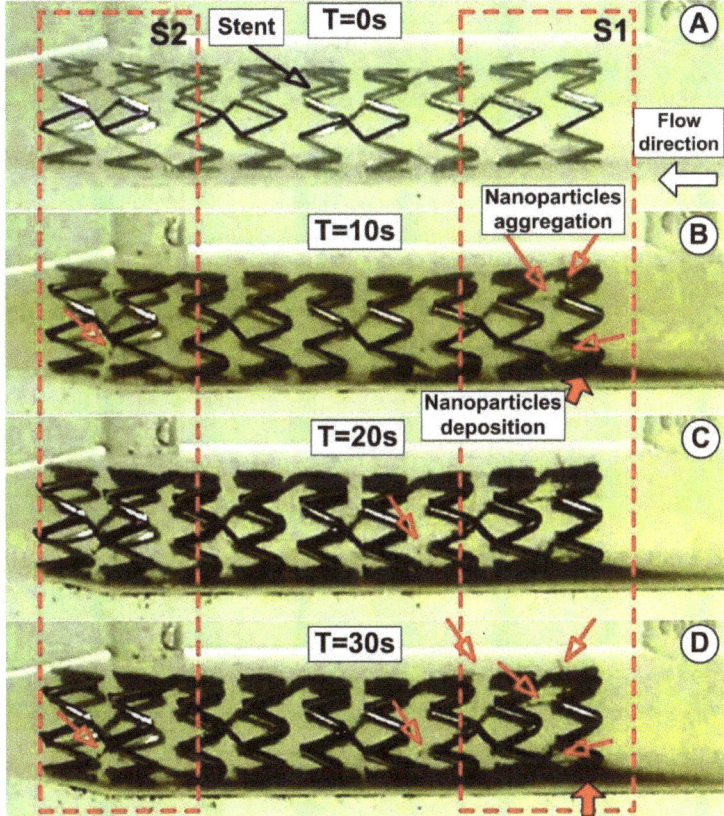

Figure 10. Magnetically induced large scales aggregation of the PEG_MNC around the targeted magnetic stent. Aggregation size evolution during different time steps of the injection period of 30 s. S1 and S2 regions where chain-like magnetic particles structure are generated (red arrows).

The PEG_MNC accumulation for the various injection time steps near the stent struts for the fixed magnet position is shown in Figure 10 (magnet position at 15 mm from the stent lower part and 18.15 mm from the upper part). As shown in Figure 10, the circulating magnetic clusters eventually completely encircle the stent struts. It is crucial to note that the magnetic cluster depositions are uneven, more robust at the last strut ring in the stent exit section, and more uniform on the first strut ring (in the stent inlet region). This occurs because the geometry of the stent and the strength of the applied external magnetic field affect local magnetic cluster deposition. Figure 9 illustrates this situation clearly, showing how the exposed portion of the struts from the distal end of the stent was typically not well covered by magnetic clusters. The cluster deposition around the stent struts is significantly influenced by local hemodynamics. The near-wall residence duration was prolonged, and the magnetic particle deposition was improved by the presence of the recirculation flow around the struts. The balance between the hydrodynamic force produced by the angular velocity of the recirculation region, the intensity of the magnetic field produced by the ferromagnetic stent, and the superimposed external magnetic field also contributes to the coverage of the struts with magnetic clusters. In particle deposition, the magnet position is crucial. The collection of particles in the bottom part is more noticeable than the accumulation of particles in the top part of the stent because the magnet is closer to the stent's lower part. The syringe's exit part moved at the same speed as the working fluid through the artery model during the model's suspension injection, which was carried out at a constant flow rate of 1.5 mL/s. Given the proximity of the PEG suspension injection point to the stent (19 × stent diameter ≅ 60 mm), the PEG_MNC begins to deposit around the proximal stent struts and at the stent inlet section nearly within one second (Figure 10B).

As observed in Figure 10D, the accumulation of particles in this location is more prominent than the accumulation of particles in the top part of the stent because the magnet is closer to the stent's bottom part (around the stent elements). Additionally, using a single magnet causes increased particle collection on the bottom wall of the artery due to the magnet's position on the stent.

Using an image-processing application (ImageJ, https://imagej.nih.gov/ij/ (accessed on 3 August 2022)), the thickness of the particles on the artery's lower wall was measured to quantify the PEG_MNC deposition at the end of the injection period of 30 s. (Figure 10D).

3.5. PEG_MNC Aggregation and Chain Formation Inside the Expanded Stent

The near-wall residence duration was prolonged, and the magnetic particle deposition was improved by the presence of the recirculation flow around the struts. However, it is crucial to note that the magnetic cluster depositions are uneven, more robust at the last strut ring in the stent exit section, and more uniform on the first strut ring (in the stent inlet region).

The acceleration of the particles in the direction of the magnetic field source is caused by the magnetic force, which is produced by an external magnetic field. Therefore, the magnetic flux concentrations close to the target location are more significant than 110 mT (120 mT at the distance of 15 mm) (Figure 3), sufficient to cause particle saturation.

In our experiment, magnetic particles create chain-like structures in a stented arterial model (Figure 11). The chains are usually oriented according to the applied external magnetic field perpendicular to the flow direction. In our case, the exception is chained Ch 2 and 4, where the geometry of the stent affects how the chains develop. Due to the strong magnetic field, the parabolic flow profile does not bend the shape of the chains.

The chain length variations corresponding to different injection times are presented in Table 3.

Table 3. Investigated chain length variations.

Time [s]	Chain 1 [mm]	Chain 2 [mm]	Chain 3 [mm]
0 (start)	0	0	0
10	0.26	0.47	0.21
20	0.84	0.58	0.27
30 (end)	0.86	0.63	0.27

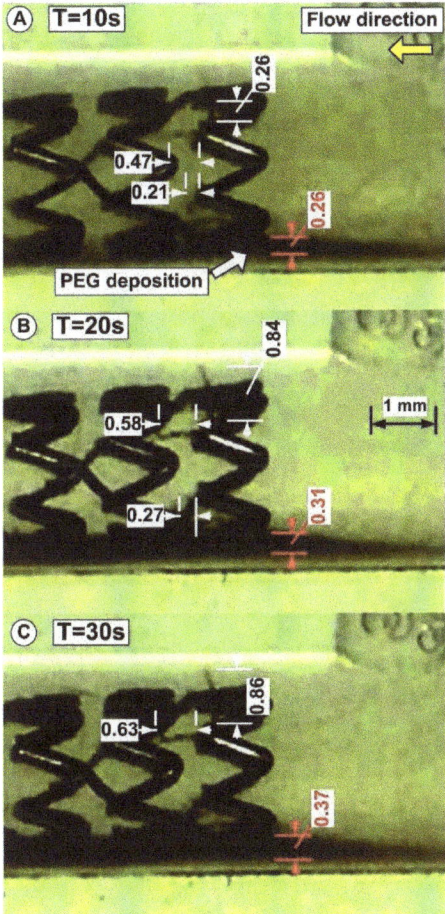

Figure 11. Detail regarding the chain-like particle structure evolution during injection time of 30 s. Data shows chain structure development and growth for chains Ch 1, Ch 2, and Ch 3 (according to Figure 9). Additionally, the figure presents the PEG_MNC deposition evolution during the injection period of 30 s (red arrows). All presented data are in mm.

3.6. Suspension Model's Aggregation Process in the Magnetic Field

At room temperature (25 °C), optical microscopy of the suspension model's aggregation process was carried out in the external magnetic field. The microscope bench was modified to accommodate a field generator based on permanent magnets that produced a 100 mT magnetic field and had an optical cell with a 0.1 mm thickness.

Randomly dispersed particles become magnetized by a perpendicular applied field due to field-induced chaining.

The microscope pictures of the PEG_MNC suspension before magnetic field application (Figure 12A) and the following 5 s of the magnetic field application are shown in Figure 12B. According to the optical microscopy pictures (Figure 12B), these particles form long linear aggregates due to dipolar magnetic interactions when there is an external magnetic field.

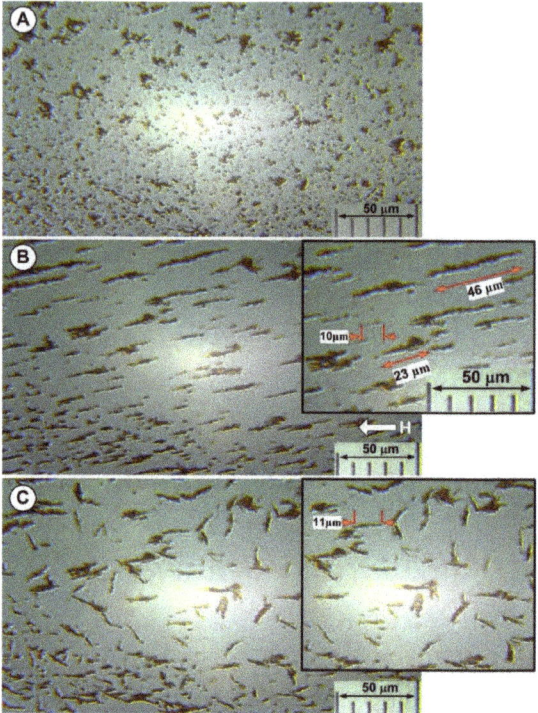

Figure 12. (**A**) PEG_MNC nanocomposite in carrier fluid without the magnetic field (optical microscopy). The figure shows the initial state with a homogeneous nanoparticle suspension when the magnetic field is switched on (t = 0); (**B**) aggregation of the PEG_MNC nanocomposite in carrier fluid under the action of the externally applied magnetic field of intensity H = 110 mT. Detail regarding the length and thickness of different chains generated. (**C**) Evolution of the PEG_MNC aggregate after turning off the magnetic field. Detail shows that the size of nanoparticle aggregates decreases after the magnetic field is turned off.

Considering that the average sizes of the multicore particles (as determined by DLS) are 102 ± 28 nm (size range: 70 ÷ 150 nm), these significant structures generated in the range of several micrometres (formed under magnetic fields) assume that the behaviours found might lead to the characteristics of the MVE discussed in the previous chapter.

To understand how the strength and length of the external magnetic field affect the size of the chain-like structures, it is crucial to study how the formations arise.

The spindles in the final stages of the process have an average thickness of $\cong$ 2 μm and a length of tens of microns (Figure 12B).

3.7. Colloidal Stability of the PEG_MNCs

The colloidal stability of the PEG_MNCs was investigated. For this purpose, an MNC aqueous dispersion (MNC-AD) was prepared using one stage of distilled water. Then, the MNC-AD was vigorously stirred using ultrasonication in several successive steps. After each ultrasonication step (~1000 J/step in 10 mL MNC-AD), most MNCs sediment within about one hour, leading to a diluted turbid supernatant.

The hydrodynamic diameter and zeta potential were measured with a Malvern Zetasizer Nano instrument. Dynamic light scattering provides three crucial details about the final PEGylated NP [43,44]: NP size, zeta potential (details on the NPs' colloidal stability and surface coating), and size distribution. The Malvern Zeta Sizer Nano Series (operated at a scattering angle of 173°) was used for the DLS measurements (1 mL of particle suspen-

Acknowledgments: For S.I. Bernad, this work was supported by the RA-TB/CFATR/LHC multiannual research program 2021-2022. For L. Vekas and V. Socoliuc, this work was supported partially by the RA-TB/CFATR/LMF multiannual research program 2021–2022 and by a grant from the Ministry of Research, Innovation and Digitization, CNCS/CCCDI-UEFISCDI, project number PN-III-P2-2.1-PED-2021-2049, within PNCDI III. For I. Craciunescu, R. Turcu, I.C. Ioncica, and E. Bernad, this work was supported partially by a grant from the Ministry of Research, Innovation and Digitization, CNCS/CCCDI-UEFISCDI, project number PN-III-P2-2.1-PED-2021-2049, within PNCDI III.

Conflicts of Interest: The authors declare no conflict of interest.

References

1. Garg, S.; Serruys, P.W. Coronary stents current status. *J. Am. Col. Cardiol.* **2010**, *56*, S1–S42. [CrossRef] [PubMed]
2. Buchanan, G.L.; Chieffo, A.; Colombo, A. Is there still a survival advantage to bypass surgery over percutaneous intervention in the modern era? *Prog. Cardiovasc. Dis.* **2015**, *58*, 335–341. [CrossRef]
3. Clare, J.; Ganly, J.; Bursill, C.A.; Sumer, H.; Kingshott, P.; de Haan, J.B. The Mechanisms of Restenosis and Relevance to Next Generation Stent Design. *Biomolecules* **2022**, *12*, 430. [CrossRef] [PubMed]
4. Jacobin-Valat, M.J.; Laroche-Traineau, J.; Larivière, M.; Mornet, S.; Sanchez, S.; Biran, M.; Lebaron, C.; Boudon, J.; Lacomme, S.; Cérutti, M.; et al. Nanoparticles functionalised with an anti-platelet human antibody for in vivo detection of atherosclerotic plaque by magnetic resonance imaging. *Nanomed. NBM* **2015**, *11*, 927–937. [CrossRef]
5. Nguyen, L.T.H.; Muktabar, A.; Tang, J.; Dravid, V.P.; Thaxton, C.S.; Venkatraman, S.; Woei, K. Engineered nanoparticles for the detection, treatment and prevention of atherosclerosis: How close are we? *Drug Discov. Today* **2017**, *22*, 1438–1446. [CrossRef]
6. Garello, F.; Svenskaya, Y.; Parakhonskiy, B.; Filippi, M. Micro/Nanosystems for Magnetic Targeted Delivery of Bioagents. *Pharmaceutics* **2022**, *14*, 1132. [CrossRef] [PubMed]
7. Moulton, K.S.; Heller, E.; Konerding, M.A.; Flynn, E.; Palinski, W.; Folkman, J. Angiogenesis inhibitors endostatin or TNP–470 reduce intimal neovascularization and plaque growth in apolipoprotein E-deficient mice. *Circulation* **1999**, *99*, 1726–1732. [CrossRef]
8. Marx, S.O.; Totary-Jain, H.; Marks, A.R. Vascular smooth muscle cell proliferation in restenosis. *Circ. Cardiovasc. Interv.* **2011**, *4*, 104–111. [CrossRef]
9. Unverdorben, M.; Vallbracht, C.; Cremers, B.; Heuer, H.; Hengstenberg, C.; Maikowski, C.; Werner, G.S.; Antoni, D.; Kleber, F.X.; Bocksch, W.; et al. Paclitaxelcoated balloon catheter versus paclitaxel-coated stent for the treatment of coronary in-stent restenosis. *Circulation* **2009**, *119*, 2986–2994. [CrossRef]
10. Sella, G.; Gandelman, G.; Teodorovich, N.; Tuvali, O.; Ayyad, O.; Abu Khadija, H.; Haberman, D.; Poles, L.; Jonas, M.; Volodarsky, I.; et al. Mid-Term Clinical Outcomes Following Drug-Coated Balloons in Coronary Artery Disease. *J. Clin. Med.* **2022**, *11*, 1859. [CrossRef]
11. Schiele, T.M.; Krotz, F.; Klauss, V. Vascular restenosis—striving for therapy. *Expert Opin. Pharmacother.* **2004**, *5*, 2221–2232. [CrossRef] [PubMed]
12. Abbasnezhad, N.; Zirak, N.; Champmartin, S.; Shirinbayan, M.; Bakir, F. An Overview of In Vitro Drug Release Methods for Drug-Eluting Stents. *Polymers* **2022**, *14*, 2751. [CrossRef] [PubMed]
13. Gazeau, F. Cooperative organization in iron oxide multicore nanoparticles potentiates their efficiency as heating mediators and MRI contrast agents. *ACS Nano.* **2012**, *6*, 10935–10949. [CrossRef]
14. Liu, X.; Wang, N.; Liu, X.; Deng, R.; Kang, R.; Xie, L. Vascular Repair by Grafting Based on Magnetic Nanoparticles. *Pharmaceutics* **2022**, *14*, 1433. [CrossRef] [PubMed]
15. George, T.A.; Hsu, C.-C.; Meeson, A.; Lundy, D.J. Nanocarrier-Based Targeted Therapies for Myocardial Infarction. *Pharmaceutics* **2022**, *14*, 930. [CrossRef] [PubMed]
16. Tombácz, E.; Turcu, R.; Socoliuc, V.; Vékás, L. Magnetic iron oxide nanoparticles: Recent trends in design and synthesis of magnetoresponsive nanosystems. *Biochem Biophys Res Commun.* **2015**, *468*, 442–453. [CrossRef]
17. Hu, M.; Butt, H.-J.; Landfester, K.; Bannwarth, M.B.; Wooh, S.; Therien-Aubin, H. Shaping the Assembly of Superparamagnetic Nanoparticles. *ACS Nano.* **2019**, *13*, 3015–3022. [CrossRef]
18. Kappes, M.; Bernhard, F.; Pfister, F.; Huber, C.; Friedrich, P.R.; Stein, R.; Braun, J.; Band, J.; Schreiber, E.; Alexiou, C.; et al. Superparamagnetic Iron Oxide Nanoparticles for Targeted Cell Seeding: Magnetic Patterning and Magnetic 3D Cell Culture. *Adv. Funct. Mater.* **2022**, 2203672. [CrossRef]
19. Gaëtan, M.; Van de Walle, A.; Perez, J.E.; Ukai, T.; Maekawa, T.; Luciani, N.; Wilhelm, C. High-Throughput Differentiation of Embryonic Stem Cells into Cardiomyocytes with a Microfabricated Magnetic Pattern and Cyclic Stimulation. *Adv. Funct. Mater.* **2020**, *30*, 2002541. [CrossRef]
20. Dutz, S.; Clement, J.H.; Eberbeck, D.; Gelbrich, T.; Hergt, R.; Müller, R.; Wotschadlo, J.; Zeisberger, M. Ferrofluids of magnetic multicore nanoparticles for biomedical applications. *J. Mag. Magn. Mater.* **2009**, *321*, 1501–1504. [CrossRef]
21. Socoliuc, V.; Avdeev, M.V.; Kuncser, V.; Turcu, R.; Tombácz, E.; Vékás, L. Ferrofluids and bio-ferrofluids: Looking back and stepping forward. *Nanoscale* **2022**, *14*, 4786–4886. [CrossRef] [PubMed]

22. Illés, E.; Szekeres, M.; Tóth, I.Y.; Farkas, K.; Földesi, I.; Szabó, Á.; Iván, B.; Tombácz, E. PEGylation of Superparamagnetic Iron Oxide Nanoparticles with Self-Organizing Polyacrylate-PEG Brushes for Contrast Enhancement in MRI Diagnosis. *Nanomaterials* **2018**, *8*, 776. [CrossRef] [PubMed]
23. Illés, E.; Szekeres, M.; Kupcsik, E.; Tóth, I.Y.; Farkas, K.; Jedlovszky Hajdú, A.; Tombácz, E. PEGylation of surfaced magnetite core-shell nanoparticles for biomedical application. *Colloids Surf. A* **2014**, *460*, 429–440. [CrossRef]
24. Tóth, I.Y.; Illés, E.; Bauer, R.A.; Nesztor, D.; Szekeres, M.; Zupkó, I.; Tombácz, E. Designed polyelectrolyte shell on magnetite nanocore for dilution-resistant biocompatible magnetic fluids. *Langmuir* **2012**, *28*, 16638–16646. [CrossRef] [PubMed]
25. Singh, D.; McMillan, J.M.; Liu, X.M.; Vishwasrao, H.M.; Kabanov, A.V.; Sokolsky-Papkov, M.; Gendelman, H.E. Formulation design facilitates magnetic nanoparticle delivery to diseased cells and tissues. *Nanomedicine* **2014**, *9*, 469–485. [CrossRef]
26. Oberoi, H.S.; Nukolova, N.V.; Kabanov, A.V.; Bronich, T.K. Nanocarriers for delivery of platinum anticancer drugs. *Adv. Drug Del. Rev.* **2013**, *65*, 1667–1685. [CrossRef]
27. Lim, E.K.; Jang, E.; Lee, K.; Haam, S.; Huh, Y.M. Delivery of cancer therapeutics using nanotechnology. *Pharmaceutics* **2013**, *5*, 294. [CrossRef]
28. Suk, J.S.; Xu, Q.; Kim, N.; Hanes, J.; Ensign, L.M. PEGylation as a strategy for improving nanoparticle-based drug and gene delivery. *Adv. Drug Deliv. Rev.* **2016**, *99*, 28–51. [CrossRef]
29. Yu, M.K.; Park, J.; Jon, S. Targeting Strategies for Multifunctional Nanoparticles Cancer Imaging and Therapy. *Theranostic* **2012**, *2*, 3–44. [CrossRef]
30. Bernad, S.I.; Craciunescu, I.; Sandhu, G.S.; Dragomir-Daescu, D.; Tombacz, E.; Vekas, L.; Turcu, R. Fluid Targeted delivery of functionalized magnetoresponsive nanocomposite particles to a ferromagnetic stent. *J. Magn. Magn. Mater.* **2021**, *519*, 167489. [CrossRef]
31. Sankar, D.S.; Hemalatha, K. Pulsatile flow of Herschel–Bulkley fluid through stenosed arteries—A mathematical model. *Int. J. Non-Linear Mech.* **2006**, *41*, 979–990, 2006. [CrossRef]
32. Bernad, S.I.; Susan-Resiga, D.; Bernad, E. Hemodynamic Effects on Particle Targeting in the Arterial Bifurcation for Different Magnet Positions. *Molecules* **2019**, *24*, 2509. [CrossRef] [PubMed]
33. Bernad, S.I.; Susan-Resiga, D.; Vekas, L.; Bernad, E.S. Drug targeting investigation in the critical region of the arterial bypass graft. *J. Magn. Magn. Mater.* **2019**, *475*, 14–23. [CrossRef]
34. Torii, R.; Wood, N.B.; Hadjiloizou, N.; Dowsey, A.W.; Wright, A.R.; Hughes, A.D.; Davies, J.; Francis, D.P.; Mayet, J.; Yang, G.; et al. Stress phase angle depicts differences in coronary artery hemodynamics due to changes in flow and geometry after percutaneous coronary intervention. *Am. J. Physiol. Heart Circ. Physiol.* **2009**, *296*, H765–H776. [CrossRef] [PubMed]
35. Uthamaraj, S.; Tefft, B.J.; Klabusay, M.; Hlinomaz, O.; Sandhu, G.; Dragomir-Daescu, D. Design and validation of a novel ferromagnetic bare metal stent capable of capturing and retaining endothelial cells. *ABME* **2014**, *24*, 2416. [CrossRef] [PubMed]
36. Tefft, B.J.; Uthamaraj, S.; Harbuzariu, A.; Harburn, J.J.; Witt, T.A.; Newman, B.; Psaltis, P.J.; Hlinomaz, O.; Holmes, D.R., Jr.; Gulati, R.; et al. Nanoparticle-Mediated Cell Capture Enables Rapid Endothelialization of a Novel Bare Metal Stent. *Tissue Eng. Part A* **2018**, *24*, 1157–1166. [CrossRef] [PubMed]
37. Cho, Y.I.; Kensey, K.R. Effects of the non-Newtonian viscosity of blood on flows in a diseased arterial vessel. Part 1: Steady flows. *Biorheology* **1991**, *28*, 241–262. [CrossRef]
38. Mandal, P.K. An unsteady analysis of non-Newtonian blood flow through tapered arteries with a stenosis. *Non. Linear Mech.* **2005**, *40*, 151–164, 2005. [CrossRef]
39. Akbar, N.S.; Nadeem, S. Carreau fluid model for blood flow through a tapered artery with a stenosis. *Ain Shams Eng. J.* **2014**, *5*, 1307–1316. [CrossRef]
40. Nowak, J.; Wiekhorst, F.; Trahms, L.; Odenbach, S. The influence of hydrodynamic diameter and core composition on the magnetoviscous effect of biocompatible ferrofluids. *J. Phys. Condens. Matter* **2014**, *26*, 176004. [CrossRef]
41. Vasilescu, C.; Latikka, M.; Knudsen, K.D.; Garamus, V.M.; Socoliuc, V.; Turcu, R.; Tombacz, E.; Susan-Resiga, D.; Ras, R.H.A.; Vekas, L. High concentration aqueous magnetic fluids: Structure, colloidal stability, magnetic and flow properties. *Soft Matter* **2018**, *14*, 6648–6666. [CrossRef] [PubMed]
42. Carreau, P.J. Rheological Equations from Molecular Network Theories. *Trans. Soc. Rheol.* **1972**, *16*, 99–127. [CrossRef]
43. Bhattacharjee, S.; Elimelech, M.; Borkovec, M. DLVO Interaction between Colloidal Particles: Beyond Derjaguin's Approximation. *Croat. Chem. Acta* **1998**, *71*, 883–903.
44. Lim, J.; Yeap, S.P.; Che, H.X.; Low, S.C. Characterization of magnetic nanoparticle by dynamic light scattering. *Nanoscale Res. Lett.* **2013**, *8*, 381. [CrossRef] [PubMed]
45. Gregory, J. Interaction of Unequal Double layers at Constant Charge. *J. Colloids Interface Sci.* **1975**, *51*, 44–51. [CrossRef]
46. Ivanov, A.O.; Novak, E.V. Phase separation of ferrocolloids: The role of van der Waals interaction. *Colloid J.* **2007**, *69*, 332. [CrossRef]
47. Ivanov, A.O.; Kuznetsova, O.B. Magnetic properties of dense ferrofluids: An influence of interparticle correlations. *Phys. Rev.* **2001**, *E 64*, 041405. [CrossRef]
48. Socoliuc, V.; Turcu, R. Large scale aggregation in magnetic colloids induced by high frequency magnetic fields. *J. Magn. Magn. Mater.* **2020**, *500*, 166348. [CrossRef]
49. Ta, H.T.; Truong, N.P.; Whittaker, A.K.; Davis, T.P.; Peter, K. The effects of particle size, shape, density and flow characteristics on particle margination to vascular walls in cardiovascular diseases. *Expert Opin. Drug Deliv.* **2018**, *15*, 33–45. [CrossRef]

50. Bae, J.E.; Huh, M.I.; Ryu, B.K.; Do, J.Y.; Jin, S.U.; Moon, M.J.; Jung, J.C.; Chang, Y.; Kim, E.; Chi, S.G.; et al. The effect of static magnetic fields on the aggregation and cytotoxicity of magnetic nanoparticles. *Biomaterials* **2011**, *32*, 9401–9414. [CrossRef] [PubMed]
51. van Vlerken, L.E.; Vyas, T.K.; Amiji, M.M. Poly(ethylene glycol)-modified nanocarriers for tumor-targeted and intracellular delivery. *Pharm. Res.* **2007**, *24*, 1405–1414. [CrossRef]
52. Kanaras, A.G.; Kamounah, F.S.; Schaumburg, K.; Kiely, C.J.; Brust, M. Thioalkylated tetraethylene glycol: A new ligand for water soluble monolayer protected gold clusters. *Chem. Commun.* **2002**, *20*, 2294–2295. [CrossRef] [PubMed]
53. Derjaguin, B.V.; Churaev, N.V.; Muller, V.M. The Derjaguin—Landau—Verwey—Overbeek (DLVO) Theory of Stability of Lyophobic Colloids. In *Surface Forces*; Springer: Boston, MA, USA, 1987. [CrossRef]
54. Hermann, J.; DiStasio, R.A., Jr.; Tkatchenko, A. First-Principles Models for van der Waals Interactions in Molecules and Materials: Concepts, Theory, and Applications. *Chem. Rev.* **2017**, *117*, 4714–4758. [CrossRef] [PubMed]
55. Jun, Y.W.; Casula, M.F.; Sim, J.H.; Kim, S.Y.; Cheon, J.; Alivisatos, A.P. Surfactant-assisted elimination of a high energy facet as a means of controlling the shapes of TiO_2 nanocrystals. *J. Am. Chem. Soc.* **2003**, *125*, 15981–15985. [CrossRef] [PubMed]
56. Förster, S.; Antonietti, M. Amphiphilic block copolymers in structure-controlled nanomaterial hybrids. *Adv. Mat.* **1998**, *10*, 195–217. [CrossRef]
57. Illés, E.; Szekeres, M.; Tóth, I.Y.; Szabó, Á.; Iván, B.; Turcu, R.; Vékás, L.; Zupkó, I.; Jaics, G.; Tombácz, E. Multifunctional PEG-carboxylate copolymer coated superparamagnetic iron oxide nanoparticles for biomedical application. *J. Mag. Magn. Mater.* **2018**, *451*, 710–720. [CrossRef]

Article

Novel Salinomycin-Based Paramagnetic Complexes—First Evaluation of Their Potential Theranostic Properties

Irena Pashkunova-Martic [1,*], Rositsa Kukeva [2], Radostina Stoyanova [2], Ivayla Pantcheva [3], Peter Dorkov [4], Joachim Friske [1], Michaela Hejl [5], Michael Jakupec [5], Mariam Hohagen [6], Anton Legin [5], Werner Lubitz [7], Bernhard K. Keppler [5], Thomas H. Helbich [1] and Juliana Ivanova [8]

[1] Department of Biomedical Imaging and Image-Guided Therapy, Division of Molecular and Structural Preclinical Imaging, Preclinical Imaging Laboratory, Medical University of Vienna & General Hospital of Vienna, Waehringer Guertel 18–20, 1090 Vienna, Austria
[2] Institute of General and Inorganic Chemistry, Bulgarian Academy of Sciences, Akad. G. Bonchev Str., bl. 11, 1113 Sofia, Bulgaria
[3] Faculty of Chemistry and Pharmacy, Sofia University "St. Kliment Ohridski", J. Bourchier Blvd., 1, 1164 Sofia, Bulgaria
[4] Chemistry Department, R&D, BIOVET Ltd., 39 Peter Rakov Str., 4550 Peshtera, Bulgaria
[5] Institute of Inorganic Chemistry, University of Vienna, Waehringer Strasse 42, 1090 Vienna, Austria
[6] Department of Inorganic Chemistry—Functional Materials, University of Vienna, Waehringer Strasse 42, 1090 Vienna, Austria
[7] BIRD-C GmbH, Dr. Bohrgasse 2–8, 1030 Vienna, Austria
[8] Faculty of Medicine, Sofia University "St. Kliment Ohridski", Kozjak Str., 1, 1407 Sofia, Bulgaria
* Correspondence: irena.pashkunova-martic@meduniwien.ac.at; Tel.: +43-1-40400-48195; Fax: +43-1-40400-48980

Citation: Pashkunova-Martic, I.; Kukeva, R.; Stoyanova, R.; Pantcheva, I.; Dorkov, P.; Friske, J.; Hejl, M.; Jakupec, M.; Hohagen, M.; Legin, A.; et al. Novel Salinomycin-Based Paramagnetic Complexes—First Evaluation of Their Potential Theranostic Properties. *Pharmaceutics* **2022**, *14*, 2319. https://doi.org/10.3390/pharmaceutics14112319

Academic Editor: Carlotta Marianecci

Received: 30 September 2022
Accepted: 26 October 2022
Published: 28 October 2022

Publisher's Note: MDPI stays neutral with regard to jurisdictional claims in published maps and institutional affiliations.

Copyright: © 2022 by the authors. Licensee MDPI, Basel, Switzerland. This article is an open access article distributed under the terms and conditions of the Creative Commons Attribution (CC BY) license (https:// creativecommons.org/licenses/by/ 4.0/).

Abstract: Combining therapeutic with diagnostic agents (theranostics) can revolutionize the course of malignant diseases. Chemotherapy, hyperthermia, or radiation are used together with diagnostic methods such as magnetic resonance imaging (MRI). In contrast to conventional contrast agents (CAs), which only enable non-specific visualization of tissues and organs, the theranostic probe offers targeted diagnostic imaging and therapy simultaneously. Methods: Novel salinomycin (Sal)-based theranostic probes comprising two different paramagnetic metal ions, gadolinium(III) (Gd(III)) or manganese(II) (Mn(II)), as signal emitting motifs for MRI were synthesized and characterized by elemental analysis, infrared spectral analysis (IR), electroparamagnetic resonance (EPR), thermogravimetry (TG) differential scanning calorimetry (DSC) and electrospray ionization mass spectrometry (ESI-MS). To overcome the water insolubility of the two Sal-complexes, they were loaded into empty bacterial ghosts (BGs) cells as transport devices. The potential of the free and BGs-loaded metal complexes as theranostics was evaluated by in vitro relaxivity measurements in a high-field MR scanner and in cell culture studies. Results: Both the free Sal-complexes (Gd(III) salinomycinate (Sal-Gd(III) and Mn(II) salinomycinate (Sal-Mn(II)) and loaded into BGs demonstrated enhanced cytotoxic efficacy against three human tumor cell lines (A549, SW480, CH1/PA-1) relative to the free salinomycinic acid (Sal-H) and its sodium complex (Sal-Na) applied as controls with IC_{50} in a submicromolar concentration range. Moreover, Sal-H, Sal-Gd(III), and Sal-Mn(II) were able to induce perturbations in the cell cycle of treated colorectal and breast human cancer cell lines (SW480 and MCF-7, respectively). The relaxivity (r_1) values of both complexes as well as of the loaded BGs, were higher or comparable to the relaxivity values of the clinically applied contrast agents gadopentetate dimeglumine and gadoteridol. Conclusion: This research is the first assessment that demonstrates the potential of Gd(III) and Mn(II) complexes of Sal as theranostic agents for MRI. Due to the remarkable selectivity and mode of action of Sal as part of the compounds, they could revolutionize cancer therapy and allow for early diagnosis and monitoring of therapeutic follow-up.

Keywords: theranostics; paramagnetic salinomycin complexes; bacterial ghosts; gadolinium; manganese; MRI

1. Introduction

Cancer theranostics, the combination of a diagnostic with a therapeutic, pave the way for personalized medicine, allowing for simultaneous and precise diagnosis and treatment of malignant diseases. Current theranostic concepts aim at (i) facilitating the observation and monitoring of anticancer drugs, (ii) providing targeted and personalized cancer therapy, and (iii) realizing simultaneous diagnosis and treatment in (early-stage) cancer patients [1]. Magnetic Resonance Imaging (MRI) is a powerful non-invasive diagnostic technique, which takes advantage of a very high spatial and temporal resolution and can provide detailed molecular/cellular information when combined with a contrast agent with high relaxivity to overcome the lack of sensitivity inherent to MRI [2]. In contrast to conventional contrast agents (CAs), which enable only non-specific visualization of tissues and organs, the theranostic probe offers targeted diagnostic imaging and therapy simultaneously [3]. Presently, there is only one clinically approved theranostic agent for the treatment of progressive prostate cancer, recently approved by the FDA [4]. Therefore, the development of efficient, tumor-directed drugs with good magnetic susceptibility and a higher safety profile remains the main objective in modern oncology.

The design of new CAs plays an essential role in contrast-enhanced MRI today. Currently used CAs are predominantly low-molecular-weight gadolinium(III) (Gd(III))-based complexes, which can provide outstanding positive MR images with high resolution [5]. Unfortunately, some investigations have shown that Gd(III) could be involved in nephrogenic systemic fibrosis (NSF), which limits its clinical applications [6]. Newer studies, however, report the accumulation of Gd(III) in various tissues (bone, brain, and kidneys) of patients who were not diagnosed with renal impairment [7,8]. Recent toxicity concerns associated with the long-term use of low-molecular-weight acyclic Gd-based contrast agents (GBCA) have resulted in the restriction of their administration by the European Medicines Agency (EMA) and have triggered risk warnings from the U.S. Food and Drug Administration (FDA) [9,10]. Therefore, novel chelators, on the one hand, as well as alternative paramagnetic centers, on the other, as part of the MRI CAs, are needed urgently.

Due to its favorable properties, such as high spin number, long electronic relaxation time, and labile water exchange, manganese(II) (Mn(II)) represents an attractive alternative metal center for the design of novel MRI CAs [11]. The natural polyether antibiotic Salinomycin (Sal) has attracted the attention of numerous scientists all over the world as a highly selective cytotoxic agent and has been the subject of intense investigations during the past few years [12–19]. The distinctive pentacyclic spiroketal ring system that characterizes the molecule, and the ample presence of oxygen atoms from different functional groups (carboxylic, carbonyl, ether, hydroxyl), makes it a potential ligand able to bind paramagnetic metal cations such as Gd(III) or Mn(II). Therefore, it is expected to provide better stability and a safer profile. Sal has been reported to form complexes with metal(II) ions such as Zn^{2+}, Cu^{2+}, Co^{2+}, and Ni^{2+}, which exerted superior anticancer activity in vitro compared to the free ligand [20,21]. To the best of our knowledge, there are no data on the coordination of the antibiotic to trivalent metal ions.

Bacterial ghost cells (BGs) represent empty cell envelopes of Gram-negative bacteria devoid of cytoplasmic content and free of nucleic acids. They are produced by the controlled expression of the plasmid-encoded lysis gene E. Protein E, which leads to the fusion of the inner and outer membranes of the bacteria and the formation of a tunnel structure through which the cytoplasmic content is expelled as a result of the osmotic pressure difference between the cytoplasm and the exterior [22]. The BGs can be used as carriers or targeting vehicles for active agents, which are specific to various types of tissue. Moreover, active agents are efficiently transported to the desired destination since it is possible to prepare BGs that contain only the desired active substance and a high degree of loading, and thus, high efficiency of the active agent can be achieved. When BGs loaded with an active agent (in particular, a diagnostic, therapeutic, or theranostically active agent) are administered, they are internalized by cancer cells, followed by degradation of the BGs within the cancer cells. Thereupon, the active agent is released into the cytoplasm of the cancer cells, inducing cell death of the cancer cells [23,24].

Herein, we report, for the first time, the evaluation of Sal complexes with the paramagnetic ions Gd(III) (Sal-Gd(III)) and Mn(II) (Sal-Mn(II)) and their incorporation into BGs as effective theranostic agents for the early diagnosis and monitoring of cancer therapy with MRI.

2. Methods
2.1. Chemicals

The commercially available pharmaceutical-grade salinomycin sodium ($C_{42}H_{69}O_{11}Na$; SalNa) was provided by Biovet Ltd. (Peshtera, Bulgaria), purity of >95%. Salinomycinic acid was prepared as previously reported [21]. Organic solvents (MeCN, MeOH, DMSO) and metal salts of analytical grade ($GdCl_3$ and $MnCl_2.4H_2O$) were purchased from Fisher Scientific (Loughborough, UK). The tetrazolium salt 3-(4,5-dimethylthiazol-2-yl)-2,5-diphenyl tetrazolium bromide (MTT) was bought from Sigma-Aldrich (Vienna, Austria).

2.2. Synthesis of Paramagnetic Complexes of Sal
2.2.1. Synthesis of Gd(III) Salinomycinate

The metal salt ($GdCl_3.6H_2O$, 0.1742 g, 0.47 mmol) was dissolved in 2 mL water. The solution was added to salinomycin sodium solution (0.2561 g, 0.33 mmol, dissolved in MeCN:MeOH = 1:5). The reaction mixture was stirred for 30 min at room temperature. After the slow evaporation of the solvents for seven days, a white precipitate was formed. The solid phase was washed with water, filtered off, and dried over P_4O_{10}. Yield: 221 mg, 82%. Anal. Calcd. for $C_{126}H_{213}O_{36}Gd$ (MW = 2458.25 g/mol): H, 8.65%; C, 61.43%; Gd, 6.39%. Found: H, 8.74%; C, 61.61%; Gd 6.28%. ESI-MS, m/z: 906.76 [$(C_{42}H_{68}O_{11})Gd$]$^+$, calcd: 906.40 (100%), 908.40 (88.0%), 904.40 (82.4%), 905.40 (63%); 1657.56 [$C_{42}H_{69}O_{11})_2Gd$]$^+$, calcd: 1656.89 (100%); 1657.90 (90.9%); 1658.90 (88%); 1654.89 (82.4%); 1659.90 (80%); 1655.89 (74.9%); 1655.89 (63%).

2.2.2. Synthesis of Mn(II) Salinomycinate

Salinomycin sodium (0.3850 g, 0.5 mmol) was dissolved in mixed solvents (20 mL MeOH + 2 mL MeCN). The solution of Mn(II) chloride ($MnCl_2.4H_2O$, 1 mmol, 198 mg in 4 mL water) was added to the ligand solution. The reaction mixture was stirred at room temperature for 30 min. The solution slowly evaporated, and the resulting light-brownish precipitate was washed with water, filtered off, and dried over P_4O_{10} for three days. Yield: 329 mg, 83%. Anal. Calcd. for $C_{84}H_{142}O_{24}Mn$ (MW = 1590.94 g/mol): H, 8.93%; C, 63.36%; Mn, 3.45%. Found: H, 9.35%; C, 63.05%; Mn, 3.24%. ESI-MS, m/z: 804.76 [$(C_{42}H_{69}O_{11})Mn$]$^+$, calcd.: 804.42 (100%); 805.43 (45.4%); 1578.55 [$(C_{42}H_{69}O_{11})_2MnNa$]$^+$, calcd: 1576.90 (100%); 1577.90 (90.9%); 1578.90 (40.8%).

The Gd(III) and Mn(II) complexes of salinomycin are insoluble in water and soluble in organic solvents, such as MeOH, MeCN, EtOH, $CHCl_3$, DMSO, and hexane.

2.2.3. Loading of Gd(III) and Mn(II) Salinomycinates into Empty Bacterial Ghosts Cells (BGs)

Non-pathogenic *Escherichia coli Nissle* 1917 (*EcN*1917) and *Escherichia coli NM*522 BGs (*NM*522) were successfully loaded by simple resuspension of BGs (10 mg BGs were first suspended in PBS, pH 7.4, 0.05 M) with each Sal-complex solutions (Sal-Gd(III), Sal-Mn(II), SalNa, and SalH) in methanol. Ten mg of each Sal-based compound were used for loading *EcN*1917 and 5 mg for *NM*522. The BGs suspensions were stirred for two hours at RT. Subsequently, the loaded BGs were collected by centrifugation at 11 300 g for 15 min, and the pellets were washed three times with Milli-Q® water. One mL aliquots were stored at −20 °C for further analyses.

Quantification of Sal-Gd(III)/Sal-Mn(II) Extracted from BGs

In order to determine the amount of Sal-complex within the BGs, 4 mg loaded BGs were resuspended in 0.5 mL of 96% ethanol (Carl Roth, Vienna, Austria), followed by 5 min of ultrasonication. Subsequently, the ethanolic extract was diluted equally with Milli-Q® H_2O (1:1) and immediately centrifuged at $11,300 \times g$ for 15 min at 4 °C. FPLC analysis was performed using an AKTÄ Purifier 10 System® (GE Healthcare, Chicago, IL, USA) equipped with a Superdex 10/300 GL column (24 mL; Cytiva, Germany), and UV and conductivity detectors. MeOH-acetate (65:35) was used as an eluent. The quantification was carried out using the peak area method applying free Sal-Mn(II) complex as an external standard.

2.3. Physical Measurements

2.3.1. Infrared Spectral Analysis (IR)

The infrared spectra of both compounds were recorded on a Specord-75IR (Carl-Zeiss, Oberkochen, Germany) in a nujol mull.

2.3.2. Electroparamagnetic Resonance Analysis (EPR)

The EPR analysis was performed on a Bruker EMX PremiumX instrument (Karlsruhe, Germany). All measurements were carried out in the X-band at a frequency of microwave electromagnetic radiation 9.45 GHz. Quantitative analysis was performed using Bruker software (Version 1.1b 119) equipped with a spin count option. For temperature variation, a variable temperature unit, ER4141VTM, was used.

2.3.3. Electrospray Ionization Mass Spectrometry (ESI-MS)

ESI-MS spectra were recorded on Waters Micromass ZQ2000 Single Quadrupole Mass spectrometer (Waters, Milford, MA, USA) in MeOH/H_2O (10% H_2O), positive mode, in the range of 0–2000 m/z.

2.4. Thermogravimetric Analysis (TGA)

Thermogravimetric analysis (TGA) and differential scanning calorimetry (DSC) was executed using a Netzsch® STA-449 F3 Jupiter instrument from 40 to 800 °C under airflow of 20 mL·min^{-1} as carrier gas with a heating rate of 10 °C min^{-1}. Simultaneous thermal analyses allow the measurement of mass changes and thermal effects in the range of 150 °C to 2400 °C. The percentage of mass loss was estimated in the temperature range of 40–800 °C.

2.5. In Vitro Relaxivity Measurements

As a reference substance, pure MeOH was used as a negative control, and gadopentetate dimeglumine (Magnevist®) and gadoteridol (ProHance®) as positive controls. Serial dilutions of BGs loaded with the salinomycin complexes were carried out in ultra-pure Milli-Q® water ranging from 0.02 to 0.1 mM for Mn(II) salinomycinate or Gd(III) salinomycinate, respectively. For the free salinomycinate complexes, six dilutions ranging from 0.01 mM to 3.3 mM (for the Gd(III) salinomycinate) and from 0.02 mM to 1.4 mM (for the Mn(II) salinomycinate) in MeOH with a volume of 0.5 mL were prepared. All probes were measured in Eppendorf safe-lock polypropylene tubes with a diameter of 0.4 cm. The tubes were placed in the center of a plastic box and measured at ambient temperature.

Relaxivity measurements were conducted on a high-field MRI scanner (9.4 Tesla, Bruker Biospec). T_1, T_2, and T_{2*} measurements of different contrast media concentrations were performed using the T_1 mapping inversion recovery RARE sequence, the T_2 mapping multislice multiecho spin echo sequence, and the T_{2*} mapping multi gradient echo sequence for calculation of relaxivities (R_1, R_2, R_{2*}).

To determine the T_1 spin-lattice relaxation times, spin echo inversion recovery (IR) sequences with inversion times from 0 to 3500 ms (TI = 0, 60, 80, 100, 150, 200, 250, 300, 400, 500, 750, 1000, 1250, 1500, 1750, 2000, 2500, 3500 ms) were used. An adiabatic pulse was applied for B1-insensitive inversion. The other parameters were TR/TE = 5000/8.1 ms; flip angle 180°; 8 turbo factor 11; FOV 180 × 180 mm; resolution matrix 192 × 192; bandwidth 260 Hz/pixel; and slice thickness 3 mm. For all measurements, the relaxation rates (R_1) and the relaxivities (r_1) were calculated for each substance. The relaxivities (r_1, r_2) were calculated as the slope of the linear regression of R_1 and R_2 as a function of the contrast agent concentration.

2.6. Cytotoxicity Tests

2.6.1. Cell Culture

Four adherently growing human cancer cell lines were used for this study: CH1/PA-1 (ovarian teratocarcinoma) cells were a gift from Lloyd R. Kelland (CRC Center for Cancer Therapeutics, Institute of Cancer Research, Sutton, UK), whereas A549 (non-small-cell lung cancer) and SW480 (colon carcinoma) cells were kindly provided by the Institute of Cancer Research, Department of Medicine I, Medical University of Vienna, Austria, and MCF-7 (mammary carcinoma) cells by the Department of Pharmaceutical Sciences, University of Vienna. The first three cell lines were grown in Eagle's minimal essential medium (MEM) supplemented with L-glutamine (4 mM), sodium pyruvate (1 mM), 1% (v/v) non-essential amino acid solution, and 10% (v/v) heat-inactivated fetal calf serum (from BioWest) in 75 cm^2 flasks at 37 °C under a humidified atmosphere containing 5% CO_2 in the air. MCF-7 cells were cultured in High Glucose Dulbecco's Minimal Essential Medium (DMEM) supplemented with 10% (v/v) fetal calf serum (FCS; from BioWest), L-glutamine (4 mM) and 0.01 mg/mL of human insulin. All cell culture media, supplements, and assay reagents were purchased from Sigma-Aldrich, and all plastic ware from Starlab unless stated otherwise.

2.6.2. MTT Assay

The antiproliferative activity of the compounds was determined by the colorimetric MTT assay (MTT = 3-(4,5-dimethyl-2-thiazolyl)-2,5-diphenyl-2H-tetrazolium bromide). The 1 × 10^3 CH1/PA-1, 2 × 10^3 SW480 and 3 × 10^3 A549 cells were seeded in 100 µL per well into 96-well microculture plates. After 24 h, the tested compounds were dissolved in DMSO (Fisher Scientific), serially diluted in supplemented MEM not to exceed a final DMSO content of 0.5% (v/v) and added in 100 µL per well. After 96 h, the drug-containing medium was replaced with 100 µL of an RPMI 1640/MTT mixture [six parts of RPMI 1640 medium, supplemented with 10% heat-inactivated fetal bovine serum (FBS), and 4 mM L-glutamine, with one part of MTT solution in phosphate-buffered saline (5 mg/mL)] per well. After incubation for four hours, the MTT-containing medium was replaced with 150 µL DMSO per well to dissolve the formazan product formed by viable cells. Optical density at 550 nm (and at a reference wavelength of 690 nm) was measured with a microplate reader (ELx808, Bio-Tek). The 50% inhibitory concentrations (IC_{50}) relative to untreated controls were interpolated from the concentration-effect curves. At least three independent experiments were performed, each in triplicate per concentration level.

2.6.3. Cell Cycle Studies—Impact of Free Sal-H, Sal-Gd(III), and Sal-Mn(II) on the Cell Cycle

Colon carcinoma cells (SW480) and breast adenocarcinoma cells (MCF-7) were seeded in 12-well plates (CytoOne, tissue culture treated) in density of 8 × 10^4 cells per well in 1 mL of the corresponding medium. After a recovery time of 24 h, cells were treated with different concentrations of Sal-H, Sal-Mn(II), and Sal-Gd(III). For this purpose, the test substances were dissolved in DMSO and diluted in a medium such that the maximum concentration of DMSO in the cells did not exceed 0.5%. For assay validation, the well-known cell cycle inhibitors etoposide and gemcitabine were applied as positive controls. Plates were incubated with test compounds for 24 h at 37 °C, 5% CO_2. Following the exposure, the medium was completely removed, and adherent cells were gently washed twice with ice-cold PBS. Propidium iodide (PI, 1.0 mg/mL) and HFS-buffer (0.1% (v/v) Triton X-100, 0.1% (w/v) sodium citrate in Milli-Q® water) were mixed for staining to yield a final PI concentration of 40 µg/mL. The ice-cold staining solution was added to the cells (500 µL/well). The staining was carried out overnight at 4 °C in the dark. To prepare the samples for measurement, the staining solution was vigorously pipetted against the surface to achieve sufficient resuspension of the cells. For each sample, 200 µL of the stained cell suspensions were added to a 96-well round-bottom FACS plate (Falcon®). The fluorescence of all samples stained with a DNA-intercalating agent was measured no longer than 24 h after staining using a flow cytometer (Guava easyCyte 8HT, Millipore®). GuavaSoft™ software was used in the InCyte-modus, and 10,000 events/probes were counted.

The flow cytometry data sets were evaluated in FlowJo software (v 10.8.1). The single-cell populations were gated based on forward and side scatters characteristics. The Watson Pragmatic model was applied as a standard method to analyze the resulting histograms: the recorded red fluorescence intensities of cells in the S phase were located between normally distributed G1/G0 and G2/M peaks. The number of cells in each phase was calculated from the model by means of integration. Means and standard deviations were calculated from at least three independently performed experiments. The figures were adjusted in GIMP freeware (v 2.10.24).

3. Results

3.1. Synthesis and Characterization

The elemental analysis demonstrated that salinomycin sodium reacts with Gd(III) and Mn(II) to form homometallic mononuclear complexes of composition [Gd($C_{42}H_{69}O_{11}$)$_3$(H_2O)$_3$] and [Mn($C_{42}H_{69}O_{11}$)$_2$(H_2O)$_2$] respectively. The ESI-MS spectra of both complexes corresponded to the elemental analysis and contained several signals due to the dissociation of water molecules, ligand anions, and/or complexation with Na^+ (ESI-MS Spectra, Supplementary Materials Sections S1 and S2).

In the IR spectrum of the Gd(III) salinomycinate, two characteristic bands at 1540 cm^{-1} and 1400 cm^{-1} corroborated the monodentate coordination mode of the carboxylate anion to Gd(III). The shift of the band for the carbonyl group from 1700 cm^{-1} to 1690 cm^{-1} compared to the IR spectrum of salinomycinic acid proved the participation of the carbonyl group of the antibiotic in weak hydrogen bonds. The strong, broadband at 3400 cm^{-1} confirmed the presence of hydrogen-bonded hydroxyl groups (IR Spectra, Supplementary Materials Section S3).

The IR spectrum of Mn(II) salinomycinate consisted of four characteristic bands at 1400 cm^{-1}, 1550 cm^{-1}, 1690 cm^{-1}, and 3400 cm^{-1}. The first two bands were assigned to symmetric and asymmetric stretching vibrations of the deprotonated carboxyl group. The difference between both bands ($\Delta \nu \leq 150$ cm^{-1}) [21] confirmed the monodentate coordination of the deprotonated carboxyl group to the paramagnetic metal center. Similar to the IR spectrum of Gd(III) salinomycinate, the shift of the band for the carbonyl group from 1700 cm^{-1} to 1690 cm^{-1}, compared to the IR spectrum of salinomycinic acid, confirmed the participation of the carbonyl group of the antibiotic in weak hydrogen bonds. The broad band at 3400 cm^{-1} was assigned to the stretching vibrations of hydrogen-bonded hydroxyl groups.

The EPR spectra of the Gd(III) salinomycinate, recorded at 100 K and 295 K, are shown in Figure 1. A series of signals, distributed from 0 to 500 mT, were registered. The spectra of Figure 1 demonstrated that, in the whole temperature range, the spectrum remained unchanged independent of the number and the positions of the signals, and the temperature lowering led to an increase in signal intensities. Upon closer observation, some prominent features of the spectra were revealed—the signals with $g_{eff} \approx 6.0$, 2.8, and 2.0. This set of signals is known as the U (ubiquitous) spectrum of Gd^{3+} ions. It should be mentioned that a characteristic feature of the U-spectrum is the almost equal intensity of the signals with $g_{eff} \approx 6.0$, 2.8 [25]. The regarded complex spectrum with the above-described arrangement of the EPR signals is due to isolated Gd^{3+} ions in a low symmetry field with the zero-field splitting parameter D > 0.3 cm^{-1}. Under such a geometry, for the Gd^{3+} ions ($^8S_{7/2}$, $4f^7$ electronic configuration) a multitude of electron transitions is possible between eight energy levels ($m_s = \pm 1/2$, $m_s = \pm 3/2$, $m_s = \pm 5/2$ and $m_s = \pm 7/2$). It is noteworthy that the considered U-spectrum is assigned to Gd^{3+} ions with a coordination number higher than six in non-crystalline materials [26].

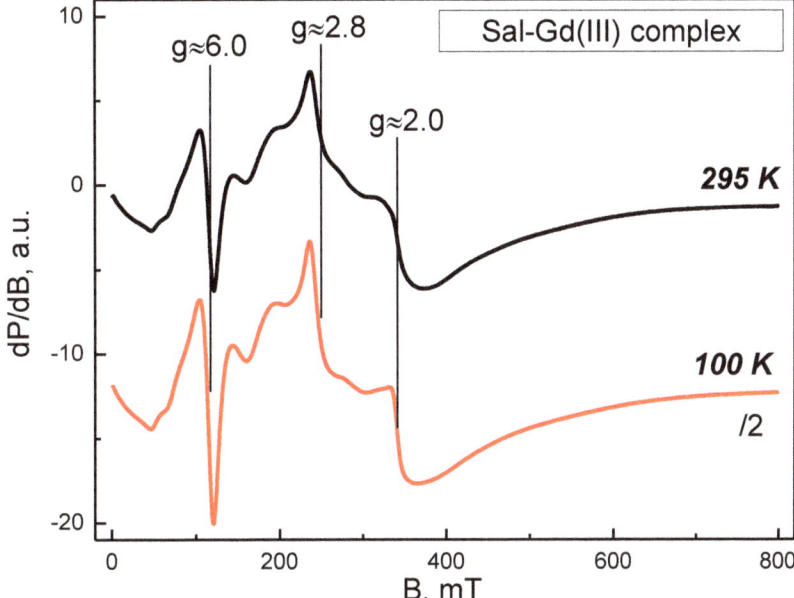

Figure 1. EPR spectra of Sal-Gd(III) complex (solid-state) at 295 and 100 K. The U-spectra signals ($g \approx 6.0$, $g \approx 2.8$, and $g \approx 2.0$) are designated.

The Mn(II) salinomycinate complex was studied by EPR analysis at 100 and 295 K (Figure 2). At both measurement temperatures, identical spectrum features were observed. In the central magnetic field region, a relatively broad signal with the following EPR parameters was registered: $g_{eff} \approx 1.99$, $\Delta H_{pp} \approx 54.0$ mT. Six clearly distinct narrow lines are superimposed on it (Figure 2, inset) at a distance of approximately 9.5 mT. At a lower magnetic field, an additional signal could be noticed with a g-factor around 5.0. The signal with $g \approx 1.99$, as well as the six narrower lines located on it, were assigned to Mn^{2+} ions, which are characterized by ground state $^6S_{5/2}$ and electron and nuclear spin number—S = 5/2, I = 5/2. The registered sextet of lines was attributed to the hyperfine structure of Mn^{2+} ions. Principally, the hyperfine structure possessed a unique value of a hyperfine splitting constant, A_{hfs}, depending on the closest surrounding Mn^{2+} ions. The experimentally determined constant (A_{hfs}) for Mn(II) salinomycinate in this study was 9.5 mT. For comparison, the hyperfine splitting constant of the aqua complex $[Mn(H_2O)_6]^{2+}$ was known to be 9.4 mT and, thus, approximated the corresponding constant found for the Mn(II) salinomycinate complex [27]. This similarity implies the coordination of Mn^{2+} to water molecules and OH groups in the studied complex.

The registration of the EPR signal with $g \approx 5$ suggests the presence of a large zero-field splitting constant, $D > 0.3$ cm^{-1}. Therefore, the EPR spectrum of the Mn(II) salinomycinate complex showed the presence of isolated Mn^{2+} ions in a low-symmetry ligand field.

3.2. Thermogravimetric Analysis (TGA)

TGA and DSC analysis of the two Sal metal complexes were studied. The heating rates were suitably controlled at 10 °C·min^{-1} under airflow of 20 mL·min^{-1}, and the weight loss was measured from ambient temperature to 800 °C, depicted in Figure 3.

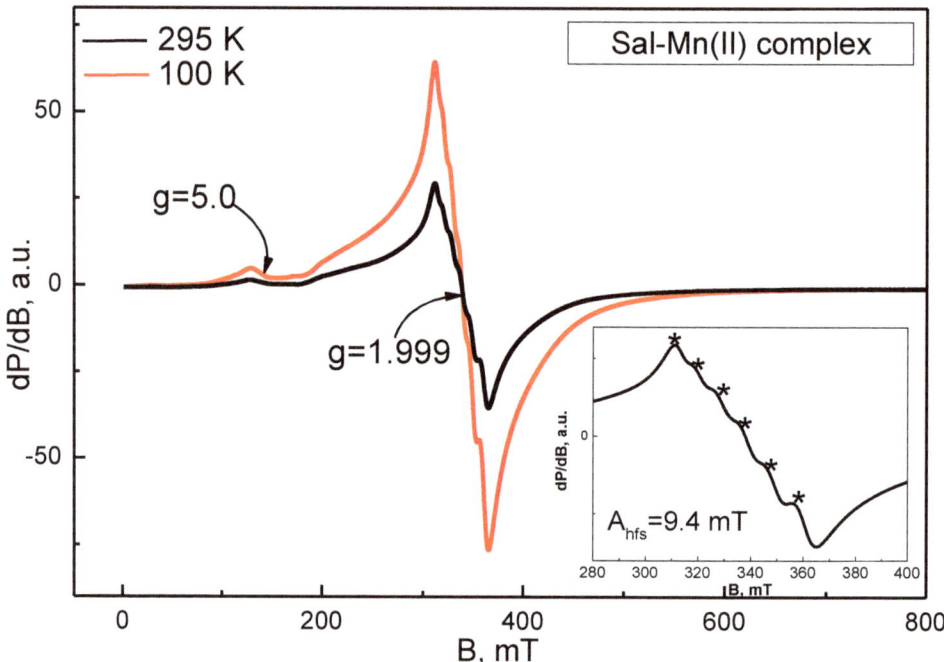

Figure 2. EPR spectra of Sal-Mn(II) complex at 295 and 100 K. The six lines of the Mn^{2+} hyperfine structure are shown in the inset.

The two Sal metal complexes were thermally stable up to 55 °C (TGA; Figure 3A,B). A mass loss in percentage was observed with a broad and flat endothermic peak at 100 °C in the DSC curve for Sal-Mn(II) (Figure 3A) and a more defined endothermic peak at 70 °C for Sal-Gd(III) (Figure 3B). This may be due to the removal of coordinated and absorbed water as depicted in the light green color (3% for the Sal-Mn(II) complex, Figure 3A and 5% for the Sal-Gd(III) one, Figure 3B). The second drop in the masses starting from 150 °C to 475 °C (Figure 3A) and from 150 °C to 440 °C (Figure 3B), highlighted in dark green, can be assigned to the two-step decomposition of the bis- or trisalinomycinates (corresponding to 73% and 72% mass loss respectively (Figure 3A,B). This thermal process is associated with two small exothermic peaks for each Sal complex at 190 °C and 370 °C and at 260 °C and 380 °C in the DSC curve (Figure 3A,B; red-orange line on the right). The final and last step, with a mass loss of 15% between 480–540 °C for the Sal-Mn(II) and 450–540 °C for the Sal-Gd(III) compound, correlated with a strong exothermic signal at 500 °C in the DSC curve. This was most probably due to the oxidation and combustion of the rest of organic matter and the formation of MnO or Gd$_2$O$_3$ as final residues. All these findings are in good correlation with published characterizations of metal complexes using TGA-DSC analyses under air atmosphere [28,29].

3.3. In Vitro MRI

The magnetic susceptibility of the free Sal-complexes and the loaded ones into BGs was determined in a dilution series in MeOH or in Milli-Q® water, respectively, and compared to two clinically used contrast agents, the linear gadopentetate dimeglumine (Magnevist®) and cyclic gadoteridol (ProHance®). A steep signal increase with increasing concentrations has been observed for the prepared contrast agents (Figure 4).

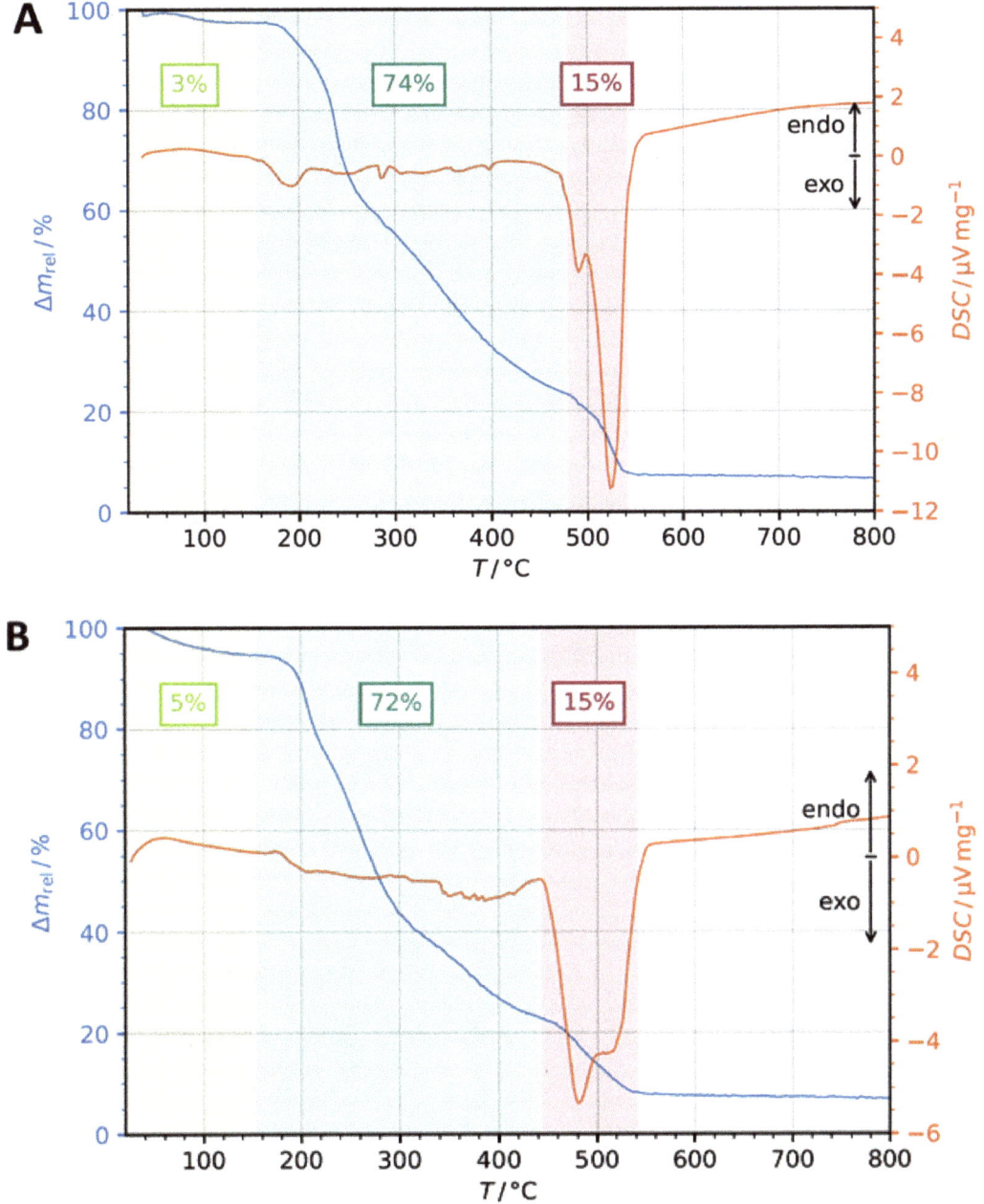

Figure 3. DSC and mass loss profiles of (**A**) Sal-Mn(II) and (**B**) Sal-Gd(III) complexes.

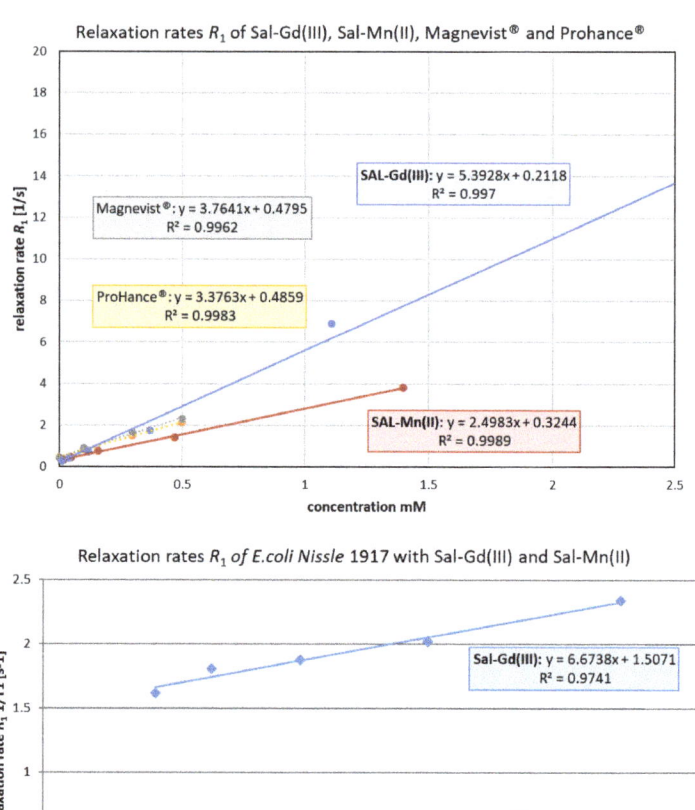

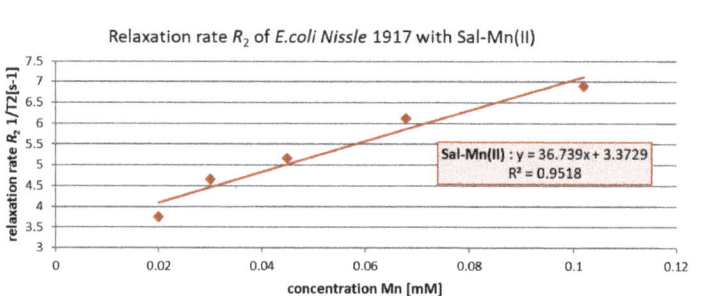

Figure 4. Plots of relaxation rates (R_1/R_2, [1/s]) to concentration (mM) curves for Sal-Gd(III), Sal-Mn(II), and *EcN1917*-BGs loaded with both Sal complexes compared to clinically applied gadopentetate dimeglumine (Magnevist®, linear, grey line) and gadoteridol (ProHance®, cyclic, yellow line). For better comparison, one point from the Sal-Gd(III) graph corresponding to 0.33 mM concentration has been omitted. A strong signal increase in MRI using either the non-loaded Sal-Gd(III) and Sal-Mn(II) or the loaded complexes was shown, comparable with or superior to the effect caused by the clinically most applied CAs, gadopentetate dimeglumine and gadoteridol. The Sal-Gd(III) compound caused a predominant T_1 effect only; thus, just the R_1 dependence with increasing concentration is shown.

4. Cell Culture Studies

4.1. Cytotoxicity Studies

The cytotoxic activity of the salinomycinate complexes alone and when loaded into *EcN1917* and *NM522* was tested against three human cell lines and showed pronounced anticancer potency (Figures 5 and 6 and Table 1). Our results for the cytotoxicity of the paramagnetic complexes of Sal with Mn(II) and Gd(III), and loaded into BGs were compared with those for the cytotoxic activity of four Pt-containing conventional chemotherapeutic drugs and are summarized in Table 3 [30,31]. Data for the cytotoxicity of salinomycin (SalH) and salinomycin-sodium (SalNa) are also given. The results revealed that both paramagnetic complexes of salinomycin are more cytotoxic against the tested tumor cell lines compared to salinomycin and its sodium complex. The effect was more pronounced on SW480 and CH1/PA-1 cell lines. Both paramagnetic complexes of salinomycin showed dose-dependent cytotoxicity against the tested tumor cell lines (Figures 5 and 6).

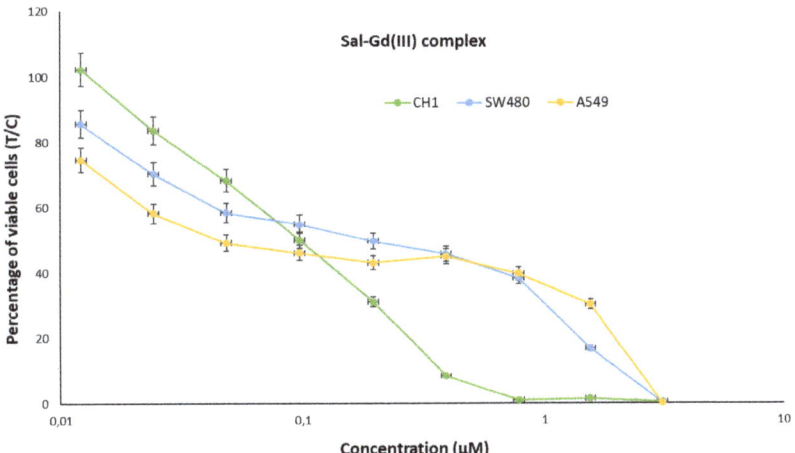

Figure 5. Concentration-effect curves of the Sal-Gd(III) complex in A549 (yellow), CH1/PA-1 (green), and SW480 (blue) cells, determined by the MTT assay after 96 h treatment. Values are normalized relative to untreated controls and represent means and standard deviations of at least three independent experiments.

Table 1. Cytotoxicity evaluation of SalH, SalNa, free Gd(III) and Mn(II) salinomycinates, and with loaded BGs: mean IC_{50} values (in µM) ± standard deviations from at least three independent MTT assays in each of three human cancer cell lines.

	Sample	Cell Line A549	SW480	CH1/PA-1
1.	Sal-H	0.23 ± 0.06	1.1 ± 0.6	0.32 ± 0.12
2.	Sal-Na	0.27 ± 0.02	0.88 ± 0.44	0.43 ± 0.11
3.	$[Mn(Sal)_2(H_2O)_2]$ = Sal-Mn(II)	0.19 ± 0.11 *	0.52 ± 0.22	0.17 ± 0.05
4.	$[Gd(Sal)_3(H_2O)_3]$ = Sal-Gd(III)	0.15 ± 0.12 *	0.36 ± 0.12	0.093 ± 0.025
5.	EcN1917 + 10 mg/mL Sal-Mn(II)	0.22 ± 0.09	0.55 ± 0.06	0.12 ± 0.03
6.	EcN1917 + 10 mg/mL Sal-Gd(III)	0.093 ± 0.043	0.28 ± 0.14	0.086 ± 0.021
7.	NM522 + 5 mg/mL Sal-Mn(II)	0.28 ± 0.17 *	0.54 ± 0.27	0.12 ± 0.02
8.	NM522 + 5 mg/mL Sal-Gd(III)	0.092 ± 0.041	0.31 ± 0.15	0.088 ± 0.008

* The bigger standard deviation is due to a flatter curve.

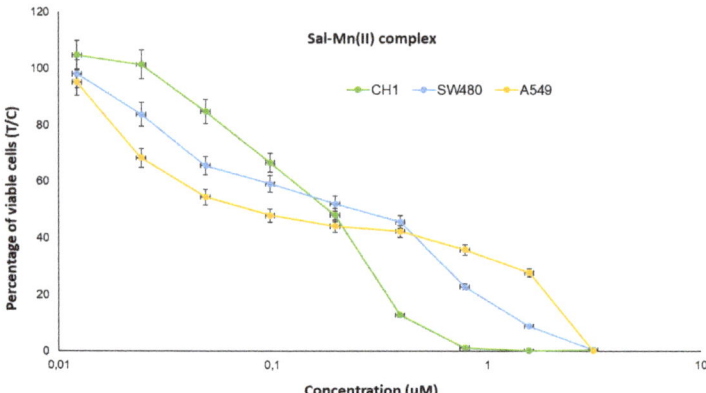

Figure 6. Concentration-effect curves of the Sal-Mn(II) complex in A549 (yellow), CH1/PA-1 (green), and SW480 (blue) cells, determined by the MTT assay after 96 h treatment. Values are normalized relative to untreated controls and represent means and standard deviations of at least three independent experiments.

After incorporation into empty bacterial envelopes, both Sal complexes retained their antitumor activity. Notably, both *NM522* formulations loaded with half the amount of the Sal complex as that for the *EcN1917* exhibited similar cytotoxic efficacy at lower concentrations (Figures 7 and 8). The non-loaded *EcN1917* and *NM522* show no activity.

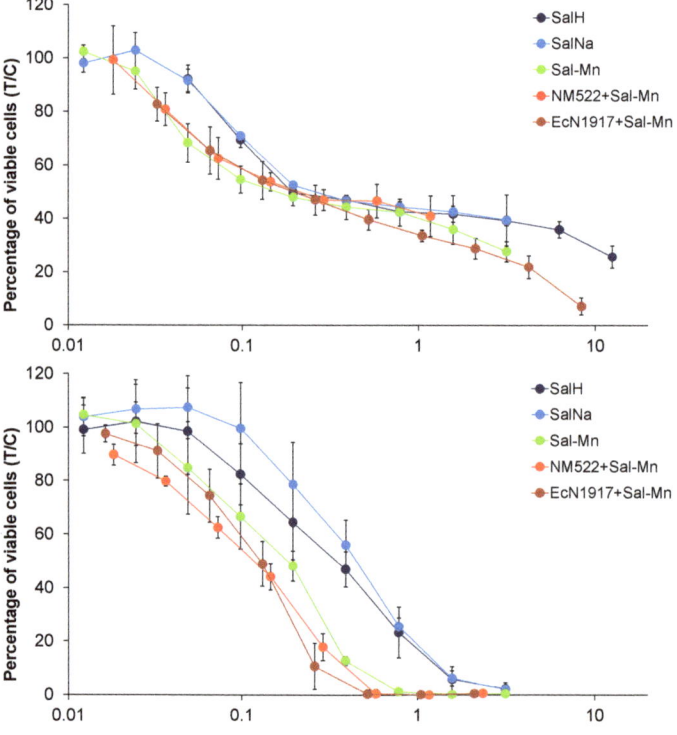

Figure 7. *Cont.*

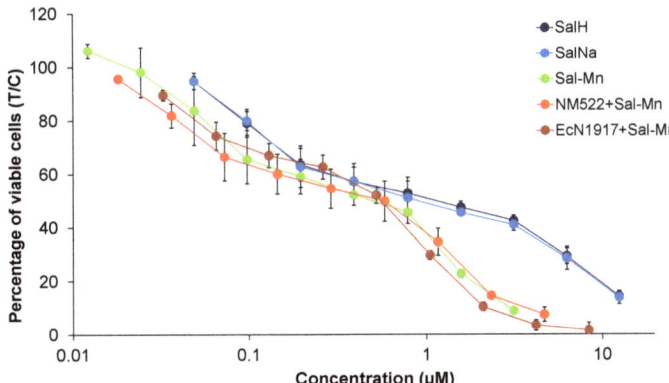

Figure 7. Concentration-effect curves of Sal-H, Sal-Na, Sal-Mn(II), *NM522* + Sal-Mn(II), and *EcN1917* + Sal-Mn(II) in A549 (**top**), CH1/PA-1 (**middle**), and SW480 (**bottom**) cells, determined by the MTT assay after 96 h treatment. Values are normalized relative to untreated controls and represent means and standard deviations of at least three independent experiments.

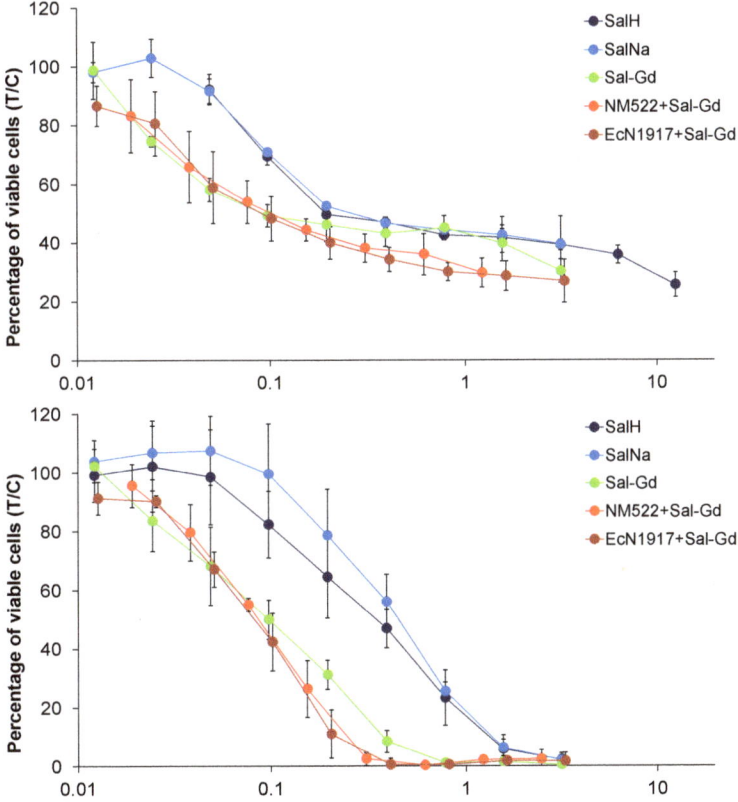

Figure 8. *Cont.*

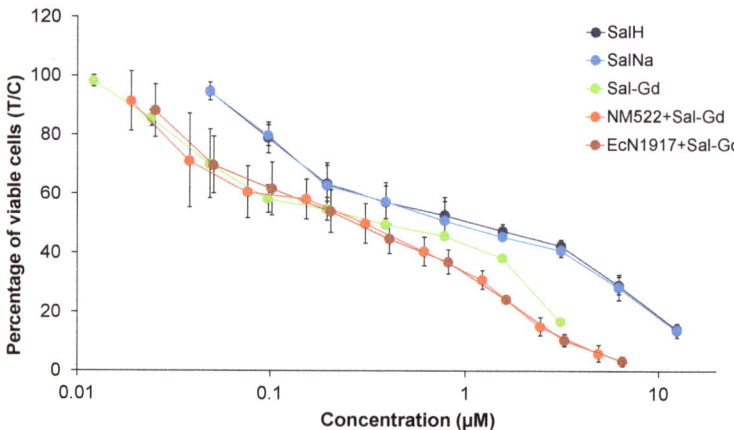

Figure 8. Concentration-effect curves of SalH, SalNa, Sal-Gd(III), *NM522* + Sal-Gd(III) and *EcN1917* + Sal-Gd(III) in A549 (**top**), CH1/PA-1 (**middle**) and SW480 (**bottom**) cells, determined by the MTT assay after 96 h treatment. Values are normalized relative to untreated controls and represent means and standard deviations of at least three independent experiments.

Surprisingly, some of the loaded BGs exerted higher cytotoxic efficacy than the free Sal-Mn(II) and Sal-Gd(III). In particular, *EcN 1917* loaded with Sal-Gd(III) possessed an IC$_{50}$ of only 0.09 µM against A549 and 0.12 µM for *Ec NM522* loaded with half the starting Sal complex as that for *EcN 1917*. The cytotoxic effect of both loaded BGs on SW480 was also more pronounced, with an IC$_{50}$ of 0.27 µM and 0.40 µM relative to the 0.36 µM of the Sal-Gd(III) alone. Tumor cell eradication of CH1/PA-1 cells was similar after the application of both free and loaded Sal-Gd(III). Control Sal-Na exerted less cytotoxic activity after entrapment into BGs. For both A549 and SW480, cancer cell line cytotoxicity decreased after loading of Sal-Mn(II), with IC$_{50}$ values of 0.33 and 0.75 µM, while the effect remained unchanged for CH1/PA-1. Altogether, the prepared free and loaded compounds possessed a superior cytotoxic efficacy on all three cancer cell lines tested (Table 1).

4.2. Cell Cycle Studies

In general, the SW480 cell line was more sensitive to tested compounds than the MCF-7 cell line. The concentrations of 4–8 µM were sufficient for altering the cell cycle distribution in colon cancer cells upon 24 h treatment (Figure S1). Salinomycinic acid (Figure S1A) and its metal complexes (Figure S1C,E) diminished the SW480 cell fraction in the G0/G1-phase (by 8–17%) and correspondingly increased the cell fraction in the S-phase (by 9–10%), followed by a G2/M-phase increase at the highest concentrations (by 7–8%). Effects on the cell cycle distribution of MCF-7 breast cancer cells were more pronounced at higher (16–32 µM) concentrations (Figure S1B,D,F). The tested compounds mainly decreased the number of cells in the G0/G1-phase (by 15–18%) and increased the number of cells in the G2/M-phase (5–13%), with a minor effect on the S-phase fraction (±6%).

The concentration-dependent effect of the individual compounds on SW480 and MCF-7 cells is exemplified in Figures S2 and S3, respectively. Sal-H was the least active compound in both cell lines, with moderate activity at the highest concentrations applied. Sal-Mn(II) and Sal-Gd(III) demonstrated a more pronounced effect on the cell cycle in both colon and breast cancer cells. The direct correlation between the number of salinomycin ligands in the complex and the potency to induce cell cycle perturbations could be observed. Sal-Gd(III) demonstrated the highest activity in both cell lines (Figures S2 and S3C).

5. Discussion

Pilot clinical trials have demonstrated that salinomycin inhibited the disease progression in patients diagnosed with invasive carcinoma. Data revealed antibiotic-induced apoptosis in the metastatic cancer cells, and no severe, long-lasting side effects were observed [32]. These promising results demonstrated the remarkable potency of salinomycin for cancer treatment. Coordination of the antibiotic to divalent metal ions can further enhance its cytotoxicity and diminish its toxicity [21,32,33]. Chelated metal ions can also provide a convenient handle for bioconjugation with other molecules via axial coordination. Herein, we explored the structures of two paramagnetic complexes of salinomycin and demonstrated, for the first time, their potency as theranostic probes. Moreover, the loading of the novel potent theranostics into BGs may provide a targeted drug delivery that can circumvent the water insolubility of the free Sal-complexes and pave the way to clinical translation.

The IR spectra of salinomycin with Mn(II) and Gd(III) reported in this study were similar to the IR spectra of $[Co(C_{42}H_{69}O_{11})_2(H_2O)_2]$, $[Ni(C_{42}H_{69}O_{11})_2(H_2O)_2]$, $[Cu(C_{42}H_{69}O_{11})_2(H_2O)_2]$, $[Zn(C_{42}H_{69}O_{11})_2(H_2O)_2]$, and $[Cd(C_{42}H_{69}O_{11})_2(H_2O)_2]$ discussed previously [33,34]. The literature spectroscopic data for M(II) salinomycinates have suggested that the organic ligand was coordinated to the metal center via a terminal deprotonated carboxyl group and a terminal secondary hydroxyl group [21,34]. A distorted octahedral molecular geometry of the complexes of salinomycin with M(II) has been proposed [21,34].

Herein, we applied a solid-state EPR spectroscopy to obtain more detailed information about the structures of salinomycin complexes with Gd(III) and Mn(II). The complex EPR spectrum of Gd(III) salinomycinate is due to Gd^{3+} ions placed in a crystal field with low symmetry and characterized by a high splitting constant at zero magnetic field [35,36]. For the Mn(II) salinomycinate, the described broad signal consists of six superimposed lines, which are attributable to the ultrafine interaction characteristics of Mn(II) ions (nuclear spin I = 5/2). The calculated superfine interaction constant A_{hfs} is ca. 9.5 mT. For comparison, the superfine interaction constant for $[Mn(H_2O)_6]^{2+}$ is 9.4 mT [27]. The additional signal at $g \approx 5.0$ suggests that the cleavage constant of Cramer doublets in a zero magnetic field is significant (above 0.3 cm^{-1}). A quantitative EPR analysis was additionally applied for the estimation of the metal ion to ligand ratio. The ratio Gd^{3+} to salinomycin was established to be 1:5. The discrepancy to the formula obtained via elemental analysis could be explained by some loss of EPR signal as a result of the large zero-field splittings constant value. The experimentally evaluated ratio Mn^{2+} to salinomycin was found to be 1:2.5 showing a good agreement with the results from elemental analysis. The presence of aqua ligands in the coordination sphere of the Sal-Mn(II) complex was also supposed by the EPR analysis.

When taken together, the data from the elemental analysis and spectroscopic studies (IR, ESI-MS, EPR) allowed us to conclude that the complexes of salinomycin with Gd(III) and Mn(II) possess a tricapped, trigonal, prismatic geometry (Figure 9) and an octahedral molecular geometry (Figure 10) respectively. The salinomycinate monoanions are coordinated to the metal center via a deprotonated carboxyl group and a terminal secondary hydroxyl group. The water molecules coordinated to the metal center stabilize the pseudo-cyclization of the complex by hydrogen bonds with the organic ligand. In addition, TGA and DCS studies confirmed coordinated water molecules for both Gd(III) and Mn(II).

Relaxivity (r_1 or r_2) in MRI is defined as the paramagnetic enhancement of the water protons relaxation rates (R_1 and R_2) caused by 1 mM contrast agent concentration. It thus directly relates to the MRI efficiency of the contrast agent. The majority of CAs are based on paramagnetic ions (predominately Gd(III)) chelated in stable complexes, which reduce the longitudinal relaxation time (T_1) of water protons in the body. Consequently, positive (bright) T_1-weighted MR images can be seen (T_1-enhancers).

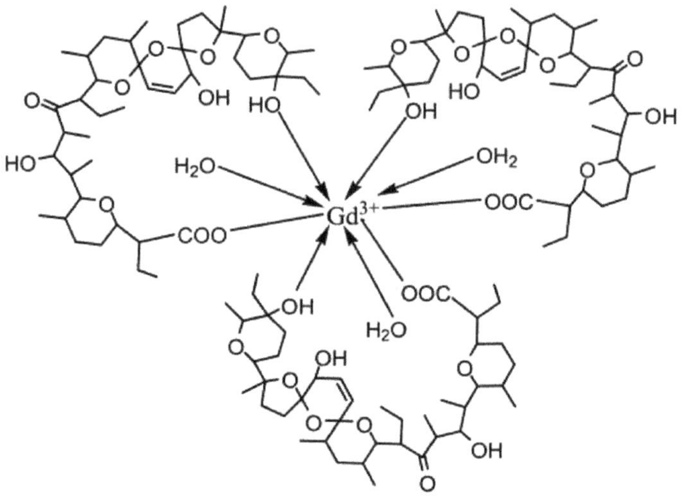

Figure 9. Proposed chemical structure of [Gd(C$_{42}$H$_{69}$O$_{11}$)$_3$(H$_2$O)$_3$].

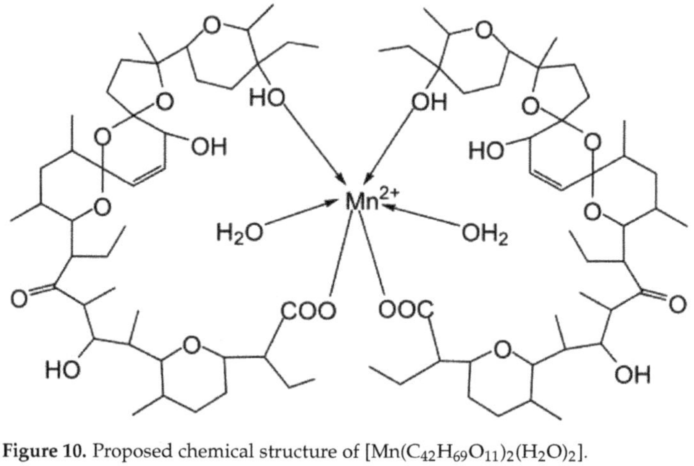

Figure 10. Proposed chemical structure of [Mn(C$_{42}$H$_{69}$O$_{11}$)$_2$(H$_2$O)$_2$].

As new MRI contrast agents, both the free and the loaded into BGs Gd(III) salinomycinate and Mn(II) salinomycinate were evaluated for their ability to modify the R_1 of water protons on a high-field scanner (9.4 Tesla; BioSpec 94/30USR, Bruker, Germany). The measured relaxivity values (r_1) of Gd(III) salinomycinate and Mn(II) salinomycinate, 5.4 and 2.5 1/mM*s, respectively, are higher than or as high as the most clinically applied probes (Figure 4). Even though the r_1 of Mn(II) salinomycinate is not as high as that of the salinomycin complex with Gd(III), Sal-Mn(II) causes an excellent contrast in MRI and may serve as a good alternative to the Gd-based contrast agents (GBCA). Moreover, Mn-based agents have shown a better diagnostic performance than Gd-based agents in certain disease areas, such as pancreatic lesions [37]. Compared to other Mn-based small molecular MRI contrast agents, Mn(II) salinomycinate (r_1 = 2.5 1/mM*s) shows a relaxivity as high as that recently reported by Wang et al. for Mn-PyC3A-3-OBn (r_1 = 2.6 1/mM·s, in Tris buffer at 1.4 T), a novel Mn(II)-based complex designed for liver-specific imaging [38]. The only Mn(II)-based complex that was clinically available for some time was the liver-specific Mangafodipir® or Teslascan® ([Mn(dpdp)]$^{4-}$), in which the relaxation enhancement arises from the Mn(II) ion released from the complex in vivo [39]. The Mn(II) complex of salinomycin, described in this study, exerts a higher relaxivity than Teslascan® generates in vivo (r_1 = 2.3 1/mM·s, at 1 T, MRI-relaxivity MR-TIP: Database) and a comparable strong signal increase compared to the aqueous solutions of Teslascan® (r_1 = 2.8 1/mM·s) [39].

All Sal-complexes loaded into BGs showed a superior signal enhancement in MRI, with r_1 = 2.578 1/mM·s and a pronounced r_2 = 36.74 1/mM·s for the bacterial cells, carrying Sal-Mn and an increased r_1 = 6.674 1/mM·s for those with entrapped Sal-Gd(III). This is comparable with the r_1 of clinically used Magnevist® and ProHance® and about ten times higher when compared to the r_2 = 3.70 [1/mM·s] reported for Mangafodipir®. The contrast efficiency of the loaded BGs with Sal-Gd(III) (r_1 = 6.674 1/mM·s) improved approximately 1.8 times compared to the clinically applied Magnevist® (r_1 = 3.76 [1/mM·s]).

The summarized proton relaxivities of our Gd(III) and Mn(II) complexes of salinomycin with regard to various published low-molecular-weight paramagnetic complexes ere shown in Table 2.

Table 2. Comparison of proton relaxivity r_1 in [1/mM·s] for Sal-Gd(III), Sal-Mn(II), and *E. coli Nissle 1917* loaded with the two Sal complexes with various low-molecular-weight Gd(III) and Mn(II) complexes.

Complex	Relaxivity Values r_1/r_2, [1/mM·s] (RT/20 °C/37 °C)	Magnetic Field Strength, [T]
Gd-DTPA^{2-} (Magnevist®)	3.76/n.d.	9.4
Gd-HPDO3A (ProHance®)	3.37/n.d.	9.4
JTIP1/[Mn(Sal)$_2$(H$_2$O)$_2$]	2.50/n.d.	9.4
JTIP3/[Gd(Sal)$_3$(H$_2$O)$_3$]	5.40/n.d.	9.4
E. coli Nissle 1917 loaded with JTIP1/[Mn(Sal)$_2$(H$_2$O)$_2$]	2.58/36.74	9.4
E. coli Nissle 1917 loaded with JTIP3/[Gd(Sal)$_3$(H$_2$O)$_3$]	6.67/n.a.	9.4
[Mn(DPDP)]$^{4-}$ (Teslascan®) [39]	2.80/3.70	4.7
Mn-PyC3A [40]	3.4/n.d.	3
Mn-EDTA-BTA [41]	3.50	0.47
Mn(HBET-NO$_2$)]$^{2-}$ [42]	2.33	1.4
Mn-CDTA^{2-} [43]	2.99	0.47
Mn-EDTA^{2-} [43]	2.12	0.47
Mn-DTPA^{3-} [43]	1.57	0.47

With regard to the T_1 values and the calculated relaxivities (r_1), it is known that, at higher field strength, the relaxivity values tend to decrease slightly [44]. Nevertheless, the table above truly illustrates the great potential of the newly prepared salinomycinates as contrast agents for MRI. In addition, r_1 is always influenced by several physical and chemical parameters, such as the strength and homogeneity of the applied magnetic field, the hydration state of the contrast agent, its molecular size, internal and anisotropic rotation, water exchange rates, and molecular tumbling [45].

Notably, the chemical composition of Gd(III) salinomycinate and Mn(II) salinomycinate comprises three and two inner-sphere water molecules, respectively, which presumably contributes to higher relaxivity values and, in turn, to an increased contrast in MRI. It is known that the interaction dynamics between water and Gd(III) or Mn(II) complexes highly affect the relaxivity, while the mechanism remains unclear. Based on Solomon-Bloembergen-Morgan (SBM) theory, the direct water coordination to the paramagnetic metal centers is the major contributor to T_1 inner-sphere relaxivity [40].

As expected, the signal enhancement in MRI caused by the Sal-complexes loaded within the bacterial cells was superior compared to the non-loaded free Sal-Gd(III) and

Sal-Mn(II), even at lower Gd(III) or Mn(II) concentrations (0.06–0.1 mM Gd(III) and 0.05–0.1 mM Mn(II) applied). This fact could be explained by a slowed molecular tumbling of the macromolecular constructs and a more effective water exchange rate provided by the aqueous media in which the BGs were resuspended.

Remarkably, the complexes of salinomycin with Gd(III) and Mn(II) and the loaded BGs exerted superior cytotoxicity on A549 compared to all conventional Pt-containing clinical chemotherapeutics (Table 3) [30,31]. All compounds demonstrated much higher antitumor activity against SW480 than did satraplatin, cisplatin, and carboplatin, and comparable cytotoxicity to oxaliplatin [30,31]. The cytotoxicity of salinomycin with Mn(II) against CH1/PA-1 was similar to that of oxaliplatin and satraplatin, more pronounced in contrast to carboplatin, and lower compared to cisplatin. The cytotoxicity of the Gd(III) complex of salinomycin against CH1/PA-1 was either much more elevated set against clinical chemotherapeutics (carboplatin, oxaliplatin) or comparable (satraplatin, cisplatin) [30,31].

Table 3. Cytotoxic activity of paramagnetic complexes of salinomycin with Gd(III) and Mn(II), loaded BGs and Pt-containing clinical chemotherapeutics: mean IC_{50} values (in µM) ± standard deviations from at least three independent MTT assays in each of three human cancer cell lines.

	Sample / Cell Line	A549	SW480	CH1/PA-1
1.	Sal-H	0.23 ± 0.06	1.06 ± 0.58	0.32 ± 0.12
2.	Sal-Na	0.27 ± 0.02	0.88 ± 0.44	0.43 ± 0.11
3.	[Mn(Sal)$_2$(H$_2$O)$_2$] = Sal-Mn(II)	0.19 ± 0.11	0.52 ± 0.22	0.17 ± 0.05
4.	[Gd(Sal)$_3$(H$_2$O)$_3$] = Sal-Gd(III)	0.15 ± 0.12	0.36 ± 0.12	0.09 ± 0.03
5.	EcN1917 + 10 mg/mL Sal-Mn(II)	0.33 ± 0.14	0.82 ± 0.10	0.19 ± 0.04
6.	EcN1917 + 10 mg/mL Sal-Gd(III)	0.09 ± 0.04	0.27 ± 0.13	0.082 ± 0.020
7.	NM522 + 5 mg/mL Sal-Mn(II)	0.75 ± 0.46	1.44 ± 0.74	0.32 ± 0.054
8.	NM522 + 5 mg/mL Sal-Gd(III)	0.12 ± 0.05	0.40 ± 0.19	0.112 ± 0.010
Satraplatin [30]	Pt(NH$_3$)(cha)Cl$_2$(OAc)$_2$	6.4 ± 0.4	1.5 ± 0.1	0.10 ± 0.12
Cisplatin [30]	Pt(NH$_3$)$_2$Cl$_2$	6.2 ± 1.2	3.3 ± 0.2	0.077 ± 0.006
Carboplatin [46]	Pt(NH$_3$)$_2$(CBDCA)	91 ± 102	37 ± 1	0.81 ± 0.17
Oxaliplatin [31]	Pt(DACH)(ox)	0.98 ± 0.21	0.29 ± 0.05	0.18 ± 0.01

The antiproliferative activity of Mn(II) salinomycinate against the more chemo-resistant A549 cell line was much more pronounced compared to that of Mn(II) with other ligands [47–51]. The IC_{50} value for [Mn(Sal)$_2$(H$_2$O)$_2$] was from eight to 984 times lower compared to IC_{50} values for the other Mn(II) complexes. This enormous difference in the cytotoxicity of Mn(II) complexes against the A549 cell line could not be attributed to the differences in the experimental protocols. We found only one study where the cytotoxicity data for Gd(III) complexes with other ligands against A549 were reported [52]. The cytotoxicity of the Sal-Gd(III) complex against the A549 cell line was superior compared to the Gd(III) complex described in the literature (the IC_{50} value was 733 times lower compared to the value for the published complex) [53].

The results demonstrate that salinomycinic acid and its metal-based complexes Sal-Gd(III) and Sal-Mn(II) may induce cell cycle perturbations in both breast cancer MCF-7 and colon cancer SW480 cells, shifting the cell number distribution from G1/G0- towards the S- and G2/M-phases. In line with our data, Niwa et al. reported that a 24 h treatment of MCF-7 cells with 20 µM salinomycin significantly increased cell numbers in the G2/M-phase at the expense of cells in the G0/G1-phase [54].

The combination of MRI contrast agents and a highly effective anticancer drug into a single construct, such as a Sal-based theranostic complex formulated within bacterial ghosts as a transport and targeting device, may offer a powerful "all in one" tool for diagnosis, therapy, and monitoring of various cancers.

6. Conclusions

In this study, we synthesized and characterized, for the first time, the Gd(III) complex of salinomycin. We provide new data about the structure and cytotoxicity of Mn(II) salinomycinate. Further loading of both Sal-complexes into *EcN*1917 and *NM*522 BGs as transport vehicles was successfully carried out. The higher payload of paramagnetic metal ions and an anticancer drug within the BGs may provide better biocompatibility, as well as low/no toxicity or undesired side effects, making these new formulations suitable for clinical applications. To the best of our knowledge, this is the first study that demonstrates the potential use of salinomycin complexes with Gd(III) and Mn(II) as theranostic agents for MRI. This is the first report that a natural, polyether ionophorous antibiotic can act as a chelator of paramagnetic metals and an anticancer drug, exerting a superior cytotoxic effect and favorable MRI properties.

Supplementary Materials: The following supporting information can be downloaded at: https://www.mdpi.com/article/10.3390/pharmaceutics14112319/s1, Section S1. ESI-MS spectra and fragmentation scheme of [Gd(C$_{42}$H$_{69}$O$_{11}$)$_3$(H$_2$O)$_3$] complex; Section S2. ESI-MS spectra and fragmentation scheme of [Mn(C$_{42}$H$_{69}$O$_{11}$)$_2$(H$_2$O)$_2$] complex; Section S3. IR spectra of SalH, Sal-Gd(III) Sal-Mn(II); Figure S1. Cell cycle alterations in SW480 (left) and MCF-7 (right) cancer cells; Figure S2. Histograms of cell cycle phase distribution in SW480 cells exposed to increasing concentrations of Sal-H (column A), Sal-Mn(II) (column B) and Sal-Gd(III) (column C) for 24 h; Figure S3. Histograms of cell cycle phase distribution in MCF-7 cells exposed to increasing concentrations of Sal-H (column A), Sal-Mn(II) (column B) and Sal-Gd(III) (column C) for 24 h.

Author Contributions: I.P.-M.: Conceptualization, project management, methodology, data curation, formal analysis, investigation, writing—original draft, writing—review & editing. J.F.: MRI data curation, formal analysis, software, validation, writing—original draft, writing—review & editing. R.K. and R.S.: EPR analysis of both complexes. I.P.: ESI-MS spectra interpretation, writing—review & editing. P.D.: synthesis of salinomycin sodium. M.H. (Michaela Hejl): Cell culture studies, data curation. M.J.: Supervision and interpretation of cell culture studies. M.H. (Mariam Hohagen): TGA measurements, data curation, interpretation of results. A.L.: Cell cycle studies, data curation, and interpretation. W.L.: Production and supply of the bacterial ghost cells. B.K.K.: Supervision, resources. T.H.H.: Supervision, resources, writing—original draft, writing—review & editing. J.I.: Synthesis of the complexes of salinomycin with Gd(III) and Mn(II), project management, methodology, data curation, formal analysis, investigation, writing—original draft, writing—review & editing. All authors have read and agreed to the published version of the manuscript.

Funding: The authors acknowledge the financial support of this study by the Bulgarian National Science Fund (grant number: KP-06-Austria-6/6.08.2019) and the Austrian Federal Ministry of Education, Science, and Research (BMBWF) (project number: BG 07/2019). Irena Pashkunova-Martic is grateful to the Austrian Science Fond (FWF, T 1145-B) for covering her personnel costs.

Institutional Review Board Statement: Not applicable.

Informed Consent Statement: Not applicable.

Data Availability Statement: All other relevant data of this study are available from the corresponding authors upon reasonable request.

Acknowledgments: The authors thank associates Vladimir Gelev and Alexander Ronacher for ESI-MS measurements. The authors are grateful to Mary McAllister for her efforts in reading and editing the manuscript.

Conflicts of Interest: Irena Pashkunova-Martic and Juliana Ivanova declare pending patent application (patent applicants: Irena Pashkunova-Martic and Juliana Ivanova, inventors: Irena Pashkunova-Martic and Juliana Ivanova, patent application №A50210/2022). All other authors declare no conflict of interest. The company had no role in the design of the study; in the collection, analyses, or interpretation of data; in the writing of the manuscript, and in the decision to publish the results.

References

1. Jeyamogan, S.; Khan, N.A.; Siddiqui, R. Application and Importance of Theranostics in the Diagnosis and Treatment of Cancer. *Arch. Med. Res.* **2021**, *52*, 131–142. [CrossRef]
2. Sahraei, Z.; Mirabzadeh, M.; Fadaei-Fouladi, D.; Eslami, N.; Eshraghi, A. Magnetic Resonance Imaging Contrast Agents: A Review of Literature. *J. Pharm. Care* **2015**, *2*, 177–182.
3. Kojima, R.; Aubel, D.; Fussenegger, M. Novel theranostic agents for next-generation personalized medicine: Small molecules, nanoparticles, and engineered mammalian cells. *Curr. Opin. Chem. Biol.* **2015**, *28*, 29–38. [CrossRef]
4. Keam, S.J. Lutetium Lu 177 Vipivotide Tetraxetan: First Approval. *Mol. Diagn. Ther.* **2022**, *26*, 467–475. [CrossRef]
5. Lohrke, J.; Frenzel, T.; Endrikat, J.; Alves, F.C.; Grist, T.M.; Law, M.; Lee, J.M.; Leiner, T.; Li, K.-C.; Nikolaou, K.; et al. 25 Years of Contrast-Enhanced MRI: Developments, Current Challenges and Future Perspectives. *Adv. Ther.* **2016**, *33*, 1–28. [CrossRef]
6. Thomsen, H.S. NSF: Still relevant. *J. Magn. Reson. Imaging* **2014**, *40*, 11–12. [CrossRef]
7. DI Gregorio, E.; Ferrauto, G.; Furlan, C.; Lanzardo, S.; Nuzzi, R.; Gianolio, E.; Aime, S. The Issue of Gadolinium Retained in Tissues: Insights on the Role of Metal Complex Stability by Comparing Metal Uptake in Murine Tissues Upon the Concomitant Administration of Lanthanum- and Gadolinium-Diethylentriamminopentaacetate. *Investig. Radiol.* **2018**, *53*, 167–172. [CrossRef]
8. Sato, T.; Ito, K.; Tamada, T.; Kanki, A.; Watanabe, S.; Nishimura, H.; Tanimoto, D.; Higashi, H.; Yamamoto, A. Tissue gadolinium deposition in renally impaired rats exposed to different gadolinium-based MRI contrast agents: Evaluation with inductively coupled plasma mass spectrometry (ICP-MS). *Magn. Reson. Imaging* **2013**, *31*, 1412–1417. [CrossRef] [PubMed]
9. U.S. Food and Drug Administration. *FDA Drug Safety Communication: FDA Warns that Gadolinium Based Contrast Agents (GBCAs) are Retained in the Body*; Requires New Class Warnings; U.S. Food and Drug Administration: Washington, DC, USA, 2017.
10. EMA's Final Opinion Confirms Restrictions on Use of Linear Gadolinium Agents in Body Scans. Available online: http://www.ema.europa.eu/docs/en_GB/document_library/Referrals_document/gadolinium_contrast_agents_31/Opinion_provided_by_Committee_for_Medicinal_Products_for_Human_Use/WC500231824.pdf (accessed on 21 July 2017).
11. Zhen, Z.; Xie, J. Development of Manganese-Based Nanoparticles as Contrast Probes for Magnetic Resonance Imaging. *Theranostics* **2012**, *2*, 45–54. [CrossRef]
12. Naujokat, C.; Fuchs, D.; Opelz, G. Salinomycin in cancer: A new mission for an old agent. *Mol. Med. Rep.* **2010**, *3*, 555–559. [CrossRef]
13. Gupta, P.B.; Onder, T.T.; Jiang, G.; Tao, K.; Kuperwasser, C.; Weinberg, R.A.; Lander, E.S. Identification of Selective Inhibitors of Cancer Stem Cells by High-Throughput Screening. *Cell* **2009**, *138*, 645–659. [CrossRef] [PubMed]
14. Schaffhausen, J. Remaining hurdles to effective cancer therapy. *Trends Pharmacol. Sci.* **2015**, *36*. [CrossRef]
15. Jiang, J.; Li, H.; Qaed, E.; Zhang, J.; Song, Y.; Wu, R.; Bu, X.; Wang, Q.; Tang, Z. Salinomycin, as an autophagy modulator– a new avenue to anticancer: A review. *J. Exp. Clin. Cancer Res.* **2018**, *37*, 26. [CrossRef] [PubMed]
16. Wang, Q.; Liu, F.; Wang, L.; Xie, C.; Wu, P.; Du, S.; Zhou, S.; Sun, Z.; Liu, Q.; Yu, L.; et al. Enhanced and Prolonged Antitumor Effect of Salinomycin-Loaded Gelatinase-Responsive Nanoparticles via Targeted Drug Delivery and Inhibition of Cervical Cancer Stem Cells. *Int. J. Nanomed.* **2020**, *15*, 1283–1295. [CrossRef] [PubMed]
17. Antoszczak, M.; Huczyński, A. Salinomycin and its derivatives—A new class of multiple-targeted "magic bullets". *Eur. J. Med. Chem.* **2019**, *176*, 208–227. [CrossRef]
18. Dewangan, J.; Srivastava, S.; Rath, S.K. Salinomycin: A new paradigm in cancer therapy. *Tumor Biol.* **2017**, *39*, 1010428317695035. [CrossRef]
19. Huczynski, A. Salinomycin—A New Cancer Drug Candidate. *Chem. Biol. Drug Des.* **2012**, *79*, 235–238. [CrossRef]
20. Momekova, D.; Momekov, G.; Ivanova, J.; Pantcheva, I.; Drakalska, E.; Stoyanov, N.; Guenova, M.; Michova, A.; Balashev, K.; Arpadjan, S.; et al. Sterically stabilized liposomes as a platform for salinomycin metal coordination compounds: Physicochemical characterization and in vitro evaluation. *J. Drug Deliv. Sci. Technol.* **2013**, *23*, 215–223. [CrossRef]
21. Ivanova, J.; Pantcheva, I.N.; Zhorova, R.; Momekov, G.; Simova, S.; Stoyanova, R.; Zhecheva, E.; Ivanova, S.; Mitewa, M. Synthesis, spectral properties, antibacterial and antitumor activity of salinomycin complexes with Co(II), Ni(II), Cu(II) and Zn(II) transition metal ions. *J. Chem Chem Eng.* **2012**, *6*, 551–562.
22. Koller, V.J.; Dirsch, V.M.; Beres, H.; Donath, O.; Reznicek, G.; Lubitz, W.; Kudela, P. Modulation of bacterial ghosts—induced nitric oxide production in macrophages by bacterial ghost-delivered resveratrol. *FEBS J.* **2013**, *280*, 1214–1225. [CrossRef]
23. Langemann, T.; Koller, V.J.; Muhammad, A.; Kudela, P.; Mayr, U.B.; Lubitz, W. The bacterial ghost platform system: Production and applications. *Bioeng. Bugs* **2010**, *1*, 326–336. [CrossRef]
24. Tabrizi, C.A.; Walcher, P.; Mayr, U.B.; Stiedl, T.; Binder, M.; McGrath, J.; Lubitz, W. Bacterial ghosts—biological particles as delivery systems for antigens, nucleic acids and drugs. *Curr. Opin. Biotechnol.* **2004**, *15*, 530–537. [CrossRef]
25. Mazur, M.; Poprac, P.; Valko, M.; Rhodes, C.J. 'U-spectrum' type of Gd(III) EPR spectra recorded at various stages of TEOS-based sol–gel process. *J. Sol-Gel Sci. Technol.* **2016**, *79*, 220–227. [CrossRef]
26. Kiran, N.; Kummara, V.K.; Ravi, N.; Lenine, D. Structure and EPR investigations on Gd3+ ions in magnesium-lead-borophosphate glasses. *J. Mol. Struct.* **2020**, *1208*, 127877. [CrossRef]
27. Gagnon, D.M.; Hadley, R.C.; Ozarowski, A.; Nolan, E.M.; Britt, R.D. High-Field EPR Spectroscopic Characterization of Mn(II) Bound to the Bacterial Solute-Binding Proteins MntC and PsaA. *J. Phys. Chem. B* **2019**, *123*, 4929–4934. [CrossRef] [PubMed]

28. Thurston, J.H.; Trahan, D.; Ould-Ely, T.; Whitmire, K.H. Toward a General Strategy for the Synthesis of Heterobimetallic Coordination Complexes for Use as Precursors to Metal Oxide Materials: Synthesis, Characterization, and Thermal Decomposition of $Bi_2(Hsal)_6 \cdot M(Acac)_3$ (M = Al, Co, V.; Fe, Cr). *Inorg. Chem.* **2004**, *43*, 3299–3305. [CrossRef] [PubMed]
29. Moreira, J.M.; Campos, G.F.; Pinto, L.M.D.C.; Martins, G.R.; Tirloni, B.; Schwalm, C.S.; de Carvalho, C.T. Copper (II) complexes with novel Schiff-based ligands: Synthesis, crystal structure, thermal (TGA–DSC/FT-IR), spectroscopic (FT-IR, UV-Vis) and theoretical studies. *J. Therm. Anal. Calorim.* **2022**, *147*, 4087–4098. [CrossRef]
30. Varbanov, H.P.; Göschl, S.; Heffeter, P.; Theiner, S.; Roller, A.; Jensen, F.; Jakupec, M.A.; Berger, W.; Galanski, M.S.; Keppler, B.K. A Novel Class of Bis- and Tris-Chelate Diam(m)inebis(dicarboxylato)platinum(IV) Complexes as Potential Anticancer Prodrugs. *J. Med. Chem.* **2014**, *57*, 6751–6764. [CrossRef]
31. Banfić, J.; Legin, A.A.; Jakupec, M.A.; Galanski, M.; Keppler, B.K. Platinum(IV) complexes featuring one or two axial ferrocene bearing ligands—synthesis, characterization, and cytotoxicity. *Eur J. Inorg Chem.* **2014**, *2014*, 484–492. [CrossRef]
32. Naujokat, C.; Steinhart, R. Salinomycin as a Drug for Targeting Human Cancer Stem Cells. *J. Biomed. Biotechnol.* **2012**, *2012*, 950658. [CrossRef]
33. Atanasov, V.; Stoykova, S.; Goranova, Y.; Nedzhib, A.; Tancheva, L.; Ivanova, J.; Pantcheva, I. Preliminary study on in vivo toxicity of monensin, salinomycin and their metal complexes. *Bul. Chem.Commun.* **2014**, *46*, 233–273.
34. Ivanova, J.; Pantcheva, I.N.; Mitewa, M.; Simova, S.; Tanabe, M.; Osakada, K. Cd(II) and Pb(II) complexes of the polyether ionophorous antibiotic salinomycin. *Chem. Cent. J.* **2011**, *5*, 52. [CrossRef] [PubMed]
35. Azzoni, C.B.; Martino, D.D. EPR study of Gd3+ doped lead oxide based glasses. *J. Mater. Sci.* **1999**, *34*, 3931–3935. [CrossRef]
36. Malchukova, E.; Boizot, B.; Ghaleb, D. Guillaume Petite β-Irradiation Effects in Gd-Doped Borosilicate Glasses Studied by EPR and Raman Spectroscopies. *J. Non Cryst Solids* **2006**, *352*, 297–303. [CrossRef]
37. Diehl, S.J.; Lehmann, K.J.; Gaa, J.; McGill, S.; Hoffmann, V.; Georgi, M. MR Imaging of Pancreatic Lesions. Comparison of manganese-DPDP and gadolinium chelate. *Investig. Radiol.* **1999**, *34*, 589–595. [CrossRef] [PubMed]
38. Wang, J.; Wang, H.; Ramsay, I.A.; Erstad, D.J.; Fuchs, B.C.; Tanabe, K.K.; Caravan, P.; Gale, E.M. Manganese-Based Contrast Agents for Magnetic Resonance Imaging of Liver Tumors: Structure–Activity Relationships and Lead Candidate Evaluation. *J. Med. Chem.* **2018**, *61*, 8811–8824. [CrossRef] [PubMed]
39. Pan, D.; Caruthers, S.D.; Senpan, A.; Schmieder, A.H.; Wickline, S.A.; Lanza, G.M. Revisiting an old friend: Manganese-based MRI contrast agents. *Wiley Interdiscip. Rev. Nanomed. Nanobiotechnology* **2011**, *3*, 162–173. [CrossRef] [PubMed]
40. Gale, E.M.; Wey, H.-Y.; Ramsay, I.; Yen, Y.-F.; Sosnovik, D.E.; Caravan, P. A Manganese-based Alternative to Gadolinium: Contrast-enhanced MR Angiography, Excretion, Pharmacokinetics, and Metabolism. *Radiology* **2018**, *286*, 865–872. [CrossRef] [PubMed]
41. Islam, K.M.; Kim, S.; Kim, H.-K.; Park, S.; Lee, G.-H.; Kang, H.J.; Jung, J.-C.; Park, J.-S.; Kim, T.-J.; Chang, Y. Manganese Complex of Ethylenediaminetetraacetic Acid (EDTA)–Benzothiazole Aniline (BTA) Conjugate as a Potential Liver-Targeting MRI Contrast Agent. *J. Med. Chem.* **2017**, *60*, 2993–3001. [CrossRef]
42. Gale, E.M.; Mukherjee, S.; Liu, C.; Loving, G.S.; Caravan, P. Structure–Redox–Relaxivity Relationships for Redox Responsive Manganese-Based Magnetic Resonance Imaging Probes. *Inorg. Chem.* **2014**, *53*, 10748–10761. [CrossRef] [PubMed]
43. Borodin, O.Y.; Sannikov, M.Y.; Belyanin, M.L.; Filimonov, V.D.; Usov, V.Y.; Rybakov, Y.L.; Gukasov, V.M.; Shimanovskii, N.L. Relaxivity of Paramagnetic Complexes of Manganese and Gadolinium. *Pharm. Chem. J.* **2019**, *53*, 635–637. [CrossRef]
44. Shen, Y.; Goerner, F.L.; Snyder, C.; Morelli, J.N.; Hao, D.; Hu, D.; Li, X.; Runge, V.M. T1 Relaxivities of Gadolinium-Based Magnetic Resonance Contrast Agents in Human Whole Blood at 1.5, 3, and 7 T. *Investig. Radiol.* **2015**, *50*, 330–338. [CrossRef] [PubMed]
45. Wahsner, J.; Gale, E.M.; Rodríguez-Rodríguez, A.; Caravan, P. Chemistry of MRI Contrast Agents: Current Challenges and New Frontiers. *Chem. Rev.* **2019**, *119*, 957–1057. [CrossRef] [PubMed]
46. Varbanov, H.P.; Valiahdi, S.M.; Kowol, C.R.; Jakupec, M.A.; Galanski, M.S.; Keppler, B.K. Novel tetracarboxylatoplatinum(iv) complexes as carboplatin prodrugs. *Dalton Trans.* **2012**, *41*, 14404–14415. [CrossRef] [PubMed]
47. Uivarosi, V.; Munteanu, A. Flavanoid complexes as promising anticancer metallodrugs. In *Flavanoids—From Biosynthesis to Human Health*; IntechOpen Ltd.: London, UK, 2017; Chapter 14; pp. 305–333.
48. Anđelković, K.; Milenković, M.R.; Pevec, A.; Turel, I.; Matić, I.Z.; Vujčić, M.; Sladić, D.; Radanović, D.; Brađan, G.; Belošević, S.; et al. Synthesis, characterization and crystal structures of two pentagonal-bipyramidal Fe(III) complexes with dihydrazone of 2,6-diacetylpyridine and Girard's T reagent. Anticancer properties of various metal complexes of the same ligand. *J. Inorg. Biochem.* **2017**, *174*, 137–149. [CrossRef] [PubMed]
49. Wang, F.-Y.; Xi, Q.-Y.; Huang, K.-B.; Tang, X.-M.; Chen, Z.-F.; Liu, Y.-C.; Liang, H. Crystal structure, cytotoxicity and action mechanism of Zn(II)/Mn(II) complexes with isoquinoline ligands. *J. Inorg. Biochem.* **2017**, *169*, 23–31. [CrossRef] [PubMed]
50. Icsel, C.; Yilmaz, V.T.; Aydinlik, Ş.; Aygun, M. New manganese(II), iron(II), cobalt(II), nickel(II) and copper(II) saccharinate complexes of 2,6-bis(2-benzimidazolyl)pyridine as potential anticancer agents. *Eur. J. Med. Chem.* **2020**, *202*, 112535. [CrossRef] [PubMed]
51. Kovala-Demertzi, D.; Staninska, M.; Garcia-Santos, I.; Castineiras, A.; Demertzis, M.A. Synthesis, crystal structures and spectroscopy of meclofenamic acid and its metal complexes with manganese(II), copper(II), zinc(II) and cadmium(II). Antiproliferative and superoxide dismutase activity. *J. Inorg. Biochem.* **2011**, *105*, 1187–1195. [CrossRef]
52. Huang, Y.; Hu, L.; Zhang, T.; Zhong, H.; Zhou, J.; Liu, Z.; Wang, H.; Guo, Z.; Chen, Q. $Mn_3[Co(CN)_6]_2@SiO_2$ Core-shell Nanocubes: Novel bimodal contrast agents for MRI and optical imaging. *Sci. Rep.* **2013**, *3*, 2647. [CrossRef] [PubMed]

53. Kaur, J.; Tsvetkova, Y.; Arroub, K.; Sahnoun, S.; Kiessling, F.; Mathur, S. Synthesis, characterization, and relaxation studies of Gd-DO3A conjugate of chlorambucil as a potential theranostic agent. *Chem. Biol. Drug Des.* **2017**, *89*, 269–276. [CrossRef]
54. Niwa, A.M.; D'epiro, G.F.R.; Marques, L.A.; Semprebon, S.C.; Sartori, D.; Ribeiro, L.R.; Mantovani, M.S. Salinomycin efficiency assessment in non-tumor (HB4a) and tumor (MCF-7) human breast cells. *Naunyn-Schmiedebergs Arch. Fur Exp. Pathol. Und Pharmakol.* **2016**, *389*, 557–571. [CrossRef] [PubMed]

Review

Iron-Based Magnetic Nanosystems for Diagnostic Imaging and Drug Delivery: Towards Transformative Biomedical Applications

Stefan H. Bossmann [1], Macy M. Payne [1], Mausam Kalita [2], Reece M. D. Bristow [3], Ayda Afshar [4] and Ayomi S. Perera [3,*]

[1] The University of Kansas Medical Centre and The University of Kansas Comprehensive Cancer Centre, 3901 Rainbow Blvd, Kansas City, MO 66160, USA
[2] Department of Radiology and Biomedical Imaging, University of California, San Francisco, CA 94143, USA
[3] Department of Chemical and Pharmaceutical Sciences, Kingston University London, Kingston upon Thames KT1 2EE, UK
[4] Department of Mechanical Engineering, University College London, Torrington Place, London WC1E 7JE, UK
* Correspondence: a.perera@kingston.ac.uk

Abstract: The advancement of biomedicine in a socioeconomically sustainable manner while achieving efficient patient-care is imperative to the health and well-being of society. Magnetic systems consisting of iron based nanosized components have gained prominence among researchers in a multitude of biomedical applications. This review focuses on recent trends in the areas of diagnostic imaging and drug delivery that have benefited from iron-incorporated nanosystems, especially in cancer treatment, diagnosis and wound care applications. Discussion on imaging will emphasise on developments in MRI technology and hyperthermia based diagnosis, while advanced material synthesis and targeted, triggered transport will be the focus for drug delivery. Insights onto the challenges in transforming these technologies into day-to-day applications will also be explored with perceptions onto potential for patient-centred healthcare.

Keywords: magnetic hyperthermia; MRI technology; patient-centred healthcare; iron oxide nanoparticles; nanotechnology

1. Introduction

1.1. Introduction to Iron Based Magnetic Nano Systems

The manipulation of magnetic properties to develop advanced technologies in biomedicine was first used clinically in the early 1980s with magnetic resonance imaging (MRI) as a diagnostic tool [1]. This technology utilized superparamagnetic nanoparticles (SPIONS) as contrasting agents to enhance the MR signals, of which various types of iron oxides were the prime agents [2]. Since then the usage of magnetic iron-based components for imaging and drug delivery has become prominent and steadily expanding areas in biomedical research. Iron is ideal for such applications as it is a ferromagnetic metal (i.e., permanently magnetic) that can form oxides such as magnetite (Fe_3O_4) and maghemite (γ-Fe_2O_3), which are considered superparamagnetic or display enhanced magnetic properties in the presence of external magnetic fields (Figure 1) [3]. It must be noted that not all forms of iron are magnetic or suitable for biomedical applications. Hematite (α-Fe_2O_3) for example, is very weakly magnetic and Wüstite (FeO) is magnetic but is unstable at ambient conditions and is thus nonviable for biomedicinal purposes.

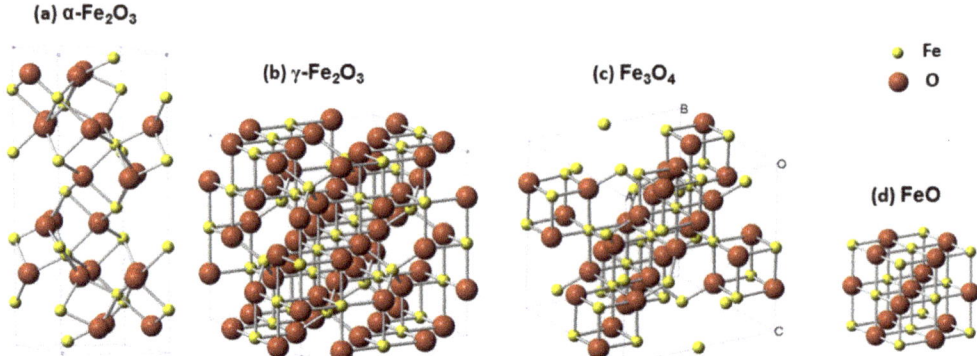

Figure 1. Representation of crystallographic unit cells of four major iron oxides species: (**a**) Hematite—α-Fe$_2$O$_3$, (**b**) maghemite—γ-Fe$_2$O$_3$, (**c**) magnetite—Fe$_3$O$_4$, and (**d**) wüstite—FeO. Image taken with permission from Zhu et al., 2016 [3].

Iron-based magnetic materials have currently become significant in many applications in nanotechnology, including biomedicine [4] and far beyond. The term nanotechnology is defined by the European commission as " ... areas of science and engineering where phenomena that take place at dimensions in the nanometre scale are utilised in the design, characterisation, production and application of materials, structures, devices and systems" [5]. This term is rapidly gaining public attention and significant global economic investments due to explosive achievement of technological advances that affect almost every area of life, ranging from electronics [6] such as smart phones and televisions to building materials [7] such as concrete. The materials used in nanotechnology applications, i.e., nanomaterials, can be broadly defined as those that have individual units with at least one Cartesian dimension within 1–100 nm [8]. This definition, however, is not universal and may vary according to the specific field and/or context in which it is used [9,10]. Iron oxide species have been utilized in various formats including nanoparticles, nanofibres [11], sol-gels [12] or as nanoparticle-incorporated composite materials such as hydrogels [13–15], core–shell structures [16], liposomes [17,18], etc., which will be discussed throughout this review with respect to materials development and modes and scope of applications.

1.2. Advantages of Iron Based Magnetic Systems

Iron oxide nanomaterials have key advantages that make them particularly attractive for biomedical applications. They have shown to have biocompatibility as proven via many toxicological studies [19] and in vivo applications such as MRI contrast agents for imaging [20], and therapeutic applications such as magnetic hyperthermia and targeted drug delivery, mostly but not limited to cancer treatment [21]. They have been successfully used in multiple clinical trials with hyperthermia based cancer treatment [22,23], that utilizes alternating magnetic fields on a magnetically responsive fluid. This allows for tissue-specific localization of heat in tissues that located deep within the body with high intensity and in a noninvasive manner to treat cancers such as prostate carcinoma and glioblastoma, among others [22], thus illustrating the potential to transform such therapeutics in to safer, more efficient procedures.

Sustainable approaches to synthesis of magnetic iron oxide nanoparticles have seen parallel and rapid growth in recent years, along with biocompatibility studies [24–27]. One strategy of avoiding or minimizing toxicity in vivo is coating and stabilizing the nanoparticle surface with biocompatible materials such as lauric acid and proteins such as serum albumin [28]. Investigations have also been carried out to customize nanoparticle size, shape, surface morphologies, crystallinity, uniformity, etc. in order to optimize biocompatibility [16,21]. Another promising approach with increasing interest is greener synthesis using biological components and species such as plant extracts, bacteria, fungi and algae. Such techniques are found to be biologically safer and can be achieved in a myriad of shapes such as cubes, tetragonal crystals, spheres, cylinder, and hexagonal rods, etc., among others [29]. Additionally, such biological synthesis has embedded processes that can replace chemicals needed for reduction, capping and stabilization in conventional synthesis, leading to cost-effectiveness [29–32].

1.3. Review Overview

This review aims to discuss key materials and mode of applications of magnetic systems that incorporate one or more nanosized materials containing iron, along two key branches: (1) diagnostic imaging and (2) drug delivery (Figure 2). Insights will be given on the enormous possibilities as well as significant challenges in transforming these materials and technologies for real world applications in biomedicine.

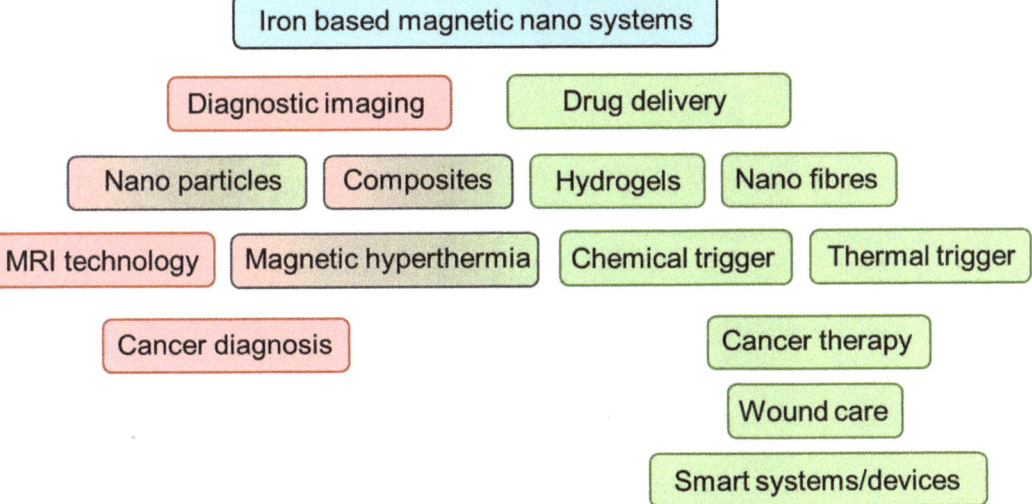

Figure 2. Overview of review topics discussed. Red depicts materials and methods and applications in diagnostic imaging and green represents the same in drug delivery. Boxes with combined red-green indicates both.

2. Diagnostic Imaging

2.1. Asset of IONPs to Different Modalities

Iron oxide nanoparticles have ready application towards magnetic resonance imaging as a contrast agent to depress the relaxation rate of tissues where they are accumulated. However, the usage of IONPs can extend beyond its application as a contrast agent for MRI alone. Correctly coupled or functionalized with fluorophores or radiotracers the usage of IONPs can be extended for use in photo-acoustic imaging (PAI), positron emission tomography (PET), and single photon emission computed tomography (SPECT), computing tomography (CT), and magnetic particle imaging (MPI) in addition to its natural contrast agent application in MRI [33]. This can allow for both qualitative and quantitative imaging by combining modalities which enable quantification of iron content in additional to anatomical information.

2.2. IONPs as MRI Contrast Agents

Iron oxide nanoparticles are common contrast agents for detecting and tracking cells. They have been shown to retain overall cell viability with rare detrimental effects on proliferation, apoptosis, or necrosis over a wide range of iron concentrations, and can label a wide variety of cell types from stem cells, cancer cells, immune cells, and more [34]. Their adaptability in size, the ability to perform surface modification, and renal clearance capabilities leads to a safer contrast agent that can remain visible for longer periods of time with fewer risks than gadolinium-based agents (GBCA) [35] images. These nanoparticles cause a decrease in the transverse (T_2) relaxation by creating magnetic field inhomogeneity [34]. This creates signal loss surrounding the iron oxide nanoparticle, commonly referred to as the "blooming artifact," which is detectable within an image. Comparison of a T_1 and T_2 contrast agent can be seen in Figure 3. Iron oxide nanoparticles are viewed indirectly where regions of signal loss vary in size depending upon their target, application, and size of iron oxide nanoparticles used.

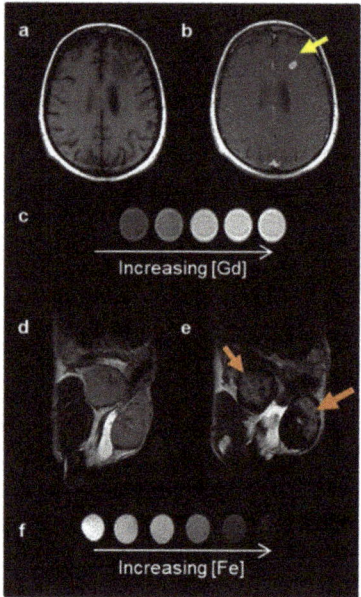

Figure 3. Two classes of MRI contrast agents. (**a**) Pre- and (**b**) post-GBCA T1-weighted MRI on a brain metastasis in a melanoma patient. (**c**) T1 contrast agents decrease the spin-lattice (T1) relaxation time, increasing signal with increasing agent concentration, and produce brighter contrast images. (**d**) Pre- and (**e**) post-IONP-based contrast agent T2-weighted MRI on inflamed mouse mammary gland tumours. (**f**) T2 contrast agents decrease the spin-spin (T2) relaxation time, decreasing signal with increased agent concentration, and produce darker contrast images. Image taken with permission from Jeon et al., 2021 [35].

2.3. Contrast Agent Application in MRI

Iron oxide nanoparticles are incredibly sensitive to the pulse sequence used in image acquisition. These nanoparticles cause a decrease in the transverse (T2) relaxation by creating magnetic field inhomogeneity [36]. This creates signal loss surrounding the iron oxide nanoparticle, commonly referred to as the "blooming artefact," which is detectable within an image (Figure 4). Regions of signal loss vary in size depending upon their target, application, and the particular size of iron oxide nanoparticle used. Iron oxide nanoparticles are viewed indirectly, where the decreased relaxivity depends on the contrast agent concentration.

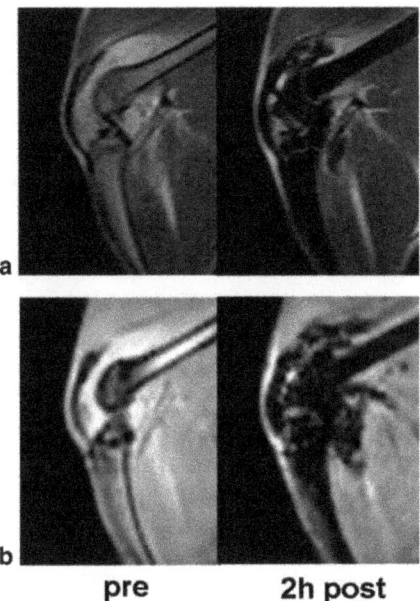

Figure 4. The T2 and T2* effect of ultra-small paramagnetic iron oxide nanoparticles: Corresponding T2-weighted (**a**) two-dimensional Spin Echo (SE) images and T2*-weighted (**b**) three-dimensional Spoiled Gradient Recalled Acquisition in the Steady State (SPGR) images of an arthritic knee joint before and two hours after injection. The USPIO results in hypointense regions with marked signal loss. Image taken with permission from Simon et al., 2006 [36].

In vivo cell tracking can be performed with either an intravenous injection of the nanoparticle or initial incubation with the cells for later injection. For innately phagocytic cells, simple co-incubation will traditionally prompt sufficient labelling; however, for other cell types, efficient labelling may require the use of additional agents such as transfection agents or electroporation [37]. However, upon cell-labelling, it is possible to utilize labelled cells for therapeutic information, such as monitoring the migration and survival of cells and confirming appropriate administration allowing for optimization of treatments to occur in real-time. Labelled cells can also be utilized to image cancer cells with the potential to detect singular cancer cells and monitor metastatic spread. The ability to monitor singular cancer cells is only possible through the blooming artefact making a 20-micron cell present as hypo-intense voids visible in 300-micron slices [36]. A common compound of interest in current clinical trials, Feraheme (Ferumoxytol) is presumed to have a half-life in humans of 24–36 h, allowing a larger window for imaging, and repeat imaging, compared to GBCAs. Ferumoxytol is the leading agent referenced in a majority of clinical trials utilizing iron oxide as an MRI contrast agent. There had been a surge in repurposing this FDA-approved drug initially developed for iron replacement in anemia for chronic kidney disease, yet it had come under scrutiny for off-label use as a contrast agent when a small population of users (not associated with clinical studies for MR imaging) suffered from a series of anaphylactic reactions [38]. Thus, it has now been labeled with an FDA black box warning which initially caused a lull or the termination of ongoing research projects. Fortunately, new safety assessments have indicated that there are less than 1% incidences of moderate to severe toxicities [32] and future research outlooks are promising and excitement over its application has grown.

A study by Corwin et al. investigated the improvement in image quality with steady state MR-angiography (SS-MRA) with ferumoxytol in comparison to conventional first pass MRA. They cited the primary advantage with iron oxide contrast agents was the possibility of administration to patients with renal failure, as with gadolinium-based contrast agents they face the risk of nephrogenic systemic fibrosis. In their analysis they were able to increase the average vessel sharpness significantly, increase resolution, and maintain a comparable signal to noise ratio of the external iliac artery with ferumoxytol assisted SS-MRA with an SNR of 42.2 for first pass MRA and 41.8 for SS-MRA. The use of ferumoxytol provided superior vessel sharpness but also allows for prolonged imaging times, as gadolinium-based contrast agents rapid clearance limits acquisition to first pass -MRA methods [38,39]. While the possibility for high levels of intravascular concentrations of the iron oxide agent can induce artifacts due to the inherent signal lass from a T_2-based contrast agent, the utilization of lower concentrations can circumvent this shortcoming.

Pan et al. further adapted the iron oxide nanoparticle through the utilization of hyaluronic acid and iron oxide nanoparticles to generate a molecular MRI probe with the ability to adapt and switch from a T_1 and T_2 contrast agent for evaluation of atherosclerotic plaques. The overall structure uses HA as the core of the nanomaterials with IONP-P nanoparticles on the surface of the HA platform. These nanoparticles are presumed to by phagocytosed by macrophages where the IONP-HP were found to target macrophages likely due to the CD44 receptor-targeted HA, and form clusters where the particle size of the IONP-HP before clustering is 153 nm and after clustering increases to 562 nm. An initial simulation of the iron oxide nanoparticles and hyaluronic acid nanoparticles (IONP-HP) tested the clusters admission into macrophages in acidic environments with pH's similar to that of the lysosomes they would inhabit by creating a series of solutions with pH values between 4 to 7.4 and incubated these solutions for a week as they formed clusters. This cluster formation would encourage the T_2 contrast effect postulated by researchers. TEM results confirmed that when pH levels decrease below 5, the nanoparticles become more compact, and the MRI results confirmed that this produces a T_2 contrast effect. MR signal demonstrated an increase with incubation time for T_1-weighted images and a decrease in T_2-weighted images with the SNR in the T_1 images decreasing slowly with time with peak signal at 12 h with no T_2 enhancement happening concurrently. The ability of the nanoparticles to present T_1 or T_2 enhancement was found to depend upon the particle size, although the aggregation of the nanoparticles was found to affect both transverse and longitudinal relaxation in MRI [40]. It is postulated that the IONP-HP clusters aggravate the proton dephasing of surrounding water molecules; thus, the T_2 contrast effect is enhanced by the variety of sizes within the cluster. Animal studies with mice to determine if MRI would be able to identify stable versus vulnerable plaques determined that MRI identification of stable vs. vulnerable plaques was possible upon obtaining T_1 and T_2 weighted images. The signal intensity of vulnerable plaques has an apparent increase at 2 h post-injection for T_1 weighted images, whereas the stable plaque does not, likely due to poor infiltration of macrophages into the stable plaques. This confirms the generation of a "switchable" MRI probe where the IONP-HP can be "turned on" through macrophages as engulfing time increases, thus providing a novel molecular MRI probe for future uses built upon a hyaluronic acid platform.

Current clinical trials include studies such as those performed at the M.D. Anderson Cancer Center which recently completed a clinical trial assessing the use of Feraheme to enable researchers in visualizing cancerous lymph nodes and perform liver imaging with MRI. It involved the intravenous administration of Feraheme at 6 mg of iron/kg and then repeat MR scans over a series of 3 days. The initial scan was taken without the application of the contrast agent and then 38 and 72 h post ferumoxytol injection. Following suite, the Allegheny Singer Research Institute conducted a study to investigate radiotherapy with iron oxide nanoparticles and MRI-Guided Linear Accelerator (MRI-Linac) for primary and metastatic hepatic cancers. Imaging was performed on a 1.5 T instrument to investigate if the use of ferumoxytol for MR-SPION radiotherapy will assist in the detection and avoidance of functional liver tissue to increase the safety of liver stereotactic body radiotherapy. Furthermore, the National Institute of Neurological Disorders and Stroke and National Institutes of Health Clinical Center recently completed a study of in vivo characterization of inflammation with ferumoxytol on 7 T instruments. Researchers focused on assessing if ferumoxytol caused prolonged changes in the brain in MRI, investigated if iron collected in the globus pallidus, and if the iron oxide contrast agent could assist in the identification of inflammation caused by multiple sclerosis. Participants partook in five clinical visits where ferumoxytol was delivered intravenously on the second with the following three including brain imaging on a 7 T MRI and blood draws. These clinical trials are a small segment of the work being done to test and establish iron oxide nanoparticles as a potential contrast agent to enhance and further the diagnostic capabilities of MRI. Future work is likely to build upon these findings as the number of available iron oxide nanoparticles increase, and enhanced surface modification of them allows for more efficient targeting and applicability.

Overall, iron oxide nanoparticles are a promising contrast agent which can allow for increased sensitivity of MRI detection of small lesions and tissue boundaries. There are a broad variety of synthesis routes possible with extensive surface modifications possible that can increase the targetability and sensitivity of these emerging contrast agent alternatives [41]. Through their increased biocompatibility and tunability it may be possible to generate a selective molecular MR probe that would enable enhanced imaging without impacting overall cell viability or exposing patients to potential toxicity allowing for more accurate diagnostic testing and analysis of disease genesis and progression.

2.4. Multimodal Imaging Applications

Blood circulation time was investigated by injections of 100 μL of IONP-HP solution into the tail vein of C57BL/6j mice. Blood samples were then collected at various time points as well as prior to injection, and the iron content was monitored through inductively coupled plasma optical emission spectrometry.

Through Equation (1), where ta and tb are the two time points of blood collection and Ca and Cb are the elevated Fe concentration in the circulation at the two time points after the injection of IONP-HP solution, the blood circulation half-life was determined. Overall cytotoxicity and biocompatibility were tested using an MTT assay with RAR264.7 cells and 15 BALB/c mice. Within the mouse group, the 15 mice were separated into two groups where one subset was injected with 80 μL of IONHP-HP and the other 80 μL of saline. Overall behaviors and body weight were carefully observed, and mice were sacrificed with inspection of the heart, liver, spleen, kidneys, and lungs was performed. For biodistribution, a third group of 10 BALB/c mice were injected with 150 μL of IONP-HP in the tail vein, and major organs were harvested at 24- and 48-h time points. The biodistribution of the IONP-HP was analysed through ICP-OES. Overall, there was no death, weight loss, or behavioural changes in mice after injection and H&E staining demonstrated no significant changes to the major organs. There were slight iron increases in the liver and kidneys at the 24-h time point.

$$t_{1/2} = \frac{t_b - t_a x 0.693}{ln C_a - ln C_b} \quad (1)$$

Initial in vitro MRI studies to examine the relaxation properties of the nanoparticles included obtaining T_1 and T_2 weighted images of different variations of the iron oxide nanoparticles, including the hyaluronic acid-iron oxide nanoparticle, with varying concentrations of iron ranging from 0.062 to 2 mm (0.062, 0.125, 0.25, 0.5, 1, and 2 mM) on a 1.5 T scanner. T_1 and T_2 weighted images were acquired through a turbo spin-echo sequence with the T_1 having a variation of repetitions times (TR) from 200 to 600 ms in 100 ms increments and an echo time (TE) of 6.5 ms, while the T_2 imaging used a constant TR of 200 ms and a variety of echo times from 13 to 91 ms in 13 ms increments. The longitudinal and transverse relaxation times were then determined for each of the nanoparticle variations, as well as the signal-to-noise ratios. The ability of the nanoparticles to present T_1 or T_2 enhancement was found to depend upon the particle size, although the aggregation of the nanoparticles was found to affect both transverse and longitudinal relaxation in MRI [15]. It is postulated that the IONP-HP clusters aggravate the proton dephasing of surrounding water molecules; thus, the T_2 contrast effect is enhanced by the variety of sizes within the cluster.

Mice were separated into vulnerable and stable groups, placed as such depending upon the feeding strategies used in their development. MR imaging of mice in the vulnerable plaque group was used to acquire T_1 weighted images of the abdomen with a 9.4 T Bruker instrument. Upon intravenous injection of IONP-HP, T_1 and T_2 weighted images of both stable and vulnerable plaque groups before and at various time points post-injection. Initially, at the 30-min time point post-injection, the vulnerable plaques showed no signal change. However, between 1- and 2-h post-injection, the vulnerable plaque mice demonstrated substantial T1 enhancement indicating the IONP-HP had been phagocytosed by the macrophages yet maintained a "loose" structure. Upon 24 h and up to 43 h post-injection, the signals became very low indicated that the IONP-HP in the macrophages had entered the vulnerable plaques and caused the contrast agent to move from T_1 to T_2 enhancement. Presumably, the switch from T_1 to T_2 mode is due to the loose structure of the IONP-HP collapsing in the secondary lysosome of macrophages. Due to the poor solubility of PAA on the surface of IONPs, the NPs form a cluster structure, thereby switching from T_1 to T_2 enhancement. Overall, MRI identification of stable vs. while iron oxide nanoparticles have firmly established a space as potential contrast agents for MRI, ongoing work has also investigated their applicability in other regions of imaging. These tactics allow for increased specificity of imaging while retaining the ability to localize regions within anatomical MRI scans. While these tactics often require additional functionalization of iron oxide nanoparticles, they pose a variety of benefits and uses.

Gholipour et al. developed a PET/MRI probe through a biotinylated thiosemicarbazone dextran-coated iron oxide nanoparticle. Biotin was used to functionalize the nanoparticle and increase delivery of the Ga-69 radiolabeled nanoparticles to tumor regions. These nanoparticles were then tested against subcutaneous 4T1 tumors in BALB/c mice. After tail vein administration, biodistribution studies were conducted and found that while there was a sizeable accumulation of the NP in the liver and spleen, the uptake in the tumor was higher than in other organs (excluding liver and spleen). Overall, the tumor uptake of the injected dose was 5.5% dose/g which was determined to be a middle ground when compared to other studies [41]. However, this multimodal nanoparticle comes with the remarkable benefits of being cheaper to synthesis due to the availability of thiosemicarbazone which was used as the chelator and the aldehyde platform which was used in conjugation of both the chelator and biotin targeting agent. When combined with the increased colloidal stability, this work demonstrates the potential application for usages of iron oxide nanoparticles as dual PET/MRI nanoparticles which can be further modified for optimal tumor targeting.

Bell et al. focused on a functionalized nanoparticle to work as a multimodal agent for optoacoustic (OA) and magnetic resonance imaging. Initial studies focused on assessing the use of iron oxide nanoparticles functionalized with indocyanine green (ICG) and Flamma® 774 as agents to induce increased T_2 relaxation rate for MR imaging while performing as a multispectral optoacoustic tomography (MSOT) contrast agent as well [42]. The FeO-OA dye nanoparticles were then tested in phantoms as both optoacoustic and MR imaging agents where subsequently the most successful was further tested in vivo against subcutaneous U87MG tumors in NCr nude mice [41]. Preliminary evidence from initial testing and phantom studies indicated the ICG capped nanoparticles had negligible contrast agent efficacy in MR imaging while the FeO-774 had representative peaks in MSOT imaging at 780 nm while retaining the relaxation properties of the iron oxide in MR imaging. After intra-tumoral injection of the FeO-774 nanoparticle the nanoparticle provided strong T_2 contrast in MR scans, remained stable in the mouse model, and remained detectable in MSOT. This study presents an alternative usage to functionalized iron oxide nanoparticles for utilization in multiple image modalities that do not require exposure to radiation.

In an additional study with multifunctional nanoparticles, Peng et al. generated macrophage laden gold nanoflowers which were embedded with ultrasmall iron oxide nanoparticles (Fe_3O_4/Au DSNFs). The iron oxide provided T_1 contrast for MR imaging due to their small size (USPIOs have been shown to act as a T_1 contrast agent) 38 while the gold NPs could be utilized for CT contrast enhancements. In vivo studies comparing macrophage laden Fe_3O_4/Au DSNFs (MA@Fe_3O_4/Au DSNFs) and Fe_3O_4/Au DSNFs in a breast tumor model of 4T1 implanted subcutaneously into ICR mice indicated improved performance from the targeted Fe_3O_4/Au DSNFs. After tail vein injection the SNR of the tumor region as 1.5–1.9 times greater with the MA@Fe_3O_4/Au DSNFs compared to the Fe_3O_4/Au alone. Furthermore, after CT the MA@Fe_3O_4/Au DSNFs reached a summit value of 46 HU, 1.2 times greater than that of the free nanoparticles. Ultimately, Peng and colleagues were able to sufficiently demonstrate the effectivity of a multifunctional nanoparticle utilizing iron oxide and gold to perform as a multimodal probe for both CT and MRI.

While the use of iron oxide nanoparticles for use as MRI contrast agents is a common application there are a variety of imaging modalities where iron oxide nanoparticles can be useful to enhance diagnostic and therapeutic capabilities. The ability to functionalize iron oxide nanoparticles with radiotracers, alternative metal nanoparticles, and targeting agents lends itself to a multitude of applications. Through the use of additional functionalization, iron oxide nanoparticles lend themselves as a convenient mechanism to overcome the shortcomings of each imaging modality independently.

2.5. Potential Socio-Economic Sustainability

Medical imaging is inherently expensive. The cost to build, procure, operate, and maintain the instrumentation is often transferred to the patients with the high cost per scan. Additionally, those responsible for the attainment and purchasing of MRI equipment in particular, have denoted their primary concern is patient and operator safety rather than factors which would induce a lower economic footprint or pertain to socio-economic sustainability [43]. This indicates that there is a large area of growth possible in development of advanced instrumentation, since cost is often a secondary concern to its safety and diagnostic capabilities. As the instrumentation and efficacy of contrast agents improve, patient outcomes are likely to improve correspondingly. Little work has been done to thoroughly investigate the cost-effectivity of iron oxide nanoparticles as contrast agents in MRI and other imaging modalities, however a litany of clinical experiments are underway to test FDA approved iron-oxide nanoparticles for alternative uses as a contrast agent. Targeting the cost effectiveness towards utilizing medical imaging for early diagnosis and intervention in patients is an arduous process which involves the quality-adjusted-life years (QALY) gained compared to the increased cost per year along with comparison of a willingness to pay threshold. These factors are subject to fluctuate depending upon the disease in question, patient age, and other comorbidity factors.

While there are studies that investigate the cost and individual impact on imaging in certain disease models, and the cost-effectivity thereof, the global accessibility and sustainability of medical imaging is more nuanced. Pertaining towards MRI, access to this modality of imaging and health management is relegated to the upper middle- and high-income countries, with 90% of the world lacking access [44]. However, there is large opportunity for growth to access these regions allowing for the full implementation of medical imaging through advancing technologies and targeting the applicability of MRI at lower field strengths. Some of the hurdles include developing instrumentation that can be used in an unshielded environment, has less energy requirements, and can be operated and analyzed via telemedicine to truly address accessibility constraints seen presently [40]. Fortunately, work has already begun to address the need for smaller and potentially portable MR spectrometers, with additional investigation into the utilization of ultra-low-field, and very-low-field MRI in these underrepresented demographics. While these instruments are likely to suffer from lower signal to noise ratios and resolutions, they will still present an additional and capable diagnostic tool that will open doors for patients as well as employment for maintenance, management, and operation of the spectrometers.

While access to qualified individuals to operate the instrumentation may be difficult initially, networks which will allow for remote training and access, combined with the possibility of harnessing Starlink capabilities to provide internet access in remote regions, may further assist in bridging the gap of accessibility for this diagnostic tool. With the development of more accessible scanners at lower field strengths, it will become more pressing to find safe and effective contrast agents that will better identify areas of malignancies to afford a comparable level of diagnostic efficacy found at higher fields.

3. Drug Delivery

Iron-based magnetic components incorporated into various materials and composites are being intensively explored for targeted and controlled release of drugs, towards development of sustainable, robust and efficient platforms. This section will focus on trends in materials development for advanced drug delivery including drug-carrying nanoparticles, composites, fibres and hydrogels with modes of release including magnetic hyperthermia and chemical and thermal triggers (Figure 5). Specific target applications will also be discussed in terms of cancer treatment, wound care and smart devices/technologies.

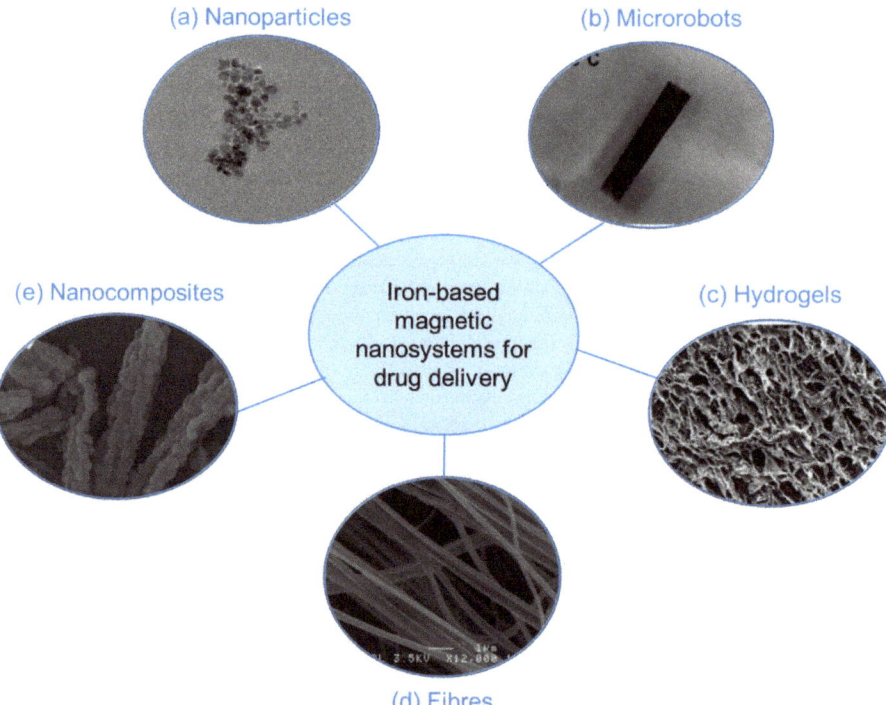

Figure 5. Various advanced platforms consisting of iron-based magnetic nanosystems for applications in drug delivery. Reproduced with permission from (**a**) Guo et al., 2018 [45]. (**b**) Fusco et al., 2019 [46]. (**c**) Perera et al., 2022 [13]. (**d**) Perera et al., 2018 Copyright © 2022 American Chemical Society [11] and (**e**) Huang et al., 2008 [47].

3.1. Advanced Materials

3.1.1. Nanoparticle-Based: Coatings, Ligands and Composite Materials

The surface of SPIONS can be readily functionalised with linkers, receptors, drug molecules, etc., which allows them to be coated with desired drugs for targeted delivery. This method has been investigated extensively for cancer treatment. Tagging the nanoparticles with receptors recognised by cancerous and tumour cells allows for their uptake by these cells, and provides convenient means of using magnetic nanoparticles (MNPs) as vehicles for drug delivery. Targeted drug delivery systems have used a variety of anti-cancer drug carriers, including magnetic iron oxide nanoparticles and natural biodegradable or non-biodegradable polymers [48]. The drugs can either be adsorbed on the surface of these carriers or enclosed inside of them and can be delivered to a given region using an external magnetic stimuli [49–52].

Doxorubicin is a widely investigated anticancer drug that can be delivered via MNP coating with multiple linker molecules such as citric acid, folic acid, chitosan, etc., in order to optimise MNP stabilisation, dispersion and effective drug delivery, typically characterised via cytotoxicity [45,53]. Doxorubicin has also been used in conjunction with lipid materials and MNPs to form magnetically sensitive lipocomplexes for in vivo drug delivery against CT26 mouse colorectal carcinoma [54–56]. Results have indicated superior anticancer activity by the magnetic lipocomplex, along with higher cell uptake, in comparison with standard doxorubicin.

Chlorambucil, another viable candidate for MNP-based delivery, is used for the treatment of chronic lymphocytic leukaemia, as well as both Hodgkin's and non-Hodgkin's lymphoma [57,58]. In a recent study, chlorambucil was incorporated into Fe_3O_4 iron oxide MNPs via a chitosan shell and showed increased efficacy in drug release onto cancer cells, compared to the non-complexed drug [59]. Another recent investigation with anticancer drug violamycin loaded onto 8–10 nm iron oxide nanoparticles, have shown increased effectiveness against to MCF-7 breast cancer cell line [53,60], indicating the potential in such SPION-based systems in drug delivery.

SPION-containing composites, i.e., materials with two or more components that display advanced chemical-physical properties to its individual components, are another area of interest for enhanced drug delivery applications with promising results obtained from graphene oxide-Fe_3O_4 and γ-Fe_2O_3-SBA-15 silica composites [47,53,61] among others. The latter study indicated the formation nanostructures with "rice like grain" consistency that displayed improved ibuprofen release in stimulated bodily fluids. A polymer-magnetic composite consisting of poly(N-vinyl-2-pyrrolidone) and Fe_3O_4 iron oxide ring-shaped nanostructure carrying doxorubicin have shown good in vivo tumour inhibition under magnetic hyperthermia [41,54]. Moreover, Zn-Al layered double hydroxide (LDH) nanosheets doped with Fe_3O_4 MNPs carrying doxorubicin were found to be effective against HepG2 cancer cell line, according to a recent study [49,62]. This material was found to be pH responsive, with more acidic conditions resulting in a larger doxorubicin release profile, thus allowing scope for chemically triggered drug release.

While promising, the composite SPION-based platforms have drawbacks of being expensive, and having complex synthesis, which will affect scale-up and real life applications. Nevertheless, their superior drug release ability warrants further research and exploration onto counteracting such effects.

3.1.2. Hydrogels Systems

A hydrogel is a 3-dimensional polymer network, which has water entrapped within its molecular space [63]. Thus, they are made predominantly of water, but nanoparticles and drugs can be incorporated into the water-polymer matrix for targeted and/or triggered drug delivery. Iron oxide MNPs have been particularly effective in hydrogel-based biomedical systems, ranging from bioseperation and tissue engineering in addition to drug delivery [60,63–65]. Recent research have significantly advanced the development of iron oxide MNP incorporated hydrogels, providing viable options to effectively transform drug delivery in to safer, more efficient and sustainable paths [13].

The polymers used to form hydrogels can be natural, synthetic, hybrid or bioinspired and need to be biocompatible for applications in drug delivery. Natural polymers such as gelatin and alginate have been used to develop hydrogels with doxorubicin loaded SPIONs and have indicated cytotoxicity via pH triggered drug release [56]. Dextran-MNP-based hydrogels have also shown to be magnetic and pH stimuli sensitive drug carrier as a dual tuneable drug delivery system [58]. Although highly biocompatible, such natural hydrogel platforms suffer from the difficulty of high scale extraction of polymers from natural sources.

Polyvinyl alcohol (PVA) is a synthetic biocompatible polymer which can be used to produce hydrogels for a wide range of uses including tissue engineering, graphing [66,67] and contact lenses [68]. One of the advantages of PVA as a hydrogel is its use in facile production of cryogels by simply dissolving in water followed by freeze/thawing [69]. A new study investigated the incorporation of acetaminophen and citric acid loaded Fe_3O_4 MNPs into various shapes of PVA hydrogels, followed by characterisation by temperature triggered drug release and magnetic hyperthermia experiments [13]. The study also looked at two different shapes, both a disc shape and a hemisphere shape, which is a novel concept (Figure 6). The study showed that there was evidence of a shape selective aspect in the magnetic hyperthermia studies, rendering a novel path for customized drug delivery for wound care applications and has potential for in vivo applications as well. Another synthetic polymer and Fe_2O_3 MNP containing hydrogel as also proven to be promising as an anti-inflammatory drug-releasing agent [70] in a recent study. This hybrid hydrogel system consisting of poly (ethylene glycol)-block-poly(propyleneglycol)-block-poly (ethylene glycol) (Pluronic P123), loaded with MNPs was loaded with diclofenac sodium and tested for pH based and temperature induced drug release, and have displayed superior results compared to its non MNP counterpart.

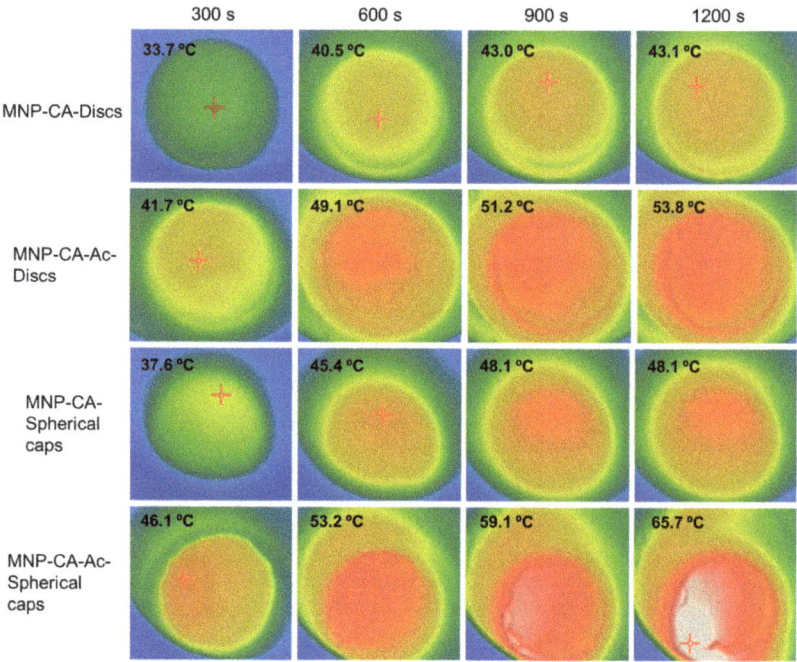

Figure 6. Thermal images indicating magnetic hyperthermia mediated temperature increase of disc shaped and spherical cap shaped gels with and without acetaminophen coating (MNP-CA and MNP-CA-Ac respectively) over time. The hyperthermia set up consisted of a copper coil and a Hikvision ds-2tp21b thermal imaging camera. The coil was heated under 19.95 V with a 14 A (AC) current at a frequency of 0.816 MHz. Gels were kept in the centre of the coil on a Teflon container and subject to heating while thermal images were taken every 10 s for 20 min. Image taken with permission from Perera et al., 2022 [13].

3.1.3. Nano-Fibre Based Materials

Fibres consisting of a polymer base along with MNPs have been investigated for numerous biomedical applications, including drug-carrying vehicles, advanced scaffolds for tissue engineering and bases for wound dressing [71–73] (Figure 7). The morphology of such nanoscale fibres resembles biological tissue, and as a result they are of particular interest in biomedicine. They also have high surface area, surface energy and can be organized into porous hierarchical structures, which makes them ideal for cell and tissue adhesion as well as the adsorption of drug molecules. Fibre dimensions can be readily customized to adapt to different applications with techniques such as electrospinning [74–79]. Various morphologies including hollow core−shell microparticle encapsulated fibres have been achieved via this method [80,81]. Furthermore, fibre composition have been optimized to incorporate components with biodegradable [82,83] properties and even those that include living tissue [84–86]. Scale-up targeting industrial applications of such materials has also been achieved [87].

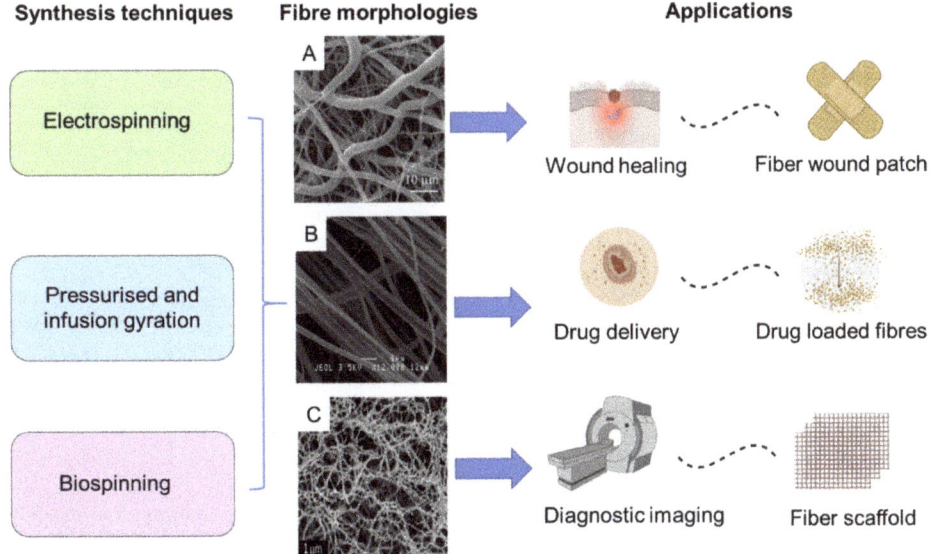

Figure 7. Diversity and versatility of MNP incorporated fibre based magnetic nanosystems in terms of synthesis, fibre morphology and scope of applications. Reproduced with permission from: (**A**)—Do Pham et al., 2021 [72]. (**B**)—Perera et al., 2018 Copyright © 2022 American Chemical Society [11]. (**C**)—Copyright © 2022 Mahmoudi et al., 2016 [73].

In spite of the breadth of research available, challenges still remain and prevent translation into real world biomedical applications, ranging from mass production issues to long-term in vivo stability. For instance, lack of control over fibre diameter, pore sizes, and morphological heterogeneities in the fibres produced via electrospinning, have led to reduced cell penetration, which is critical for long-term use as tissue scaffolds [80,82,88,89]. Nanoscale fibre diameters have also shown to be difficult to attain, unlike those in microscale [79] or have resulted in low yields [78]. Moreover, weaker mechanical strength and toxic and/or non-biocompatible components used during synthesis have limited applications in biomedicine [79]. Biospinning, has been used as an alternative technique to produce fibres with greater mechanical strength for scaffolds for tendons or bones [81]. However, this method is hampered by high cost, difficulty in scale-up, longer production times, and lack of customizability. Melt spinning is another alternative and produces fibres by extruding a heated polymer through a spinneret with textural control for cell applications [90–93]. This too has shown to be non-viable due to high costs on energy and equipment, and limitations in cell-penetration. One solution to address the latter issue was to encapsulate cells via interfacial complexation, which is cheaper [94,95], yet this too was found to be non-scalable and the fibres produced were morphologically heterogeneous, which were significant disadvantages. A recent study detailed a technique based on infusion-gyration to produce polyvinyl alcohol (PVA) and Fe_3O_4 MNP incorporated fibres, in a fast and cost-effective manner, with controllable sizes and the potential for scale-up. The fibres produced via this method were also found to be remotely actuated, rendering potential for significant advancements in drug delivery for patient-centred wound care and tissue engineering.

3.2. Advanced Applications

3.2.1. Magnetic Hyperthermia for Cancer Treatment

Hyperthermia, a phenomenon broadly defined in biomedicine as elevating tissue temperatures by means of external stimuli beyond normal physiological values, has been used for treatment of various diseases including Rheumatic conditions [96,97] and immunosuppression in management of pain and inflammation [97–99]. Cancer treatment however, is by far the most-explored area in hyperthermia-mediated treatments with applications in drug delivery [100–103]. It compliments conventional treatment methods including surgery, radiation, immunotherapy or chemotherapy. Hyperthermia treatments can be administered across either locally or the whole body, which advantageous for targeted drug delivery [104,105]. Typical temperature ranges used for cancer treatment via hyperthermia fall into two major categories of temperatures higher than 46 °C and temperatures within 41–46 °C.

Nanocomposites consisting of polymer matrices with SPIONS are gaining prominence in hyperthermia-related research on drug delivery [106,107]. They are superior to conventional methods which have drawbacks such as: (1) excessive heating of surrounding tissue causing cell/tissue damage; (2) under-heating of target areas located deep in the body or inside hard bone tissue; and (3) limited heat penetration resulting in recurrent tumour growth or incomplete removal [92,108]. The use of SPIONs ensures convenient thermal triggering of the nanocomposites via external magnetic fields or radiation to convert dissipated magnetic energy into heat, and are hence, an effective source of inducing hyperthermia [92,103]. Additionally, SPION-based delivery systems have advantages that include: (1) easy absorption into cancer cells due to their small sizes; (2) efficient delivery via multiple routes such as injection, liposomes, etc.; (3) ability to be functionalized with drugs or target-specific binding agents to increase selectivity and efficiency of treatment; (4) cost effective and sustainable due to requiring less trigger energy as a result of high heating efficiency.

An effective drug delivery strategy must encompass the above features but go beyond them and be able to interact with complex cellular functions in new ways [93,109]. Moreover, it needs to be biodegradable, biocompatible, and comfortable for patients with minimal adverse effects both during and after drug administration [95,110]. Furthermore, a high drug loading capability is desired, as well as a simple and cost-effective synthesis process [111,112]. Such desirable characteristics can be achieved through the use of SPIONS, which have proven to be effective in delivering a variety of drugs to a specific target in the human body via sustained or controlled release, in recent studies [113,114]. Several clinical studies have been conducted and have shown promise [22,23], however, MNP- or SPION-based magnetic hyperthermia are yet to enter into real world, applications in healthcare. Nonetheless, their beneficial features offer the possibility to develop advanced and multidimensional approaches to non-invasive and precise drug delivery, and hence, have viable potential to effectively transform the field.

3.2.2. Wound Care Applications

Wound care applications that utilize magnetic nanoparticle integrated hyperthermia are emerging as a promising area for enhanced, safer and less invasive drug release to treat surface wounds. These have been used for controlled release of broad spectrum antimicrobials [46,115–117], as well as providing various cues for neural regeneration [102,103,118,119]. Additionally, magnetic fields have shown to modulate mechanosensitive ion channels in cells [120] at low frequencies (<100 Hz), providing further evidence for the potential of magnetic systems in wound care. Furthermore, general heating in therapeutic ranges achievable with these approaches have been shown to act as gene expression triggers [118,121], while systems containing MNPs in conjugation with enzymes [107] and synthetic vesicles [108] have shown promise in enhanced wound healing. Hence, it is likely that a multimodal approach that involves heat-mediated drug release together with systems such as magnetic hydrogels would have a number of useful applications in regenerative medicine.

The best wound dressings should serve multiple purposes, including preventing acute or chronic infection, preserving a balanced environment for moisture and gas exchange, absorbing extrudates and blood from wounds, and stimulating cell migration and proliferation, which will aid in wound healing [109,110,112,114,122–125]. All these characteristics can simultaneously be present in nanofibre wound dressings. Nanofibre dressings enable for both oxygen permeability and wound protection against bacterial invasion owing to their small pore sizes. Furthermore, because of the ease with which chemical and biological molecules can be encapsulated during the spinning process, nanofibres can have wide appeal as viable vehicles for targeted and localized drug delivery [117,126]. In such cases, therapeutic agents have been readily incorporated into nanofibres for controlled and efficient release.

3.2.3. Magnetically Actuated Smart Devices and Microrobots

Actuation of the drug-carrying platform by an external magnetic field (i.e., magnetic actuation), is a scarcely explored area but has enormous potential in biomedicine. Such systems can potentially lead to remote controlled, precise, and safer pathways of drug delivery. A recent study indicates that shape switching magnetic hydrogel bilayers can be used to develop tubular microrobots by coupling a thermoresponsive hydrogel nanocomposite with a poly(ethylene glycol)diacrylate (PEGDA) layer [46]. The magnetic response has been achieved by using graphene oxide or silica-coated superparamagnetic iron oxide nanoparticles dispersed within the thermoresponsive hydrogel matrix, leading to magnetic actuation capability. Results have indicated that such magnetic composite systems can be optimized via shape (ex. helical microrobots) to enhance drug release and motility. Other studies have shown that core–shell ZnNCs encapsulated within mesoporous silica particles to carry and release drugs under magnetic hyperthermia, as remotely-controlled mechanised nanosystem [118,127].

Magnetic actuation can be taken a step further in to the development of smart devices and lead to advances in the rapidly evolving field of micro-robotics with a focus on biomedical applications [45,119,128–131]. Magnetically actuated miniature robots are able to access obscure regions of the human body and are capable of penetrating and manipulating matter as small as subcellular entities. Research and development of these systems have expanded rapidly over the past two decades due to high potential application in bioengineering and biomedicine [132,133]. While there are various methods of obtaining actuation in miniature robots, magnetic actuation offers a safe but effective approach to remotely control such systems via dynamic magnetic fields. Recent technologies and systems include soft ferromagnetic robot for surgery [133,134], magnetic and optical oxygen sensor for in situ intraocular sensing [135], magnetic microgrippers for biopsy [122,136], magnetic spore-based microrobots for remote detection of toxins [123,137], fluorescent magnetic microrobots for single-cell manipulation [124,138], magnetotactic bacteria swarms for targeted delivery [125,139], and magnetic scaffold to culture cells for tissue engineering [126,140].

A summary of key applications of iron based magnetic nanomaterials and methods of incorporation are depicted below (Table 1).

Table 1. Summary of applications, methods of incorporation and examples of iron based magnetic nanomaterials.

Application	Method	Examples	Example Reference(s)
MRI Contrasting agents	Nanoparticles and fluids	Iron oxide nanoparticles and hyaluronic acid nanoparticles (IONP-HP)	Pan et al. [40]
Cancer treatments	Coating	Doxorubicin, Violamycin	Guo et al. [45], Marcu et al. [53]
	Ligand	Chlorambucil-Chitosan Shell	Rozman et al. [57]
	Composite-Coating	LDH-Fe_3O_4 (doxorubicin)	Chai et al. [62]
	Composite	Poly(N-vinyl-2-pyrrolidone)-Fe_3O_4 iron oxide ring-shaped nanostructure	Wang et al. [48]
	Hydrogels	Dextran-MNP-based hydrogel	Zeng et al. [58]
Wound cleaning	Composite-Coating	γ-Fe_2O_3-SBA-15 silica (ibuprofen)	Huang et al. [47]
	Hydrogels	PVA-Fe_3O_4 (acetaminophen and citric acid), Poly (ethylene glycol)-block-poly(propyleneglycol)-block-poly (ethylene glycol) (Pluronic P123) hybrid system with Fe_2O_3	Perera et al. [13], Pandey et al. [70]
	Fibre	PVA Fe_3O_4 MNP incorporated fibres	Perera et al. [11]
Magnetic smart devices and microrobots	Hydrogel-Coating	Hydrogel nanocomposite with a poly(ethylene glycol)diacrylate (PEG-DA) layer	Fusco et al. [46]

4. Outlook

4.1. Challenges

The use of iron-containing nanoparticles can be challenging, although the human body has mastered the balancing act between using iron as co-factor in vital enzymes and massive inflammation caused by reactive oxygen species (ROS). Because iron(II/III) can easily change its redox state, it is typically catalysing or facilitating one electron oxidation- or reduction-chemistry, thus leading to the formation of radicals (Figure 8) [12,74,141]. Complexation with glutathione and a plethora of other (bio)organic (macro)molecules is able to shift the redox transition of Fe(II) to Fe(III) and vice versa over a broad potential range [127,142]. The human body stores iron as Fe(II) in ferritins and hemosiderin to tightly regulate its availability and delivers stored iron throughout the body by transferrin and other transport proteins [129,130,143]. The total amount of iron stored in the human body is 600 to 1000 mg in adult males and 200 to 300 mg in adult females [129,130]. One of the problems with iron-containing nanoparticles for diagnostics and treatment is that the iron concentration that is administered can exceed the total storage capacity of the human body [131,144]. Smaller nanoparticles can be filtered from circulation by means of renal excretion. The threshold for renal clearance is a particle diameter of about 5.5 nm [133], depending on the chemical structure of the nanoparticle. Larger nanoparticles can be excreted by means of hepatobiliary elimination via bile ducts and intestines [145,146]. Nanoparticles that are being taken up by Kupffer cells undergo long-time retention and slow degradation. Depending on the location of the iron-containing nanoparticles and the existence of protective coatings, biocorrosion occurs within 24 to 96 h. If nanoparticle uptake by Kupffer cells is the dominant pathways, iron has to effectively excreted to avoid iron overload. Otherwise, systemic damage is observed. The major iron-induced pathways leading to the release of hydroxyl radicals (HO) and reactive oxygen species (ROS) in the cells, membrane damage, protein damage and aggregation, mitochondrial damage resulting in cytochrome C release and apoptosis, as summarized in Figure 7 [12,95,143]. Therefore, iron overload has to be avoided, resulting in restrictions of using iron-containing nanoparticles as contrast agents in several diseases, such as chronic liver diseases [136]. It should also be mentioned that iron overload (hemochromatosis) from using iron-containing nanoparticles as MRI contrast agents and blood transfusions is additive. Besides genetic mutations, the blood transfusions are considered the major source of iron overload observed in the clinic [147]. MRI imaging is a suitable method to quantitatively detect iron overload that can be treated by means of chelation therapy [148].

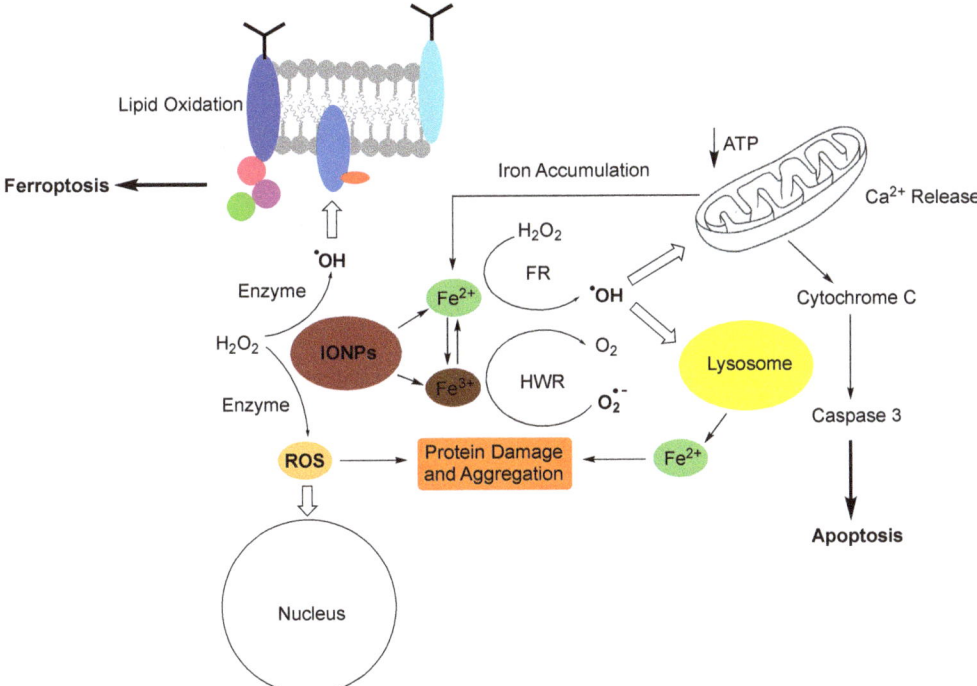

Figure 8. Pathways to iron-induced apoptosis (Ref. [127]) or ferroptosis (Refs. [149,150]). Small Iron-containing nanoparticles (IONPs) create oxidative stress by releasing Fe(II) and Fe(III) cations and by presenting a large assessable surface for redox cycling. In the presence of cellular hydrogen peroxide (H_2O_2), Fe(II) enables the Fenton reaction, which created either the hydroxyl radical (HO·, shown), or higher valent iron-species Ref. [151] (not shown). Fe(III) is reduced by the superoxide anion (O_2^-) in the Haber-Weiss reaction (Refs. [138,151]). Oxidative stress leads to membrane damage, protein damage and aggregation, nuclear damage, mitochondrial damage leading to cytochrome C release and apoptosis (Ref. [150]). Ferroptosis is triggered by radical-induced membrane damage. It is noteworthy that mitochondria do not play a part in ferroptosis (Ref. [137]).

4.2. Size-Dependence of Iron-Nanoparticle Toxicity

The bioavailability of iron-nanoparticles is strongly dependent on their size [131,144]. Fe-induced apoptosis or ferroptosis are only observed in organs, which can be reached by the nanoformulations. In general, the ROS generated by large Fe-nanoparticles (d > 5.5 nm) is minimal, indicating that cellular uptake (except for Kupffer cells in liver and spleen) is minimal. In sharp contrast, ultra-small Fe-nanoparticles (d < 5.5 nm) cause significant ROS because of their larger particle surface and faster corrosion kinetics, compared to larger nanoparticles. Although many types of ultrasmall nanoparticles induce the generation of ROS (e.g., Au, SiO_2, and Fe_3O_4), only Fe_3O_4 (and other Fe-containing nanoparticles) catalyze the Fenton/Haber-Weiss-Reactions [127,131,137], that generate significantly higher concentrations of hydroxyl radicals (HO.) and higher-valent iron-species. These conditions favour lipid peroxidation and, ultimately, ferroptosis. When using ultrasmall iron oxide nanoparticles are utilized for MRI contrast, Fe-toxicity is closely associated with the injection rate. A slower injection rate will provide more time for binding of bioavailable Fe(II) to ferritin. A typical example of Fera-heme that is injected at a rate of up to 30 mg (Fe)/s in the clinic [138]. In mouse experiments, ultrasmall Fe-nanoparticles were preferentially taken up by the heart, followed by liver, spleen, and lung. Uptake lead in all organs to massive ROS-induced inflammation. Whereas effective uptake by liver and spleen could have been anticipated, because the reticuloendothelial system is located in these organs, effective uptake by the heart was somewhat surprising. If the dosing of Fe-nanoparticles was too high, virtually all mice died of heart failure caused by inflammation [139]. Fe-nanoparticle uptake by the lung is mainly caused by the very high available surface and is potentially life-threatening as well. In contract to virtually all other organs, the iron concentration in the mouse kidneys was found to be low. This is a clear indication that iron from ultrasmall nanoparticles is taken up by the surrounding tissue and does, at least initially, not reach the kidneys within 24 h.

4.3. Environmental Considerations

The environmental impact of nanoparticles is under discussion. Whereas it is well established that small nanoparticles are a considerable health risk in high concentrations, larger nanoparticles (d > 10 nm) in lower concentrations generally do not pose a significant risk. However, there is not sufficient data on long-term exposure [140]. In our oxidative atmosphere, virtually all nanoparticles undergo oxidation, with the exception of metal oxides and silicates. These oxidation products can be enriched in the environment, for instance in reductive regions, such as fluvial sediments and sludge from sewage treatment plants [152]. It should also be noted that noble metal nanoparticles have slower oxidation kinetics, which allows them to pass through the filters of a sewage treatment plants into the aquifer where they constitute a health risk [142]. Iron and other d-block metal containing small nanoparticles undergo rapid corrosion in the environment, which releases the metals [143,147]. Depending on the pH-conditions in aquifers, metal hydroxide precipitation can occur [144]. It should also be noted that inhaling small nanoparticles can lead to significant lung inflammation, as discussed above [153].

4.4. Potential Impact on Real-Life Practices: Probable Trends

There is agreement in the field that we are one to two decades away from the onset of ultra-high field MRI beyond 10 T in the clinic [146]. This field strength will be required for truly contrast-free MRI imaging of most diseases. Until then, the use of contrast agents will be mandatory to enhance signal-to-noise and to shorten the residence time of patients in the MR imager. Compared to gadolinium(III) compounds, which exhibit significant nephrotoxicity [149], Fe-containing nanoparticles can be better managed in the clinic, albeit they also pose some risk of inflammation [147].

In addition to their clinical use as MRI contrast agents, Fe-nanoparticles are magnetic and, therefore, suitable for magnetically aided drug delivery (including chemotherapy), and angiogenic therapy, in which a massive inflammation is caused in the (micro) blood vessels of the tumour microenvironment to cut off the tumour from nutrient and oxygen supply [150]. Ferroptosis is a most promising treatment strategy of virtually all solid tumours that is driven by iron-dependent lipid peroxidation [137,138,151]. Although it appears that the onset of ferroptosis does not require a mitochondrial contribution, lipid peroxidation will heavily influence mitochondrial morphology, bioenergetics, and metabolism [138]. Ferroptosis is favoured in cancer cells vs. normal cells because of the metabolic dysfunction of the former. When combined with targeting strategies of overexpressed clusters of differentiation at the surface of cancer cells, the use of ultra-small Fe-nanoparticles may constitute a promising new strategy to eradicate primary tumours and metastases alike [124,138].

Fe-nanoparticles can also be utilized to target and enhance the effects of hyperthermia. They can be applied in conjunction with A/C-magnetic [151] and radiofrequency (RF)-hyperthermia [154,155] alike. Since a pro-inflammatory tumour microenvironment is required for a successful immune response, which can be triggered by hyperthermia [156], the use of Fe-nanoparticles is potentially synergetic with hyperthermia and are thus likely to be utilized in such applications in the near future.

5. Conclusions

Nanotechnology can play a major role in transforming the biomedical field with safer, effective, more advanced, and socioeconomically sustainable materials and technologies. This review focuses on selected recent and significand trends on magnetic nanosystems that incorporate iron-based materials, for diagnostic MRI technology and drug delivery for showcasing this fast-approaching transformation. Key characteristics, advantages and versatility of magnetic iron oxide nanoparticles are emphasised. Principle features of MRI technology, and how it can benefit by the utilization of iron oxide nanoparticles as contrast agents, is critically discussed along with key developments. Novel material development with advances in mode and scope if application in drug delivery are also discussed with respect to iron oxide nanomaterials. Finally, a look at challenges to these applications along with size-dependent iron-nanoparticle toxicity, environmental considerations and potential on real-life practices including probable trends are highlighted.

Author Contributions: Conceptualization A.S.P., S.H.B.; investigation A.S.P., S.H.B., M.M.P., M.K., R.M.D.B., A.A.; writing—original draft preparation, A.S.P., S.H.B., M.M.P., M.K., R.M.D.B., A.A.; writing—review and editing A.S.P., S.H.B.; supervision, A.S.P., S.H.B.; funding acquisition A.S.P., S.H.B. All authors have read and agreed to the published version of the manuscript.

Funding: This work was supported by a HEFCE-GCRF P2034-117 for ASP and the Department of Chemical and Pharmaceutical Sciences, Kingston University, as well as NIH 1R01EB028848-01 and NSF EFRI CEE 1933321 for SHB and MMP.

Conflicts of Interest: The authors declare no conflict of interest.

References

1. Edelman, R.R. The History of MR Imaging as Seen through the Pages of Radiology. *Radiology* **2014**, *273*, S181–S200. [CrossRef] [PubMed]
2. Lodhia, J.; Mandarano, G.; Ferris, N.; Eu, P.; Cowell, S. Development and use of iron oxide nanoparticles (Part 1): Synthesis of iron oxide nanoparticles for MRI. *Biomed. Imaging Interv. J.* **2010**, *6*, e12. [CrossRef] [PubMed]
3. Zhu, W.; Winterstein, J.; Maimon, I.; Yin, Q.; Yuan, L.; Kolmogorov, A.N.; Sharma, R.; Zhou, G. Atomic Structural Evolution during the Reduction of α-Fe2O3 Nanowires. *J. Phys. Chem. C* **2016**, *120*, 14854–14862. [CrossRef] [PubMed]
4. Ramos, A.P.; Cruz, M.A.E.; Tovani, C.B.; Ciancaglini, P. Biomedical applications of nanotechnology. *Biophys. Rev.* **2017**, *9*, 79–89. [CrossRef] [PubMed]
5. The Appropriateness of Existing Methodologies to Assess the Potential Risks Associated with Engineered and Adventitious Products of Nanotechnologies. Available online: https://ec.europa.eu/health/ph_risk/committees/04_scenihr/docs/scenihr_o_003b.pdf (accessed on 16 August 2022).
6. Payal; Pandey, P. Role of Nanotechnology in Electronics: A Review of Recent Developments and Patents. *Recent Pat. Nanotechnol.* **2022**, *16*, 45–66. [CrossRef] [PubMed]
7. Sanchez, F.; Sobolev, K. Nanotechnology in concrete—A review. *Constr. Build. Mater.* **2010**, *24*, 2060–2071. [CrossRef]
8. Malhotra, B.D.; Ali, M.A. Nanomaterials in Biosensors: Fundamentals and Applications. In *Nanomaterials for Biosensors*; William Andrew: Norwich, NY, USA, 2018; pp. 1–74. [CrossRef]
9. Nanomaterials definition matters. *Nat. Nanotechnol.* **2019**, *14*, 193. [CrossRef]
10. Definition of a Nanomaterial. Available online: https://ec.europa.eu/environment/chemicals/nanotech/faq/definition_en.htm (accessed on 16 August 2022).
11. Perera, A.S.; Zhang, S.; Homer-Vanniasinkam, S.; Coppens, M.-O.; Edirisinghe, M. Polymer–Magnetic Composite Fibers for Remote-Controlled Drug Release. *ACS Appl. Mater. Interfaces* **2018**, *10*, 15524–15531. [CrossRef]

12. Caetano, B.L.; Guibert, C.; Fini, R.; Fresnais, J.; Pulcinelli, S.H.; Ménager, C.; Santilli, C.V. Magnetic hyperthermia-induced drug release from ureasil-PEO-γ-Fe2O3 nanocomposites. *RSC Adv.* **2016**, *6*, 63291–63295. [CrossRef]
13. Perera, A.S.; Jackson, R.J.; Bristow, R.M.D.; White, C.A. Magnetic cryogels as a shape-selective and customizable platform for hyperthermia-mediated drug delivery. *Sci. Rep.* **2022**, *12*, 9654. [CrossRef]
14. Shirakura, T.; Kelson, T.J.; Ray, A.; Malyarenko, A.E.; Kopelman, R. Hydrogel Nanoparticles with Thermally Controlled Drug Release. *ACS Macro Lett.* **2014**, *3*, 602–606. [CrossRef] [PubMed]
15. Li, J.; Mooney, D.J. Designing hydrogels for controlled drug delivery. *Nat. Rev. Mater.* **2016**, *1*, 16071. [CrossRef]
16. Ling, D.; Hyeon, T. Chemical Design of Biocompatible Iron Oxide Nanoparticles for Medical Applications. *Small* **2013**, *9*, 1450–1466. [CrossRef] [PubMed]
17. López-Noriega, A.; Hastings, C.L.; Ozbakir, B.; O'Donnell, K.E.; O'Brien, F.J.; Storm, G.; Hennink, W.E.; Duffy, G.P.; Ruiz-Hernández, E. Hyperthermia-Induced Drug Delivery from Thermosensitive Liposomes Encapsulated in an Injectable Hydrogel for Local Chemotherapy. *Adv. Healthc. Mater.* **2014**, *3*, 854–859. [CrossRef] [PubMed]
18. May, J.P.; Li, S.-D. Hyperthermia-induced drug targeting. *Expert Opin. Drug Deliv.* **2013**, *10*, 511–527. [CrossRef]
19. Vakili-Ghartavol, R.; Momtazi-Borojeni, A.A.; Vakili-Ghartavol, Z.; Aiyelabegan, H.T.; Jaafari, M.R.; Rezayat, S.M.; Arbabi Bidgoli, S. Toxicity assessment of superparamagnetic iron oxide nanoparticles in different tissues. *Artif. Cells Nanomed. Biotechnol.* **2020**, *48*, 443–451. [CrossRef]
20. Nosrati, H.; Salehiabar, M.; Fridoni, M.; Abdollahifar, M.-A.; Manjili, H.K.; Davaran, S.; Danafar, H. New Insight about Biocompatibility and Biodegradability of Iron Oxide Magnetic Nanoparticles: Stereological and In Vivo MRI Monitor. *Sci. Rep.* **2019**, *9*, 7173. [CrossRef]
21. Xie, W.; Guo, Z.; Gao, F.; Gao, Q.; Wang, D.; Liaw, B.S.; Cai, Q.; Sun, X.; Wang, X.; Zhao, L. Shape-, size- and structure-controlled synthesis and biocompatibility of iron oxide nanoparticles for magnetic theranostics. *Theranostics* **2018**, *8*, 3284–3307. [CrossRef]
22. Etemadi, H.; Plieger, P.G. Magnetic Fluid Hyperthermia Based on Magnetic Nanoparticles: Physical Characteristics, Historical Perspective, Clinical Trials, Technological Challenges, and Recent Advances. *Adv. Ther.* **2020**, *3*, 2000061. [CrossRef]
23. Liu, X.; Zhang, Y.; Wang, Y.; Zhu, W.; Li, G.; Ma, X.; Zhang, Y.; Chen, S.; Tiwari, S.; Shi, K.; et al. Comprehensive understanding of magnetic hyperthermia for improving antitumor therapeutic efficacy. *Theranostics* **2020**, *10*, 3793–3815. [CrossRef]
24. Ali, A.; Zafar, H.; Zia, M.; Ul Haq, I.; Phull, A.R.; Ali, J.S.; Hussain, A. Synthesis, characterization, applications, and challenges of iron oxide nanoparticles. *Nanotechnol. Sci. Appl.* **2016**, *9*, 49–67. [CrossRef]
25. Wu WHe, Q.; Jiang, C. Magnetic Iron Oxide Nanoparticles: Synthesis and Surface Functionalization Strategies. *Nanoscale Res. Lett.* **2008**, *3*, 397. [CrossRef]
26. Samrot, A.; Sahithya, C.; Selvarani A, J.; Purayil, S.; Ponnaiah, P. A review on synthesis, characterization and potential biological applications of superparamagnetic iron oxide nanoparticles. *Curr. Res. Green Sustain. Chem.* **2021**, *4*, 100042. [CrossRef]
27. Laurent, S.; Forge, D.; Port, M.; Roch, A.; Robic, C.; Vander Elst, L.; Muller, R. Magnetic Iron Oxide Nanoparticles: Synthesis, Stabilization, Vectorization, Physicochemical Characterizations, and Biological Applications. *Chem. Rev.* **2008**, *108*, 2064–2110. [CrossRef]
28. Janko, C.; Zaloga, J.; Pöttler, M.; Dürr, S.; Eberbeck, D.; Tietze, R.; Lyer, S.; Alexiou, C. Strategies to optimize the biocompatibility of iron oxide nanoparticles—"SPIONs safe by design". *J. Magn. Magn. Mater.* **2017**, *431*, 281–284. [CrossRef]
29. Kaushal, P.; Verma, N.; Kaur, K.; Sidhu, A.K. Green Synthesis: An Eco-friendly Route for the Synthesis of Iron Oxide Nanoparticles. *Front. Nanotechnol.* **2021**, *3*, 655062. [CrossRef]
30. Saif, S.; Tahir, A.; Chen, Y. Green Synthesis of Iron Nanoparticles and Their Environmental Applications and Implications. *Nanomaterials* **2016**, *6*, 209. [CrossRef]
31. Ge, S.; Shi, X.; Sun, K.; Li, C.; Uher, C.; Baker, J.R.; Banaszak Holl, M.M.; Orr, B.G. Facile Hydrothermal Synthesis of Iron Oxide Nanoparticles with Tunable Magnetic Properties. *J. Phys. Chem. C* **2009**, *113*, 13593–13599. [CrossRef]
32. Perez De Berti, I.O.; Cagnoli, M.V.; Pecchi, G.; Alessandrini, J.L.; Stewart, S.J.; Bengoa, J.F.; Marchetti, S.G. Alternative low-cost approach to the synthesis of magnetic iron oxide nanoparticles by thermal decomposition of organic precursors. *Nanotechnology* **2013**, *24*, 175601. [CrossRef]
33. Alphandéry, E. Iron oxide nanoparticles as multimodal imaging tools. *RSV Adv.* **2019**, *9*, 40577–40587. [CrossRef]
34. Makela, A.V.; Murrell, D.H.; Parkins, K.M.; Kara, J.; Gaudet, J.M.; Foster, P.J. Cellular Imaging With MRI. *Top Magn. Reson. Imaging* **2016**, *25*, 177–186. [CrossRef]
35. Jeon, M.; Halbert, M.V.; Stephen, Z.R.; Zhang, M. Iron Oxide Nanoparticles as T(1) Contrast Agents for Magnetic Resonance Imaging: Fundamentals, Challenges, Applications, and Prospectives. *Adv. Mater.* **2021**, *33*, e1906539. [CrossRef] [PubMed]
36. Simon, G.H.; von Vopelius-Feldt, J.; Wendland, M.F.; Fu, Y.; Piontek, G.; Schlegel, J.; Chen, M.H.; Daldrup-Link, H.E. MRI of arthritis: Comparison of ultrasmall superparamagnetic iron oxide vs. Gd-DTPA. *J. Magn. Reson. Imaging* **2006**, *23*, 720–727. [CrossRef] [PubMed]
37. Peng, Y.; Wang, X.; Wang, Y.; Gao, Y.; Guo, R.; Shi, Z.; Cao, X. Macrophage-Laden Gold Nanoflowers Embedded with Ultrasmall Iron Oxide Nanoparticles for Enhanced Dual-Mode CT/MR Imaging of Tumors. *Pharmaceutics* **2021**, *13*, 995. [CrossRef] [PubMed]
38. Chen, C.; Ge, J.; Gao, Y.; Chen, L.; Cui, J.; Zeng, J.; Gao, M. Ultrasmall superparamagnetic iron oxide nanoparticles: A next generation contrast agent for magnetic resonance imaging. *Wiley Interdisc. Rev. Nanomed. Nanobiotechnol.* **2022**, *14*, e1740. [CrossRef]

39. Corwin, M.T.; Fananapazir, G.; Chaudhari, A.J. MR Angiography of Renal Transplant Vasculature with Ferumoxytol: Comparison of High-Resolution Steady-State and First-Pass Acquisitions. *Acad. Radiol.* **2016**, *23*, 368–373. [CrossRef]
40. Elhalawani, H.; Awan, M.J.; Ding, Y.; Mohamed, A.S.R.; Elsayes, A.K.; Abu-Gheida, I.; Wang, J.; Hazle, J.; Gunn, G.B.; Lai, S.Y.; et al. Data from a terminated study on iron oxide nanoparticle magnetic resonance imaging for head and neck tumors. *Sci. Data* **2020**, *7*, 63. [CrossRef]
41. Pan, C.; Lin, J.; Zheng, J.; Liu, C.; Yuan, B.; Akakuru, O.U.; Zubair Iqbal, M.; Fang, Q.; Hu, J.; Chen, J.; et al. An intelligent T(1)-T(2) switchable MRI contrast agent for the non-invasive identification of vulnerable atherosclerotic plaques. *Nanoscale* **2021**, *13*, 6461–6474. [CrossRef]
42. Gul, S.; Khan, S.B.; Rehman, I.U.; Khan, M.A.; Khan, M.I. A Comprehensive Review of Magnetic Nanomaterials Modern Day Theranostics. *Front. Mater.* **2019**, *6*, 179. [CrossRef]
43. Bell, G.; Balasundaram, G.; Attia, A.B.E.; Mandino FOlivo, M.; Parkin, I.P. Functionalised iron oxide nanoparticles for multimodal optoacoustic magnetic resonance imaging. *J. Mater. Chem. B* **2019**, *7*, 2212–2219. [CrossRef]
44. Lindgreen, A.; Antioco, M.; Harness, D.; van der Sloot, R. Purchasing and Marketing of Social and Environmental Sustainability for High-Tech Medical Equipment. *J. Bus. Ethics* **2009**, *85*, 445–462. [CrossRef]
45. Guo, L.; Chen, H.; He, N.; Deng, Y. Effects of surface modifications on the physicochemical properties of iron oxide nanoparticles and their performance as anticancer drug carriers. *Chin. Chem. Lett.* **2018**, *29*, 1829–1833. [CrossRef]
46. Fusco, S.; Huang, H.-W.; Peyer, K.E.; Peters, C.; Häberli, M.; Ulbers, A.; Spyrogianni, A.; Pellicer, E.; Sort, J.; Pratsinis, S.E.; et al. Shape-Switching Microrobots for Medical Applications: The Influence of Shape in Drug Delivery and Locomotion. *ACS Applied Mater. Interfaces* **2015**, *7*, 6803–6811. [CrossRef]
47. Huang, S.; Yang, P.; Cheng, Z.; Li, C.; Fan, Y.; Kong, D.; Lin, J. Synthesis and Characterization of Magnetic FexOy@SBA-15 Composites with Different Morphologies for Controlled Drug Release and Targeting. *J. Phys. Chem. C* **2008**, *112*, 7130–7137. [CrossRef]
48. Wang, X.; Qi, Y.; Hu, Z.; Jiang, L.; Pan, F.; Xiang, Z.; Xiong, Z.; Jia, W.; Hu, J.; Lu, W. Fe$_3$O$_4$@PVP@DOX magnetic vortex hybrid nanostructures with magnetic-responsive heating and controlled drug delivery functions for precise medicine of cancers. *Adv. Compos. Hybrid Mater.* **2022**, *5*, 1–13. [CrossRef]
49. Ebadi, M.; Bullo, S.; Buskaran, K.; Hussein, M.Z.; Fakurazi, S.; Pastorin, G. Dual-Functional Iron Oxide Nanoparticles Coated with Polyvinyl Alcohol/5-Fluorouracil/Zinc-Aluminium-Layered Double Hydroxide for a Simultaneous Drug and Target Delivery System. *Polymers* **2021**, *13*, 855. [CrossRef]
50. Ali, I.; Lone, M.N.; Suhail, M.; Mukhtar, S.D.; Asnin, L. Advances in Nanocarriers for Anticancer Drugs Delivery. *Curr. Med. Chem.* **2016**, *23*, 2159–2187. [CrossRef]
51. Hossen, S.; Hossain, M.K.; Basher, M.K.; Mia, M.N.H.; Rahman, M.T.; Uddin, M.J. Smart nanocarrier-based drug delivery systems for cancer therapy and toxicity studies: A review. *J. Adv. Res.* **2019**, *15*, 1–18. [CrossRef]
52. Senapati, S.; Mahanta, A.K.; Kumar, S.; Maiti, P. Controlled drug delivery vehicles for cancer treatment and their performance. *Signal Transduct. Target. Ther.* **2018**, *3*, 7. [CrossRef]
53. Marcu, A.; Pop, S.; Dumitrache, F.; Mocanu, M.; Niculite, C.M.; Gherghiceanu, M.; Lungu, C.P.; Fleaca, C.; Ianchis, R.; Barbut, A.; et al. Magnetic iron oxide nanoparticles as drug delivery system in breast cancer. *Appl. Surf. Sci.* **2013**, *281*, 60–65. [CrossRef]
54. Din, F.U.; Aman, W.; Ullah, I.; Qureshi, O.S.; Mustapha, O.; Shafique, S.; Zeb, A. Effective use of nanocarriers as drug delivery systems for the treatment of selected tumors. *Int. J. Nanomed.* **2017**, *12*, 7291–7309. [CrossRef] [PubMed]
55. Choi, G.E.; Kang, M.S.; Kim, Y.J.; Yoon, J.J.; Jeong, Y.I. Magnetically Responsive Drug Delivery Using Doxorubicin and Iron Oxide Nanoparticle-Incorporated Lipocomplexes. *J. Nanosci. Nanotechnol.* **2019**, *19*, 675–679. [CrossRef] [PubMed]
56. Jahanban-Esfahlan, R.; Derakhshankhah, H.; Haghshenas, B.; Massoumi, B.; Abbasian, M.; Jaymand, M. A bio-inspired magnetic natural hydrogel containing gelatin and alginate as a drug delivery system for cancer chemotherapy. *Int. J. Biol. Macromol.* **2020**, *156*, 438–445. [CrossRef] [PubMed]
57. Rozman, C.; Montserrat, E. Chronic Lymphocytic Leukemia. *New Engl. J. Med.* **1995**, *333*, 1052–1057. [CrossRef] [PubMed]
58. Zeng, N.; He, L.; Jiang, L.; Shan, S.; Su, H. Synthesis of magnetic/pH dual responsive dextran hydrogels as stimuli-sensitive drug carriers. *Carbohydr. Res.* **2022**, *520*, 108632. [CrossRef]
59. Hussein-Al-Ali, S.H.; Hussein, M.Z. Chlorambucil-Iron Oxide Nanoparticles as a Drug Delivery System for Leukemia Cancer Cells. *Int. J. Nanomed.* **2021**, *16*, 6205–6216. [CrossRef]
60. Liang, Y.-Y.; Zhang, L.-M.; Jiang, W.; Li, W. Embedding Magnetic Nanoparticles into Polysaccharide-Based Hydrogels for Magnetically Assisted Bioseparation. *ChemPhysChem* **2007**, *8*, 2367–2372. [CrossRef]
61. Yang, X.; Zhang, X.; Ma, Y.; Huang, Y.; Wang, Y.; Chen, Y. Superparamagnetic graphene oxide–Fe$_3$O$_4$ nanoparticles hybrid for controlled targeted drug carriers. *J. Mater. Chem.* **2009**, *19*, 2710–2714. [CrossRef]
62. Chai, J.; Ma, Y.; Guo, T.; He, Y.; Wang, G.; Si, F.; Geng, J.; Qi, X.; Chang, G.; Ren, Z.; et al. Assembled Fe$_3$O$_4$ nanoparticles on ZnAl LDH nanosheets as a biocompatible drug delivery vehicle for pH-responsive drug release and enhanced anticancer activity. *Appl. Clay Sci.* **2022**, *228*, 106630. [CrossRef]
63. Ahmed, E.M. Hydrogel: Preparation, characterization, and applications: A review. *J. Adv. Res.* **2015**, *6*, 105–121. [CrossRef]
64. Liu, T.-Y.; Hu, S.-H.; Liu, K.-H.; Liu, D.-M.; Chen, S.-Y. Preparation and characterization of smart magnetic hydrogels and its use for drug release. *J. Magn. Magn. Mater.* **2006**, *304*, e397–e399. [CrossRef]

65. Stylios, G.K.; Wan, T.Y. Investigating SMART Membranes and Coatings by In Situ Synthesis of Iron Oxide Nanoparticles in PVA Hydrogels. *Adv. Sci. Technol.* **2009**, *60*, 32–37.
66. Lee, K.Y.; Mooney, D.J. Hydrogels for Tissue Engineering. *Chem. Rev.* **2001**, *101*, 1869–1880. [CrossRef] [PubMed]
67. Alexandre, N.; Ribeiro, J.; Gärtner, A.; Pereira, T.; Amorim, I.; Fragoso, J.; Lopes, A.; Fernandes, J.; Costa, E.; Santos-Silva, A.; et al. Biocompatibility and hemocompatibility of polyvinyl alcohol hydrogel used for vascular grafting—In vitro and in vivo studies. *J. Biomed. Mater. Res. Part A* **2014**, *102*, 4262–4275.
68. Hyon, S.-H.; Cha, W.-I.; Ikada, Y.; Kita, M.; Ogura, Y.; Honda, Y. Poly(vinyl alcohol) hydrogels as soft contact lens material. *J. Biomater. Sci. Polym. Ed.* **1994**, *5*, 397–406. [CrossRef]
69. Adelnia, H.; Ensandoost, R.; Shebbrin Moonshi, S.; Gavgani, J.N.; Vasafi, E.I.; Ta, H.T. Freeze/thawed polyvinyl alcohol hydrogels: Present, past and future. *Eur. Polym. J.* **2022**, *164*, 110974. [CrossRef]
70. Pandey, D.K.; Kuddushi, M.; Kumar, A.; Singh, D.K. Iron oxide nanoparticles loaded smart hybrid hydrogel for anti-inflammatory drug delivery: Preparation and characterizations. *Colloids Surf. A Physicochem. Eng. Asp.* **2022**, *650*, 129631. [CrossRef]
71. Leung, V.; Ko, F. Biomedical applications of nanofibers. *Polym. Adv. Technol.* **2011**, *22*, 350–365. [CrossRef]
72. Do Pham, D.D.; Jenčová, V.; Kaňuchová, M.; Bayram, J.; Grossová, I.; Šuca, H.; Urban, L.; Havlíčková, K.; Novotný, V.; Mikeš, P.; et al. Novel lipophosphonoxin-loaded polycaprolactone electrospun nanofiber dressing reduces Staphylococcus aureus induced wound infection in mice. *Sci. Rep.* **2021**, *11*, 17688. [CrossRef]
73. Mahmoudi, M.; Zhao, M.; Matsuura, Y.; Laurent, S.; Yang, P.C.; Bernstein, D.; Ruiz-Lozano, P.; Serpooshan, V. Infection-resistant MRI-visible scaffolds for tissue engineering applications. *BioImpacts BI* **2016**, *6*, 111–115. [CrossRef]
74. Suwantong, O. Biomedical applications of electrospun polycaprolactone fiber mats. *Polym. Adv. Technol.* **2016**, *27*, 1264–1273. [CrossRef]
75. Liu, H.; Ding, X.; Zhou, G.; Li, P.; Wei, X.; Fan, Y. Electrospinning of Nanofibers for Tissue Engineering Applications. *J. Nanomater.* **2013**, *2013*, 495708. [CrossRef]
76. Rodríguez, K.; Gatenholm, P.; Renneckar, S. Electrospinning cellulosic nanofibers for biomedical applications: Structure and in vitro biocompatibility. *Cellulose* **2012**, *19*, 1583–1598. [CrossRef]
77. Zhang, Y.; Lim, C.T.; Ramakrishna, S.; Huang, Z.-M. Recent development of polymer nanofibers for biomedical and biotechnological applications. *J. Mater. Sci. Mater. Med.* **2005**, *16*, 933–946. [CrossRef] [PubMed]
78. Zafar, M.; Najeeb, S.; Khurshid, Z.; Vazirzadeh, M.; Zohaib, S.; Najeeb, B.; Sefat, F. Potential of Electrospun Nanofibers for Biomedical and Dental Applications. *Materials* **2016**, *9*, 73. [CrossRef]
79. Tamimi, E.; Ardila, D.C.; Haskett, D.G.; Doetschman, T.; Slepian, M.J.; Kellar, R.S.; Vande Geest, J.P. Biomechanical Comparison of Glutaraldehyde-Crosslinked Gelatin Fibrinogen Electrospun Scaffolds to Porcine Coronary Arteries. *J. Biomech. Eng.* **2015**, *138*, 011001–01100112. [CrossRef]
80. Agarwal, S.; Wendorff, J.H.; Greiner, A. Use of electrospinning technique for biomedical applications. *Polymer* **2008**, *49*, 5603–5621. [CrossRef]
81. Gosline, J.M.; Guerette, P.A.; Ortlepp, C.S.; Savage, K.N. The mechanical design of spider silks: From fibroin sequence to mechanical function. *J. Exp. Biol.* **1999**, *202 Pt 23*, 3295–3303. [CrossRef]
82. Ashammakhi, N.; Ndreu, A.; Piras, A.M.; Nikkola, L.; Sindelar, T.; Ylikauppila, H.; Harlin, A.; Gomes, M.E.; Neves, N.M.; Chiellini, E.; et al. Biodegradable Nanomats Produced by Electrospinning: Expanding Multifunctionality and Potential for Tissue Engineering. *J. Nanosci. Nanotechnol.* **2007**, *7*, 862–882. [CrossRef]
83. Teo, W.-E.; He, W.; Ramakrishna, S. Electrospun scaffold tailored for tissue-specific extracellular matrix. *Biotechnol. J.* **2006**, *1*, 918–929. [CrossRef]
84. Townsend-Nicholson, A.; Jayasinghe, S.N. Cell Electrospinning: A Unique Biotechnique for Encapsulating Living Organisms for Generating Active Biological Microthreads/Scaffolds. *Biomacromolecules* **2006**, *7*, 3364–3369. [CrossRef] [PubMed]
85. Jayasinghe, S.N.; Irvine, S.; McEwan, J.R. Cell electrospinning highly concentrated cellular suspensions containing primary living organisms into cell-bearing threads and scaffolds. *Nanomedicine* **2007**, *2*, 555–567. [CrossRef] [PubMed]
86. Yan, S.; Li, X.; Dai, J.; Wang, Y.; Wang, B.; Lu, Y.; Shi, J.; Huang, P.; Gong, J.; Yao, Y. Electrospinning of PVA/sericin nanofiber and the effect on epithelial-mesenchymal transition of A549 cells. *Mater. Sci. Eng. C* **2017**, *79*, 436–444. [CrossRef]
87. Persano, L.; Camposeo, A.; Tekmen, C.; Pisignano, D. Industrial Upscaling of Electrospinning and Applications of Polymer Nanofibers: A Review. *Macromol. Mater. Eng.* **2013**, *298*, 504–520. [CrossRef]
88. Guimarães, A.; Martins, A.; Pinho, E.D.; Faria, S.; Reis, R.L.; Neves, N.M. Solving cell infiltration limitations of electrospun nanofiber meshes for tissue engineering applications. *Nanomedicine* **2010**, *5*, 539–554. [CrossRef]
89. Martins, A.; Araújo, J.V.; Reis, R.L.; Neves, N.M. Electrospun nanostructured scaffolds for tissue engineering applications. *Nanomedicine* **2007**, *2*, 929–942. [CrossRef]
90. Gomes, M.E.; Azevedo, H.S.; Moreira, A.R.; Ellä, V.; Kellomäki, M.; Reis, R.L. Starch–poly(ε-caprolactone) and starch–poly(lactic acid) fibre-mesh scaffolds for bone tissue engineering applications: Structure, mechanical properties and degradation behaviour. *J. Tissue Eng. Regen. Med.* **2008**, *2*, 243–252. [CrossRef] [PubMed]
91. Sinclair, K.D.; Webb, K.; Brown, P.J. The effect of various denier capillary channel polymer fibers on the alignment of NHDF cells and type I collagen. *J. Biomed. Mater. Res. Part A* **2010**, *95*(4), 1194–1202. [CrossRef] [PubMed]
92. Tanaka, K.; Ito, A.; Kobayashi, T.; Kawamura, T.; Shimada, S.; Matsumoto, K.; Saida, T.; Honda, H. Heat immunotherapy using magnetic nanoparticles and dendritic cells for T-lymphoma. *J. Biosci. Bioeng.* **2005**, *100*, 112–115. [CrossRef]

93. Singh, R.; Lillard, J.W., Jr. Nanoparticle-based targeted drug delivery. *Exp. Mol. Pathol.* **2009**, *86*, 215–223. [CrossRef]
94. Wan, A.C.A.; Liao, I.C.; Yim, E.K.F.; Leong, K.W. Mechanism of Fiber Formation by Interfacial Polyelectrolyte Complexation. *Macromolecules* **2004**, *37*, 7019–7025. [CrossRef]
95. Tamargo, J.; Le Heuzey, J.Y.; Mabo, P. Narrow therapeutic index drugs: A clinical pharmacological consideration to flecainide. *Eur. J. Clin. Pharmacol.* **2015**, *71*, 549–567. [CrossRef] [PubMed]
96. Jeziorski, K. Hyperthermia in rheumatic diseases. A promising approach? *Reumatol. Rheumatol.* **2018**, *56*, 316–320. [CrossRef] [PubMed]
97. Brenner, M.; Braun, C.; Oster, M.; Gulko, P.S. Thermal signature analysis as a novel method for evaluating inflammatory arthritis activity. *Ann. Rheum. Dis.* **2006**, *65*, 306–311. [CrossRef] [PubMed]
98. Otremski, I.; Erling, G.; Cohen, Z.; Newman, R.J. The effect of hyperthermia (42.5 °C) on zymosan-induced synovitis of the knee. *Rheumatology* **1994**, *33*, 721–723. [CrossRef]
99. Schmidt, K.L.; Simon, E. Thermotherapy of Pain, Trauma, and Inflammatory and Degenerative Rheumatic Diseases. In *Thermotherapy for Neoplasia, Inflammation, and Pain*; Kosaka, M., Sugahara, T., Schmidt, K.L., Simon, E., Eds.; Springer: Tokyo, Japan, 2001; pp. 527–539.
100. Jordan, A.; Scholz, R.; Wust, P.; Fähling, H.; Roland, F. Magnetic fluid hyperthermia (MFH): Cancer treatment with AC magnetic field induced excitation of biocompatible superparamagnetic nanoparticles. *J. Magn. Magn. Mater.* **1999**, *201*, 413–419. [CrossRef]
101. Kumar, C.S.S.R.; Mohammad, F. Magnetic nanomaterials for hyperthermia-based therapy and controlled drug delivery. *Adv. Drug Deliv. Rev.* **2011**, *63*, 789–808. [CrossRef]
102. Huang, H.; Delikanli, S.; Zeng, H.; Ferkey, D.M.; Pralle, A. Remote control of ion channels and neurons through magnetic-field heating of nanoparticles. *Nat. Nanotechnol.* **2010**, *5*, 602–606. [CrossRef]
103. Funnell, J.L.; Balouch, B.; Gilbert, R.J. Magnetic Composite Biomaterials for Neural Regeneration. *Front. Bioeng. Biotechnol.* **2019**, *7*, 179. [CrossRef]
104. Falk, M.H.; Issels, R.D. Hyperthermia in oncology. *Int. J. Hyperthermia.* **2001**, *17*, 1–18. [CrossRef]
105. Behrouzkia, Z.; Joveini, Z.; Keshavarzi, B.; Eyvazzadeh, N.; Aghdam, R.Z. Hyperthermia: How Can It Be Used? *Oman Med. J.* **2016**, *31*, 89–97. [CrossRef] [PubMed]
106. Chang, D.; Lim, M.; Goos, J.A.C.M.; Qiao, R.; Ng, Y.Y.; Mansfeld, F.M.; Jackson, M.; Davis, T.P.; Kavallaris, M. Biologically Targeted Magnetic Hyperthermia: Potential and Limitations. *Front. Pharmacol.* **2018**, *9*, 831. [CrossRef] [PubMed]
107. Ziv-Polat, O.; Topaz, M.; Brosh, T.; Margel, S. Enhancement of incisional wound healing by thrombin conjugated iron oxide nanoparticles. *Biomaterials* **2010**, *31*, 741–747. [CrossRef] [PubMed]
108. Li, X.; Wang, Y.; Shi, L.; Li, B.; Li, J.; Wei, Z.; Lv, H.; Wu, L.; Zhang, H.; Yang, B.; et al. Magnetic targeting enhances the cutaneous wound healing effects of human mesenchymal stem cell-derived iron oxide exosomes. *J. Nanobiotechnol.* **2020**, *18*, 113. [CrossRef] [PubMed]
109. Afshar, A.; Yuca, E.; Wisdom, C.; Alenezi, H.; Ahmed, J.; Tamerler, C.; Edirisinghe, M. Next-generation Antimicrobial Peptides (AMPs) incorporated nanofibre wound dressings. *Med. Devices Sens.* **2021**, *4*, e10144. [CrossRef]
110. Khil, M.S.; Cha, D.I.; Kim, H.Y.; Kim, I.S.; Bhattarai, N. Electrospun nanofibrous polyurethane membrane as wound dressing. Journal of biomedical materials research. *Part B Appl. Biomater.* **2003**, *67*, 675–679. [CrossRef]
111. Patra, J.K.; Das, G.; Fraceto, L.F.; Campos, E.V.R.; Rodriguez-Torres, M.d.P.; Acosta-Torres, L.S.; Diaz-Torres, L.A.; Grillo, R.; Swamy, M.K.; Sharma, S.; et al. Nano based drug delivery systems: Recent developments and future prospects. *J. Nanobiotechnology* **2018**, *16*, 71. [CrossRef]
112. Zhou, Y.; Yang, D.; Chen, X.; Xu, Q.; Lu, F.; Nie, J. Electrospun Water-Soluble Carboxyethyl Chitosan/Poly(vinyl alcohol) Nanofibrous Membrane as Potential Wound Dressing for Skin Regeneration. *Biomacromolecules* **2008**, *9*, 349–354. [CrossRef]
113. Liu, J.F.; Jang, B.; Issadore, D. Use of magnetic fields and nanoparticles to trigger drug release and improve tumor targeting. *Wiley Interdiscip. Rev. Nanomed. Nanobiotechnol.* **2019**, *11*, e1571. [CrossRef]
114. Jayakumar, R.; Prabaharan, M.; Sudheesh Kumar, P.T.; Nair, S.V.; Tamura, H. Biomaterials based on chitin and chitosan in wound dressing applications. *Biotechnol. Adv.* **2011**, *29*, 322–337. [CrossRef]
115. Gao, F.; Li, X.; Zhang, T.; Ghosal, A.; Zhang, G.; Fan, H.M.; Zhao, L. Iron nanoparticles augmented chemodynamic effect by alternative magnetic field for wound disinfection and healing. *J. Control. Release* **2020**, *324*, 598–609. [CrossRef] [PubMed]
116. Ibelli, T.; Templeton, S.; Levi-Polyachenko, N. Progress on utilizing hyperthermia for mitigating bacterial infections. *Int. J. Hyperth.* **2018**, *34*, 144–156. [CrossRef] [PubMed]
117. Weng, L.; Xie, J. Smart electrospun nanofibers for controlled drug release: Recent advances and new perspectives. *Curr. Pharm. Des.* **2015**, *21*, 1944–1959. [CrossRef] [PubMed]
118. Thomas, C.R.; Ferris, D.P.; Lee, J.-H.; Choi, E.; Cho, M.H.; Kim, E.S.; Stoddart, J.F.; Shin, J.-S.; Cheon, J.; Zink, J.I. Noninvasive Remote-Controlled Release of Drug Molecules in Vitro Using Magnetic Actuation of Mechanized Nanoparticles. *J. Am. Chem. Soc.* **2010**, *132*, 10623–10625. [CrossRef]
119. Xu, T.; Yu, J.; Yan, X.; Choi, H.; Zhang, L. Magnetic Actuation Based Motion Control for Microrobots: An Overview. *Micromachines* **2015**, *6*, 1346–1364. [CrossRef]
120. Hughes, S.; El Haj, A.J.; Dobson, J. Magnetic micro- and nanoparticle mediated activation of mechanosensitive ion channels. *Med. Eng. Phys.* **2005**, *27*, 754–762. [CrossRef]

121. Deckers, R.; Quesson, B.; Arsaut, J.; Eimer, S.; Couillaud, F.; Moonen, C.T.W. Image-guided, noninvasive, spatiotemporal control of gene expression. *Proc. Natl. Acad. Sci. USA* **2009**, *106*, 1175–1180. [CrossRef]
122. Leong, T.G.; Randall, C.L.; Benson, B.R.; Bassik, N.; Stern, G.M.; Gracias, D.H. Tetherless thermobiochemically actuated microgrippers. *Proc. Natl. Acad. Sci. USA* **2009**, *106*, 703–708. [CrossRef]
123. Zhang, Y.; Zhang, L.; Yang, L.; Vong, C.I.; Chan, K.F.; Wu, W.K.K.; Kwong, T.N.Y.; Lo, N.W.S.; Ip, M.; Wong, S.H.; et al. Real-time tracking of fluorescent magnetic spore-based microrobots for remote detection of *C. diff* toxins. *Sci. Adv.* **2019**, *5*, eaau9650. [CrossRef]
124. Steager, E.B.; Selman Sakar, M.; Magee, C.; Kennedy, M.; Cowley, A.; Kumar, V. Automated biomanipulation of single cells using magnetic microrobots. *Int. J. Robot. Res.* **2013**, *32*, 346–359. [CrossRef]
125. Felfoul, O.; Mohammadi, M.; Taherkhani, S.; de Lanauze, D.; Zhong Xu, Y.; Loghin, D.; Essa, S.; Jancik, S.; Houle, D.; Lafleur, M.; et al. Magneto-aerotactic bacteria deliver drug-containing nanoliposomes to tumour hypoxic regions. *Nat. Nanotechnol.* **2016**, *11*, 941–947. [CrossRef] [PubMed]
126. Kim, S.; Qiu, F.; Kim, S.; Ghanbari, A.; Moon, C.; Zhang, L.; Nelson, B.J.; Choi, H. Fabrication and Characterization of Magnetic Microrobots for Three-Dimensional Cell Culture and Targeted Transportation. *Adv. Mater.* **2013**, *25*, 5863–5868. [CrossRef] [PubMed]
127. Koppenol, W.H.; Hider, R.H. Iron and redox cycling. Do's and don'ts. *Free Radic. Biol. Med.* **2019**, *133*, 3–10. [CrossRef] [PubMed]
128. Floyd, S.; Pawashe, C.; Sitti, M. An untethered magnetically actuated micro-robot capable of motion on arbitrary surfaces. In Proceedings of the 2008 IEEE International Conference on Robotics and Automation, Pasadena, CA, USA, 19–23 May 2008; pp. 419–424.
129. Saito, H. Metabolism of Iron Stores. *Nagoya J. Med. Sci.* **2014**, *76*, 235–254. [PubMed]
130. Kohgo, Y.; Ikuta, K.; Ohtake, T.; Torimoto, Y.; Kato, J. Body iron metabolism and pathophysiology of iron overload. *Int. J. Hematol.* **2008**, *88*, 7–15. [CrossRef]
131. Yarjanli, Z.; Ghaedi, K.; Esmaeili, A.; Rahgozar, S.; Zarrabi, A. Iron oxide nanoparticles may damage to the neural tissue through iron accumulation, oxidative stress, and protein aggregation. *BMC Neurosci.* **2017**, *18*, 51. [CrossRef]
132. Yang, Z.; Zhang, L. Magnetic Actuation Systems for Miniature Robots: A Review. *Adv. Intell. Syst.* **2020**, *2*, 2000082. [CrossRef]
133. Longmire, M.; Choyke, P.L.; Kobayashi, H. Clearance properties of nano-sized particles and molecules as imaging agents: Considerations and caveats. *Nanomedicine (lond)* **2008**, *3*, 703–717. [CrossRef]
134. Kim, Y.; Parada, G.A.; Liu, S.; Zhao, X. Ferromagnetic soft continuum robots. *Sci. Robot.* **2019**, *4*, eaax7329. [CrossRef] [PubMed]
135. Ergeneman, O.; Dogangil, G.; Kummer, M.P.; Abbott, J.J.; Nazeeruddin, M.K.; Nelson, B.J. A Magnetically Controlled Wireless Optical Oxygen Sensor for Intraocular Measurements. *IEEE Sens. J.* **2008**, *8*, 29–37. [CrossRef]
136. Zhou, Q.; Wei, Y. For Better or Worse, Iron Overload by Superparamagnetic Iron Oxide Nanoparticles as a MRI Contrast Agent for Chronic Liver Diseases. *Chem. Res. Toxicol.* **2017**, *30*, 73–80. [CrossRef] [PubMed]
137. Hirschhorn, T.; Stockwell, B.R. The development of the concept of ferroptosis. *Free Radic. Biol. Med.* **2019**, *133*, 130–143. [CrossRef] [PubMed]
138. Bossmann, S.H.; Oliveros, E.P.D.; Göb, S.; Siegwart, S.; Dahlen, E.P.; Payawan, L.M.; Straub, M.R.; Wörner, M.; Braun, A.M. New Evidence against Hydroxyl Radicals as Reactive Intermediates in the Thermal and Photochemically Enhanced Fenton Reactions. *J. Phys. Chem. A* **1998**, *102*, 5542–5550. [CrossRef]
139. Wu, L.; Wen, W.; Wang, X.; Huang, D.; Cao, J.; Qi, X.; Shen, S. Ultrasmall iron oxide nanoparticles cause significant toxicity by specifically inducing acute oxidative stress to multiple organs. *Part. Fibre Toxicol.* **2022**, *19*, 24. [CrossRef] [PubMed]
140. Hannah, W.; Thompson, P.B. Nanotechnology, risk and the environment: A review. *J. Environ. Monit.* **2008**, *10*, 291–300. [CrossRef] [PubMed]
141. Hartwig, A. Metal ions between essentiality and toxicity. *Chem. Unserer Zeit* **2000**, *34*, 224–231. [CrossRef]
142. Kühr, S.; Schneider, S.; Meisterjahn, B.; Schlich, K.; Hund-Rinke, K.; Schlechtriem, C. Silver nanoparticles in sewage treatment plant effluents: Chronic effects and accumulation of silver in the freshwater amphipod Hyalella azteca. *Environ. Sci. Eur.* **2018**, *30*, 7. [CrossRef]
143. Pereira, M.C.; Oliveira, L.C.A.; Murad, E. Iron oxide catalysts: Fenton and Fenton-like reactions—A review. *Clay Min.* **2012**, *47*, 285–302. [CrossRef]
144. Holleman, A.F.; Wiberg, N. *d- and f-Block Elements, Lanthanides, Actinides, Transactinides*; De Gruyter: Berlin, Germany, 2016.
145. Poon, W.; Zhang, Y.N.; Ouyang, B.; Kingston, B.R.; Wu, J.L.Y.; Wilhelm, S.; Chan, W.C.W. Elimination Pathways of Nanoparticles. *ACS Nano* **2019**, *13*, 5785–5798. [CrossRef]
146. Vachha, B.; Huang, S.Y. MRI with ultrahigh field strength and high-performance gradients: Challenges and opportunities for clinical neuroimaging at 7 T and beyond. *Eur. Radiol. Exp.* **2021**, *5*, 35. [CrossRef]
147. Malhotra, N.; Lee, J.S.; Liman, R.A.D.; Ruallo, J.M.S.; Villaflores, O.B.; Ger, T.R.; Hsiao, C.D. Potential Toxicity of Iron Oxide Magnetic Nanoparticles: A Review. *Molecules* **2020**, *25*, 3159. [CrossRef] [PubMed]
148. Wood, J.C. Use of magnetic resonance imaging to monitor iron overload. *Hematol. Oncol. Clin. North Am.* **2014**, *28*, 747–764. [CrossRef] [PubMed]
149. Mallio, C.A.; Rovira, À.; Parizel, P.M.; Quattrocchi, C.C. Exposure to gadolinium and neurotoxicity: Current status of preclinical and clinical studies. *Neuroradiology* **2020**, *62*, 925–934. [CrossRef] [PubMed]

150. Liang, P.; Ballou, B.; Lv, X.; Si, W.; Bruchez, M.P.; Huang, W.; Dong, X. Monotherapy and Combination Therapy Using Anti-Angiogenic Nanoagents to Fight Cancer. *Adv. Mater.* **2021**, *33*, e2005155. [CrossRef]
151. Hassannia, B.; Vandenabeele, P.; Vanden Berghe, T. Targeting Ferroptosis to Iron Out Cancer. *Cancer Cell* **2019**, *35*, 830–849. [CrossRef]
152. Mayes, W.M.; Jarvis, A.P.; Burke, I.T.; Walton, M.; Gruiz, K. *Trace and Rare Earth Element Dispersal Downstream of the Ajka Red Mud Spill*; Heavy metal pollution analysis; International Mine Water Association: Aachen, Germany, 2011; pp. 29–34.
153. Srinivas, A.; Rao, P.J.; Selvam, G.; Goparaju, A.; Murthy, P.B.; Reddy, P.N. Oxidative stress and inflammatory responses of rat following acute inhalation exposure to iron oxide nanoparticles. *Hum. Exp. Toxicol.* **2012**, *31*, 1113–1131. [CrossRef]
154. Chao, Y.; Chen, G.; Liang, C.; Xu, J.; Dong, Z.; Han, X.; Wang, C.; Liu, Z. Iron Nanoparticles for Low-Power Local Magnetic Hyperthermia in Combination with Immune Checkpoint Blockade for Systemic Antitumor Therapy. *Nano Lett.* **2019**, *19*, 4287–4296. [CrossRef]
155. McWilliams, B.T.; Wang, H.; Binns, V.J.; Curto, S.; Bossmann, S.H.; Prakash, P. Experimental Investigation of Magnetic Nanoparticle-Enhanced Microwave Hyperthermia. *J. Funct. Biomater.* **2017**, *8*, 21. [CrossRef]
156. Payne, M.; Bossmann, S.H.; Basel, M.T. Direct treatment versus indirect: Thermo-ablative and mild hyperthermia effects. *Wiley Interdiscip. Rev. Nanomed. Nanobiotechnol.* **2020**, *12*, e1638. [CrossRef]

Review

Magnetic Solid Nanoparticles and Their Counterparts: Recent Advances towards Cancer Theranostics

Mónica Cerqueira [1,2,3], Efres Belmonte-Reche [3], Juan Gallo [3], Fátima Baltazar [1,2,*] and Manuel Bañobre-López [3,*]

[1] Life and Health Sciences Research Institute (ICVS), Campus de Gualtar, University of Minho, 4710-057 Braga, Portugal; monica.cerqueira@inl.int
[2] ICVS/3B's—PT Government Associate Laboratory, 4805-017 Guimarães, Portugal
[3] Advanced (Magnetic) Theranostic Nanostructures Lab, Nanomedicine Unit, International Iberian Nanotechnology Laboratory, Avenida Mestre José Veiga, 4715-330 Braga, Portugal; efres.belmonte@inl.int (E.B.-R.); juan.gallo@inl.int (J.G.)
* Correspondence: fbaltazar@med.uminho.pt (F.B.); manuel.banobre@inl.int (M.B.-L.)

Abstract: Cancer is currently a leading cause of death worldwide. The World Health Organization estimates an increase of 60% in the global cancer incidence in the next two decades. The inefficiency of the currently available therapies has prompted an urgent effort to develop new strategies that enable early diagnosis and improve response to treatment. Nanomedicine formulations can improve the pharmacokinetics and pharmacodynamics of conventional therapies and result in optimized cancer treatments. In particular, theranostic formulations aim at addressing the high heterogeneity of tumors and metastases by integrating imaging properties that enable a non-invasive and quantitative assessment of tumor targeting efficiency, drug delivery, and eventually the monitoring of the response to treatment. However, in order to exploit their full potential, the promising results observed in preclinical stages need to achieve clinical translation. Despite the significant number of available functionalization strategies, targeting efficiency is currently one of the major limitations of advanced nanomedicines in the oncology area, highlighting the need for more efficient nanoformulation designs that provide them with selectivity for precise cancer types and tumoral tissue. Under this current need, this review provides an overview of the strategies currently applied in the cancer theranostics field using magnetic nanoparticles (MNPs) and solid lipid nanoparticles (SLNs), where both nanocarriers have recently entered the clinical trials stage. The integration of these formulations into magnetic solid lipid nanoparticles—with different composition and phenotypic activity—constitutes a new generation of theranostic nanomedicines with great potential for the selective, controlled, and safe delivery of chemotherapy.

Keywords: solid lipid nanoparticles; magnetic nanoparticles; magnetic solid lipid nanoparticles; cancer theranostics; MRI-contrast agents

1. Introduction

Cancer is a malignant disease involving uncontrolled and rapid growth of aberrant and nonfunctional cells as a result of epigenetic and genetic modifications. These have the capacity to metastasize to distant organs of the body [1]. This heterogeneous disease ranks as a principal public health concern worldwide [2]. In total, 18.1 million new cancer cases were diagnosed in 2018, whilst 9.6 million deaths were related to the disease. Moreover, a 60% incidence increase in new global cancer cases is expected to occur over the next two decades, according to the World Health Organization (WHO) [3].

The main tool for an efficient cancer treatment is an early diagnosis, as according to WHO reports, 30% of patients could have successfully been considered cured if diagnosed at an early stage of the disease. When the tumor is identified early (in the first stages), combinations of surgery, chemotherapy, and radiotherapy are usually viable options as treatments with higher success rates and less side effects [4]. However, the latter occurrence of the symptoms leads quite often to a cancer diagnosis at more advanced stages—stage three or four. Then, the subscripted cancer treatment will be dependent on the type and stage of the tumor/s, in addition to the patient's condition—older and weaker patients are normally spared treatments due to their aggressiveness—where late diagnosis (and/or surgical tumor inaccessibility) limits the treatment of cancers to chemotherapy and immunotherapy [4].

Several research fields are focused on finding anticancer drugs that achieve a selective phenotypic cytotoxic effect on cancer cells. These should, at the same time, stop or slow down tumor growth whilst being less toxic (or ideally innocuous) to healthy tissues [5]. Chemotherapeutic agents obtain different mechanisms of action depending on their pharmacophore structure and other moieties (its chemical structure). Hence, chemotherapeutics can be classified as alkylating agents (e.g., cisplatin and cyclophosphamide), anti-metabolites (e.g., methotrexate and fluorouracil), anthracyclines with DNA-binding antibiotics (e.g., doxorubicin (DOX)), topoisomerase inhibitors (e.g., etoposide), and microtubule stabilizers (e.g., paclitaxel, docetaxel) [4,6]. Although usually effective, the main drawback of these drugs is their selectivity issues, as they can usually have a phenotypic effect on the much more abundant healthy tissue as well. This can cause short and then long-term health sequels in patients and even death [6–9].

When administered intravenously, chemotherapeutics are systemically distributed and therefore can potentially reach all organs. Given its nature as a blood detoxifier—converting xenobiotics into waste products—the liver is usually specially affected by the non-selective action of the drugs [10]. Systemic distribution also reduces the in situ concentration of the compounds in the tumor area. They may therefore require a higher posology to achieve the desired effect, compromising their narrow therapeutic margins [5,11,12]. The poor pharmacokinetics, specificity, and the generation of cancer multidrug-resistance (MDR) can further reduce their therapeutic margins [5–7,13]. Altogether, the treatments available and the current poor success rates associated with them require smart targeted strategies to achieve chemotherapeutic selectivity in addition to better early diagnosis and in situ therapies.

Nanotechnology has evolved into a multidisciplinary field, having revolutionized many scientific and nonscientific areas since 1970, including: applied physics, materials chemistry, chemistry mechanics, robotics, medicine, and biological and electrical engineering [14]. In the bioscience and medicine fields, nanomaterials have a wide range of applications. In cancer therapy, for example, they have been used as diagnostic tools and as drug delivery formulations [15,16]. Their nanoscale size (1–100 nm) makes them ideal candidates for surface nano-engineering and the production of functionalized nanostructures [17]. Hence, they are currently being applied as drug delivery systems (DDS), sensors, and tissue engineering catalyzers, amongst others [18]. Due to their unique physical and optical properties and chemical stability, nanoparticles can grant selectivity to drugs for specific body/organ/tissue targeting and even for individual recognition and targeting of single cancer cells [15,19]. Hence, the nanoparticles' characteristics can benefit the bioactivity of therapeutic compounds through the reduction of the concentration needed for a certain phenotypic outcome, potentially increasing their therapeutic margins and pharmacokinetic properties and altogether reducing their potential harmful secondary effects in healthy tissues (Figure 1) [14,18,19].

Many nanoformulations have been investigated pre-clinically, yet only a minority have advanced to clinic stages [20]. Currently, those approved by the U.S. FDA and European Medicines Agency (EMA) [21] include: Abraxane [22], Doxil [23], and Patisiran/ONPATTRO [24]. These formulations respond to the need for creating new systems that efficiently improve drug selectivity and delivery and that help promote an accurate and safer treatment of cancer.

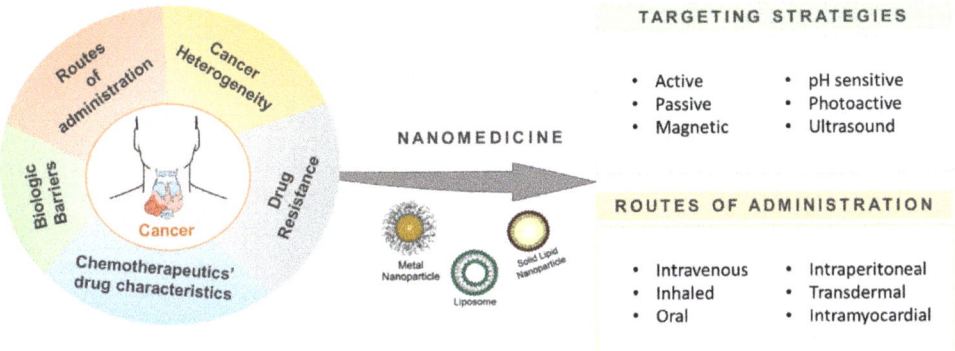

Figure 1. Nanomedicine applications in cancer therapy. Nanoparticles, as drug delivery systems, can enhance the drug targeting to specific body/organ/tissue or even single cancer cells through different targeting strategies (e.g., active/passive, endogenously/exogenously responsive) and different routes of administration (intravenous, oral, or intraperitoneal, among others).

Within the cancer field, magnetic nanoparticles (MNPs) have gained interest as highly functionalized tools that can be applied to diagnosis, monitorization, and therapy. Their relative straightforward synthesis, functionalization, purification, and characterization, together with their usually good biodegradability and diagnostic platform potential, confer major advantages for their use in cancer theranostics [25–35]. Recently, NanoTherm®, a new platform for the intermittent glioblastoma treatment multiform, was approved by the EMA and evidences the potential these systems have in cancer diagnosis and therapy [36]. Another type of nanoparticle, which is based on solid lipid nanoparticles (SLNs), has also been studied abundantly and is currently applied in cancer therapy. Here, SLNs have been used as a drug delivery system that has the potential to control the release of the loaded chemotherapy and decrease their toxicity with an enhancement of biocompatibility in comparison to inorganic or polymeric nanoparticles [37–40].

In this review, we provide an overview of recent developments to fight cancer using MNPs and SLNs, alone or in combination, to yield magnetic solid lipid nanoparticles (mSLNs), where we highlight their performance and potential application in diagnosis, drug delivery, and other therapeutic approaches such as magnetic hyperthermia and theranostics. Special focus will be paid to those reports offering results at the advanced preclinical stage, both in vitro and in vivo.

2. Magnetic Nanoparticles

MNPs are being widely studied nowadays in many areas (such as in the biomedical field), because they offer a plethora of opportunities [25]. Their physicochemical properties, superparamagnetic behavior, small size, and capability to promote biological interactions at the cellular and molecular level [25,26] allow MNPs to be employed as drug delivery systems [28,29], magnetic resonance imaging contrast enhancers [30], and hyperthermia inducers [31] for the treatment of cancer.

A key component of these MNPs is the metal used in their formulations. Thus, they are usually ferrites (MFe_2O_4, $Ni_aZn_{(1-a)}Fe_2O_4$, $Mn_aZn_{(1-a)}Fe_2O_4$) [41], metal alloys (FeCo, alnico, and permalloy), or iron-based magnetic oxides (hematite (α-Fe_2O_3), magnetite (Fe_3O_4), and maghemite (γ-Fe_2O_3)) [31]. The most commonly used nanoparticles in the biomedical field are superparamagnetic iron oxide nanoparticles (SPIONs), such as Fe_3O_4 and γ-Fe_2O_3, which present high biocompatibility and lower toxicity compared to other metal structures (e.g., quantum dots, gold nanoparticles, and carbon nanotubes (CNTs) may present lower biodegradation and body-elimination issues [25], together with increased cytotoxicity [32,41]). Their superparamagnetic properties enable a degree of control through the application of an alternating magnetic field (AMF). Here, selective application of the AMF can force the MNPs to generate local heat and promote the direct tumor ablation and/or the drug release into the desired region, ultimately avoiding invasive diagnostic and therapeutic techniques [32,33].

MNP performance is dependent on their composition, morphology, surface coating, and size of the inorganic core, all of which influence their in vivo behavior [25] and potential toxicity [41]. Studies performed in a mouse model with MNPs coated with DMSA (dimercaptosuccinic acid) revealed accumulation in the liver, spleen, and lungs without side effects [34]. Hence, the functionalization of the formulations' surface with targeted ligands can be a strategy to reduce toxicity in untargeted organs, whilst also increasing the therapeutic efficacy in targeted ones [41].

2.1. Magnetic Nanoparticles as Drug Delivery Systems

MNPs have become an interesting vehicle for drug delivery in the cancer therapy field. The MNPs' design and formulation are part of an interdisciplinary scientific communication where bio-physicochemical interactions between MNPs and cells are optimized. As described by Hung et al. [41], an efficient DDS should: (i) have the capacity to load the appropriate drug/active compound, (ii) improve the biocompatibility, stability, and protect the drug and its bioactivity, and (iii) promote drug delivery at the required site with low toxicity for the healthy cells/tissues, [41].

As several MNP production methods have been currently described in the literature, the process can be selected based on the ultimate purpose/objective of the MNPs, which for most is the maximization of the desired phenotypic effect on cancer. On the one hand, the co-precipitation of salts with stabilizing polymer/s, hydro/solvothermal procedures, thermal decomposition, and reverse microemulsions can be considered the traditional methods of MNPs synthesis [27]. On the other hand, newer strategies include microfluidic and biogenic synthesis [36]. In either case, the resulting MNPs are usually constituted by a magnetic core–shell encapsulated by a polymer coating [42], where chemotherapeutics are loaded into (Figure 2). In this manner, the chemotherapeutics also help improve their colloidal stability and pharmacokinetic properties for the posterior systemic administration [43]. The drug loading can also be performed by several methodologies [27,42], although the methodology most employed makes use of the direct encapsulation of the drug or its absorption in the MNPs through physical or chemical interactions. The drug loading efficiency is here dependent on both the properties and compatibilities of the chemotherapy with the MNP and its coating [1]. Hence, MNP coating selection and optimization is the common strategy to effectively load hydrophobic [44] or hydrophilic drugs into the nanoformulations [43]. Different coatings may also have different feasibilities for the formulation administration [45]. Altogether, an effective coating selection will promote the correct loading of the drugs, prevent the nanoparticle agglomeration, and promote an efficient and controlled release at the target site. Typical coatings include lipids, surfactants, or polymers (such as dextran or polyethylene glycol (PEG)). These organic surfactants and polymers enhance the biocompatibility of the nanoparticles and promote opsonization resistance. This expands their systemic circulation time and increases the fraction of nanoparticles that ultimately reach the target (tumor cells) [25,46]. Furthermore, coatings can also lower unwanted cytotoxicity in healthy tissues. For example, for iron oxide nanoparticles coated with PEG, Ruiz and co-workers demonstrated an enhanced residence time and reduced liver and spleen particle accumulation when compared to its uncoated counterpart [35].

MNPs loaded with active agents (chemotherapy, DNA, RNA, or antibodies) can further improve their therapeutic effects and margin whilst grating a degree of control over their release in the biological environment [46–50]. Additional selectivity and modulation of the MNP response can be achieved by functionalizing the MNPs [47]. For example, functionalized MNPs have already been prepared to be sensitive to internal metabolic factors of the tumor, such as pH, hypoxia, specific enzymes, and the Warburg effect [25,46–49]. MNP formulations have also been prepared to be sensitive to an external stimulus to be subjected over the tumor area, such as light or temperature [16,25,30,51]. For the latter, MNPs under either near-infrared (NIR) light or an alternating magnetic field (as the external stimuli) have been found to further modulate the release of the loaded drug [31,52]. Hence, the stimuli can provide an additional level of control over the drug release equilibrium [25,47].

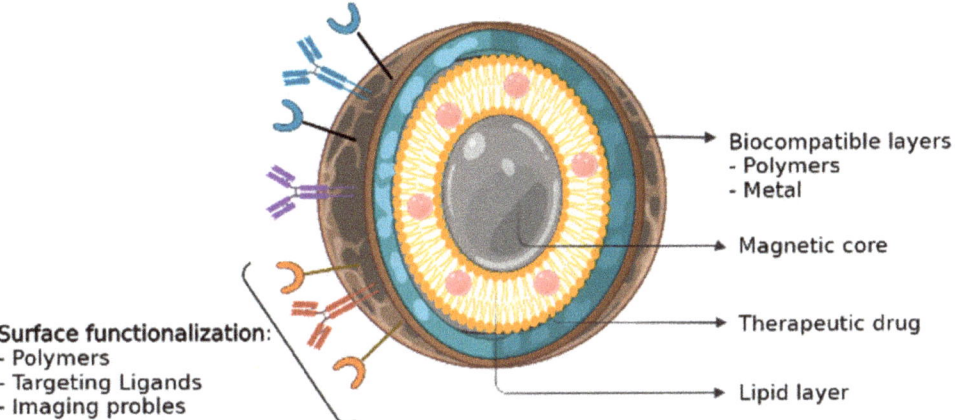

Figure 2. Magnetic nanoparticle (MNP) structure. MNPs are usually constituted by a magnetic core–shell encapsulated by a biocompatible coating [42], where chemotherapeutics are loaded into.

The sum of all of these characteristics makes MNPs very interesting tools for the safe and selective targeting of cancer, in addition to their theranostic capabilities [48–50].

2.2. Magnetic Nanoparticles in Cancer Diagnostics

The WHO's 2018 world cancer report predicted an increase by 2040 of 60% in cancer incidence. Currently, early detection is the most effective way to increase the probability for successfully overcoming most cancers. These malignancies ideally require a non-invasive, fast, and precise diagnostic system able to provide the position, size, and characteristics of the main tumor, in addition to that of other metastatic bodies [53].

A diagnostic tool used in clinic for tumor detection is magnetic resonance imaging (MRI). MRI is a non-invasive, safe, and painless technique that uses magnetism and radio pulses to produce images of different internal tissues and organs from different angles and perspectives. The result is usually a clear depiction of soft tissues, including some tumors [41].

MRI is based on the properties of some atoms to absorb energy in the form of radio waves when under a magnetic field. Such an event causes a spin polarization that can induce a signal in a radio frequency coil that can then be detected by a nearby antennae/detector. Usually, hydrogen nuclei consisting of a single proton are used to create the signals. Hydrogen is naturally abundant in all forms of life and hence can be used to create a macroscopic polarization of hydrogen-rich tissues (rich in water and fats). The pulses of radio waves excite the nuclear spin energy transition whilst the magnetic field gradient localizes their polarization in space. After the excitation, the technique measures the relaxation of the hydrogen in the longitudinal (T_1-spin-lattice relaxation) and transverse planes (T_2-spin-spin relaxation) [33,53,54]. The image formed here is dependent on the tissue's local atomic density and the association of hydrogen to other atoms. Furthermore, the pulse sequence can generate different contrasts between tissues, as can specific agents that increase the capabilities of MRI. These agents shorten the relaxation times of the nearby tissue, thus overcoming sensitivity limitations of the technique. These can be categorized by their planar outcome, T_1 and/or T_2 effects (longitudinal or transverse effect on relaxation time of water protons, respectively [55]). Similarly, longitudinal and transverse relaxivity (r_1 and r_2) are a measure of the goodness of a contrast agent for T_1- and T_2-weighted MR imaging, respectively, and indicate the concentration of contrast agent (mM) that is needed to shorten the relaxation time by one second.

MNPs are a type of MRI contrast agents with multifunctional properties that are considered interesting probes for their co-localization in specific tissues, such as some tumors. Guldris and co-workers [56–58] and Keasberry et al. [59] reported that proper designs of iron oxide MNPs can significantly enhance the diagnostic capability of MRI when compared to other nanostructured Fe-based contrast agents currently available. The most common magnetic labels used in vivo are based on gadolinium (Gd) complexes and iron oxide magnetic nanoparticles (Fe_3O_4). The latter has already been successfully used in clinical diagnosis as an MRI contrast agent (e.g., Abdoscan®, Resovist®, Feridex®) [60]. Additionally, and in opposition to Gd complexes, iron-based contrast agents have the potential to be used in T_1- or T_2-weighted imaging with better biocompatibility and safety [54]. Likewise, manganese oxide nanoparticles are of growing interest as an alternative to the Gd chelates as T_1 contrast agents [61,62].

To date, several works in the literature have attempted the optimization of MNPs as MRI contrast agents to improve their imaging capabilities for cancer diagnosis. Tse and co-workers reported the synthesis of a prostate specific membrane antigen (PSMA)-targeting iron oxide using a solvent evaporation method, which when directly injected into the prostate induced negative contrast visualization in the MRI [63]. The authors noted the great applicability of the MNPs for the detection and localization of prostate cancer as the result of the great increase in image contrast in in vivo experiments. Similarly, Salimi et al. synthesized iron oxide magnetic nanoparticles coated with a fourth generation polyamidoamine dendrimer (G_4@IONPs). These G_4@IONPs MNPs, which were synthesized via a co-precipitation method, significantly shortened the transverse relaxation times (T_2) in in vivo MRI imaging of the mice's liver after the intravenous administration of the G_4@IONPs MNPs [64]. Gonzalez-Rodrigues et al. followed a different approach and synthesized multifunctional graphene oxide magnetite (GO-Fe_3O_4) loaded with doxorubicin to obtain a formulation with dual magnetic resonance and fluorescence imaging capabilities [65]. The synthesis was here achieved via a coupling reaction between 3-aminopropyltriethoxysilane (APTES)-Fe_3O_4 nanoparticles and GO in the presence of the coupling agents N'-ethylcarbodiimide hydrochloride (EDC) and N-hydroxysuccinimide (NHS). These GO-Fe_3O_4 MNPs exhibited a high r_2/r_1 ratio and great potential to be used as a negative MRI contrast agent in vitro in both cervical and breast cancers cell lines (HeLa and MCF-7, respectively). The authors also reported the use of MRI to study the DOX release from the nanocarrier, together with the translocation of the GO-Fe_3O_4 into the cancer cells [65]. In their study, the MRI analysis provided extensive information regarding the drug's spatial-temporal release and the consequent evaluation of the overall therapeutic efficiency. Another study was conducted by Gallo and co-workers using eco-friendly synthesis of MnO_2_CQDs (carbon quantum dots), which showed OFF–ON responsiveness in the presence of redox stimuli for dual MRI/fluorescence imaging applicability [66].

2.3. Magnetic Nanoparticles for Cancer Treatment

Hyperthermia. The use of heat as a treatment for cancer was first tested in 1898 by Frans Westermark, who used hot water in an intracavitary spiral tube to treat advanced cervical cancer [67]. In 1957, Gilchrist et al. administered magnetic nanoparticles for the first time with the intention of generating induction heating capable of selectively killing lymphatic metastases [68]. The authors delivered 5 mg of Fe_3O_4 per gram of lymph nodes tissue and then applied an alternating magnetic field (AMF) of 15.9–19.1 kAm^{-1} at 1.2 MHz to obtain a temperature rise of 14 °C. The results of the experiments showed a significant cancer cell death rate without side effects to surrounding tissues [68]. Since then, different methods have been developed to deliver heat as a system for cancer ablation.

This effect, known as hyperthermia or overheating, is a phenomenon where an abnormal higher body temperature occurs (higher than the normal corporal temperature of 37 °C) [69]. This effect can have a variety of origins, including a natural immunological defense mechanism (fever), designed to increase the body's temperature when suffering an infection [69]. Similarly, overheating can be employed for cancer therapy purposes [70]. Conventional hyperthermia, such as radiofrequency or microwave, is here applied as an adjuvant therapy, ultimately exposing tissues to higher temperatures (up 42 °C) that promote cancer cells apoptosis [71]. As mentioned before, cancer is characterized by an intensification of the cells metabolism rate, amongst other changes, that combined with a disorganized vascular system [1] results in an increased sensitivity to hyperthermia (since the ability to disperse heat is diminished) [68,72]. Additionally, hyperthermia increases the susceptibility of cancer cells to other treatments, including chemotherapy and radiotherapy [72]. However, the main problem of classical hyperthermia is the lack of homogeneity in the heat distribution profile, which can cause harmful side effects in the bordering healthy tissues. Such problems highlight the need to control the temperature increase [73].

An alternative that can allow the control of the temperature is the use of tough, magnetic nanoparticles as generators of local heat in specific areas. When an external AMF is applied to generate heat, the approach is called magnetic hyperthermia [74]. Magnetic hyperthermia is a non-invasive treatment where, in the presence of an AMF, magnetic material can transform electromagnetic radiation into thermal energy. Nearby cancer cells heat up to ideally result in tumor ablation [51,75]. Furthermore, intravenous administration of MNPs allows their accumulation on tumorous tissues via passive (by the enhanced permeability and retention (EPR) effect) and potentially active mechanisms (where the MNP surface possesses specific ligands for the surface receptors present in cancer cells) [76]. This accumulation can enable the repetition of posterior AMF treatments with no further MNP administration [33,75]. Additionally, the incorporation of chemotherapeutic drugs inside the formulation allows a synergistic combination of magnetic hyperthermia and chemotherapy, which can overcome some of the concerns related to the magneto-thermal conversion efficiency in vivo (such as degradation of magnetic susceptibility or their inherent absorption under AMF) [77].

Rego et al. evaluated the performance of aminosilane-coated superparamagnetic iron oxide nanoparticles as a magnetic hyperthermia treatment in a glioblastoma tumor model. A C6 cell model was evaluated in vitro, whilst Wistar rats were implanted by stereotaxis with C6 cells via stereotaxis for their in vivo evaluation. The authors applied an AMF of 874 kHz and 200 Gauss (20 mT) and observed a 52% and 32.8% in vitro and in vivo cancer cell death, respectively [78]. It is important to highlight that the allowed electromagnetic field that can be applied to living organs should not exceed an upper limit given by the product $H \cdot f = 4.5 \times 10^8$ $Am^{-1}s^{-1}$ (according to the Brezovich criterion [79]) or $H \cdot f = 5 \times 10^9$ $Am^{-1}s^{-1}$ (according to Herg et al. [80]).

Similarly, in a recent study, Kandasamy et al. synthesized hydrophilic and surface-functionalized superparamagnetic iron oxide nanoparticles (SPIONs). The synthesized SPIONs were functionalized in situ with short-chained molecules, including 1,4-diaminobenzene (14DAB), 4-aminobenzoic acid (4ABA), and 3,4-diaminobenzoic acid (34DABA). Moreover, their combination with terephthalic acid (TA)/2-aminoterephthalic acid (ATA)/trimesic acid (TMA)/pyromellitic acid (PMA) molecules was explored. The results showed that only the 4DAB-, 4ABA-, 34DABA-, and 4ABA-TA-coated SPIONS presented higher magnetization values than free SPIONS. More specifically, 34DABA-coated SPIONs-based aqueous ferrofluid (AFF, 0.5 mg mL^{-1}) showed a faster thermal response and achieved the therapeutic temperature of 42 °C, ultimately having a higher cytotoxic efficiency (61–88%) in HepG2 liver cancer cells [81]. Table 1 summarizes other representative biological studies that have applied MNPs hyperthermia in cancer context.

Table 1. Studies using magnetic nanoparticles (MNPs) for magnetic hyperthermia treatment in cancer.

MNP (Particle Size) + Surface Modification	Treatment + Cancer Model	Results	Ref
SPIONs (250 nm) were coated with targeted CXCR4.	Treatment: 869 kHz and 20 kA·m^{-1} for the first 30 min of the experiment, followed by another 30 min at 554 kHz, and 24 kA·m^{-1}. Cancer model: glioblastoma (LN229) and normal kidney cells (HK-2).	In vitro, the targeted treatment conjugated with MH strategy showed a lethal outcome of, approximately, 100% for LN229 cancer cells after 72 h of treatment. The safety profile of NPs was confirmed by the minimal cytotoxicity observed in control group (JK cells—HK-2 cell line).	[82]
IONPs (not specified) were coated with DMSA and conjugated with Gem and the pseudo-peptide NucAnt (N6L).	Treatment: H = 15.4 kA m^{-1}; f = 435 kHz. Cancer model: pancreatic cancer model (BxPC-3 and PANC-1 cancer cell lines). Athymic nude mice were subcutaneously injected with 2×10^6 BxPC-3 cells.	Combined chemotherapy and treatment with NPs-based MH showed increased cytotoxicity and cell death in vitro (~90% of viable cells compared to approximately 10% when no MH was applied). In vivo, Gem MNPs and the hyperthermia therapy managed to cause an almost complete tumor remission in mice xenografts (at day 28) when compared to the groups receiving only the mono-modal MNP therapy or just the hyperthermia.	[83]
IONPs (46 nm). Fe$_3$O$_4$@Au MNPs were prepared and loaded with C225.	Treatment: I = 30 A; f = 230 kHz. Cancer model: glioblastoma cancer model (U251 cancer cell line). Male and female Balb/c nu/nu nude mice were subcutaneously injected with 2×10^6 U251 cells.	The combined triple therapy decreased, in vitro, cell viability with a high rate of apoptosis via caspase-3, caspase-8, and caspase-9 expression upregulation. In vivo, a significant tumor growth inhibition (approximately 95% of tumor remission) was measured compared to the control groups.	[84]
SPIONs (100 nm) were modified with anti-CD44 antibody.	Treatment: I = 50 A; f = 237 kHz. Cancer model: head and neck squamous cell carcinoma stem cells model (Cal-27 cancer cell line). Male Balb/c nude mice were subcutaneously injected with 5×10^7 Cal-27 cells.	CD44-SPIONPs exhibited good biocompatibility and a programmed cell death in cancer stem cells after an AMF application. In vivo, 33.43% of tumor growth inhibition was observed on the treated group.	[85]
^{225}Ac SPIONs (10 nm) were attached the attachment of CEPA and transtuzumab to the surface.	Treatment: magnetic flux density from 100 to 300 G and frequency range of 386–633 kHz. Cancer model: ovarian cancer model (SKOV-3 cancer cell line).	^{225}Ac@Fe$_3$O$_4$-CEPA-trastuzumab showed a high cytotoxic effect towards SKOV-3 ovarian cancer cells expressing the HER2 receptor, in vitro.	[17]

IONPs: iron oxide nanoparticles; DMSA: dimercaptosuccinic acid; C225: cetuximab; MNPs: magnetic nanoparticles; Gem: gemcitabine; SPIONs: superparamagnetic iron oxide nanoparticles; AMF: alternating magnetic field; MH: magnetic hyperthermia; ^{225}Ac: actinium-225; CEPA: 3-phosphonopropionic acid; NPs: nanoparticles; CXCR4: chemokine cell surface receptor 4.

Chemotherapeutic drug delivery. Chemotherapeutic agents target cells at different phases of cell cycle, which directly or indirectly inhibit the uncontrolled growth of cancer cells [86]. However, the small molecules' lack of specificity and selectivity towards the cancer tissue can also promote damage to healthy cells, as stated earlier [6–10]. MNPs as a drug delivery system are a potential solution for the delivery of drugs to the desired specific sites. These systems can promote a controlled drug release over time, which provides more efficient therapy for the patient [33] without promoting an overdosage of the drug and associated side effects [87,88]. The drug release from MNPs could present a constant profile (ultimately maintaining a constant concentration for a certain time) or a sigmoidal drug release, reaching a maximum concentration [88]. The use of MNPs as a chemotherapeutic vehicle has been studied [33] since the 1980s, and since then different formulations have been described that incorporate drugs such as DOX [89], paclitaxel (PTX) [90], and methotrexate (MTX) [91] as safer and potential alternatives for the treatment of different cancer types.

In MNPs, these therapeutics can be found either as part of the coating of the nanoparticles (maintained through interactions formed with the surface-active functional groups of the MNPs) or encapsulated/embedded inside them. Both approaches, and especially the latter, can help protect the healthy cells and tissues against the bioactivity of the chemotherapeutic drugs needed to combat cancer. The specific activation of the magnetic nanocarriers under particular conditions after reaching the cancer area can then promote the release of the loaded drugs in the tumor microenvironment. For instance, AMF-generated heat (magnetic hyperthermia) and pH (as the tumor microenvironment has a lower pH than normal physiological values [92]) [93] have been successfully employed as MNP-activation stimuli. Reports of MNPs sensitive to both stimuli have also been reported by Yu et al. [94]. Here, $Fe_3O_4@SiO_2$ coated with mPEG-poly(l-asparagine) MNPs showed sensitivity to both stimuli (temperature and pH) and as a result displayed an increased DOX release in the tumor region [94]. Similarly, a recent work developed nanocarriers based on an Fe/Mg-carbonate apatite (Fe/Mg-CA) nanoparticles formulation, where different concentrations of Fe^{+3} and Mg^{+2} were used under specific pH to trigger the release of the loaded DOX. The biodistribution study was performed ex vivo; here, both nanoparticles promoted the accumulation of DOX in breast tumors whilst also causing a bigger cytotoxic effect on the cancer and a half-life circulation improvement when compared to the free drug [89].

Applying an AMF as a stimulus for the activation of the drug-loaded MNPs can create a synergistic cytotoxic effect on cancer, where the sum of the parts (the chemotherapy and the magnetic hyperthermia) can cause a bigger phenotypic effect than the individual treatments, as demonstrated by diverse research groups. For example, for the treatment of primary central nervous system lymphoma (PCNSL), Dai et al. [91] used six experimental groups (control, Fe_3O_4, MTX, Fe_3O_4@MTX, Fe_3O_4 with hyperthermia, and Fe_3O_4@MTX with hyperthermia) and observed an increase in the apoptosis rate in vitro for the combinatorial treatment when compared to the other groups used. In their in vivo evaluation, the same combination managed to inhibit more the tumor growth when compared to the rest of groups used, as well as managed to decrease the overall tumor cell numbers as measured by H&E staining (hematoxylin and eosin staining). Their results highlight the advantages of this dual treatment in oncology [91]. Other examples are shown in Table 2, which summarizes other similar studies involving chemotherapeutics with or without the application of magnetic hyperthermia or photothermic conditions.

Table 2. Studies using magnetic nanoparticles (MNPs) as drug delivery systems for cancer therapy.

MNP (Particle Size) + Surface Modification	Treatment + Cancer Model	Results	Ref
SPIONs (12 nm). SPIONs were coated with a DMSA, MF66, and covalently functionalized with (i) DOX (MF66-DOX), (ii) pseudopeptide NuCant (MF66-N6L), and (iii) with both (MF66-DOX-N6L).	Treatment: DOX + AMF (H = 15.4 kA/m; f = 435 kHz). Cancer model: breast cancer model (BT474 cell line). Female athymic nude mice were subcutaneously injected (on rear backside) with 2.0×10^6 BT474 cells.	The thermo-chemotherapeutic treatment favors the tumor regression in 50% comparatively to control group in vivo (between day 6 and day 17). MF66-DOX-N6L plus hyperthermia application increased their internalization in cancer cells and enhanced in 90% the cytotoxic effect in vitro, comparatively to control group.	[95]

Table 2. *Cont.*

MNP (Particle Size) + Surface Modification	Treatment + Cancer Model	Results	Ref
IONPs (112 nm). MnFe$_2$O$_4$ MNPs were synthesized and were encapsulated in PTX loaded thioether-containing ω-hydroxyacid-co-poly(d,l-lactic acid) (TEHA-co-PDLLA).	Treatment: PTX + AMF (25 mT; f = 765 kHz). Cancer model: colorectal cancer model (Caco-2 cell line) + human mesenchymal stem cells derived from adipose tissue.	In vitro experiments showed that NPs were able to sustain PTX release for up 18 days. Moreover, NPs showed great anticancer activity in a dose-dependent manner with low toxicity toward the primary human stem cells derived from adipose tissue.	[96]
IONPs (122 nm). IONPs were modified with a layer of di-carboxylate polyethylene glycol and carboxylate-methoxy polyethylene glycol. Then, IONPs were coated with silica, obtaining PEGylated silica-coated IONs (PS-IONs).	Treatment: DOX + CDDP. Cancer model: breast cancer model (MCF7 cell line); mouse fibroblast cell line (L929).	NPs showed a dual stimuli-triggered release behavior. A release rate of 69% and 84%, for DOX and CDDP, respectively, was measured during the first 30 h in an acidic environment under photothermal conditions. PS-IONs demonstrated potent antitumor activity in vitro, which was significantly enhanced when exposed to low-power near-IR laser irradiation.	[97]
IONPs (non-mentioned). Surface modification is not mentioned.	Treatment: ferumoxytol. Cancer model: mouse mammary tumor virus—polyoma middle T antigen—MMTV-PyMT; MDA-MB-468). Human fibrosarcoma cells (HT1080); murine macrophages (RAW264.7); human dermal fibroblasts (PCS-201-012); human umbilical vein endothelial cells (HUVECs). Female FVB/N were injected with 2.3 × 10^6 MMTV-PyMT cancer cells.	Ferumoxytil NPs caused tumor growth inhibition by increasing caspase-3 activity. Moreover, macrophages exposed to the NPs enhanced mRNA transcription associated with pro-inflammatory Th1-type responses. In vivo, IONs significantly inhibited the growth of subcutaneous adenocarcinomas compared to controls (tumor size reduction of 53% at day 21), as well as the development of liver metastasis. Additionally, NPs allowed its use as T_2-weighted image for tumor imaging.	[98]
IONPs (20 nm). Surface modification is not mentioned.	Treatment: AT. Cancer model: lung cancer model (A549 and H1975) and human normal lung epithelial cells (BEAS2B); mouse normal liver cells (AML12); rat normal liver cells (BRL3A). Male athymic nude mice were subcutaneously injected with 5 × 10^5 A549 and H1975 into the dorsal flanks.	AT-MNPs demonstrated inhibition in cancer viability (less than 50% viable cells), whilst displaying no toxicity in vivo. AT-MNP treatment intensified the non-small-cell lung cancer apoptosis, activating the caspase-3 route and downregulating the anti-apoptotic proteins Bcl2 and BclXL, in addition to upregulating the proapoptotic Bax and Bad signals.	[99]
SPIONs (165 nm). Surface modification is not mentioned.	Treatment: MTX + AMF (H023.9 kA/m, f = 410 kHz). Cancer model: human bladder cancer cell line (T24). Male SCID (BALB/cJHanHsd-Prkdc) were subcutaneously injected with 2 × 10^6 T24 cancer cells dorsally between the hindlegs.	The results revealed that the relapse-free destruction of tumors was superior when the combination of chemotherapy and magnetic hyperthermia was used (13 days post-treatment versus 15 days post-treatment under monotherapy). The authors also observed an impairment of proapoptotic signaling, cell survival, and cell cycle pathways.	[100]

SPIONs: superparamagnetic iron oxide nanoparticles; DMSA: dimercaptosuccinic acid; DOX: doxorubicin; AMF: alternating magnetic field; IONPs: iron oxide nanoparticles; MNPs: magnetic nanoparticles; AT: actein; PTX: paclitaxel; CDDP: cisplatin; MTX: methotrexate.

2.4. Magnetic Nanoparticles for Theranostic Applications

MNPs have the potential to be used as theranostic platforms in the cancer research field. A theranostic platform combines diagnostic and therapeutic capabilities in the same formulation, enabling efficient tumor targeting, treatment, and therapy response monitoring (or image-guided therapeutics, the visualization of tissue images before, during, and after the treatment) [33]. This combination can help tailor the therapy requirements for each patient within an individualized therapeutic strategy design, with a greater probability of a positive outcome and, at the same time, reduced side effects (Figure 3) [27].

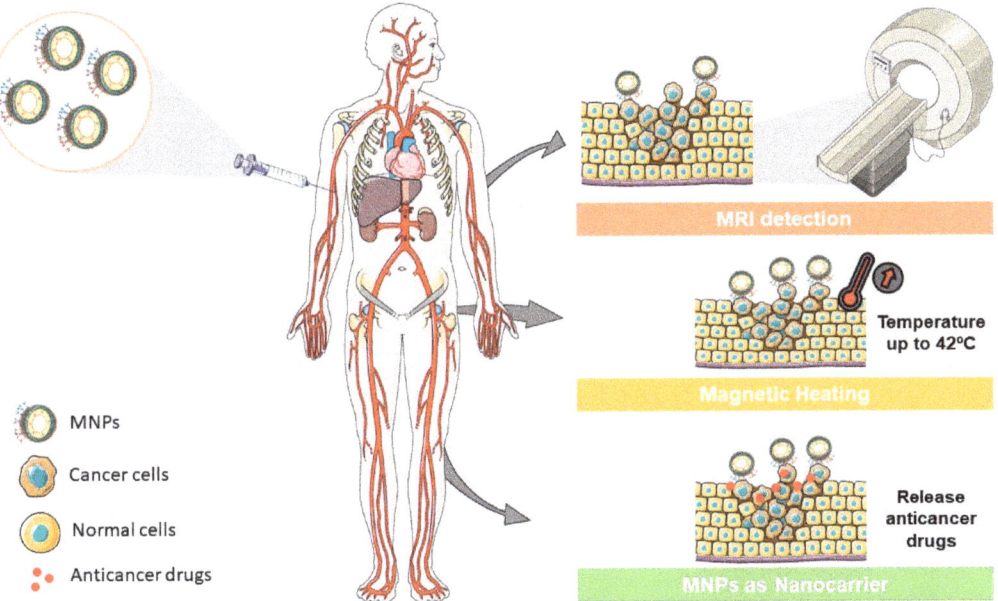

Figure 3. MNP applications in different cancer areas. MNPs could be used as (i) contrast agents to enhance the MRI detection in cancer diagnosis, as (ii) generators for magnetic heating in specific regions such as solid tumors, and as (iii) nanocarriers to deliver specific drugs in cancer treatment.

Following this path, Abedi et al. [101] synthesized MNP as theranostic platforms by combining modified magnetic mesoporous silica nanoparticles (MMSNs) with imidazoline groups (MMSN-Imi) conjugated with cisplatin (Cis-Pt). The nanoparticles displayed high r_2/r_1 reflexivity values and a growth inhibition of ovarian carcinoma cells through apoptosis and necrosis induction, confirming their theranostic applicability in cancer treatment and control [97]. Zhang et al. [98] followed a similar approach by designing an LDH-Fe_3O_4-HA (hyaluronic acid) core–shell loaded with encapsulated DOX. The functionalized surface of the Fe_3O_4 nanoparticles granted good colloidal stability and cytocompatibility to the nanoformulation, whilst also displaying high r_1 values and control over its drug release in a pH-dependent manner. The nanoparticles in in vitro phenotypic activity managed to selectively target B16 melanoma cells. The authors also evaluated the nanoparticles' theranostic efficiency in vivo, using B16 melanoma tumor-bearing C57BL/6 mice through intravenous injection. In vivo, the data showed a reduction of tumor growth in addition to an enhanced MRI contrast in the functionalized nanoparticle-treated group [102]. Table 3 shows other recent studies where theranostic magnetic platforms were designed, synthesized, and evaluated.

Table 3. Studies using magnetic nanoparticles (MNPs) for cancer theranostics.

MNP (Particle Size) + Composition	Treatment + Cancer Model	Results	Ref
MnO_2 NPs (107 nm) loaded with poly(N-vinylcaprolactam) nanogels (PVCL NGs) (DOX/MnO_2@PVCL NG).	Treatment: DOX Cancer model: melanoma cancer model (B16 cancer cell line). In vivo: mouse model of subcutaneous B16 melanoma.	NPs showed interesting biocompatibility properties in addition to redox responsiveness in tumoral tissues. In an in vivo tumor model (with relatively high concentration of GSH), a release of Mn^{+2} from DOX/MnO_2@PVCL NG occurred that enhanced T_1-weighted MRI. In parallel, the DOX release from the NPs inhibited the tumor growth (1 versus 14 relative tumor growth for dual-treatment and control, respectively).	[103]
Fe_3O_4 IONPs (200–300 nm) were synthesized and functionalized with PDA, PEG, and cRGD (Fe_3O_4@PDA-PEG-cRGD).	Treatment: DOX + photothermal effect (1 W/cm^2). Cancer model: colon cancer model (HCT-116 cancer cell line). Male nude mice were subcutaneously injected with HCT-116 cells (5×10^6/mL).	In vitro and in vivo, NPs were capable of targeting tumor cells and promoting the drug internalization. The cytotoxic effect was also significant (survival rate of 25.6% comparatively to control group) whilst the nanocarriers displayed good thermal stability and photothermal conversion efficiency, pH responsiveness, and an enhancement of T_2-MRI contrast. In vivo, the authors observed a decrease in tumor growth around 67% when compared the dual-treatment with the control.	[104]
IONPs (26 nm) were coated with casein (CION) and functionalized with the tumor-targeting ATF of urokinase plasminogen activator and the antitumor drug CDDP (ATF-CNIO-CDDP).	Treatment: CDDP. Cancer model: pancreatic cancer model (MIA PaCa-2 cancer cell line). Female nu/nu mice were injected with 1×10^6 MIA PaCa-2 cells (orthotopic pancreatic tumor model).	NPs promote a T_2-MRI contrast, combined with an improvement of therapeutic effectiveness (0.75 g versus 1.5 g of tumor weight for treated group and control, respectively) and a decrease on harmful side effects in comparison to the free drug.	[105]
SPIONs (260 nm) were coated with FA and ACPP (F/A-PLGA@DOX/SPIO).	Treatment: DOX. Cancer model: human non-small cell lung cancer model (A549 cell line). Normal liver cell (L02 cell line). Male BALB/c nude mice were subcutaneously injected with 3×10^7 A549 cells into the right-rear leg.	F/A-PLGA@DOX/SPIO induced apoptosis in the cancer cells, accelerating the overproduction of ROS. MRI was used to track the NPs in cancer cells (T_2-weighted MRI). In vivo, a reduction in tumor growth was observed (around 67% comparatively to control group), NPs showed a good biocompatibility and long plasma stability, with a capability to induce tumor necrosis, whilst no significant damage or inflammation was detected in healthy organs.	[106]
SPIONs (6 nm) were coated with dextran (FeDC-E NPs).	Treatment: erlotinib. Cancer model: lung cancer model (CL1-5-F4 cancer cell line). Male BALB/c nude mice were subcutaneously injected with 2.5×10^6 of CL1-5-F4 cells.	Theranostic NPs showed a significant therapeutic effect with targeting properties against invasive and migrative cancer cells. These NPs enabled their localization using T_2-weighted MRI. EGFR–ERK–NF-κB signaling pathways were suppressed when after tumors treatment.	[107]

SPIONs: superparamagnetic iron oxide nanoparticles; IONPs: iron oxide nanoparticles; MNPs: magnetic nanoparticles; NPs: nanoparticles; DOX: doxorubicin; PDA: polydopamine; PEG: poly(ethylene glycol); cRGD: cyclic arginine-glycine-aspartate motif; ATF: amino-terminal fragment; GSH: glutathione; MRI: magnetic resonance imaging; CDDP: cisplatin; FA: folic acid; ACPP: activable cell-penetrating peptide; ROS: reactive oxygen species.

To date, several MNPs are in the early stages of clinical trials or in a pre-clinal phase, while different designs have already made it into the clinics for medical imaging and the therapeutic application of solid tumors, such as Feridex IV® (liver and spleen), Lumiren® (bowel), Combidex® (lymph node metastases), and NanoTherm® [36,108].

3. Solid Lipid Nanoparticles

SLNs were first remarked upon in the early 1990s [74,109–111] as an upgraded alternative of the polymeric, inorganic, and liposomal nanoparticles traditionally used until then as carriers [40]. SLNs are colloidal nanoparticles composed of a lipid matrix, solid at both room and body temperatures [112], and surfactants used as stabilizing and solvating agents (Figure 4) [113]. Different lipid and surfactant compositions can control the size, polydispersity, surface charge, stability, and drug release profile of the formulation [106]. The selection of the lipids can also influence the biodegradability, stability, and affinity by drugs and other elements (metals, dyes, etc.). Commonly, fatty acids such as mono-, di-, and triglycerides, fatty alcohols, and waxes are used for the preparation of SLNs [114]. The small size of the formulations (ranging from 10 to 1000 nm), the large surface-to-volume ratio, and the high drug encapsulation efficiency are the key advantages of SLNs. Additionally, these formulations can potentiate the therapeutic effectiveness of hydrophobic pharmaceuticals [36] by improving their bioavailability, protection from biodegradation and clearance by the reticuloendothelial system (RES), and controlling the drug release rate [115].

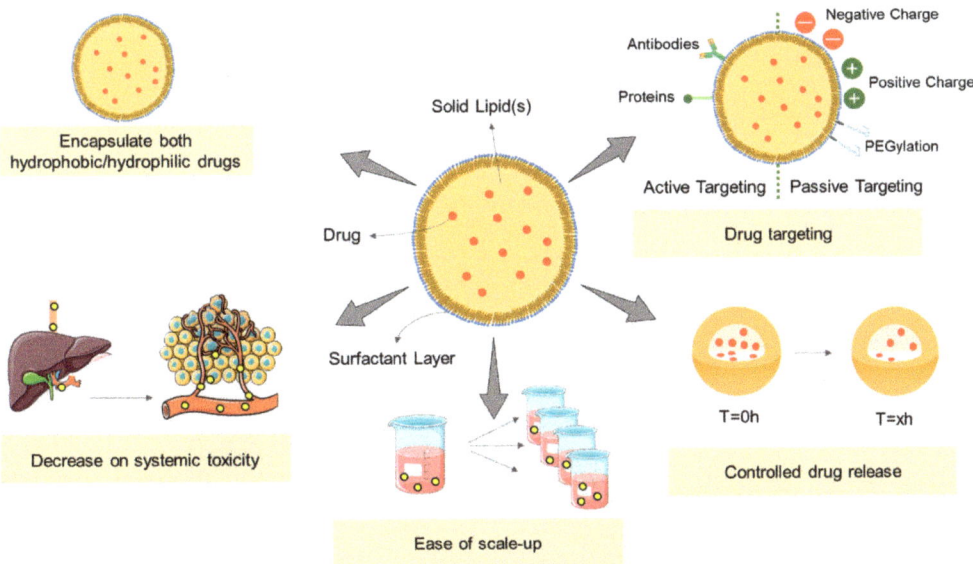

Figure 4. Highlight the applications of SLNs and their major advantages. SLNs could be used as a drug carrier for both hydrophobic and hydrophilic drugs, capable of controlling the drug release, avoiding the "burst effect", and additionally promoting a target delivery that decreases the systemic toxicity. These nanocarriers could be easily scaled up in a cost-effective manner. Adapted from [116].

3.1. SLNs as Drug Delivery Systems

The design of the SLN is the starting point for its development as a potential nanocarrier. For the synthesis of the SLNs, a high-pressure homogenization technique (HPH) methodology has been developed and amply used because of its easiness, efficacy, and relatively low cost [37]. Microemulsions, solvent emulsification method, solvent evaporation or diffusion, and double emulsion techniques have also been used for the preparation of the formulations [37,38,113,117]. However, some of these techniques have drawbacks and limitations, including—for HPH methodology—the mechanic stress applied to the final formulation. Similarly, other techniques depend on a recrystallization step that can reduce the effectiveness of the drug loading (which, however, can be overcome using a heterogenous lipid phase [92]) [118].

SLNs formulations are already approved by the FDA and included in the "Generally Recognized As Safe" (GRAS) list. They are recognized as safe to be administered via different routes including intranasal [119], by inhalation [120], intravenous [121], subcutaneous [122], rectal [123], oral [124], ocular [125], and intramuscular [126]. SLNs' design empowers the biodistribution pharmacokinetics of the intended drugs, improving the drug treatment effectiveness by overcoming the MDR [127]. Additionally, the possibility to modify the SLNs' surface enhances the capability to overcome biological barriers to target cancer cells with minimal side effects [38] and decrease the initial rapid drug release, called the "burst effect" [37] (major drawback of the drug delivery systems since they could expose the patient to a drug overdose [128]). Identical to what happens with MNPs, coating the SLNs with PEG avoids the rapid immune system cell uptake of these nanocarriers and increases their circulation time [37,38,118,128]. The effect of the functionalization with a PEG-coating was evaluated by Arduino et al., who observed an enhanced ability of the formulation to cross the blood–brain barrier and, consequently, the accumulation of the encapsulated drugs in the brain [129]. Dhiman et al. applied a different approach by synthesizing PEGylated SLNs to enhance the pharmacological profile of the drug in a pathological cardiac hypertrophic model. Their data showed an increase in the circulation time of the PEG-coated nanoparticles and a significant preclusion of the cardiac hypertrophy when compared to the free drug [130].

The therapeutic effect of the encapsulated drug is potentially more efficient when the SLNs selectively deliver the drug to its specific site of action. However, the effective accumulation of nanoparticles in solid tumors depends also on the tumors' microenvironment characteristics as well as the nanoparticles' physicochemical properties. It has been debated that the EPR effect can hypothetically cause the passive accumulation of nanoparticles, liposomes, or other carriers and macromolecules in tumors because of the enhanced vascular permeability and poor lymphatic drainage surrounding the tumors [131]. This is a consequence of the tumor's growth requirements, which demands and consumes a high and continuous supply of nutrients and oxygen to be able to sustain its uncontrolled proliferation (Figure 5). To accomplish this, the malignant cells secrete proteins and growth factors, such as fibroblast growth factor (FGF) and vascular endothelial growth factor (VEGF), to induce new blood vessels in a process called angiogenesis, which is one of the hallmarks of cancer [132,133]. The rapid generation of new capillaries in addition to a lack of vasculature supportive tissue (basal membrane) can form an abnormal vessel architecture, with endothelium gaps of diameters between 200 nm to 2 μm of size [134]. Due to this situation, the circulating nanoparticles can easily reach the tumor region through the gaps located in the surrounded blood vessels because of their characteristic small sizes compared to the pore size (<200 nm) [46,131,134–136]. In conjugation with an enhanced permeability, an enhanced retention can also be observed due to the deficiency of the lymphatic system. This is because the nanoparticles (characterized by a larger hydrodynamic size) are incapable of returning to the surrounding capillaries, which ultimately increases their retention time in the tumor [136–138].

SLNs can also accumulate in the tumor regions through active delivery mechanisms. For this, the SLNs' surface are functionalized with ligands that can selectively recognize overexpressed receptors on the surface of cancer cells and, ultimately, be translocated inside the cells [136,138]. Consequently, the selective delivery of the pharmacologically active compounds to the tumor can reduce the toxicity and harmful side effects on other healthy cells [46,137,138]. Using these ideas, Rosière and co-workers [139] developed an SLN based on a folate-conjugated copolymer of PEG and chitosan (F-PEG-HTCC) with paclitaxel encapsulated within. In vitro studies with the functionalized SLN showed a decrease of the IC$_{50}$ (half-maximum inhibitory concentration) in overexpressed folate receptor (FR) cell lines in comparison with healthy cell lines with a normal expression of FR. In vivo studies were conducted using female CD1 and BALB/c mice intrapulmonary implanted with M109-HiFR lung cells. Developed nanoparticles were administered to mice through the endotracheal route to perform pharmacokinetic studies. Data demonstrated an enhanced penetrability and prolonged lung residence of the drug-loaded SLNs [139]. Hyaluronic acid is another ligand commonly used as a functionalization moiety for active targeting, as several tumor types are characterized by the overexpression of its receptors (CD44 and CD168). In vitro results obtained by Campos et al. [140] showed enhanced targeting cellular uptake with time/dose-controlled delivery when using a chitosan and hyaluronan (HA)-coated SLN. Their results pointed to an improvement of the chemotherapeutic efficiency [140]. Similarly, SLNs loaded with methotrexate and functionalized with carbohydrates (fucose) were synthesized by Garg and co-workers [141]. In vitro results showed an increase in cytotoxicity against the MCF-7 cancer cell line in comparison to the free drug. Furthermore, in vivo studies were performed using DMBA-induced breast cancer in female Wistar rats. Nanoparticles were intravenously injected into rats and results showed an accumulation of the functionalized SLNs in the tumor microenvironment, which ultimately was associated with an increase in the efficiency of the antitumor treatment.

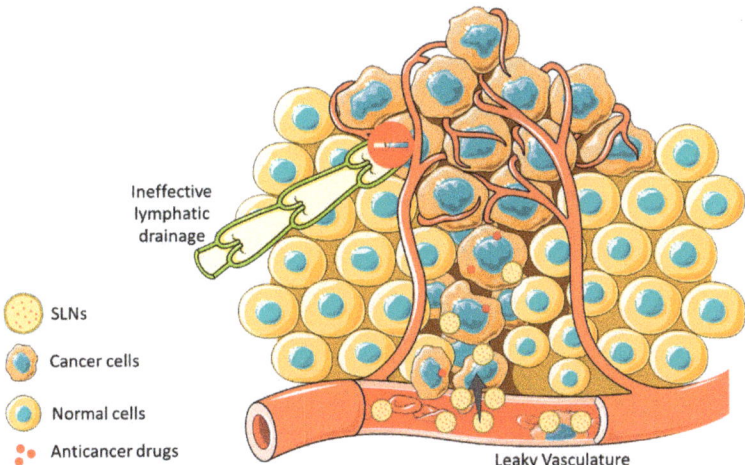

Figure 5. Schematic illustration of the EPR effect and nanoparticles uptake through size across cancerous tissues. EPR effect promotes an increased accumulation of nanoparticles in cancer cells facing normal cells, due to the leaky vasculature within the tumor region being allied to a dysfunctional lymphatic system.

3.2. Solid Lipid Nanoparticles in Cancer Treatment

As drug nanocarriers, SLNs enable the encapsulation of hydrophobic and hydrophilic drugs (a detailed review on the hydrophilic drug encapsulation can be consulted in [142]) through three potentially distinct manners [76,138]. These can be: (i) dispersed homogeneously in the lipid matrix, (ii) dispersed throughout the shell (surfactant layer), and (iii) incorporated in the core (Figure 6). Several studies have already verified the efficient incorporation of different chemotherapeutic drug types [143–146] and their evaluation in a wide range of cancers.

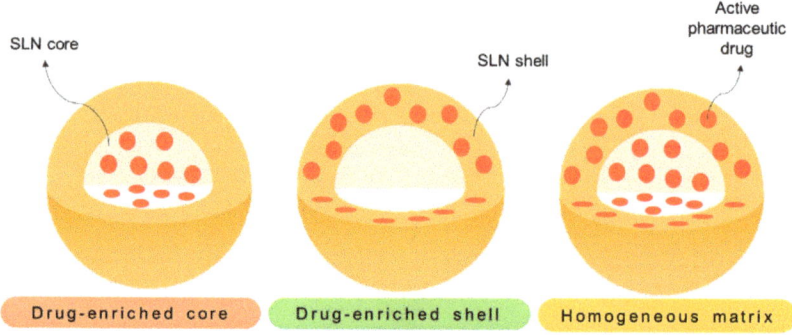

Figure 6. Different representative models of SLNs. On the different models, the drug distribution is represented across (i) the core (drug-enriched core), (ii) the surfactant shell (drug-enriched shell), and (iii) through the core and shell (homogeneous matrix).

For breast cancer, Xu and colleagues [147] studied the applicability of paclitaxel-loaded SLNs in a drug-resistant breast cancer cell line (MCF-7), whilst Eskiler et al. observed an enhanced anticancer activity of tamoxifen (Tam)-loaded SLNs by inducing apoptosis in both MCF-7 and MCF-7 Tam-resistant cell lines [148]. In the latter, a healthy breast control cell line (MCF-10A) was also used and showed no damage after treatment, validating their use as selective formulations that can even overcome Tam resistance.

Glioma (brain cancer) has also been targeted with SLNs in some studies to improve the treatment outcome. Marslin et al. used an SLN encapsulated with albendazole (ABZ) [149] and observed an in vitro biphasic release of the drug, where 82% of ABZ was released in 24 h, in addition to an increase of its cytotoxicity and drug uptake in U-87 MG cells compared to the free drug [149].

In a similar approach for lung cancer studies, docetaxel (DTX)-loaded SLNs showed, in in vitro studies, a better controlled drug release and an overall activity gain of 100-fold in comparison with the free-drug-treated control in 4T1 cells. Considering the improvement in cellular uptake, SLN-DTX significantly accumulated in cancer cells associated with an induction of cellular apoptosis. Subsequent in vivo studies showed a reduction of tumor growth with the SLNs treatment, without a detectable systemic toxicity in the mice model employed [150].

Other examples can be found in Table 4, which summarizes recent SLN preparations and uses them as potential cancer treatments.

Table 4. Solid lipid nanoparticles (SLNs) as drug delivery systems for cancer therapy.

SLN (Particle Size) + Surface Modification/Loading	Drug + Cancer Model	Results	Ref
SLNs (200 nm). Surface modification is not mentioned.	Drug: DOX. Cancer model: murine malignant melanoma (B16F10 cells). C57BL/6 mice (12–16 weeks old) were intravenously injected with 1×10^5 B16F10 cells.	In vivo, mice treated with SLNs-DOX, obtained, approximately, a 60% reduction of tumor area when compared to mice treated with free DOX. No significant differences were found in the survival rates or body weight between different treatment groups, indicating no detectable SLPs-DOX in vivo toxicity during the timeframe of these tests.	[151]

Table 4. Cont.

SLN (Particle Size) + Surface Modification/Loading	Drug + Cancer Model	Results	Ref
PTX-SLN (<200 nm). Surface modification is not mentioned.	Drug: PTX. Cancer model: breast cancer model (MCF-7 cancer cell line).	Xu et al. observed an enhanced anticancer activity of PTX-SLNs, which significantly increased the intracellular uptake (almost 10 ng more of PTX *per* mg of protein comparatively to control) of the drug when compared to the free drug. The results demonstrated that the use of SLNs could efficiently avoid the multidrug resistance mechanisms observed in breast cancer cells.	[147]
SLN-TMZ (279 nm). Surface modification is not mentioned.	Drug: TMZ. Cancer model: melanoma cancer model (JR8 and A2058 cell lines; B16-F10 mouse melanoma cell line). Female C57BL6/J mice were subcutaneously injected with 1×10^6 B16-F10 cells.	NPs showed in vitro and in vivo their ability to target tumor cells and promote drug internalization, reducing the therapeutic dosage needed to be administered in the in vivo model. Here, SLN-TMZ also displayed a higher mice survival rate compared to that obtained using the free drug (increasing from 50 to 100%). Moreover, the in vitro tumor angiogenesis was found to be inhibited (HUVEC method).	[152]
Chol-CUR-SLN (170 nm). Surface modification is not mentioned.	Drug: CUR. Cancer model: breast cancer model (MDA-MB-231 cell line).	In vitro results showed that Chol-CUR-SLN efficiently targeted and accumulated in cancer cells. It also exhibited a higher inhibitory effect on cell viability (20% of higher cytotoxicity in comparison to free drug) and proliferation when compared to free CUR. Chol-CUR-SLN significantly improved the induction of apoptosis (63.87% versus 55.4%) in MDA-MB-231 cells, compared to free CUR.	[153]
SLN-MTX (300 nm) loaded with an ApoE mimicking chimera peptide to actively target the brain.	Drug: MTX. Cancer model: glioblastoma cancer model (F98/Fischer glioblastoma human primary culture).	A reduction of tumor growth (relative tumor growth of approximately 4 versus 10 for treated and control groups, respectively) was observed with SLN-MTX. Moreover, an increase of apoptosis was noted, demonstrating that the developed SLN could be an alternative to conventional therapy.	[154]
TAT PTX/TOS-CDDP SLNs (100 nm) modified with DSPE-PEG and TAT for co-delivery of PTX and TOS-CDDP.	Drug: PTX + TOS-CDDP. Cancer model: cervical cancer model (HeLa cancer cell line). BALB/c nude mice were subcutaneously injected with 1×10^6 of HeLa cells.	TAT PTX/TOS-CDDP SLNs had a slower drug release in comparison with PTX/TOS-CDDP SLNs. Here, the drug release was greatly affected by a lower pH. The in vitro cellular uptake study also showed that tumor cells could uptake more efficiently the TAT PTX/TOS-CDDP SLNs when compared with other SLNs. Moreover, these nanoparticles showed a synergistic effect in the suppression of tumor growth in vivo (inhibition rate of 72.2%) with lower toxicity (calculated by the bodyweight loss during the experiment). Moreover, the formulation increased the drug accumulation in tumor tissue in comparison to the administration of the free drug.	[155]

Table 4. *Cont.*

SLN (Particle Size) + Surface Modification/Loading	Drug + Cancer Model	Results	Ref
c-SLN (200 nm). Surface modification is not mentioned.	Drug: FA+ ASP. Cancer model: pancreatic cancer model (PaCa-2 and Panc-1 cell lines). Male SCID mice were subcutaneously injected with 1×10^6 PaCa-2 cells.	In vitro studies demonstrated that NPs with the conjugated treatment effectively inhibited cell growth, inducing apoptosis. The use of the dual treatment loaded in the SLNs presented significantly better results in cell viability assays when compared to the cells treated with the free drugs. The in vivo studies presented a tumor growth suppression of 45% compared to the control group. However, this result was not statistically significant. By performing the immunohistochemistry analysis, an increased expression of pro-apoptotic proteins was detected.	[156]

SLNs: solid lipid nanoparticles; NPs: nanoparticles; PTX: paclitaxel; TMZ: temozolomide; CUR: curcumin; Chol: cholesterol; ApoE: very low-density lipoprotein receptor binding; MTX: methotrexate; DSPE: 1,2-distearoyl-sn-glycero-3-phosphorylethanolamine; PEG: poly(ethylene glycol); TAT: trans-activating transcriptional activator; TOS-CDDP: α-tocopherol succinate-cisplatin prodrug; c-SLN: chitosan-coated solid lipid nanoparticle; FA: ferulic acid; ASP: aspirin; DOX: doxorubicin; HUVEC: human umbilical vein endothelial cells.

SLN formulations represent an advanced nanocarrier system suitable to provide safer and more efficient anticancer treatments, since they are able to overcome many of the limitations of a free-drug administration. However, SLNs with therapeutic properties are still in the initial stages of research and show very limited clinical translation. Large-scale manufacturing processes (able to preserve the stability of drugs), sterilization, and other fabrication technical issues are still challenges that need to be overcome before commercially available SLN products become a reality [157]. For example, an optimization of the SLNs design is still required when using recrystallization synthetic procedures where a drug expulsion from the system can occur, reducing the drug loading capacity [158–160], and where the lack of interactions between the drug and the lipid matrix, as well as their chemical nature and state, could also contribute to the poor drug encapsulation [158]. Furthermore, some studies noted a relatively high percentage (70–99.9%) of water content in the dispersion [37,161]. Despite these particular limitations, SLNs constitute simple, scalable, and cost-efficient drug carriers able not only to encapsulate one or several drug candidates and enable multidrug co-delivery approaches but also to provide a functionalization platform towards specific targeting and accumulation in the tumor region, thus offering an enhanced therapeutic index and reduced systemic toxicity. Beyond the encapsulation of anticancer drugs [146–150,152–156], SLNs have already been used to encapsulate siRNA [162,163], DNA [162], platelet aggregation inhibitors [164], and magnetic particles [164,165]. The latter will be further discussed in the next section.

4. Magnetic Solid Lipid Nanoparticles

As aforementioned, SLNs present a broad variety of advantages for the treatment of cancer. Several research groups have focused on the development of these new nanoplatforms, trying to exploit and maximize their benefits [164–167]. However, somewhat surprisingly, the magnetic material incorporation in the SLNs was not explored until quite recently.

Different metals and metal derivatives such as iron oxide, gold, and gadolinium [74,83,95–99] have been incorporated in the nanoformulations producing novel platforms with great potential in cancer therapy and tissue imaging. In particular, encapsulated iron oxide and gadolinium have been studied abundantly as magnetic delivery systems that can be guided to tumor regions and/or activated for controlled drug release and cell ablation (magnetic hyperthermia) via an external magnetic field or by endogenous stimuli such as pH changes [168–171]. In particular, iron oxide nanoparticles are considered biocompatible and safe materials and are the gold standard magnetic nanoparticles in medical research, despite the fact that they are able to cause cytotoxicity from the generation of ROS species via the Fenton reaction, which can lead to the damage of DNA, lipids, proteins, and carbohydrates [171,172].

Magnetic solid lipid nanoparticles (mSLNs) represent a new class of functional nanoplatforms that usually consist of inorganic magnetic nanoparticles incorporated in solid lipid nano-matrices and which have great applicability in the medical field [173,174]. For example, Igartua et al. [173] synthesized a colloidal lipid nanoparticle loaded with magnetite using a warm emulsions methodology. The preliminary small size and high entrapment efficiency of the mSLNs managed to fuse the benefits of both types of nanocarriers (SLNs and MNPs) and overcome their independent application issues. mSLNs have shown an enhanced colloidal and chemical stability and caused lower toxicity in vitro compared to the MNPs alone, as described by Müller and colleagues [175], and in vivo using a immunocompetent mice model as described by García-Hevia L. and co-workers [176]. Other groups developed mSLNs constituted with polylactide/glycolide (PLA/GA) and loaded with several different quantities of magnetite to show a controlled drug release via magnetic heating up to 42 °C [177].

mSLN synthesis can be achieved through different methodologies, including emulsification ultrasonic dispersion [178], emulsification–diffusion followed by sonication [179], chemical co-precipitation [165,180], and solvent evaporation [181]. The characterization of the resulting mSLNs allows for the elucidation of the structure of the formulation, where the metals can be embedded in the core and/or surface as described by several authors [179–182]. On the one hand, the metal nanoparticles can be embedded in the lipidic core, where the MNPs' hydrophobic surface shows chemical affinity by the lipid matrix to yield mSLNs. For the mSLN surface, different surfactants can be used during the synthesis to confer colloidal stability and solvation in water. A schematic representation of mSLNs can be seen in Figure 7.

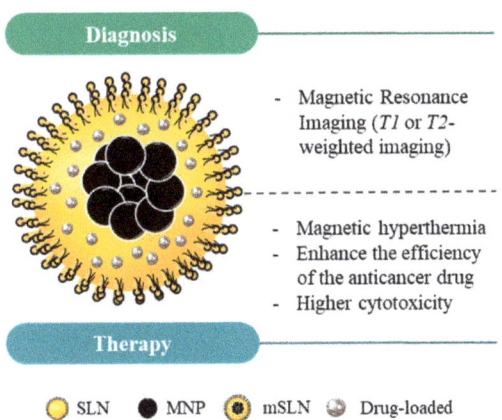

Figure 7. Schematic structure of magnetic solid lipid nanoparticles (mSLNs) and their application in cancer theranostics. Due to the properties of magnetic nanoparticles (MNPs), mSLNs can be used for diagnostic purposes (e.g., MRI application) and cancer therapy via magnetic hyperthermia. Moreover, magnetic hyperthermia in mSLNs offers an extra level of control over the drug release into the region of interest, ultimately increasing the cytotoxicity for cancer cells in comparison with SLNs or MNPs alone.

On the other hand, the metal nanoparticles can be confined in the mSLN surface. Hsu and Su [172] synthesized a new platform that conjugated magnetic heating with a controlled release of the encapsulated drugs (tetracaine) using lipid matrices with γ-Fe_2O_3

particles on their surface. γ-Fe$_2$O$_3$ could then be energized using an external magnetic field, generating enough heat to induce direct thermotherapy as well as to stimulate the release of the loaded drugs in the surrounding tissues. They applied an alternating magnetic field of 60 kA/m at 25 kHz to obtain an increase in temperature of 13 °C in 20 min (up to absolute values of 50 °C). Approximately 35% of the encapsulated tetracaine was released from the mSLNs in 20 min of exposure to the alternating magnetic field [172].

Another example of MNPs loaded in SLNs with applicability in controlled drug release was explored by Pang et al. Here, MNPs were first coated with oleic acid and then loaded in the SLNs. Ibuprofen was chosen as a model drug to be also loaded within the mSLNs due to its well-known pharmacological properties. They observed a drug encapsulation efficiency of 80%, and the interaction between the encapsulated MNPs with magnetic hyperthermia application promoted a controlled release from the nanoformulation. They concluded that magnetite-loaded SLNs are viable alternatives as drug delivery systems [178]. Moreover, Oliveira and colleges developed mSLNs with PTX encapsulated via the emulsification–diffusion method. The data showed a 67% encapsulation efficiency, as well as an in vitro drug release rate increase when the temperature was raised from 25 to 43 °C by magnetic hyperthermia. They concluded that the lipid layer played a key role in the controlled drug release mechanism in response to a temperature increase. Similarly, they demonstrated that PTX-loaded mSLNs are promising systems to increase the drug bioavailability, potentially improving future cancer treatments [179]. Using the same approach, Abidi et al. observed a gradual release of albendazole from mSLNs, which reach 84% after 36 h. Their data confirmed these mSLNs as fast and high-efficiency drug delivery systems [183].

Recently, Ahmadifard and co-workers also developed chitosan-coated mSLNs, loaded with letrozole (LTZ), via a modified solvent evaporation–ultrasonic combination method. With this system, 90.1% of the drug was encapsulated, whereas 50% was released after application of a low-frequency pulsed magnetic field (LFPME) at 50 Hz for 1 h, in comparison with the non-LFPME application where the same amount of drug was released in 12 h. Similar to previous reports, their results demonstrated a promising strategy to induce a localized temperature through a magnetic field and a control of chemotherapy treatment in drug-resistant cancers via LTZ release from a nano delivery system [180].

Ghiani et al. synthesized a novel nano-sized contrast agent composed of gadolinium (III) complexes on the surface of solid lipid nanoparticles with a particle size around 50 nm. The developed paramagnetic solid lipid nanoparticles (pSLNs) demonstrated good stability. For MRI studies, IGROV-1 ovarian carcinoma-bearing BALB/c nu/nu mice were used. In vivo MRI revealed an enhancement of the T_1 signal in the tumor region, in particular when folate, used as a targeting ligand, was used to functionalize the nanoparticles' surface (through intravenous injection). Biodistribution studies in C57BL/6 mice showed an accumulation of pSLNs in the liver, highlighting the need for adjusting the approach in order to enhance the rate of hepatic clearance [184].

A recent published work by Rocha et al. describe the synthesis of a novel hybrid magnetic nanocomposite (mHNCs-DOX) which simultaneously incorporates a chemotherapeutic drug (DOX), superparamagnetic iron oxide NPs as a T_2-contrast agent (Fe$_3$O$_4$) and paramagnetic manganese oxide NPs (MnO) as a T_1-MRI contrast agent [185]. Dual T_1/T_2 MRI performance and additional thermo-chemotherapy capability were observed in vitro in triple-negative breast carcinoma cells (Hs578t cancer cell line) [185]. Table 5 further summarizes representative studies involving mSLNs for cancer treatment/theranostics.

Altogether, the mSLNs have been demonstrated to be promising tools because of their good biocompatibility [171,172,179], improvement of thermo-responsiveness compared to SLNs [168], efficiency in targeting tumors [174,181], and their high drug encapsulation efficiency. Furthermore, these nanosystems allow the application of magnetic hyperthermia as a means to provide thermal therapy and control drug release [164,172,181], in addition to being used as MRI contrast agents [174,181]. Still, there are only few studies involving tests in vivo, highlighting the need to validate the performance of these nanocarriers in more biological complex systems.

Table 5. Magnetic solid lipid nanoparticles (mSLNs) as drug delivery systems and theranostic agents against cancer.

mSLN (Particle Size) + Surface Modification	Drug + Cancer Model	Results	Ref
Wax-mSLNs (200 nm). Surface modification is not mentioned.	Drug: DOX. Cancer model: murine melanoma B16f10, Hs578t, and Dox-resistance cell lines (t84 and HCT-15).	Efficacy studies showed that DOX delivery in combination with 1 h of MH promoted a significant cytotoxic effect in vitro in melanoma cell lines compared to a treatment in which no MH was supplied (~5% vs. ~50%, respectively, when using 1 μg DOX/mL of DOX-mSLNs). Similar results were obtained in 3D in vitro using melanoma spheroids. The same dual treatment approach was applied to DOX-resistant cell lines obtaining approximately 40% of cell viability reduction.	[186]
Wax-mSLNs (250–300 nm). Surface modification is not mentioned.	Drug: OncoA. Cancer model: human lung carcinoma cell line (A549 cell line).	mSLNs showed an outstanding performance as a T_2-contrast agent in MRI ($r_2 > 800$ mm^{-1} s^{-1}). In vitro, the combination of co-loaded MNPs and OncoA with MH greatly decreased the cell viability (virtually 0% vs. 53% when performed without MH application) at the same 40 μg OncoA/mL and 25 μg Fe/mL doses).	[187]
Wax-mSLNs (200 nm). Surface modification is not mentioned.	Drug: DOX. MH: 224 kHz, 13 A, 27.6 W for 1 h for in vitro 174.5 kHz, 23 mT for 1 h for in vivo. Cancer model: murine malignant melanoma cells (B16F10 cell line); C57BL/6 mice (8–10 weeks old) were subcutaneously injected in interscapular region of mice with 5×10^5 B16F10 cells.	mSLNs-DOX showed higher cytotoxicity activity than free DOX in the whole range of DOX concentration tested both in vitro and in vivo. In vitro, a remarkable enhanced cytotoxicity was obtained when cells were exposed to the combination of chemotherapy (0.5 μ/mL) and 1 h MH (40% of viable cells vs. 85% without MH). Under a higher incubation concentration of mLNVs-DOX (1 μg DOX/mL), the results showed a cytotoxicity virtually to 100% under a combination of mLNVs-DOX with MH. In vivo, the dual treatment promoted the slowest tumor growth and smallest tumor volume, which was on average 3 and 2.1-fold smaller than the saline and free-DOX groups. Regarding imaging capability, T_2-MRI relaxation times of animal tumors treated with mSLNs were on average over 15% shorter than those of control animals injected only with saline.	[176]
Sor-mag-SLN (250 nm). Surface modification is not mentioned.	Drug: Sor. Cancer model: liver cancer model (HepG2 cell line).	The nanocarriers showed a loading efficiency of 90% and stability in an aqueous environment. Moreover, the developed nanoparticles presented a good cytocompatibility with a high antiproliferative effect against the cancer cells (40% higher in comparison to control group). This effect was associated with the capability of these nanocarriers to be specifically accumulated in the tumor region and the application of a local AMF.	[188]

Table 5. Cont.

mSLN (Particle Size) + Surface Modification	Drug + Cancer Model	Results	Ref
Mag-SLN (150 nm). Surface modification is not mentioned.	Cancer model: myeloid leukemia cancer model (HL-60/wt cell lines; L-60/adr with MRP1 = ABCC1 over-expression; HL-60/vinc with P-glycoprotein = ABCB1 over-expression), leukemia cancer model (Jurkat T-cells), and glioblastoma cancer model (U251 cell line).	The developed nanoparticles showed promising results in the context of cancer therapy, in particular against drug-resistant cell lines. The mag-SLN revealed higher cytotoxicity against resistance cell lines in comparison to DOX alone when under an AMF. Moreover, the data showed that the cells treated with a dual treatment presented an increase of nuclei fragmentation and condensed chromatin. The mag-SLNs plus MH presented apoptotic and necrotic activities. The authors proposed that the production of ROS was the cause of the higher cytotoxicity observed in the cells treated with the particles.	[168]
LMNV (100 nm). Surface modification is not mentioned.	Drug: TMZ. Cancer model: glioblastoma cancer model (U-87 cell line) and brain-endothelial cell model (bEnd.3 cell lines, an immortalized mouse BEC line).	In vitro results showed that lipid-based magnetic nanovectors presented a good loading capacity with a sustained release profile of the encapsulated chemotherapeutic drug. Moreover, a complete drug release was observed after the exposure to (i) low pH (4.5), (ii) increased concentration of hydrogen peroxide (50 µM), and (iii) increased temperature achieved through the application of an AMF. The authors noted that these nanovectors could be used as a potential hyperthermia agent, since they managed to increase apoptotic levels and decrease proliferative rates when a magnetic field of 20 mT and 750 kHz was applied, increasing the temperature to 43 °C. During in vitro tests, the capacity of LMNVs to cross the BBB was observed, where after 24 h of exposure, 40% of LMNVs were able to translocate inside the glioblastoma cells.	[189]
Gd(III)-loaded pSLNs were modified with with cellular receptors, DSPE-PEG2000-folate.	Cancer model: murine macrophage model (Raw 264.7 cell line), lymphoma cancer model (U937 cell line), and human ovarian adenocarcinoma (IGROV-1 cell line). Female Balb/C nu/nu were subcutaneously injected with 1×10^7 of IGROV-1 cells.	The data showed that pSLNs could effectively internalize in in vitro and in vivo models. Moreover, the authors detected the nanoparticles' T_1-MRI signal, at least after 30 min post-injection. The cytotoxic studies showed a decrease in cell viability when the loaded Gd(III) concentration increased within the pSLN (below 50% of viable cells). The results also demonstrated that Gd(III)-loaded pSLNs could efficiently target the cancer cells and due to the EPR effect in conjunction with its targeting properties allowed a higher internalization capacity. Moreover, they could be used as a molecular imaging tool. A macrophage uptake experiment in vivo showed that the nanoparticles could avoid the macrophage internalization and circulate for at least 6 h, increasing altogether the tumor uptake. However, the authors noted an excessive accumulation in the liver with slow elimination rates after performing the biodistribution study.	[184]

Table 5. Cont.

mSLN (Particle Size) + Surface Modification	Drug + Cancer Model	Results	Ref
Sor-Mag-SLNs (300 nm). Surface modification is not mentioned.	Drug: Sor. Cancer model: liver cancer model (HepG2 cell line).	The results showed an increase of the cytotoxic effects of sorafenib. Using an external magnetic field, it was possible to guide and improve the drug effect in the desired area. Quantitative evaluation of cell mortality indicated 95% of cell death compared to the control (5%). Moreover, the authors mentioned that the nanocarriers could be an effective approach to reduce the undesired side effects of chemotherapeutic drugs and improve their pharmacokinetic properties.	[190]
Nut-Mag-SLNs (180 nm) were loaded with fluorescenin-PEG-DSPE (FITC-PEG-DSPE).	Drug: Nut. Cancer model: glioblastoma cancer model (U-87 cancer cell line) and brain endothelial cell model (bEnd.3 cell lines, an immortalized mouse BEC line).	Nut-Mag-SLNs presented a good colloidal stability and could efficiently cross an in vitro blood–brain barrier model. The authors observed that the nanovectors were magnetically activated, enabling their pass through the BBB, and could also deliver the drug loads to glioblastoma cells. Moreover, they observed an enhanced antitumor activity as they obtained a 50% reduction in the metabolic activity with lower drug concentrations. Increased pro-apoptotic activity was also noted. These nanocarriers presented several advantages compared to the free drug in overcoming several limitations in glioblastoma treatments, for instance, (i) Nut-Mag-SLNs could cross the BBB, (ii) Nut-Mag-SLNs had the ability to be magnetically guided to the tumor region, and (iii) the nanoparticles showed a powerful inhibition of cancer cell proliferation while increasing the pro-apoptotic activity.	[181]
mSLNs (180 nm). Surface modification is not mentioned.	Cancer model: colon cancer model (HT-29 cell line).	By applying magnetic hyperthermia, results showed that mSLNs could constantly maintain the maximum temperature achieved (46 °C, in 40 min) during 1 h of exposure to a magnetic field (250 kHz and 4 kA/m). These results translated into a decrease in cell viability after magnetic treatment (up to 52% comparatively to 100% of control group). Interestingly, no cytotoxic effect was observed if only one (but not both) of the components was used alone for treatment.	[165]

Mag-SLN (mSLN): magnetic solid lipid nanoparticles; Sor: sorafenib; MRP1: multidrug resistance-associated protein 1; TMZ: temozolomide; BBB: blood–brain barrier; pSLNs: paramagnetic solid lipid nanoparticles; AMF: alternating magnetic field; DSPE: 1,2-distearoyl-sn-glycero-3-phosphorylethanolamine; PEG: poly(ethylene glycol); EPR effect: enhanced permeability and retention effect; Nut: Nutlin; DOX: doxorubicin; OncoA: oncocalyxone A.

5. Conclusions

In the last decades, medical nanoformulations have gained value in the biomedical field. Over these years, different materials have been used to form nanoparticle-based carriers including inorganics, organics, hydrogels, micelles, dendrimers, solid lipids, and other materials or combinations of them. Depending on the material, a variety of properties for diverse purposes can be achieved. Cancer theranostics is a ceaselessly growing field and clear target of nanoparticle applications, where numerous nanomaterial-related fabrication and functionalization techniques have been developed with relative success.

In this review, we analyzed the state-of-the-art MNPs, SLNs, and mSLNs, including their features, advantages, and disadvantages, as well as the most recent works concerning their application in several cancer types. The main objective in this area has been to improve cancer diagnosis and treatment by maximizing the efficiency of contrast agents and therapeutic agents.

Author Contributions: Conceptualization, F.B. and M.B.-L.; methodology, M.C.; investigation, M.C.; writing—original draft preparation, M.C. and E.B.-R.; writing—review and editing, J.G., F.B. and M.B.-L.; visualization, M.C.; supervision, F.B. and M.B.-L.; project administration, F.B. and M.B.-L.; funding acquisition, F.B., J.G. and M.B.-L. All authors have read and agreed to the published version of the manuscript.

Funding: This research was funded by the Portuguese Foundation for Science and Technology (Fundação para a Ciência e a Tecnologia—FCT) and the European Regional Development Fund (ERDF) through NORTE 2020 (2014–2020 North Portugal Regional Operational Program) under the project NORTE-01-0145-FEDER-031142 "Local specific treatment of triple-negative-breast-cancer through externally triggered target-less drug carriers (MagtargetON)", and by 2014–2020 INTERREG Cooperation Programme Spain–Portugal (POCTEP) through the project 0624_2IQBIONEURO_6_E.

Informed Consent Statement: Not applicable.

Conflicts of Interest: The authors declare no conflict of interest.

References

1. Hanahan, D.; Weinberg, R.A. Hallmarks of cancer: The next generation. *Cell* **2011**, *144*, 646–674. [CrossRef] [PubMed]
2. Wicki, A.; Witzigmann, D.; Balasubramanian, V.; Huwyler, J. Nanomedicine in cancer therapy: Challenges, opportunities, and clinical applications. *J. Control. Release* **2015**, *200*, 138–157. [CrossRef] [PubMed]
3. World Health Organization. Global cancer data. *Int. Agency Res. Cancer* **2018**, *263*, 1–3.
4. Palumbo, M.O.; Kavan, P.; Miller, W.H.; Panasci, L.; Assouline, S.; Johnson, N.; Cohen, V.; Patenaude, F.; Pollak, M.; Jagoe, R.T.; et al. Systemic cancer therapy: Achievements and challenges that lie ahead. *Front. Pharmacol.* **2013**, *4*, 57. [CrossRef]
5. Lee, J.J.; Saiful Yazan, L.; Che Abdullah, C.A. A review on current nanomaterials and their drug conjugate for targeted breast cancer treatment. *Int. J. Nanomed.* **2017**, *12*, 2373–2384. [CrossRef]
6. Pokhriyal, R.; Hariprasad, R.; Kumar, L.; Hariprasad, G. Chemotherapy Resistance in Advanced Ovarian Cancer Patients. *Biomark. Cancer* **2019**, *11*, 1179299X19860815. [CrossRef]
7. Nurgali, K.; Jagoe, R.T.; Abalo, R. Editorial: Adverse Effects of Cancer Chemotherapy: Anything New to Improve Tolerance and Reduce Sequelae? *Front. Pharmacol.* **2018**, *9*, 245. [CrossRef]
8. Hile, E.S.; Fitzgerald, G.K.; Studenski, S.A. Persistent Mobility Disability after Neurotoxic Chemotherapy. *Phys. Ther.* **2010**, *90*, 1649–1657. [CrossRef]
9. Aleman, B.M.P.; van den Belt-Dusebout, A.W.; Bruin, M.L.d.; van't Veer, M.B.; Baaijens, M.H.A.; de Boer, J.P.; Hart, A.A.M.; Klokman, W.J.; Kuenen, M.A.; Ouwens, G.M.; et al. Late cardiotoxicity after treatment for Hodgkin lymphoma. *Blood* **2007**, *109*, 1878–1886. [CrossRef]
10. Grigorian, A.; O'Brien, C.B. Hepatotoxicity Secondary to Chemotherapy. *J. Clin. Transl. Hepatol.* **2014**, *2*, 95–102. [CrossRef]
11. Housman, G.; Byler, S.; Heerboth, S.; Lapinska, K.; Longacre, M.; Snyder, N.; Sarkar, S. Drug Resistance in Cancer: An Overview. *Cancers* **2014**, *6*, 1769–1792. [CrossRef] [PubMed]
12. Chidambaram, M.; Manavalan, R.; Kathiresan, K. Nanotherapeutics to Overcome Conventional Cancer Chemotherapy Limitations. *J. Pharm. Pharm. Sci.* **2011**, *14*, 67–77. [CrossRef] [PubMed]
13. Trock, B.J.; Leonessa, F.; Clarke, R. Multidrug Resistance in Breast Cancer: A Meta-analysis of MDR1/gp170 Expression and Its Possible Functional Significance. *J. Natl. Cancer Inst.* **1997**, *89*, 917–931. [CrossRef] [PubMed]
14. Wang, M.; Thanou, M. Targeting nanoparticles to cancer. *Pharmacol. Res.* **2010**, *62*, 90–99. [CrossRef] [PubMed]
15. Mehnert, W.; Mäder, K. Solid lipid nanoparticles: Production, characterization and applications. *Adv. Drug Deliv. Rev.* **2012**, *64*, 83–101. [CrossRef]
16. Tietze, R.; Zaloga, J.; Unterweger, H.; Lyer, S.; Friedrich, R.P.; Janko, C.; Pöttler, M.; Dürr, S.; Alexiou, C. Magnetic nanoparticle-based drug delivery for cancer therapy. *Biochem. Biophys. Res. Commun.* **2015**, *468*, 463–470. [CrossRef]
17. Cędrowska, E.; Pruszyński, M.; Gawęda, W.; Żuk, M.; Krysiński, P.; Bruchertseifer, F.; Morgenstern, A.; Karageorgou, M.-A.; Bouziotis, P.; Bilewicz, A. Trastuzumab Conjugated Superparamagnetic Iron Oxide Nanoparticles Labeled with 225Ac as a Perspective Tool for Combined α-Radioimmunotherapy and Magnetic Hyperthermia of HER2-Positive Breast Cancer. *Molecules* **2020**, *25*, 1025. [CrossRef]
18. Lippacher, A.; Müller, R.; Mäder, K. Preparation of semisolid drug carriers for topical application based on solid lipid nanoparticles. *Int. J. Pharm.* **2001**, *214*, 9–12. [CrossRef]
19. Park, J.W. Liposome-based drug delivery in breast cancer treatment. *Breast Cancer Res.* **2002**, *4*, 95–99. [CrossRef]

20. Anselmo, A.C.; Mitragotri, S. An overview of clinical and commercial impact of drug delivery systems. *J. Control. Release* **2014**, *190*, 15–28. [CrossRef]
21. Anselmo, A.C.; Mitragotri, S. Nanoparticles in the clinic: An update. *Bioeng. Transl. Med.* **2019**, *4*, e10143. [CrossRef] [PubMed]
22. Green, M.R.; Manikhas, G.M.; Orlov, S.; Afanasyev, B.; Makhson, A.M.; Bhar, P.; Hawkins, M.J. Abraxane®, a novel Cremophor®-free, albumin-bound particle form of paclitaxel for the treatment of advanced non-small-cell lung cancer. *Ann. Oncol.* **2006**, *17*, 1263–1268. [CrossRef] [PubMed]
23. Alibolandi, M.; Abnous, K.; Mohammadi, M.; Hadizadeh, F.; Sadeghi, F.; Taghavi, S.; Jaafari, M.R.; Ramezani, M. Extensive preclinical investigation of polymersomal formulation of doxorubicin versus Doxil-mimic formulation. *J. Control. Release* **2017**, *264*, 228–236. [CrossRef] [PubMed]
24. Adams, D.; Gonzalez-Duarte, A.; O'Riordan, W.D.; Yang, C.-C.; Ueda, M.; Kristen, A.V.; Tournev, I.; Schmidt, H.H.; Coelho, T.; Berk, J.L.; et al. Patisiran, an RNAi Therapeutic, for Hereditary Transthyretin Amyloidosis. *N. Engl. J. Med.* **2018**, *379*, 11–21. [CrossRef]
25. Gobbo, O.L.; Sjaastad, K.; Radomski, M.W.; Volkov, Y.; Prina-Mello, A. Magnetic Nanoparticles in Cancer Theranostics. *Theranostics* **2015**, *5*, 1249–1263. [CrossRef]
26. Sun, C.; Lee, J.S.H.; Zhang, M. Magnetic nanoparticles in MR imaging and drug delivery. *Adv. Drug Deliv. Rev.* **2008**, *60*, 1252–1265. [CrossRef]
27. Xie, W.; Guo, Z.; Gao, F.; Gao, Q.; Wang, D.; Liaw, B.-S.; Cai, Q.; Sun, X.; Wang, X.; Zhao, L. Shape-, size- and structure-controlled synthesis and biocompatibility of iron oxide nanoparticles for magnetic theranostics. *Theranostics* **2018**, *8*, 3284–3307. [CrossRef]
28. Dobson, J. Magnetic nanoparticles for drug delivery. *Drug Dev. Res.* **2006**, *67*, 55–60. [CrossRef]
29. Pankhurst, Q.A.; Connolly, J.; Jones, S.K.; Dobson, J. Applications of magnetic nanoparticles in biomedicine. *J. Phys. D Appl. Phys.* **2003**, *36*, R167–R181. [CrossRef]
30. Tietze, R.; Alexiou, C. Improving cancer imaging with magnetic nanoparticles: Where are we now? *Nanomedicine* **2017**, *12*, 167–170. [CrossRef]
31. Chang, D.; Lim, M.; Goos, J.A.; Qiao, R.; Ng, Y.Y.; Mansfeld, F.M.; Jackson, M.; Davis, T.P.; Kavallaris, M. Biologically Targeted Magnetic Hyperthermia: Potential and Limitations. *Front. Pharmacol.* **2018**, *9*, 831. [CrossRef] [PubMed]
32. Calero, M.; Chiappi, M.; Lazaro-Carrillo, A.; Rodríguez, M.J.; Chichón, F.J.; Crosbie-Staunton, K.; Prina-Mello, A.; Volkov, Y.; Villanueva, A.; Carrascosa, J.L. Characterization of interaction of magnetic nanoparticles with breast cancer cells. *J. Nanobiotechnol.* **2015**, *13*, 16. [CrossRef]
33. Lima-Tenório, M.K.; Pineda, E.A.G.; Ahmad, N.M.; Fessi, H.; Elaissari, A. Magnetic nanoparticles: In vivo cancer diagnosis and therapy. *Int. J. Pharm.* **2015**, *493*, 313–327. [CrossRef] [PubMed]
34. Zamay, G.S.; Zamay, T.N.; Lukyanenko, K.A.; Kichkailo, A.S. Aptamers Increase Biocompatibility and Reduce the Toxicity of Magnetic Nanoparticles Used in Biomedicine. *Biomedicines* **2020**, *8*, 59. [CrossRef] [PubMed]
35. Ruiz, A.; Hernández, Y.; Cabal, C.; González, E.; Veintemillas-Verdaguer, S.; Martínez, E.; Morales, M.P. Biodistribution and pharmacokinetics of uniform magnetite nanoparticles chemically modified with polyethylene glycol. *Nanoscale* **2013**, *5*, 11400–11408. [CrossRef] [PubMed]
36. Mukherjee, S.; Liang, L.; Veiseh, O. Recent Advancements of Magnetic Nanomaterials in Cancer Therapy. *Pharmaceutics* **2020**, *12*, 147. [CrossRef] [PubMed]
37. Tapeinos, C.; Battaglini, M.; Ciofani, G. Advances in the design of solid lipid nanoparticles and nanostructured lipid carriers for targeting brain diseases. *J. Control. Release* **2017**, *264*, 306–332. [CrossRef]
38. Geszke-Moritz, M.; Moritz, M. Solid lipid nanoparticles as attractive drug vehicles: Composition, properties and therapeutic strategies. *Mater. Sci. Eng. C* **2016**, *68*, 982–994. [CrossRef]
39. Mukherjee, S.; Ray, S.; Thakur, R.S. Solid lipid nanoparticles: A modern formulation approach in drug delivery system. *Indian J. Pharm. Sci.* **2009**, *71*, 349–358. [CrossRef]
40. Muller, H.R.; Shegokar, R.; Keck, C.M. 20 Years of Lipid Nanoparticles (SLN & NLC): Present State of Development & Industrial Applications. *Curr. Drug Discov. Technol.* **2011**, *8*, 207–227. [CrossRef]
41. Huang, J.; Li, Y.; Orza, A.; Lu, Q.; Guo, P.; Wang, L.; Yang, L.; Mao, H. Magnetic Nanoparticle Facilitated Drug Delivery for Cancer Therapy with Targeted and Image-Guided Approaches. *Adv. Funct. Mater.* **2016**, *26*, 3818–3836. [CrossRef] [PubMed]
42. Sanson, C.; Diou, O.; Thévenot, J.; Ibarboure, E.; Soum, A.; Brûlet, A.; Miraux, S.; Thiaudière, E.; Tan, S.; Brisson, A.; et al. Doxorubicin Loaded Magnetic Polymersomes: Theranostic Nanocarriers for MR Imaging and Magneto-Chemotherapy. *ACS Nano* **2011**, *5*, 1122–1140. [CrossRef] [PubMed]
43. Furlani, E.P. Magnetic Biotransport: Analysis and Applications. *Materials* **2010**, *3*, 2412. [CrossRef]
44. Zhou, H.; Qian, W.; Uckun, F.M.; Wang, L.; Wang, Y.A.; Chen, H.; Kooby, D.; Yu, Q.; Lipowska, M.; Staley, C.A.; et al. IGF1 Receptor Targeted Theranostic Nanoparticles for Targeted and Image-Guided Therapy of Pancreatic Cancer. *ACS Nano* **2015**, *9*, 7976–7991. [CrossRef]
45. Alavijeh, A.A.; Barati, M.; Barati, M.; Dehkordi, H.A. The Potential of Magnetic Nanoparticles for Diagnosis and Treatment of Cancer Based on Body Magnetic Field and Organ-on-the-Chip. *Adv. Pharm. Bull.* **2019**, *9*, 360–373. [CrossRef]
46. Sun, T.M.; Zhang, Y.S.; Pang, B.; Hyun, D.C.; Yang, M.X.; Xia, Y.N. Engineered Nanoparticles for Drug Delivery in Cancer Therapy. Angew Chemie-Internationa. *Angew. Chem. Int. Ed.* **2014**, *53*, 12320–12364. [CrossRef]

47. Latorre, A.; Couleaud, P.; Aires, A.; Cortajarena, A.L.; Somoza, Á. Multifunctionalization of magnetic nanoparticles for controlled drug release: A general approach. *Eur. J. Med. Chem.* **2014**, *82*, 355–362. [CrossRef]
48. Cheng, M.; Ma, D.; Zhi, K.; Liu, B.; Zhu, W. Synthesis of Biotin-Modified Galactosylated Chitosan Nanoparticles and Their Characteristics in Vitro and in Vivo. *Cell. Physiol. Biochem.* **2018**, *50*, 569–584. [CrossRef]
49. Price, D.N.; Stromberg, L.; Kunda, N.K.; Muttil, P. In Vivo Pulmonary Delivery and Magnetic-Targeting of Dry Powder Nano-in-Microparticles. *Mol. Pharm.* **2017**, *14*, 4741–4750. [CrossRef]
50. Khalid, M.K.; Asad, M.; Henrich-Noack, P.; Sokolov, M.; Hintz, W.; Grigartzik, L.; Zhang, E.; Dityatev, A.; Van Wachem, B.; Sabel, B.A. Evaluation of Toxicity and Neural Uptake In Vitro and In Vivo of Superparamagnetic Iron Oxide Nanoparticles. *Int. J. Mol. Sci.* **2018**, *19*, 2613. [CrossRef]
51. Zhu, L.; Zhou, Z.; Mao, H.; Yang, L. Magnetic nanoparticles for precision oncology: Theranostic magnetic iron oxide nanoparticles for image-guided and targeted cancer therapy. *Nanomedicine* **2017**, *12*, 73–87. [CrossRef]
52. Li, J.; Zhang, W.; Ji, W.; Wang, J.; Wang, N.; Wu, W.; Wu, Q.; Hou, X.; Hu, W.; Li, L. Near infrared photothermal conversion materials: Mechanism, preparation, and photothermal cancer therapy applications. *J. Mater. Chem. B* **2021**, *9*, 7909–7926. [CrossRef] [PubMed]
53. Condeelis, J.; Weissleder, R. In Vivo Imaging in Cancer. *Cold Spring Harb. Perspect. Biol.* **2010**, *2*, a003848. [CrossRef] [PubMed]
54. Zhao, S.; Yu, X.; Qian, Y.; Chen, W.; Shen, J. Multifunctional magnetic iron oxide nanoparticles: An advanced platform for cancer theranostics. *Theranostics* **2020**, *10*, 6278–6309. [CrossRef] [PubMed]
55. Ray, S.; Li, Z.; Hsu, C.-H.; Hwang, L.-P.; Lin, Y.-C.; Chou, P.-T.; Lin, Y.-Y. Dendrimer- and copolymer-based nanoparticles for magnetic resonance cancer theranostics. *Theranostics* **2018**, *8*, 6322–6349. [CrossRef] [PubMed]
56. Guldris, N.; Argibay, B.; Kolen'Ko, Y.V.; Carbó-Argibay, E.; Sobrino, T.; Campos, F.; Salonen, L.M.; Bañobre-López, M.; Castillo, J.; Rivas, J. Influence of the separation procedure on the properties of magnetic nanoparticles: Gaining in vitro stability and T1–T2 magnetic resonance imaging performance. *J. Colloid Interface Sci.* **2016**, *472*, 229–236. [CrossRef] [PubMed]
57. Guldris, N.; Gallo, J.; García-Hevia, L.; Rivas, J.; Bañobre-López, M.; Salonen, L.M. Orthogonal Clickable Iron Oxide Nanoparticle Platform for Targeting, Imaging, and On-Demand Release. *Chem. Eur. J.* **2018**, *24*, 8624–8631. [CrossRef]
58. Guldris, N.; Argibay, B.; Gallo, J.; Iglesias-Rey, R.; Carbó-Argibay, E.; Kolen'Ko, Y.V.; Campos, F.; Sobrino, T.; Salonen, L.M.; Bañobre-López, M.; et al. Magnetite Nanoparticles for Stem Cell Labeling with High Efficiency and Long-Term in Vivo Tracking. *Bioconjugate Chem.* **2017**, *28*, 362–370. [CrossRef]
59. Keasberry, N.A.; Bañobre-López, M.; Wood, C.; Stasiuk, G.J.; Gallo, J.; Long, N.J. Tuning the relaxation rates of dual-mode T1/T2 nanoparticle contrast agents: A study into the ideal system. *Nanoscale* **2015**, *7*, 16119–16128. [CrossRef]
60. Frantellizzi, V.; Conte, M.; Pontico, M.; Pani, A.; Pani, R.; De Vincentis, G. New Frontiers in Molecular Imaging with Superparamagnetic Iron Oxide Nanoparticles (SPIONs): Efficacy, Toxicity, and Future Applications. *Nucl. Med. Mol. Imaging* **2020**, *54*, 65–80. [CrossRef]
61. García-Hevia, L.; Bañobre-López, M.; Gallo, J. Recent Progress on Manganese-Based Nanostructures as Responsive MRI Contrast Agents. *Chem. Eur. J.* **2019**, *25*, 431–441. [CrossRef] [PubMed]
62. Bañobre-López, M.; Garcia-Hevia, L.; Cerqueira, M.F.; Rivadulla, F.; Gallo, J. Tunable Performance of Manganese Oxide Nanostructures as MRI Contrast Agents. *Chem. Eur. J.* **2018**, *24*, 1295–1303. [CrossRef]
63. Tse, B.W.-C.; Cowin, G.J.; Soekmadji, C.; Jovanovic, L.; Vasireddy, R.S.; Ling, M.-T.; Khatri, A.; Liu, T.; Thierry, B.; Russell, P.J. PSMA-targeting iron oxide magnetic nanoparticles enhance MRI of preclinical prostate cancer. *Nanomedicine* **2015**, *10*, 375–386. [CrossRef] [PubMed]
64. Salimi, M.; Sarkar, S.; Saber, R.; Delavari, H.; Alizadeh, A.M.; Mulder, H.T. Magnetic hyperthermia of breast cancer cells and MRI relaxometry with dendrimer-coated iron-oxide nanoparticles. *Cancer Nanotechnol.* **2018**, *9*, 7. [CrossRef] [PubMed]
65. Gonzalez-Rodriguez, R.; Campbell, E.; Naumov, A. Multifunctional graphene oxide/iron oxide nanoparticles for magnetic targeted drug delivery dual magnetic resonance/fluorescence imaging and cancer sensing. *PLoS ONE* **2019**, *14*, e0217072. [CrossRef]
66. Gallo, J.; Vasimalai, N.; Fernandez-Arguelles, M.T.; Bañobre-López, M. Green synthesis of multimodal 'OFF–ON' activatable MRI/optical probes. *Dalton Trans.* **2016**, *45*, 17672–17680. [CrossRef]
67. Westermark, E. A case of ureteral implantation into the bladder. *J. Obstet. Women's Dis.* **1895**, *9*, 677–678. [CrossRef]
68. Gilchrist, R.K.; Medal, R.; Shorey, W.D.; Hanselman, R.C.; Parrot, J.C.; Taylor, C.B. Selective Inductive Heating of Lymph Nodes. *Ann. Surg.* **1957**, *146*, 596–606. [CrossRef]
69. Wasserman, D.D.; Healy, M. *Cooling Techniques for Hyperthermia*; StatPearls Publishing: Treasure Island, FL, USA, 2018.
70. Liu, X.; Zhang, Y.; Wang, Y.; Zhu, W.; Li, G.; Ma, X.; Chen, S.; Tiwari, S.; Shi, K.; Zhang, S.; et al. Comprehensive understanding of magnetic hyperthermia for improving antitumor therapeutic efficacy. *Theranostics* **2020**, *10*, 3793–3815. [CrossRef]
71. Kudr, J.; Haddad, Y.; Richtera, L.; Heger, Z.; Cernak, M.; Adam, V.; Zitka, O. Magnetic Nanoparticles: From Design and Synthesis to Real World Applications. *Nanomaterials* **2017**, *7*, 243. [CrossRef]
72. Huff, T.B.; Tong, L.; Zhao, Y.; Hansen, M.N.; Cheng, J.-X.; Wei, A. Hyperthermic effects of gold nanorods on tumor cells. *Nanomedicine* **2007**, *2*, 125–132. [CrossRef] [PubMed]
73. Li, Z.; Kawashita, M.; Araki, N.; Mitsumori, M.; Hiraoka, M.; Doi, M. Magnetite nanoparticles with high heating efficiencies for application in the hyperthermia of cancer. *Mater. Sci. Eng. C* **2010**, *30*, 990–996. [CrossRef]

74. Schwarz, C.; Mehnert, W.; Lucks, J.S.; Müller, R.H. Solid lipid nanoparticles (SLN) for controlled drug delivery. I. Production, characterization and sterilization. *J. Control. Release* **1994**, *30*, 83–96. [CrossRef]
75. Sanz, B.; Calatayud, M.P.; Torres, T.E.; Fanarraga, M.L.; Ibarra, M.R.; Goya, G.F. Magnetic hyperthermia enhances cell toxicity with respect to exogenous heating. *Biomaterials* **2017**, *114*, 62–70. [CrossRef] [PubMed]
76. Plassat, V.; Wilhelm, C.; Marsaud, V.; Ménager, C.; Gazeau, F.; Renoir, J.-M.; Lesieur, S. Anti-Estrogen-Loaded Superparamagnetic Liposomes for Intracellular Magnetic Targeting and Treatment of Breast Cancer Tumors. *Adv. Funct. Mater.* **2010**, *21*, 83–92. [CrossRef]
77. Hervault, A.; Thanh, N.T.K. Magnetic nanoparticle-based therapeutic agents for thermo-chemotherapy treatment of cancer. *Nanoscale* **2014**, *6*, 11553–11573. [CrossRef]
78. Rego, G.N.D.A.; Mamani, J.B.; Souza, T.K.F.; Nucci, M.P.; Da Silva, H.R.; Gamarra, L.F. Therapeutic evaluation of magnetic hyperthermia using Fe3O4-aminosilane-coated iron oxide nanoparticles in glioblastoma animal model. *Einstein* **2019**, *17*, eAO4786. [CrossRef]
79. Brezovich, I.A. Low frequency hyperthermia: Capacitive and ferromagnetic thermoseed methods. *Med. Phys. Monogr.* **1988**, *16*, 82–111.
80. Hergt, R.; Dutz, S. Magnetic particle hyperthermia—Biophysical limitations of a visionary tumour therapy. *J. Magn. Magn. Mater.* **2007**, *311*, 187–192. [CrossRef]
81. Kandasamy, G.; Sudame, A.; Luthra, T.; Saini, K.; Maity, D. Functionalized Hydrophilic Superparamagnetic Iron Oxide Nanoparticles for Magnetic Fluid Hyperthermia Application in Liver Cancer Treatment. *ACS Omega* **2018**, *3*, 3991–4005. [CrossRef]
82. Vilas-Boas, V.; Espiña, B.; Kolen'Ko, Y.V.; Bañobre-López, M.; Brito, M.; Martins, V.; Duarte, J.A.; Petrovykh, D.Y.; Freitas, P.; Carvalho, F. Effectiveness and Safety of a Nontargeted Boost for a CXCR4-Targeted Magnetic Hyperthermia Treatment of Cancer Cells. *ACS Omega* **2019**, *4*, 1931–1940. [CrossRef]
83. Sanhaji, M.; Göring, J.; Couleaud, P.; Aires, A.; Cortajarena, A.L.; Courty, J.; Prina-Mello, A.; Stapf, M.; Ludwig, R.; Volkov, Y.; et al. The phenotype of target pancreatic cancer cells influences cell death by magnetic hyperthermia with nanoparticles carrying gemicitabine and the pseudo-peptide NucAnt. *Nanomed. Nanotechnol. Biol. Med.* **2019**, *20*, 101983. [CrossRef]
84. Lu, Q.; Dai, X.; Zhang, P.; Tan, X.; Zhong, Y.; Yao, C.; Song, M.; Song, G.; Zhang, Z.; Peng, G.; et al. Fe$_3$O$_4$@Au composite magnetic nanoparticles modified with cetuximab for targeted magneto-photothermal therapy of glioma cells. *Int. J. Nanomed.* **2018**, *13*, 2491–2505. [CrossRef] [PubMed]
85. Su, Z.; Liu, D.; Chen, L.; Zhang, J.; Ru, L.; Chen, Z.; Gao, Z.; Wang, X. CD44-Targeted Magnetic Nanoparticles Kill Head And Neck Squamous Cell Carcinoma Stem Cells In An Alternating Magnetic Field. *Int. J. Nanomed.* **2019**, *14*, 7549–7560. [CrossRef] [PubMed]
86. DeVita, V.T.; Lawrence, T.S.; Rosenberg, S.A. *DeVita, Hellman, and Rosenberg's Cancer: Principles & Practice of Oncology*, 10th ed.; Lippincott Williams & Wilkins: Philadelphia, PA, USA, 2015; ISBN 9781469894553.
87. Ding, C.; Tong, L.; Feng, J.; Fu, J. Recent Advances in Stimuli-Responsive Release Function Drug Delivery Systems for Tumor Treatment. *Molecules* **2016**, *21*, 1715. [CrossRef]
88. Belanova, A.A.; Gavalas, N.; Makarenko, Y.M.; Belousova, M.M.; Soldatov, A.V.; Zolotukhin, P.V. Physicochemical Properties of Magnetic Nanoparticles: Implications for Biomedical Applications In Vitro and In Vivo. *Oncol. Res. Treat.* **2018**, *41*, 139–143. [CrossRef] [PubMed]
89. Haque, S.T.; Karim, E.M.; Abidin, S.A.Z.; Othman, I.; Holl, M.M.B.; Chowdhury, E.H. Fe/Mg-Modified Carbonate Apatite with Uniform Particle Size and Unique Transport Protein-Related Protein Corona Efficiently Delivers Doxorubicin into Breast Cancer Cells. *Nanomaterials* **2020**, *10*, 834. [CrossRef] [PubMed]
90. Ganipineni, L.P.; Ucakar, B.; Joudiou, N.; Bianco, J.; Danhier, P.; Zhao, M.; Bastiancich, C.; Gallez, B.; Danhier, F.; Préat, V. Magnetic targeting of paclitaxel-loaded poly(lactic-co-glycolic acid)-based nanoparticles for the treatment of glioblastoma. *Int. J. Nanomed.* **2018**, *13*, 4509–4521. [CrossRef] [PubMed]
91. Dai, X.; Yao, J.; Zhong, Y.; Li, Y.; Lu, Q.; Zhang, Y.; Tian, X.; Guo, Z.; Bai, T. Preparation and Characterization of Fe$_3$O$_4$@MTX Magnetic Nanoparticles for Thermochemotherapy of Primary Central Nervous System Lymphoma in vitro and in vivo. *Int. J. Nanomed.* **2019**, *14*, 9647–9663. [CrossRef] [PubMed]
92. Paliwal, R.; Paliwal, S.R.; Kenwat, R.; Das Kurmi, B.; Sahu, M.K. Solid lipid nanoparticles: A review on recent perspectives and patents. *Expert Opin. Ther. Pat.* **2020**, *30*, 179–194. [CrossRef]
93. Belyanina, I.; Kolovskaya, O.; Zamay, S.; Gargaun, A.; Zamay, T.; Kichkailo, A. Targeted Magnetic Nanotheranostics of Cancer. *Molecules* **2017**, *22*, 975. [CrossRef] [PubMed]
94. Yu, S.; Wu, G.; Gu, X.; Wang, J.; Wang, Y.; Gao, H.; Ma, J. Magnetic and pH-sensitive nanoparticles for antitumor drug delivery. *Colloids Surf. B Biointerfaces* **2013**, *103*, 15–22. [CrossRef] [PubMed]
95. Piehler, S.; Dähring, H.; Grandke, J.; Göring, J.; Couleaud, P.; Aires, A.; Cortajarena, A.L.; Courty, J.; Latorre, A.; Somoza, Á.; et al. Iron Oxide Nanoparticles as Carriers for DOX and Magnetic Hyperthermia after Intratumoral Application into Breast Cancer in Mice: Impact and Future Perspectives. *Nanomaterials* **2020**, *10*, 1016. [CrossRef]
96. Christodoulou, E.; Nerantzaki, M.; Nanaki, S.; Barmpalexis, P.; Giannousi, K.; Dendrinou-Samara, C.; Angelakeris, M.; Gounari, E.; Anastasiou, A.D.; Bikiaris, D.N. Paclitaxel Magnetic Core-Shell Nanoparticles Based on Poly(lactic acid) Semitelechelic Novel Block Copolymers for Combined Hyperthermia and Chemotherapy Treatment of Cancer. *Pharmaceutics* **2019**, *11*, 213. [CrossRef] [PubMed]

97. Khafaji, M.; Zamani, M.; Vossoughi, M.; Zad, A.I. Doxorubicin/Cisplatin-Loaded Superparamagnetic Nanoparticles as A Stimuli-Responsive Co-Delivery System For Chemo-Photothermal Therapy. *Int. J. Nanomed.* **2019**, *14*, 8769–8786. [CrossRef] [PubMed]
98. Zanganeh, S.; Hutter, G.; Spitler, R.; Lenkov, O.; Mahmoudi, M.; Shaw, A.; Pajarinen, J.S.; Nejadnik, H.; Goodman, S.; Moseley, M.; et al. Iron oxide nanoparticles inhibit tumour growth by inducing pro-inflammatory macrophage polarization in tumour tissues. *Nat. Nanotechnol.* **2016**, *11*, 986–994. [CrossRef] [PubMed]
99. Wang, M.-S.; Chen, L.; Xiong, Y.-Q.; Xu, J.; Wang, J.-P.; Meng, Z.-L. Iron oxide magnetic nanoparticles combined with actein suppress non-small-cell lung cancer growth in a p53-dependent manner. *Int. J. Nanomed.* **2017**, *12*, 7627–7651. [CrossRef] [PubMed]
100. Stapf, M.; Teichgräber, U.; Hilger, I. Methotrexate-coupled nanoparticles and magnetic nanochemothermia for the relapse-free treatment of T24 bladder tumors. *Int. J. Nanomed.* **2017**, *12*, 2793–2811. [CrossRef]
101. Abedi, M.; Abolmaali, S.S.; Abedanzadeh, M.; Farjadian, F.; Samani, S.M.; Tamaddon, A.M. Core–Shell Imidazoline–Functionalized Mesoporous Silica Superparamagnetic Hybrid Nanoparticles as a Potential Theranostic Agent for Controlled Delivery of Platinum(II) Compound. *Int. J. Nanomed.* **2020**, *15*, 2617–2631. [CrossRef]
102. Zhang, N.; Wang, Y.; Zhang, C.; Fan, Y.; Li, D.; Cao, X.; Xia, J.; Shi, X.; Guo, R. LDH-stabilized ultrasmall iron oxide nanoparticles as a platform for hyaluronidase-promoted MR imaging and chemotherapy of tumors. *Theranostics* **2020**, *10*, 2791–2802. [CrossRef]
103. Xu, F.; Zhu, J.; Lin, L.; Zhang, C.; Sun, W.; Fan, Y.; Yin, F.; Van Hest, J.C.M.; Wang, H.; Du, L.; et al. Multifunctional PVCL nanogels with redox-responsiveness enable enhanced MR imaging and ultrasound-promoted tumor chemotherapy. *Theranostics* **2020**, *10*, 4349–4358. [CrossRef] [PubMed]
104. Fan, X.; Yuan, Z.; Shou, C.; Fan, G.; Wang, H.; Gao, F.; Rui, Y.; Xu, K.; Yin, P. cRGD-Conjugated Fe_3O_4@PDA-DOX Multifunctional Nanocomposites for MRI and Antitumor Chemo-Photothermal Therapy. *Int. J. Nanomed.* **2019**, *14*, 9631–9645. [CrossRef] [PubMed]
105. Mao, H.; Qian, W.; Wang, L.; Wu, H.; Zhou, H.; Wang, A.Y.; Yang, L.; Chen, H.; Huang, J. Functionalized milk-protein-coated magnetic nanoparticles for MRI-monitored targeted therapy of pancreatic cancer. *Int. J. Nanomed.* **2016**, *11*, 3087–3099. [CrossRef] [PubMed]
106. Gao, P.; Mei, C.; He, L.; Xiao, Z.; Chan, L.; Zhang, D.; Shi, C.; Chen, T.; Luo, L. Designing multifunctional cancer-targeted nanosystem for magnetic resonance molecular imaging-guided theranostics of lung cancer. *Drug Deliv.* **2018**, *25*, 1811–1825. [CrossRef]
107. Ali, A.; Hsu, F.-T.; Hsieh, C.-L.; Shiau, C.-Y.; Chiang, C.-H.; Wei, Z.-H.; Chen, C.-Y.; Huang, H.-S. Erlotinib-Conjugated Iron Oxide Nanoparticles as a Smart Cancer-Targeted Theranostic Probe for MRI. *Sci. Rep.* **2016**, *6*, 36650. [CrossRef]
108. Thakor, A.S.; Jokerst, J.V.; Ghanouni, P.; Campbell, J.L.; Mittra, E.; Gambhir, S.S. Clinically Approved Nanoparticle Imaging Agents. *J. Nucl. Med.* **2016**, *57*, 1833–1837. [CrossRef]
109. Müller, R.H.; Maassen, S.; Schwarz, C.; Mehnert, W. Solid lipid nanoparticles (SLN) as potential carrier for human use: Interaction with human granulocytes. *J. Control. Release* **1997**, *47*, 261–269. [CrossRef]
110. Almeida, A.J.; Runge, S.; Müller, R.H. Peptide-loaded solid lipid nanoparticles (SLN): Influence of production parameters. *Int. J. Pharm.* **1997**, *149*, 255–265. [CrossRef]
111. Westesen, K.; Siekmann, B. Investigation of the gel formation of phospholipid-stabilized solid lipid nanoparticles. *Int. J. Pharm.* **1997**, *151*, 35–45. [CrossRef]
112. Kumar, M.; Kakkar, V.; Mishra, A.K.; Chuttani, K.; Kaur, I.P. Intranasal delivery of streptomycin sulfate (STRS) loaded solid lipid nanoparticles to brain and blood. *Int. J. Pharm.* **2014**, *461*, 223–233. [CrossRef]
113. Wissing, S.A.; Kayser, O.; Müller, R.H. Solid lipid nanoparticles for parenteral drug delivery. *Adv. Drug Deliv. Rev.* **2004**, *56*, 1257–1272. [CrossRef] [PubMed]
114. Katouzian, I.; Esfanjani, A.F.; Jafari, S.M.; Akhavan, S. Formulation and application of a new generation of lipid nano-carriers for the food bioactive ingredients. *Trends Food Sci. Technol.* **2017**, *68*, 14–25. [CrossRef]
115. Din, F.U.; Aman, W.; Ullah, I.; Qureshi, O.S.; Mustapha, O.; Shafique, S.; Zeb, A. Effective use of nanocarriers as drug delivery systems for the treatment of selected tumors. *Int. J. Nanomed.* **2017**, *12*, 7291–7309. [CrossRef] [PubMed]
116. Satapathy, M.K.; Yen, T.-L.; Jan, J.-S.; Tang, R.-D.; Wang, J.-Y.; Taliyan, R.; Yang, C.-H. Solid Lipid Nanoparticles (SLNs): An Advanced Drug Delivery System Targeting Brain through BBB. *Pharmaceutics* **2021**, *13*, 1183. [CrossRef] [PubMed]
117. Venishetty, V.K.; Komuravelli, R.; Kuncha, M.; Sistla, R.; Diwan, P.V. Increased brain uptake of docetaxel and ketoconazole loaded folate-grafted solid lipid nanoparticles. *Nanomed. Nanotechnol. Biol. Med.* **2013**, *9*, 111–121. [CrossRef]
118. Shegokar, R.; Singh, K.K.; Müller, R.H. Production & stability of stavudine solid lipid nanoparticles—From lab to industrial scale. *Int. J. Pharm.* **2011**, *416*, 461–470. [CrossRef]
119. Cunha, S.; Amaral, M.H.; Lobo, J.M.S.; Silva, A.C. Lipid Nanoparticles for Nasal/Intranasal Drug Delivery. *Crit. Rev. Ther. Drug Carr. Syst.* **2017**, *34*, 257–282. [CrossRef]
120. Bi, R.; Shao, W.; Wang, Q.; Zhang, N. Solid lipid nanoparticles as insulin inhalation carriers for enhanced pulmonary delivery. *J. Biomed. Nanotechnol.* **2009**, *5*, 84–92. [CrossRef]
121. Fang, Y.-P.; Chuang, C.-H.; Wu, P.-C.; Huang, Y.-B.; Tzeng, C.-C.; Chen, Y.-L.; Gao, M.-Y.; Tsai, M.-J.; Tsai, Y.-H. Amsacrine analog-loaded solid lipid nanoparticle to resolve insolubility for injection delivery: Characterization and pharmacokinetics. *Drug Des. Dev. Ther.* **2016**, *10*, 1019–1028. [CrossRef]

122. Reddy, L.H.; Sharma, R.K.; Chuttani, K.; Mishra, A.K.; Murthy, R.S.R. Influence of administration route on tumor uptake and biodistribution of etoposide loaded solid lipid nanoparticles in Dalton's lymphoma tumor bearing mice. *J. Control. Release* **2005**, *105*, 185–198. [CrossRef]
123. Mohamed, R.A.; Abass, H.A.; Attia, M.A.; Heikal, O.A. Formulation and evaluation of metoclopramide solid lipid nanoparticles for rectal suppository. *J. Pharm. Pharmacol.* **2013**, *65*, 1607–1621. [CrossRef] [PubMed]
124. Basha, S.K.; Dhandayuthabani, R.; Muzammil, M.S.; Kumari, V.S. Solid lipid nanoparticles for oral drug delivery. *Proc. Mater. Today* **2019**, *36*, 313–324. [CrossRef]
125. Seyfoddin, A.; Shaw, J.; Al-Kassas, R. Solid lipid nanoparticles for ocular drug delivery. *Drug Deliv.* **2010**, *17*, 467–489. [CrossRef] [PubMed]
126. Hassett, K.J.; Benenato, K.E.; Jacquinet, E.; Lee, A.; Woods, A.; Yuzhakov, O.; Himansu, S.; Deterling, J.; Geilich, B.M.; Ketova, T.; et al. Optimization of Lipid Nanoparticles for Intramuscular Administration of mRNA Vaccines. *Mol. Ther. Nucleic Acids* **2019**, *15*, 1–11. [CrossRef]
127. Moon, J.H.; Moxley, J.W., Jr.; Zhang, P.; Cui, H. Nanoparticle approaches to combating drug resistance. *Futur. Med. Chem.* **2015**, *7*, 1503–1510. [CrossRef]
128. Brazel, C.S.; Huang, X. The Cost of Optimal Drug Delivery: Reducing and Preventing the Burst Effect in Matrix Systems. In *Carrier-Based Drug Delivery*; ACS Symposium Series; American Chemical Society: Washington, DC, USA, 2004; Volume 879, pp. 267–282.
129. Arduino, I.; Depalo, N.; Re, F.; Magro, R.D.; Panniello, A.; Margiotta, N.; Fanizza, E.; Lopalco, A.; Laquintana, V.; Cutrignelli, A.; et al. PEGylated solid lipid nanoparticles for brain delivery of lipophilic kiteplatin Pt(IV) prodrugs: An in vitro study. *Int. J. Pharm.* **2020**, *583*, 119351. [CrossRef]
130. Dhiman, S.; Mishra, N.; Sharma, S. Development of PEGylated solid lipid nanoparticles of pentoxifylline for their beneficial pharmacological potential in pathological cardiac hypertrophy. *Artif. Cells Nanomed. Biotechnol.* **2016**, *44*, 1901–1908. [CrossRef]
131. Stylianopoulos, T. EPR-effect: Utilizing size-dependent nanoparticle delivery to solid tumors. *Ther. Deliv.* **2013**, *4*, 421–423. [CrossRef]
132. Fouad, Y.A.; Aanei, C. Revisiting the hallmarks of cancer. *Am. J. Cancer Res.* **2017**, *7*, 1016–1036.
133. Liu, Y.; Sun, D.; Fan, Q.; Ma, Q.; Dong, Z.; Tao, W.; Tao, H.; Liu, Z.; Wang, C. The enhanced permeability and retention effect based nanomedicine at the site of injury. *Nano Res.* **2020**, *13*, 564–569. [CrossRef]
134. Greish, K. Enhanced permeability and retention (EPR) effect for anticancer nanomedicine drug targeting. In *Cancer Nanotechnology*; Humana Press: New York, NY, USA, 2010; Volume 624, pp. 25–37. [CrossRef]
135. Golombek, S.K.; May, J.-N.; Theek, B.; Appold, L.; Drude, N.; Kiessling, F.; Lammers, T. Tumor targeting via EPR: Strategies to enhance patient responses. *Adv. Drug Deliv. Rev.* **2018**, *130*, 17–38. [CrossRef] [PubMed]
136. Cordero, L.B.; Alkorta, I.; Arana, L.; Cordero, L.B.; Alkorta, I.; Arana, L. Application of Solid Lipid Nanoparticles to Improve the Efficiency of Anticancer Drugs. *Nanomaterials* **2019**, *9*, 474. [CrossRef] [PubMed]
137. Natfji, A.A.; Ravishankar, D.; Osborn, H.M.I.; Greco, F. Parameters Affecting the Enhanced Permeability and Retention Effect: The Need for Patient Selection. *J. Pharm. Sci.* **2017**, *106*, 3179–3187. [CrossRef] [PubMed]
138. Bertrand, N.; Wu, J.; Xu, X.; Kamaly, N.; Farokhzad, O.C. Cancer nanotechnology: The impact of passive and active targeting in the era of modern cancer biology. *Adv. Drug Deliv. Rev.* **2014**, *66*, 2–25. [CrossRef]
139. Rosière, R.; Van Woensel, M.; Gelbcke, M.; Mathieu, V.; Hecq, J.; Mathivet, T.; Vermeersch, M.; Van Antwerpen, P.; Amighi, K.; Wauthoz, N. New folate-grafted chitosan derivative to improve delivery of paclitaxel-loaded solid lipid nanoparticles for lung tumor therapy by inhalation. *Mol. Pharm.* **2018**, *15*, 899–910. [CrossRef]
140. Campos, J.; Varas-Godoy, M.; Haidar, Z.S. Physicochemical characterization of chitosan-hyaluronan-coated solid lipid nanoparticles for the targeted delivery of paclitaxel: A proof-of-concept study in breast cancer cells. *Nanomedicine* **2017**, *12*, 473–490. [CrossRef]
141. Garg, N.K.; Singh, B.; Jain, A.; Nirbhavane, P.; Sharma, R.; Tyagi, R.K.; Kushwah, V.; Jain, S.; Katare, O.P. Fucose decorated solid-lipid nanocarriers mediate efficient delivery of methotrexate in breast cancer therapeutics. *Colloids Surf. B Biointerfaces* **2016**, *146*, 114–126. [CrossRef]
142. Mirchandani, Y.; Patravale, V.B.; Brijesh, S. Solid lipid nanoparticles for hydrophilic drugs. *J. Control. Release* **2021**, *335*, 457–464. [CrossRef]
143. Khallaf, R.A.; Salem, H.F.; Abdelbary, A. 5-Fluorouracil shell-enriched solid lipid nanoparticles (SLN) for effective skin carcinoma treatment. *Drug Deliv.* **2016**, *23*, 3452–3460. [CrossRef]
144. Kadari, A.; Pooja, D.; Gora, R.H.; Gudem, S.; Kolapalli, V.R.M.; Kulhari, H.; Sistla, R. Design of multifunctional peptide collaborated and docetaxel loaded lipid nanoparticles for antiglioma therapy. *Eur. J. Pharm. Biopharm.* **2018**, *132*, 168–179. [CrossRef]
145. Dudhipala, N.; Puchchakayala, G. Capecitabine lipid nanoparticles for anti-colon cancer activity in 1,2-dimethylhydrazine-induced colon cancer: Preparation, cytotoxic, pharmacokinetic, and pathological evaluation. *Drug Dev. Ind. Pharm.* **2018**, *44*, 1572–1582. [CrossRef]
146. Zheng, G.; Zheng, M.; Yang, B.; Fu, H.; Li, Y. Improving breast cancer therapy using doxorubicin loaded solid lipid nanoparticles: Synthesis of a novel arginine-glycine-aspartic tripeptide conjugated, pH sensitive lipid and evaluation of the nanomedicine in vitro and in vivo. *Biomed. Pharmacother.* **2019**, *116*, 109006. [CrossRef] [PubMed]

147. Xu, W.; Bae, E.J.; Lee, M.-K. Enhanced anticancer activity and intracellular uptake of paclitaxel-containing solid lipid nanoparticles in multidrug-resistant breast cancer cells. *Int. J. Nanomed.* **2018**, *13*, 7549–7563. [CrossRef] [PubMed]
148. Eskiler, G.G.; Cecener, G.; Dikmen, G.; Egeli, U.; Tunca, B. Solid lipid nanoparticles: Reversal of tamoxifen resistance in breast cancer. *Eur. J. Pharm. Sci.* **2018**, *120*, 73–88. [CrossRef] [PubMed]
149. Marslin, G.; Siram, K.; Liu, X.; Khandelwal, V.K.M.; Shen, X.; Wang, X.; Franklin, G. Solid Lipid Nanoparticles of Albendazole for Enhancing Cellular Uptake and Cytotoxicity against U-87 MG Glioma Cell Lines. *Molecules* **2017**, *22*, 2040. [CrossRef]
150. Da Rocha, M.C.O.; Da Silva, P.B.; Radicchi, M.A.; Andrade, B.Y.G.; De Oliveira, J.V.; Venus, T.; Merker, C.; Estrela-Lopis, I.; Longo, J.P.F.; Báo, S.N. Docetaxel-loaded solid lipid nanoparticles prevent tumor growth and lung metastasis of 4T1 murine mammary carcinoma cells. *J. Nanobiotechnol.* **2020**, *18*, 43. [CrossRef]
151. Valdivia, L.; García-Hevia, L.; Bañobre-López, M.; Gallo, J.; Valiente, R.; Fanarraga, M.L. Solid Lipid Particles for Lung Metastasis Treatment. *Pharmaceutics* **2021**, *13*, 93. [CrossRef]
152. Clemente, N.; Ferrara, B.; Gigliotti, C.L.; Boggio, E.; Capucchio, M.T.; Biasibetti, E.; Schiffer, D.; Mellai, M.; Annovazzi, L.; Cangemi, L.; et al. Solid Lipid Nanoparticles Carrying Temozolomide for Melanoma Treatment. Preliminary In Vitro and In Vivo Studies. *Int. J. Mol. Sci.* **2018**, *19*, 255. [CrossRef]
153. Rompicharla, S.V.K.; Bhatt, H.; Shah, A.; Komanduri, N.; Vijayasarathy, D.; Ghosh, B.; Biswas, S. Formulation optimization, characterization, and evaluation of in vitro cytotoxic potential of curcumin loaded solid lipid nanoparticles for improved anticancer activity. *Chem. Phys. Lipids* **2017**, *208*, 10–18. [CrossRef]
154. Battaglia, L.; Muntoni, E.; Chirio, D.; Peira, E.; Annovazzi, L.; Schiffer, D.; Mellai, M.; Riganti, C.; Salaroglio, I.C.; Lanotte, M.; et al. Solid lipid nanoparticles by coacervation loaded with a methotrexate prodrug: Preliminary study for glioma treatment. *Nanomedicine* **2017**, *12*, 639–656. [CrossRef]
155. Liu, B.; Han, L.; Liu, J.; Han, S.; Chen, Z.; Jiang, L. Co-delivery of paclitaxel and TOS-cisplatin via TAT-targeted solid lipid nanoparticles with synergistic antitumor activity against cervical cancer. *Int. J. Nanomed.* **2017**, *12*, 955–968. [CrossRef]
156. Thakkar, A.; Chenreddy, S.; Wang, J.; Prabhu, S. Ferulic acid combined with aspirin demonstrates chemopreventive potential towards pancreatic cancer when delivered using chitosan-coated solid-lipid nanoparticles. *Cell Biosci.* **2015**, *5*, 46. [CrossRef] [PubMed]
157. Montoto, S.S.; Muraca, G.; Ruiz, M.E. Solid Lipid Nanoparticles for Drug Delivery: Pharmacological and Biopharmaceutical Aspects. *Front. Mol. Biosci.* **2020**, *7*, 587997. [CrossRef] [PubMed]
158. Mishra, V.; Bansal, K.K.; Verma, A.; Yadav, N.; Thakur, S.; Sudhakar, K.; Rosenholm, J.M. Solid lipid nanoparticles: Emerging colloidal nano drug delivery systems. *Pharmaceutics* **2018**, *10*, 191. [CrossRef] [PubMed]
159. Sun, J.; Bi, C.; Chan, H.M.; Sun, S.; Zhang, Q.; Zheng, Y. Curcumin-loaded solid lipid nanoparticles have prolonged in vitro antitumour activity, cellular uptake and improved in vivo bioavailability. *Colloids Surf. B Biointerfaces* **2013**, *111*, 367–375. [CrossRef]
160. Jain, V.; Gupta, A.; Pawar, V.K.; Asthana, S.; Jaiswal, A.K.; Dube, A.; Chourasia, M.K. Chitosan-Assisted Immunotherapy for Intervention of Experimental Leishmaniasis via Amphotericin B-Loaded Solid Lipid Nanoparticles. *Appl. Biochem. Biotechnol.* **2014**, *174*, 1309–1330. [CrossRef]
161. Naseri, N.; Valizadeh, H.; Zakeri-Milani, P. Solid Lipid Nanoparticles and Nanostructured Lipid Carriers: Structure, Preparation and Application. *Adv. Pharm. Bull.* **2015**, *5*, 305–313. [CrossRef]
162. Jiang, S.; Eltoukhy, A.A.; Love, K.T.; Langer, R.; Anderson, D.G. Lipidoid-Coated Iron Oxide Nanoparticles for Efficient DNA and siRNA delivery. *Nano Lett.* **2013**, *13*, 1059–1064. [CrossRef]
163. Jin, J.; Bae, K.H.; Yang, H.; Lee, S.J.; Kim, H.; Kim, Y.; Joo, K.M.; Seo, S.W.; Park, T.G.; Nam, D.-H. In Vivo Specific Delivery of c-Met siRNA to Glioblastoma Using Cationic Solid Lipid Nanoparticles. *Bioconjugate Chem.* **2011**, *22*, 2568–2572. [CrossRef]
164. Oumzil, K.; Ramin, M.A.; Lorenzato, C.; Hémadou, A.; Laroche, J.; Jacobin-Valat, M.J.; Mornet, S.; Roy, C.-E.; Kauss, T.; Gaudin, K.; et al. Solid Lipid Nanoparticles for Image-Guided Therapy of Atherosclerosis. *Bioconjugate Chem.* **2016**, *27*, 569–575. [CrossRef]
165. De Escalona, M.M.; Sáez-Fernández, E.; Prados, J.C.; Melguizo, C.; Arias, J.L. Magnetic solid lipid nanoparticles in hyperthermia against colon cancer. *Int. J. Pharm.* **2016**, *504*, 11–19. [CrossRef] [PubMed]
166. Truzzi, E.; Bongio, C.; Sacchetti, F.; Maretti, E.; Montanari, M.; Iannuccelli, V.; Vismara, E.; Leo, E. Self-Assembled Lipid Nanoparticles for Oral Delivery of Heparin-Coated Iron Oxide Nanoparticles for Theranostic Purposes. *Molecules* **2017**, *22*, 963. [CrossRef] [PubMed]
167. Świętek, M.; Brož, A.; Tarasiuk, J.; Wroński, S.; Tokarz, W.; Kozieł, A.; Błażewicz, M.; Bačáková, L. Carbon nanotube/iron oxide hybrid particles and their PCL-based 3D composites for potential bone regeneration. *Mater. Sci. Eng. C* **2019**, *104*, 109913. [CrossRef] [PubMed]
168. Świętek, M.; Panchuk, R.; Skorokhyd, N.; Černoch, P.; Finiuk, N.; Klyuchivska, O.; Hrubý, M.; Molčan, M.; Berger, W.; Trousil, J.; et al. Magnetic Temperature-Sensitive Solid-Lipid Particles for Targeting and Killing Tumor Cells. *Front. Chem.* **2020**, *8*, 205. [CrossRef] [PubMed]
169. Mody, V.V.; Cox, A.G.; Shah, S.; Singh, A.; Bevins, W.; Parihar, H. Magnetic nanoparticle drug delivery systems for targeting tumor. *Appl. Nanosci.* **2014**, *4*, 385–392. [CrossRef]
170. Wydra, R.J.; Oliver, C.E.; Anderson, K.W.; Dziubla, T.D.; Hilt, J.Z. Accelerated generation of free radicals by iron oxide nanoparticles in the presence of an alternating magnetic field. *RSC Adv.* **2015**, *5*, 18888–18893. [CrossRef] [PubMed]

171. Abakumov, M.A.; Semkina, A.S.; Skorikov, A.S.; Vishnevskiy, D.A.; Ivanova, A.V.; Mironova, E.; Davydova, G.A.; Majouga, A.G.; Chekhonin, V.P. Toxicity of iron oxide nanoparticles: Size and coating effects. *J. Biochem. Mol. Toxicol.* **2018**, *32*, e22225. [CrossRef]
172. Hsu, M.-H.; Su, Y.-C. Iron-oxide embedded solid lipid nanoparticles for magnetically controlled heating and drug delivery. *Biomed. Microdevices* **2008**, *10*, 785–793. [CrossRef]
173. Igartua, M.; Saulnier, P.; Heurtault, B.; Pech, B.; Proust, J.E.; Pedraz, J.L.; Benoit, J.P. Development and characterization of solid lipid nanoparticles loaded with magnetite. *Int. J. Pharm.* **2001**, *233*, 149–157. [CrossRef]
174. Rostami, E.; Kashanian, S.; Azandaryani, A.H.; Faramarzi, H.; Dolatabadi, J.E.N.; Omidfar, K. Drug targeting using solid lipid nanoparticles. *Chem. Phys. Lipids* **2014**, *181*, 56–61. [CrossRef]
175. Müller, R.H.; Maaßen, S.; Weyhers, H.; Specht, F.; Lucks, J.S. Cytotoxicity of magnetite-loaded polylactide, polylactide/glycolide particles and solid lipid nanoparticles. *Int. J. Pharm.* **1996**, *138*, 85–94. [CrossRef]
176. García-Hevia, L.; Casafont, Í.; Oliveira, J.; Terán, N.; Fanarraga, M.L.; Gallo, J.; Bañobre-López, M. Magnetic lipid nanovehicles synergize the controlled thermal release of chemotherapeutics with magnetic ablation while enabling non-invasive monitoring by MRI for melanoma theranostics. *Bioact. Mater.* **2022**, *8*, 153–164. [CrossRef] [PubMed]
177. Babincová, M.; Čičmanec, P.; Altanerová, V.; Altaner, C.; Babinec, P. AC-magnetic field controlled drug release from magnetoliposomes: Design of a method for site-specific chemotherapy. *Bioelectrochemistry* **2002**, *55*, 17–19. [CrossRef]
178. Pang, X.; Cui, F.; Tian, J.; Chen, J.; Zhou, J.; Zhou, W. Preparation and characterization of magnetic solid lipid nanoparticles loaded with ibuprofen. *Asian J. Pharm. Sci.* **2009**, *4*, 132–137.
179. Oliveira, R.R.; Carrião, M.; Pacheco, M.T.; Branquinho, L.C.; de Souza, A.L.R.; Bakuzis, A.; Lima, E.M. Triggered release of paclitaxel from magnetic solid lipid nanoparticles by magnetic hyperthermia. *Mater. Sci. Eng. C* **2018**, *92*, 547–553. [CrossRef]
180. Ahmadifard, Z.; Ahmeda, A.; Rasekhian, M.; Moradi, S.; Arkan, E. Chitosan-coated magnetic solid lipid nanoparticles for controlled release of letrozole. *J. Drug Deliv. Sci. Technol.* **2020**, *57*, 101621. [CrossRef]
181. Grillone, A.; Battaglini, M.; Moscato, S.; Mattii, L.; Fernández, C.D.J.; Scarpellini, A.; Giorgi, M.; Sinibaldi, E.; Ciofani, G. Nutlin-loaded magnetic solid lipid nanoparticles for targeted glioblastoma treatment. *Nanomedicine* **2019**, *14*, 727–752. [CrossRef]
182. Andreozzi, E.; Wang, P.; Valenzuela, A.; Tu, C.; Gorin, F.; Dhenain, M.; Louie, A. Size-Stable Solid Lipid Nanoparticles Loaded with Gd-DOTA for Magnetic Resonance Imaging. *Bioconjugate Chem.* **2013**, *24*, 1455–1467. [CrossRef]
183. Abidi, H.; Ghaedi, M.; Rafiei, A.; Jelowdar, A.; Salimi, A.; Asfaram, A.; Ostovan, A. Magnetic solid lipid nanoparticles co-loaded with albendazole as an anti-parasitic drug: Sonochemical preparation, characterization, and in vitro drug release. *J. Mol. Liq.* **2018**, *268*, 11–18. [CrossRef]
184. Ghiani, S.; Capozza, M.; Cabella, C.; Coppo, A.; Miragoli, L.; Brioschi, C.; Bonafè, R.; Maiocchi, A. In vivo tumor targeting and biodistribution evaluation of paramagnetic solid lipid nanoparticles for magnetic resonance imaging. *Nanomed. Nanotechnol. Biol. Med.* **2017**, *13*, 693–700. [CrossRef]
185. Rocha, C.V.; da Silva, M.C.; Bañobre-López, M.; Gallo, J. (Para)magnetic hybrid nanocomposites for dual MRI detection and treatment of solid tumours. *Chem. Commun.* **2020**, *56*, 8695–8698. [CrossRef] [PubMed]
186. Jiménez-López, J.; García-Hevia, L.; Melguizo, C.; Prados, J.; Bañobre-López, M.; Gallo, J. Evaluation of Novel Doxorubicin-Loaded Magnetic Wax Nanocomposite Vehicles as Cancer Combinatorial Therapy Agents. *Pharmaceutics* **2020**, *12*, 637. [CrossRef] [PubMed]
187. De Moura, C.L.; Gallo, J.; García-Hevia, L.; Pessoa, O.D.L.; Ricardo, N.M.P.S.; López, M.B. Magnetic Hybrid Wax Nanocomposites as Externally Controlled Theranostic Vehicles: High MRI Enhancement and Synergistic Magnetically Assisted Thermo/Chemo Therapy. *Chem. Eur. J.* **2020**, *26*, 4531–4538. [CrossRef] [PubMed]
188. Grillone, A.; Riva, E.R.; Mondini, A.; Forte, C.; Calucci, L.; Innocenti, C.; Fernandez, C.D.J.; Cappello, V.; Gemmi, M.; Moscato, S.; et al. Active Targeting of Sorafenib: Preparation, Characterization, and In Vitro Testing of Drug-Loaded Magnetic Solid Lipid Nanoparticles. *Adv. Healthc. Mater.* **2015**, *4*, 1681–1690. [CrossRef]
189. Tapeinos, C.; Marino, A.; Battaglini, M.; Migliorin, S.; Brescia, R.; Scarpellini, A.; Fernández, C.D.J.; Prato, M.; Drago, F.; Ciofani, G. Stimuli-responsive lipid-based magnetic nanovectors increase apoptosis in glioblastoma cells through synergic intracellular hyperthermia and chemotherapy. *Nanoscale* **2019**, *11*, 72–88. [CrossRef]
190. Grillone, A.; Riva, E.R.; Moscato, S.; Sacco, R.; Mattoli, V.; Ciofani, G. Targeted delivery of anti-cancer drug sorafenib through magnetic solid lipid nanoparticles. In *10th Annual TechConnect World Innovation Conference and Expo, Held Jointly with the 18th Annual Nanotech Conference and Expo, and the 2015 National SBIR/STTR Conference*; Taylor and Francis Inc.: Oxfordshire, UK, 2015.

Review

Micro/Nanosystems for Magnetic Targeted Delivery of Bioagents

Francesca Garello [1], Yulia Svenskaya [2], Bogdan Parakhonskiy [3] and Miriam Filippi [4,*]

1. Molecular and Preclinical Imaging Centers, Department of Molecular Biotechnology and Health Sciences, University of Torino, Via Nizza 52, 10126 Torino, Italy; francesca.garello@unito.it
2. Science Medical Center, Saratov State University, 410012 Saratov, Russia; yulia_svenskaya@mail.ru
3. Faculty of Bioscience Engineering, Ghent University, Coupure Links 653, B-9000 Ghent, Belgium; bogdan.parakhonskiy@ugent.be
4. Soft Robotics Laboratory, Department of Mechanical and Process Engineering, ETH Zurich, 8092 Zurich, Switzerland
* Correspondence: miriam.filippi@srl.ethz.ch

Abstract: Targeted delivery of pharmaceuticals is promising for efficient disease treatment and reduction in adverse effects. Nano or microstructured magnetic materials with strong magnetic momentum can be noninvasively controlled via magnetic forces within living beings. These magnetic carriers open perspectives in controlling the delivery of different types of bioagents in humans, including small molecules, nucleic acids, and cells. In the present review, we describe different types of magnetic carriers that can serve as drug delivery platforms, and we show different ways to apply them to magnetic targeted delivery of bioagents. We discuss the magnetic guidance of nano/microsystems or labeled cells upon injection into the systemic circulation or in the tissue; we then highlight emergent applications in tissue engineering, and finally, we show how magnetic targeting can integrate with imaging technologies that serve to assist drug delivery.

Keywords: magnetic targeting; micro-systems; nano-systems; drug delivery; nanoparticles; microparticles; targeted delivery; magnetic guidance; theranostics; imaging

1. Introduction

Over the last decades, therapeutics have been developed in the form of proteins, peptides, nucleic acids, small molecules, and even cells [1–4]. Designing effective strategies for targeted delivery of these bio-active agents (bioagents) has emerged as a cornerstone for future pharmaceutical research [5]. Large efforts in drug development are currently dedicated to engineering systems that can spatiotemporally control drug activity and address specific therapeutic needs [3,4,6–9]. In particular, drug carriers have to overcome biological barriers at systemic, microenvironmental, and cellular levels that are heterogeneous across patient populations and diseases. Micro and nano-sized delivery systems have largely proved to be advantageous in controlled drug delivery [5,10–12], and are optimized in increasingly specified ways and a more personalized fashion, confirming that controlling drug dosing and targeting has become essential in the era of precision medicine [5].

Among the multitude of micro/nanocarriers developed thus far, those endowed with magnetic responsiveness stand out from the crowd as they display remarkable abilities stemming from their magnetic nature. These magnetic carriers (MCs) can be formulated by enriching or combining conventional constituents of micro/nanocarriers (lipids, proteins, polymers, or inorganic materials) with magnetically active components with ferromagnetic, paramagnetic, or superparamagnetic nature. For example, one magnetic component that is typically used is iron oxide, extensively applied in the form of particulate material, like iron oxide nanoparticles (IONPs).

The resulting MCs are popularly customized for disease detection and treatment, as they can be loaded with large amounts of drugs, nucleic acids, or imaging contrast agents [13], and chemically modified to present targeting vectors to localize into the target sites with high precision. In addition, under the application of external magnetic fields, they can be physically displaced allowing one to remotely control their position and movements. Moreover, MCs can locally release energy in the form of heat that can destruct the cells in the surroundings, a strategy known as "hyperthermic therapy" that is widely applied in cancer treatment. Interestingly, in recent years, it has been also demonstrated that, when exposed to external magnetic fields, MCs can apply localized mechanical stimuli on cells and control their behavior by affecting the molecular signaling involved in various cellular processes, such as proliferation, adhesion, and differentiation. Finally, MCs can be visualized via in vivo imaging techniques. MCs can serve as contrast agents in magnetic resonance imaging (MRI), but as they can be functionalized with diverse imaging tags, they can be also visualized by other imaging techniques. Therefore, the magnetic nature of the MCs offers the unique chance to remotely control the localization and therapeutic action of bioagents, while visually monitoring the targeted delivery process. In addition, the micro and nano-capacities of these systems make it possible to load therapeutics of various types, ranging from small molecules to nucleic acids and cells. Sensitive bioagents can be protected by encapsulation within internal compartments of larger systems, like microcapsules. Nanosized MCs, like nanoparticles, can be surrounded with surface coatings that provide accessible sites for conjugation with functional molecules.

Such versatility has promoted the use of MCs in a plethora of advanced biomedical applications that include genetic engineering, gene therapy, cell therapy, microrobotics, and tissue engineering. For these applications, MCs have been engineered as multifunctional platforms that combine mechanisms of controlled drug release, activatable cytotoxicity, spatial guidance, and multimodal imaging. In particular, the magnetic guidance can minimize drug dissemination to unspecific tissues, and can therefore substantially improve the bioagent efficiency and decrease the required dosage.

Even though the potential benefits of MCs are remarkable, certain critical aspects have to be considered in view of their full integration into the biomedical practice. For example, the constituent materials that impart magnetic properties to these systems are typically associated with biocompatibility issues. Moreover, the magnetic responsiveness and controllability of MCs depend not only on the nature of the magnetic component but also on the design parameters (such as the carrier geometry and size). The understanding and prediction of the magnetic behavior of MCs, therefore, requires the simultaneous assessment of their composition, molecular structure, and overall morphology. The magnetic and morphological characterization must then be put into the context of using MCs within biological environments. Only this way, one can identify those parameters that affect the magnetic properties and potentially alter the efficacy of the MCs in biomedical applications.

MCs can become the next generation of multimodal systems for advanced magnetically targeted delivery as they can act under the remote, high penetration, and non-invasive guidance mediated by magnetic forces. However, to fully realize the MCs' potential, their movement, affinity to the target sites, and cargo release ability have to be assessed while taking into account the specific magnetic behavior of the carrier. The design of magnetic devices has been extensively reviewed by Liu et al., and will not be discussed here [14]. In this review, we will discuss the utility of MCs in the magnetically-aided targeted delivery of various bioagents (Figure 1). We will first introduce the most widely used categories of MCs, and briefly illustrate their properties and association with diverse types of bioagents. Second, we will provide an overview of the application of MCs to drug delivery, highlighting those advantageous features that have caused the recent breakthrough of MCs in this field. Furthermore, we will describe the combination of MCs with external magnetic forces for the spatial guidance of cells and microrobotic systems, and discuss the imaging trackability of MCs within the context of drug delivery. Finally, we will critically point out the core challenges that have to be addressed to make these systems effective in specific medical applications and translatable to human use and propose future directions in advancing control delivery of pharmaceutics via MCs.

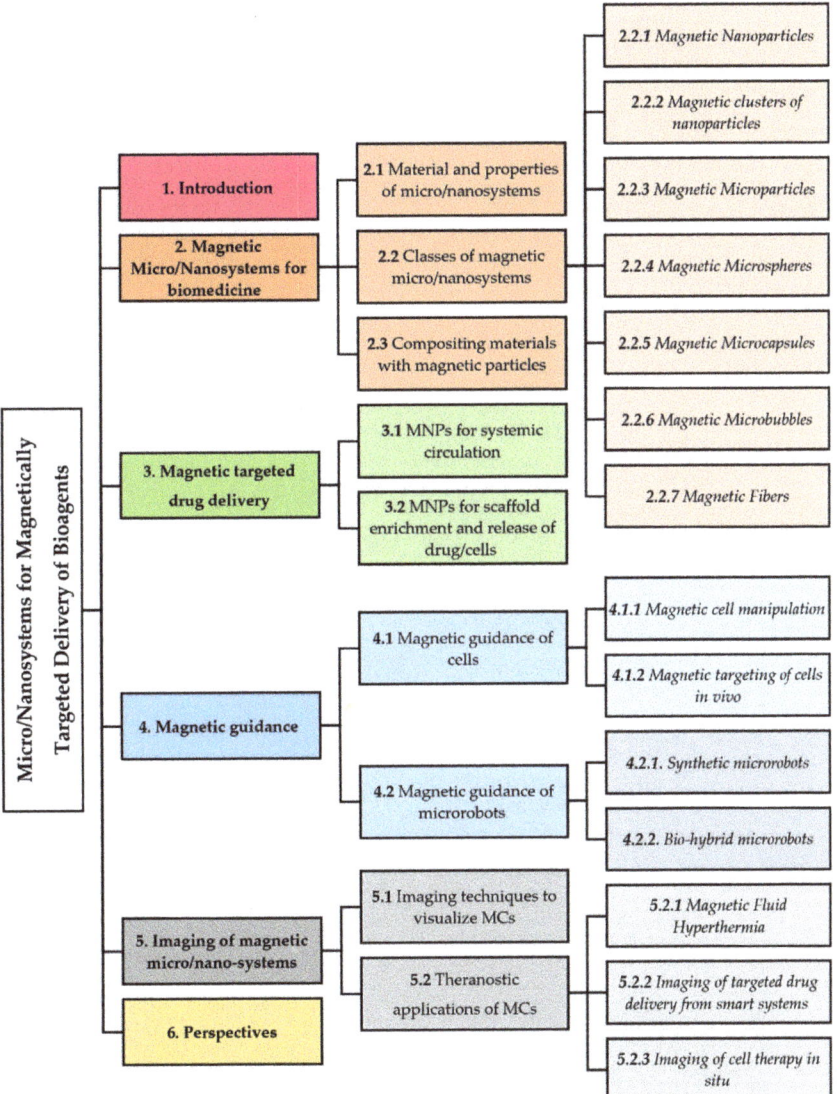

Figure 1. Schematic diagram of the review contents.

2. Magnetic Micro/Nanosystems for Biomedicine

Micro/nano-carriers can store large payloads of drugs, and protect them during vehiculation within physiological systems [11,15,16]. These platforms can have a myriad of different morphologies, in which drugs can be physically entrapped in the matrix, encapsulated in inner cavities, or attached to the platform's structure via different mechanisms of adhesion. Hollow architectures offer large inner compartments where bioagents can be stored. These systems include, for instance, microspheres or nanovesicles made of lipids, dendrimers [17–20], polymers [21], proteins [22,23], natural membranes, or other structures of biological origin (e.g., exosomes, extracellular vesicles, or yeast and bacterial capsules) [24–27]. Solid particles typically consist of inorganic materials, polymers, or proteins, and can entrap drugs within the matrix or carry them as attached to the surface (Figure 2).

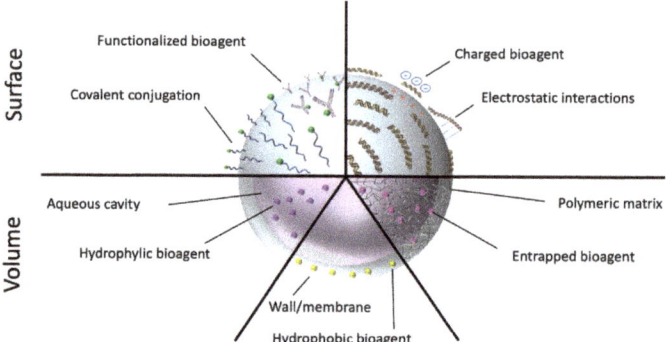

Figure 2. **Drug loading on drug delivery platforms.** Drugs can be loaded onto nano- and microcarriers via attachment to the surface or incorporated into the whole carrier volume. Bioagents can be covalently or non-covalently conjugated to the surface, or entrapped into the internal cavities or the constituent materials of the particles.

Albeit with some differences depending on the shape and composition of the carriers, their nano- or micro-size guarantees an optimal interaction with various biological entities present in the biosystems, which range from small molecules to macromolecules, organelles, and cells (Figure 3) [28]. Nevertheless, the size difference between nano- and microparticles proffers several effects on the drug loading, permeability across the biological membranes, intracellular processing, and kinetics of physiological distribution and tissue retention [29]. On the one hand, being much smaller than human cells, nanosystems can penetrate the mucosa, safely circulate into the bloodstream, and interact with cellular and subcellular targets. As such, nanomaterials can be applied to intracellular or intranuclear drug delivery, as well as to modulation of biochemical cell pathways by interacting with membrane molecules. On the other hand, microparticles could be most useful as reservoir systems for controlled drug release and protection of bioagents from enzymatic degradation.

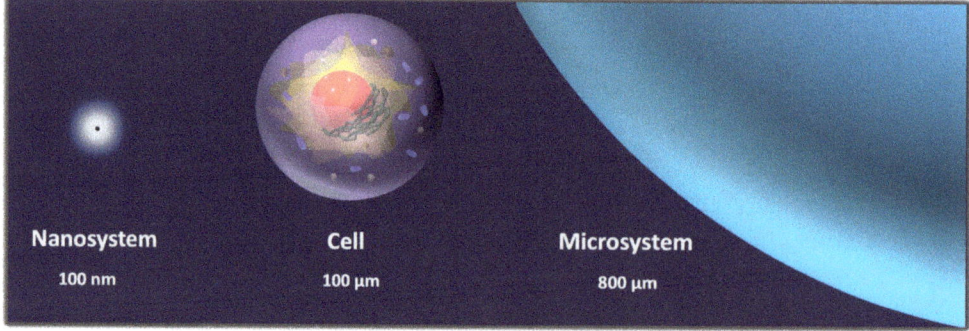

Figure 3. **Nano to micro-scale.** Comparison among the typical size of nanosystems, microsystems, and cells: nanoparticles have diameters comprised between 1 and 100 nm; microparticles have diameters ranging from 1 μm to 1000 μm, whereas the diameter of cells can vary from 10 to 100 μm.

Nano/micro drug carriers can be provided through all administration routes, however, there are some limitations for the systemic administration of microcarriers that depend on the size. When drug carriers are released into the bloodstream, they move in the body following kinetics and distribution patterns that mostly depend on the specific physicochemical properties of the carrier (passive targeting) [30]. However, nano/micro-platforms can also carry and accurately deliver their content to desired destinations (active targeting). Targeted carriers can be produced by molecular functionalization of the nano/microsystems with targeting vectors that home preferentially into target tissues [10,15,30]. Such targeting moieties provide the systems with the ability to localize into precise areas of the body, leading to improved therapeutic efficacy and reduction in off-target toxicity [3,31]. Stimuli-responsive or biodegradable materials can be used to exert spatiotemporal control over the delivery of the bioactive agents [9,32–35]. Nevertheless, these systems have be to designed through criteria that depend on the unique structure of the target tissue and organs [4].

The functionality of micro/nano-carriers can be augmented with remote controllability by imparting magnetic responsiveness to the platform structure. Carriers can be magnetized by using magnetic materials as structural bulk components or additive constituents. Magnetic targeting is based on the general assumption that MCs can be spatially controlled within living beings via magnetic fields. In particular, when the MCs are intravenously/endovascularly administered, they can accumulate within the area to which the magnetic field is applied. In addition, the local magnetic agglomeration can provide detectable imaging signals to track the local drug delivery. Structural features of the carriers, physiological parameters of the environment, and technical parameters of the experimental setting determine how effectively the MCs accumulate. Particle size, surface characteristics, as well as field strength, and features of the circulation system (like the blood flow rate) can influence the distribution of MCs at various extents. Local magnetic fields generated by permanent magnets are applied to facilitate the extravasation of carriers into the targeted area.

In the following sections, the most commonly used in MC design magnetic materials will be identified, and the classes and properties of the resulting carriers will be described. Innovative and more conventional applications of MCs will be exemplified to show how the versatility of MCs spans various domains of biomedical research and practice. Finally, the utility of MCs in targeted delivery will be discussed in view of their biophysical and magnetic properties.

2.1. Material and Properties of Micro/Nanosystems

Biomedical MCs can be obtained by incorporating a magnetically active phase in the structure of the delivery platform. As a key aspect, MCs display a magnetic fingerprint with peculiar features, such as saturation magnetization, magnetic anisotropy, and magnetic behavior in the presence of applied magnetic fields [36–39]. These features greatly vary according to the particle size and magnetic structure that can account for a single- or multi-domains organization [37,40–43]. For a better understanding of magnetism in micro and nanosystems, the reader is referred to extensive, specialized reviews on the topic [38,44]. In the following paragraphs, we will briefly survey the composition of MCs and their relevant properties for biomedical applications.

Among magnetizing materials, the magnetic spinel ferrites $M_xFe_{3-x}O_4$ (M = Mg, Mn, Fe, Co, Ni, Cu, Zn) have been extensively used to produce drug delivery systems that could overcome the major limitations found in nonmagnetic carriers, namely the poor site-specific localization and the high recognition and clearance by the effector cells of the reticuloendothelial system. Various preparations of microspheres with porous or hollow/porous architectures, as well as solid nano-sized particles were formulated from ferrites. In fact, the high magnetism of ferrites imparts high magnetic responsiveness to carriers, so that these compounds can create micro/nano-objects that can be controlled during navigation within blood vessels. Ferrite-based carriers can be irradiated with microwaves and produce heat that triggers the release of the loaded drugs [45]. Nevertheless, iron oxide, typically formulated as IONPs, has been the most extensively involved in the production of biomedical micro and nanosystems. Magnetite (Fe_3O_4) and maghemite

(γ-Fe$_2$O$_3$) are biocompatible forms of iron oxide that have been intensively exploited to produce biomedical IONPs. These nanoparticulate materials are composed of a magnetic core further surrounded by external coatings that cover the surface of the nanoparticle to increase its biocompatibility or provide suitable sites for functionalization. Coatings typically consist of inorganic materials, phospholipids, or biocompatible polymers. For example, the dextran-based coating has been employed in a variety of MCs and many of these formulations have reached advanced clinical trials or are already marketed. The magnetic core has paramagnetic or superparamagnetic properties, which render these objects visible by MRI. The popularity of IONPs derives highly tunable properties (such as size and surface characteristics) that simplify the engineering of systems customized for specific applications or endowed with smart behavior, and multifunctionality. Moreover, iron oxide is generally regarded as a biocompatible material suitable for cell experiments and safe use in living beings. In fact, iron is an essential element in human or animal bodies. As such, within relatively wide ranges of dosage and exposure, iron oxide can be efficiently processed through the physiological biochemical activities of the cells and enter the healthy metabolism of organisms. The stability of IONPs varies depending on the specific features of the system (e.g., the robustness of the coating) and the conditions of the biological environment in which the nanosystem is used. However, in general, IONPs are considered stable particles that can remain inside the cells for long time ranges (even months). As a consequence, after being administered to cells and living beings, IONPs can be safely decomposed, but typically through slow degradation kinetics. In recent years, experimental evidence has raised novel doubts in regards to toxicity. Cellular damage was observed in response to contact with IONPs and attributed to the generation of reactive species of oxygen (ROS).

2.2. Classes of Magnetic Micro/Nanosystems

The magnetic systems, which are being explored for magnetic targeting, can be categorized into various classes according to their size and morphology. In regards to drug loading/release and magnetic manipulation, the most promising classes of MCs are: nanoparticles, microspheres, microcapsules, microbubbles, fibers, nanoparticle clusters, and nanocomposite matrices (Figure 4). The following paragraphs will survey the properties of MC classes and how these characteristics can be exploited for the optimal use of MCs as bioprobes in therapeutic delivery.

2.2.1. Magnetic Nanoparticles

Magnetic nanoparticles (MNPs) are magnetic materials on the nanometer scale that have remarkable properties that evoked enormous interest in life sciences in the last decade. In particular, IONPs have emerged as diagnostic and therapeutic tools as their fundamental properties as nanomagnets, defined at the nanoscale, can be extremely useful in biological applications. IONPs are composed of magnetite or its oxidized form maghemite and have diameters ranging between about 1 and 100 nm. According to their physical properties, IONPs show different magnetic natures. Ferromagnetic particles are characterized by a permanent mean magnetic moment [46]. This comes together with a large effective magnetic anisotropy, which suppresses the thermally activated motion of the magnetic moments within the particle core. These features distinguish the ferromagnetic particles from the superparamagnetic ones. Superparamagnetism is found in MNPs with particle size smaller than a certain critical size (typically particle diameter less than 30 nm in IONPs). Below this size limit, IONPs are single-domain; namely, they are composed of one magnetic domain only. In such a condition, the overall magnetization of the MNP acts as a single giant magnetic moment that derives from summing the totality of individual magnetic moments carried by the MNP atoms. This assumption is referred to as "macro-spin approximation". The magnetization of the superparamagnetic IONPs (SPIONs) randomly flips direction under the influence of temperature, but it can also be aligned using an external magnet. In the absence of an external magnetic field, the measured magnetization of superparamagnetic nanoparticles is null (superparamagnetic state). External magnetic fields can magnetize the nanoparticles similarly to a paramagnet, but with a larger magnetic susceptibility. Once the external magnetic field is removed, the SPIONs lose their magnetization. The explanation is that, in superparamagnetism, thermal excitation causes the dipole moment of a single-domain particle to fluctuate so rapidly that no magnetic moment for macroscopic time scales can be observed. As such, superparamagnetic particles are non-magnetic in the absence of external magnetic fields but do develop a mean magnetic moment when magnetic fields are applied.

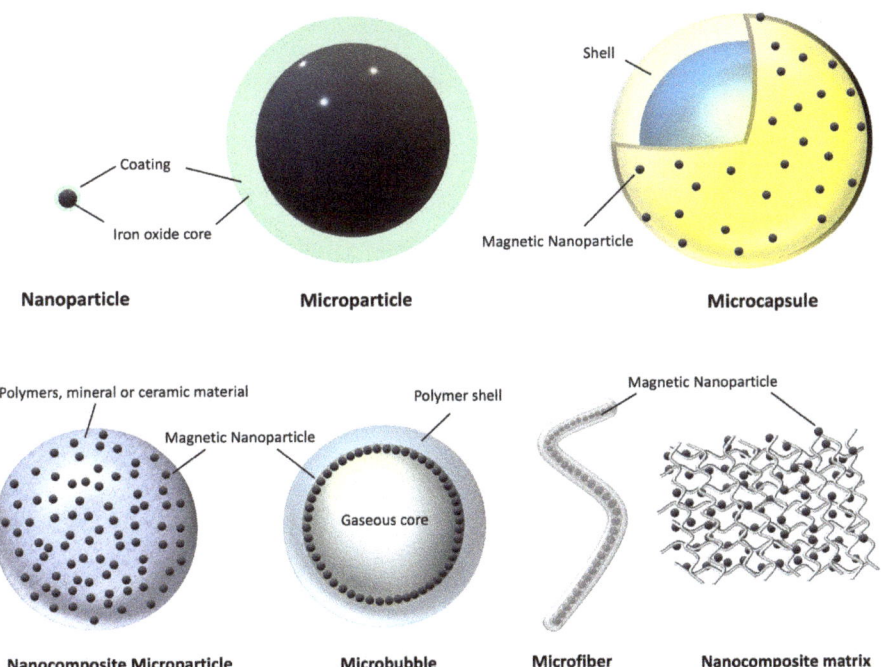

Figure 4. Different morphologies of micro/nanosystems for the magnetic targeted delivery of bioagents. Magnetic carriers can be manufactured to have different structural characteristics and classified according to their size and morphological features into: nanoparticles, microparticles, microcapsules, nanocomposite particles and matrices, microbubbles, and microfibers.

Superparamagnetism is one of the most attractive properties in the use of MNPs. In fact, the reversible magnetization ability of SPIONs has proven to be very useful to guide and manipulate them from distance through magnetic fields. This ability was transferred to cell applications, making it feasible to manipulate cells that have associated with MNPs or have internalized them. As such, SPIONs have been used for: labeling cells; separating them from complex samples or preparations; spatially guiding them; conditioning the release of drugs; remotely controlling actuatable microsystems (microactuators).

Whereas MNPs manufactured from other highly magnetic materials (like Co and Ni) are limited in their biological applicability by issues of toxicity and easy oxidization, the IONPs have instead found extensive application in gene and drug delivery, bioimaging (MRI and magnetic particle imaging), and magnetic ferrofluid hyperthermia, thus emerging in medical diagnosis and therapy. Moreover, IONPs are also used in the production of terabit magnetic storage devices and sensors, as well as for catalysis, separation of cells and biomolecules (protein separation, DNA/RNA fishing), and superparamagnetic relaxometry.

IONPs can be produced via different wet chemical, microbiological, and dry processes that can be classified as chemical, biological and physical methods [47,48]. Each technique provides manufacturers with different abilities to tailor the geometry, stability, crystallinity, dispersion trends, and magnetic properties of the nanoparticles [41]. Chemical synthesis can occur in many ways in techniques such as the chemical coprecipitation, electrochemical method, sol-gel method, flow injection method, supercritical fluid method, hydrothermal reaction, thermal and sonochemical decomposition, and nanoreactor techniques [43,47,49]. Whereas biological synthesis is based on the activity of fungi, bacteria, plants, and proteins [50,51], physical synthetic methods include: pulsed laser ablation, laser-induced pyrolysis, aerosol, power ball milling, gas-phase deposition, and electron beam lithography [47]. Physical methods are often complex procedures that guarantee only limited control of the size of particles in the nanometer range [52]. Although ensuring low cost, reproducibility, scalability, and high yield, the biological methods are time-consuming [51,53,54]. For these reasons, chemical synthesis has been the most widely employed approach as it provides a simple and reproducible route to synthesize IONPs with controlled size, composition, and morphology [55,56]. For example, in the co-precipitation method, IONPs are generated from the co-precipitation of Fe^{2+} and Fe^{3+} by the addition of a base, a process that can be controlled by varying the type of salt used, or adjusting the conditions such as the $Fe^{3+/2+}$ ions ratio, pH, and ionic strength [41].

Coating the IONPs primarily aims at protecting the core from oxidation, so that magnetism can be maintained for longer times. In addition, the nanoparticulate is also protected from aggregation, biodegradation, structural alteration, or co-assembly with components of biological systems. SPIONs are typically coated with biocompatible polymers of synthetic or natural origin, that include dextran, starch, polyethylene glycol (PEG), polyvinyl alcohol (PVA), and poly(L-lysine) (PLL) among others. These polymers offer chemical binding sites to attach drugs or receptors/ligands enabling them to recognize molecules exposed on the membrane of specific cell types. IONP surface functionalization through organic and inorganic corona shells, grafted ligands, and selective-site moieties facilitates the bonding of the nanomagnets to natural biomolecules [48]. Moreover, certain coatings enable avoid accelerated blood clearance of injectable nanomaterials and extend their circulation times, resulting in improved performances in both disease imaging and treatment. In fact, coatings can be made of some polymers escaping from the identification by effectors of the immune system. Alternatively, biomimetic delivery systems are also available, that exploit the cell membrane-based coating technology.

Even though SPIONs are generally regarded as safe biomaterials, in the past many formulations approved for clinical use were withdrawn shortly after approval [57–60]. In fact, it has to be mentioned that the cellular response to the exposure to these particles can modulate cell behavior in complex ways, and an understanding of the underlying mechanisms is still required from a generic point of view, namely about the bioeffects that are dependent on common constituents (iron oxide). In addition, the specific features of the various formulations also have to be carefully considered, as the size and surface functionalization of SPIONs can strikingly affect the toxicity. Therefore, studies focusing on the biological distribution and metabolic pathways of the MNPs in cells or bodies will lead the future evolution of research in this field [61].

2.2.2. Magnetic Clusters of Nanoparticles

Since the poor magnetization shown by individual SPIONs is unfavorable to dragging them through the body, clustering strategies were proposed to enhance the magnetic responsiveness. When combined with other materials, MNPs can be grouped to form nanometric assemblies with higher magnetic controllability and responsiveness. For instance, multilayered magnetovesicles carrying cargos in the hollow cavities can be prepared by self-assembly of IONPs grafted with poly(styrene)-b-poly(acrylic acid) block copolymers [62]. In such platforms, IONPs can be packed at a high density, substantially increasing their magnetization and the transverse relaxivity rate, an essential property that induces MRI to be dynamically monitored by magnetic resonance and ultrasound dual-modal imaging. Importantly, the H_2S microbubbles were able to ablate the tumor tissue when exposed to higher acoustic intensity. The H_2S molecules localized at a high concentration during this process could then diffuse as gasotransmitters into the inner tumor regions providing further antitumor functions.

2.2.7. Magnetic Fibers

Magnetically responsive drug delivery systems can also be designed with non-spherical particle morphology. One anisotropic design that recently became popular for specific drug delivery applications is the fiber. Magnetic fibers are fabricated by combining polymeric fibers with MNPs. One example is found in the work of Perera et al., where PVA fibers with rough 100–300 nm diameter and filled with MNPs were synthesized via infusion gyration [125]. Controlled release of drugs could be achieved by magnetic actuation. A model drug (acetaminophen) was uploaded via impregnation and rapidly released by applying magnetic fields, with 90% cumulative release achieved in 15 min only during magnetic actuation.

Drug release on demand from a fiber-shaped delivery system was triggered in response to varying pH and temperatures [126]. In such a case, magnetic bioactive glasses complexed with the antitumor drug Cisplatin were doped into the chitosan-grafted poly(ε-caprolactone) (PCL) nanofibers. Compared with control fibers kept under physiological conditions, the fibers exposed to a pH value of 5.5 and a temperature of 43 °C displayed augmented Cisplatin release rates. Following these observations, MG-63 osteosarcoma cells were treated with simultaneous chemotherapy and hyperthermia by exposure to an alternating magnetic field that activated the magnetic fibers.

Core-shell nanofibers of approximatively 300 nm in diameter were developed by coaxial electrospinning of N-carboxymethyl chitosan-PVA and PCL for the core and the shell, respectively [127]. The antitumor drug DOX and nickel ferrite nanoparticles were incorporated into the nanofibers for extended and controlled release of DOX against MCF-7 breast cancer. DOX release was studied under physiological and acidic pH in the absence or presence of applied magnetic fields, respectively, revealing that the maximum cytotoxicity on cancer cells (83%) could be obtained at day seven by magnetically stimulating nanofibers with chitosan/PCL/nickel ferrite 10 wt.% composition. Another group fabricated a magnetic nanofibrous mat-based bandage for the treatment of skin cancer by non-invasive induction of hyperthermia [128]. PCL fibers incorporating magnetite nanoparticles were obtained by electrospinning and then magnetically stimulated with an external alternating current magnetic field. The fibers locally dissipated heat energy, thus rapidly augmenting the surrounding temperature up to 45 °C. This bandage killed both parental and DOX hydrochloride-resistant HeLa cells, demonstrating enhanced DOX activity caused by increased temperatures. Mice with chemically induced skin tumors fully recovered within one month after five hyperthermic sessions of 15 min.

Due to their peculiar morphology, fibers have attracted considerable interest for drug release within specific environments and applications. Fibers are promising for the treatment of tubular or ductal anatomical structures (e.g., lung airways, blood vessels) or the construction of implantable tissue grafts. For instance, they can be used in controlled pulmonary drug delivery as they may travel deep into the lung airways. In this location, fibers facilitate the accumulation of pharmaceuticals at the target site and enable their sustained release [129]. In this context, electrospun superparamagnetic polymer-based biodegradable microrods were recently proposed. These fibers consisted of a polymeric phase of poly(l-lactide) (PLLA) and polyethylene oxide (PEO), enriched with oleic acid-coated magnetite nanoparticles. Microrods were prepared by exposing the electrospun fibers to ultraviolet (UV) irradiation and sonication, before incorporating methyl 4-hydroxybenzoate as a model drug. The authors showed that the magnetic properties of the microrods were responsible for inducing hyperthermia under AMF. Moreover, airborne microrods could be manipulated by direct current magnetic fields, thus enhancing targeted delivery to lower airways.

Scaffolds for release on demand of bioagents have been typically generated as porous structures but, in such an architecture the drugs freely diffuse, especially during long-term studies. Having good potential for better drug retention and controllable behavior, activatable magnetic fibers could have preferable morphology and profile to construct scaffolds for tissue engineering or cancer therapies. Wang et al. recently developed scaffolds from hollow fibers composed of alginate/IONPs by coaxial 3D printing [130]. Drugs, proteins, or living cells were encapsulated in the core part of the fibers that was characterized by a lower alginate gel concentration. Magnetic actuation of the fibers resulted in physical deformation of the scaffold under magnetic field exposure and subsequent release of loaded therapeutics. Moreover, due to their elongated structure, the fibers acted as diffusion barriers and minimized the uncontrolled diffusion of bioagents from the non-stimulated scaffolds. Under intermittently magnetic stimulation, a formulation with optimized composition (10 wt.% of alginate, 13% of IONPs) and crosslinking conditions (0.1 M $CaCl_2$, 10 s) displayed repeated on-demand release capability in both cell and animal experiments.

2.3. Compositing Materials with Magnetic Particles

Magnetic systems at both nano- and micro-scale can be used to create composite matrices with various applications [131–134]. Magnetic nano- and micromaterials are typically added to polymeric hydrogels or scaffolds to generate multifunctional materials that combine the properties of the polymers and the magnetic materials. Briefly, to fabricate magneto-polymeric materials, ferri, and/or ferromagnetic particles can be mixed or embedded in a polymeric matrix, and then different strategies can be followed to incorporate the particle phase either transiently or permanently [133,135].

Magnetic microcomposites include magnetic microparticles, and as such, they incorporate magnetic materials clustered in microsize sites of accumulation. Incorporating micro MCs, however, imparts a high magnetization to composites that display microscale structural heterogeneities, which might lead to gross alterations of the mechanical behavior and affect their suitability for specific applications. For instance, in additive manufacturing (e.g., deposition-based printing) a careful evaluation of the ink's mechanical properties is necessary to maintain shape fidelity and printability of the deposited material. Spangeberg et al. realized a magnetic bioink based on alginate and methylcellulose (3 and 9%, respectively) that was incorporated with magnetite microparticles [136]. While shear-thinning properties remain unaltered, viscosity and magnetization of the composite material strongly increased with the magnetic microsystem concentration. Upon optimizing the magnetic microparticle concentration (25% w/w), scaffolds were fabricated, which could deform under the application of an external magnetic field and represent a promising platform for cell mechanical stimulation and improvement of stem cell differentiation in cell-laden scaffolds.

Due to the high integrability of nanomaterials with macromolecular structures like polymers, nanocomposites display structural homogeneity, monophasic mechanical behavior, and improved physico-chemical properties that let nanocomposites emerge in biomedical applications, such as reinforcement of engineered tissue grafts [131,134,137,138]. In particular, the magnetic responsiveness renders magnetic nanocomposites particularly attractive to realize controllable cell culture models and improve cell functionality within biological tissue artifacts [138]. To include MNPs into polymeric matrices, different strategies can be applied, which include: (*i*) co-precipitation during the matrices synthesis [139,140]; (*ii*) adsorption on pre-formed matrices [141]; (*iii*) adsorption during the freezing process [142]; (*iv*) hydrothermal method [143]; (*v*) sol-gel method [144]. These approaches are based on different technical principles and have their advantages and disadvantages. For instance, the procedure for co-precipitation is a facile and convenient synthetic method, with low environmental impact, as it is carried out in aqueous solution with relatively low reaction temperatures. However, this procedure can affect the synthesis process and requires special conditions, such as high stirring speeds, controlled atmospheres, reaction medium and drying temperatures. On the other hand, the adsorption method is more predictable but limited concerning the loading amount and affected by the physical and chemical properties of the matrices, such as surface chemistry, charges as well as pore size. The recently developed method of freezing incorporation (method iii) is labor-intensive but provides four times higher loading efficiency than the coprecipitation or adsorption method. In simultaneous adsorption-freezing, the repetitive adsorption of the magnetite particles occurs in a water-based solution at a negative temperature. In such conditions, the ice crystallization provides additional pressure onto MNPs, which improves their adsorption efficiency. The hydrothermal method (iv), which uses an autoclave or microwave to reach high pressure and temperature, allows the preparation of a stoichiometric compound with appropriate crystallinity. Although this method is rather specific, the high temperature can affect the matrix stability. In the sol-gel method (v), the gel molecules stick together with the MNPs and the porous matrices. In addition to being fast and efficient, this strategy does not require any specific equipment. However, additional compounds (such as the gel molecules) are present in the final composition, which can influence the surface chemistry properties of the particles.

Besides responding to magnetic fields, the MNPs can enhance the matrix properties, by conferring for instance additional structural stability. For example, adding MNPs to the calcium carbonate particles can stabilize the unstable calcium carbonate in the vaterite phase and delay its recrystallization to calcite from hours to several days [145].

3. Magnetic Targeted Drug Delivery

Magnetic drug targeting can be realized through different approaches [121]. In the majority of cases, an external magnetic field is applied to control the localization and the functional activity (e.g., drug release) of the MCs [63,104,146–149]. External magnetic fields can be generated from various kinds of stationary magnets or via electromagnetic coils [121]. For clinical implementation, one has to take into account that, according to the guidelines provided by the International Commission of Non-ionizing Radiation Protection [150], patients could not be subjected to the magnetic flux densities that exceed 400 mT in any part of their body. In clinical practice, one to two T magnets are used to achieve local magnetic field intensities of around 400 mT. In preclinical research, magnetic drug delivery is realized in the presence of applied magnetic fields with intensities varying between 0.1 T and 1.5 T [151,152]. The distance between the magnet and the target site for delivery, as well as the geometry of the magnets, crucially affect the effectiveness of delivery [63,148]. One of the major challenges in magnetic drug delivery is delivering therapeutics within deep tissues in vivo. In particular, one common approach consists of placing stationary magnets on the skin. This way, the MCs can be efficiently guided and controlled within superficial tissue layers, but the target sites below a few centimeters (5 cm) under the skin remain not accessible. In order to drag and focus the MCs into deep tissue, a sequence of actuations can be realized by dynamically controlling magnets according to schemes defined by first-principles magneto-statics and ferrofluid transport models [153]. Such a strategy pushed magnetic nanoparticles to a deep target location.

Nevertheless, our ability to control the MCs from distance through external magnetic fields remains limited. As an alternative, magnetized implantable scaffolds can be used to attract the MCs into the target organs. In particular, nanodrugs can be delivered in a targeted fashion through implants endowed with high magnetism. Recently, poly(lactic-co-glycolic acid) (PLGA) composited with Fe_3O_4 was used to produce durable, biocompatible, and highly magnetic implant scaffolds that can locally attract magnetized chemotherapeutics and destroy cancer cells [154]. Although challenging, such approaches can facilitate more precise treatments and show promise in bone cancer, obesity, and deterioration of the inner ear [155].

3.1. MNPs for Systemic Circulation

MNPs are widely designed for the site-specific delivery of therapeutic molecules in the presence of an external magnetic field [156,157]. Chemotherapeutics [158,159], anti-inflammatory drugs [160], immunosuppressants [161], antibiotics [162,163], and antiviral agents [164] can be targeted on the diseased tissue by their use owing to sufficient surface area of MNPs and the possibility to control the payload release with or without a specific trigger. Drug molecules are generally incorporated into the carrying matrix via adsorption and encapsulation approaches or directly linked to the surface of MNPs through the covalent bonding or electrostatic attraction [157,165]. Liberation of the drug then occurs through its leakage from the matrix or as a result of the nanocarrier degradation. This process can be triggered by alternating magnetic fields [166,167] or such stimuli as pH [168], enzymatic destruction [169], and heat [170,171] utilizing linkers that are sensitive to them. Realization of such an approach should guarantee a drastic enhancement of the local drug concentration in the targeted site resulting in therapeutic efficacy increasing.

Site-specific targeting of the payload is provided either passively based on the vascular bursts [172] and enhanced permeability and retention (EPR) effect of pathologic tissue [173,174] or initiated actively via specific ligand attachment [175,176] or exploiting superparamagnetic properties of the system (magnetic targeting) [14,177]. The latter is an important advantage of MNP-based delivery systems that sets them apart from other nanotechnology-based drugs. The success of such addressing depends on the particle size, surface chemistry, magnetic polarization, and magnetophoretic movability under an external magnetic field [156]. In addition, it requires a good magnet system to guide the drug carriers to the target site [14]. The transport of MNPs in the bloodstream is defined by a dynamic equilibrium between the magnetic and hydrodynamic forces acting on them. Thus, such physical properties as the strength and gradients of the magnetic field, the volumetric and magnetic properties (magnetization) of the carriers, as well as the hydrodynamic parameters (blood flow, hematocrit, viscosity, and the particle concentration in the blood) significantly affects the efficacy of magnetic navigation [151].

The possibility of magnetic targeting of drugs has been widely demonstrated utilizing MNP-based carries systemically injected in vivo in different animal models. However, to the moment, no magnetic targeting protocol has successfully passed the regulatory approval to be translated to clinics. Only one MNP-based nanosystem (DOX-loaded MCs) was studied in patients and did not demonstrate convincing results, thus, the study was terminated in phase II/III of clinical development [178]. According to the clinicaltrials.gov database (the website was accessed on 26 February 2022), there is one ongoing clinical trial intended to evaluate the safety, efficacy, and tolerability of intratumorally injected SPIONs accompanied by the spinning magnetic field application in combination with neoadjuvant chemotherapy in osteosarcoma patients (identifying number NCT04316091). However, in this study, MNPs are administered separately from the drug, and their systemic circulation is excluded from consideration. At the same time, the ability to safely and effectively deliver therapeutic moieties to the deep tissue using MNP-based carriers and external magnetic fields remains limited. The main challenges associated with this addressing approach include the design and selection of an appropriate MC considering specific clinical indications, as well as magnet design for deep and precise addressing through the tissue taking into account the anatomical barriers [152].

A ligand-receptor-mediated targeting strategy is based on the conjugation of the MNP-based delivery systems with such segments as antibodies, aptamers, endogenous proteins (like, transferrin), hyaluronic acid, folate, and targeting peptides [175,179]. The carrier functionalization with targeting ligands enables their recognition by specific receptors that are overexpressed on the surface of target cells, and thus provides the enhanced accumulation at the desired sites. However, when the modified particles (as well as the non-modified ones) are introduced into a complex biological environment of the circulatory system, the targeting often fails due to rapid clearance by macrophages of the reticuloendothelial system (RES) [180]. This protective mechanism is so potent that the majority of the injected nanocarriers are generally cleared after a few passes through the liver and spleen [181].

The process of MNPs blood clearance depends on the injection conditions (e.g., particle injection dose and repetition time of administration), nanoparticle properties (e.g., particle size, charge, and coating structure), and animal model (mice strain, absence/presence of tumors, tumor size) was comprehensively studied by Zelepukin et al. (Figure 5) [182]. To register the MCs circulation kinetics, they placed the animal's tail inside a coil generating a magnetic field on two frequencies and measured the magnetic response of the injected MCs at a combinatorial frequency detecting the particle concentration in tail veins and arteries. This research demonstrated that the injection of low MNP doses (less than 1 mg per mouse meaning 50 µg/g tissue) granted an almost constant half-life time (1–1.6 min), while further dosage enhancement substantially prolonged the particle circulation up to 45 ± 14 min for the 10 mg per mouse dose. This effect was attributed to the saturation of the RES in the organism, which could not eliminate such large nanoparticle doses. Multiple successive administrations of MNPs also resulted in prolonged circulation, starting from the second injection (each additional dose enhanced the effect). At the same time, a one-day pause between the administrations evened out the difference in particle circulation time compared to a single injection. The negatively charged particles demonstrated a slight prolongation of the clearance process. Comparison of the circulation time for 50-, 100- and 250-nm particles coated with glucuronic acid demonstrated a well-established trend: the smaller was the carrier size, the longer its circulation occurred. However, the time difference in clearance between the 50- and 100-nm particles substantially exceeded that between 250- and 100-nm. The effect was again explained by the limited number of receptors on the surface of macrophages that bound nanoparticles, as well as by the overall saturation of the endocytosis process.

Surface modification of the drug-delivering carriers with various coatings (PEG or other hydrophilic polymers) is commonly used to decrease their opsonization and prevent recognition by the immune system, thus inhibiting the non-specific distribution of these carriers [183]. For instance, it was shown that the PEGylated single-core MNPs, which are used in magnetic particle imaging (LS-008 tracer), persisted in the bloodstream of the rats with a half-life of 4.2 ± 0.5 h when injected at the dose of 5 mg of Fe/kg [184]. The PEG-silane coating of SPIONs provided further prolongation of the blood circulation half-life (up to 7 h) [185]. It should be noted that the coating material has a strong effect on blood kinetics depending on the ionic character of the coating material as it influences the particle uptake by macrophages. Obviously, the higher efficiency of particle capture by macrophages is observed, the faster those are cleared from the blood. It was well-demonstrated that phagocytic activity inhibited by decreasing the ionic strength of the coating material deposited on the SPIONs' surface: the starch, polyacrylic acid, and carboxydextran provided the higher uptake (thus shorter circulation in the blood) of SPIONs than PEG coating [186].

Despite the highly valuable advantage of prolonged blood circulation for stealth nanoparticles, the surface coating might bring some negative aspects, like the development of immune response enhancing opsonization for repeated injections (so-called, accelerated blood clearance) [187], inhibiting of the endosomal escape [188] and complication of the cellular penetration of the carriers resulting in a significant loss of delivery system activity [189]. Even though one can guarantee that targeting receptors are not disguised by the protein corona, the coatings can reduce targeting efficiency by hiding the receptor in its own bulky structure [181]. The alternative approaches towards the blood-circulation time enhancement have been proposed so far. Namely, cellular hitchhiking meaning the use of various cells as the particle carriers [190] or the RES blockade (saturation) induced before the MNPs injection [191,192] are exploited to delay the blood clearing and increase targeting capability. Although the suppression of mononuclear phagocyte function is generally achieved by injecting large doses of various nanoparticles [193,194], it can be blocked by the organism's own intact cells as well. In particular, Nikitin et al. elegantly demonstrated that the pre-induced macrophage saturation with antibody-sensitized erythrocytes (so-called, cytoblockade of mononuclear phagocyte system) increased the circulation half-life of nanoparticles by up to 32-fold [181]. Such cytoblockade was also shown to significantly enhance the therapeutic potential for magnetically-guided drug delivery systems.

The long-term fate of MNPs needs to be addressed when studying their in vivo behavior and effectiveness [156,195]. Comprehensive reviews on their pharmacokinetics, biodistribution, and toxicity have been previously published [196,197]. Briefly, upon the intravenous injection, the MNPs are mostly accumulated within the liver and spleen, which is logical as these organs are responsible for blood purification [156]. The behavior of MNPs in living organisms is significantly influenced by their surface functionalization and physicochemical properties, like the hydrodynamic size, polydispersity index, shape, aspect ratio, stability, and surface charge. In addition, the surface roughness, ζ-potential, and the size of nanoparticles also determine the thickness and composition of the plasma protein corona forming at their surface that, in turn, defines such pharmacokinetic parameters as plasma half-life, distribution, and elimination of the carriers [196,198]. Recently, Zhang et al. have proposed a magnetothermal regulation approach towards the control of protein corona formation on iron oxide nanocarriers in vivo to improve their pharmacokinetics and therapeutic efficacy [199]. They demonstrated that the relative levels of major opsonins and dysopsonins in the corona could be tuned quantitatively by means of heat induction mediated by MNPs under AMF.

Biodegradability of the MNPs is another important issue to be investigated when designing the MNP-based drug carriers aiming at further clinical translation [177], and safe biomedical application of nanoparticles requires a clear understanding of their metabolism [200].

A broad study of the one-year fate of 17 types of MNPs performed by Zelepukin et al. demonstrated a long-lasting effect of the injected dose, hydrodynamic size, ζ-potential, surface coating, and internal architecture of MNPs on their degradation half-life [195]. The authors introduced a magnetic spectral approach for non-invasive in vivo monitoring of MC degradation based on magnetic particle quantification (MPQ) technique (Figure 5). Faster particle degradation was observed for the MNPs of smaller size and negative ζ-potential. In addition, the type and structure of the surface coating were demonstrated again to play a crucial role in their biodegradability.

3.2. MNPs for Scaffold Enrichment and Release of Drug/Cells

MCs at the nano- and microscale are useful for drug delivery, as conceived as injectable formulations and cell-interactive materials. For instance, drug-loaded MNPs can be systemically administered via intravenous injection and then magnetically guided to a target site through locally applied magnetic fields. Once at the desired site, the drug liberation can be triggered by different mechanisms. Shen et al. used NIR irradiation to release the antitumor drug DOX and the calcium ion channel blocker verapamil from injectable MNPs. The nanosystems were systemically administered and then targeted to the tumor by exposing them for a few hours to a magnetic field (about 2000 Oe) laid outside the skin in close proximity to the tumor (Figure 6A) [201]. Larger MCs realized at the microscale are also useful in drug release application, and their advantageous potential relies on two aspects. First, magnetic microsystems retain higher magnetization than MNPs, resulting in more controllable during remote navigation. The responsiveness to stationary magnetic fields of the magnetic microsystems is in general expected to be large enough to counteract the hemodynamic forces during circulatory distribution. Second, microsystems support higher payloads of bioagents and can optimally interact with cell membranes, adhering and adapting to the surface morphology of cells (Figure 6B) [202].

In addition to injectable formulation, the MCs can be used in targeted delivery as embedded in matrices that act as static platforms to control drug release or accumulation [203]. The MCs are typically used to enrich the matrices that serve as scaffolds for tissue repair [132]. The resulting nanocomposites can be magnetized under the application of magnetic fields and have increasingly been applied to tissue regeneration, especially for hard tissue repair [2,204]. This interest is justified by multiple reasons. In general, enriching the scaffolds with nanoparticulate material positively affects cell adhesion, viability, and proliferation [205]. This aspect is often reported for IONPs but not unique to them, as many other kinds of nanomaterials can provide topographical cues that modulate cell behavior [206–209]. However, using MNPs makes it possible to have remote control over cellular processes [131]. The dispersed MNPs entrapped in the scaffold matrix can be stimulated by a static or alternated external magnetic field, and the effects that such stimulation has on the cells can be different [131,138,210,211]. Magnetic cell stimulation is typically mediated by the action of mechanical forces that apply to the cell membrane and activate specific molecular signaling pathways in a process known as mechano-transduction [131,212]. For remote cell mechanostimulation through externally applied magnetic fields and magnetic nano-transducers, magnetic fields of moderate intensities (ranging from 1mT to 1T) are sufficient to trigger specific cell responses, which can be used to guide the process of tissue repair [131]. Common magnetic materials for tissue regeneration include: nanoparticles, polymers, bioceramic materials, and alloys [213,214]. Magnetic actuation of the scaffolds can improve their performance in tissue repair, impacting positively on new tissue formation. This especially applies to bone tissue engineering [213], in which novel methods combining scaffolds and stem cells with MNPs and magnetic fields have shown the ability to enhance cell osteogenic differentiation, as well as bone regeneration and angiogenesis by 2–3 folds over the controls [213,215,216]. In fact, magnetic actuation activates essential signaling pathways in osteogenesis, such as MAPK, integrin, BMP, and NF-κB. Augmenting bone repair and regeneration efficacy, magnetic actuation holds clinical potential for dental, craniofacial, and orthopedic treatments.

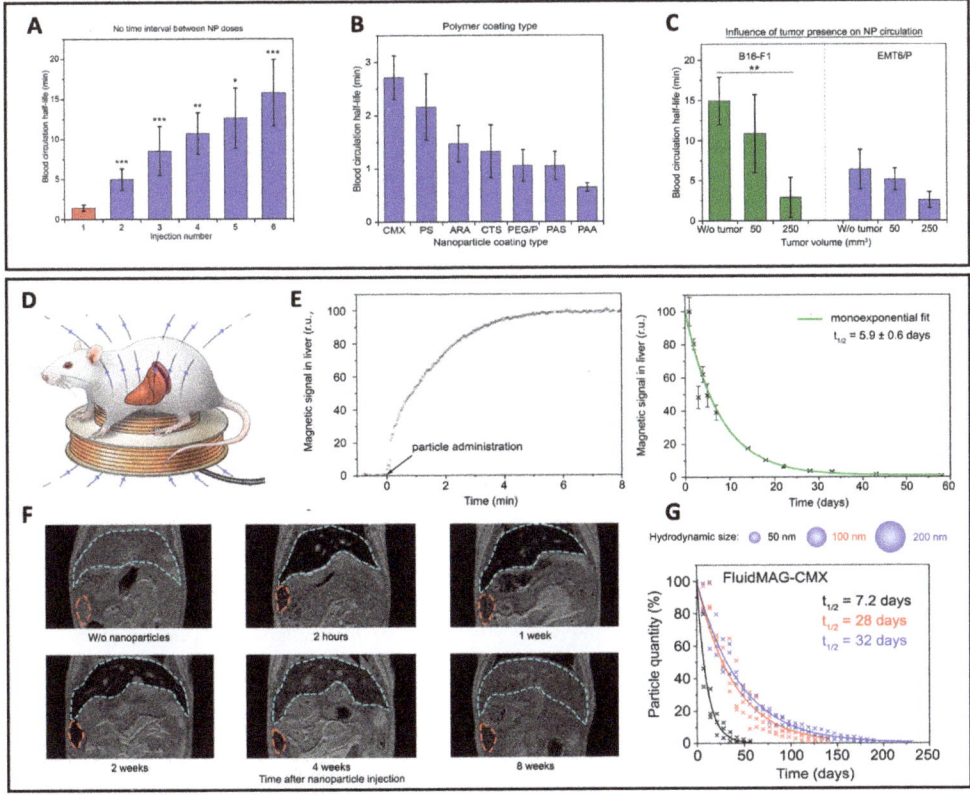

Figure 5. Circulation and distribution of MCs injected systemically. MPQ was applied to test

the blood circulation and the organ distribution of magnetic particles. (**A**) Influence of repetitive injections of 100-nm MCs on their circulation. Multiple injections of MCs (300 µg) were provided immediately after the clearance of the previous dose. (**B**) Dependence of the half-life time of 100-nm nanoparticles on polymer coating. The data are provided for a dose of 300 µg nanoparticles per mouse. (**C**) Influence of the presence and size of tumors (B16-F1 in C57Bl/6 and EMT6/P in BALB/c mice) on clearance of 50-nm MCs from the bloodstream. The data are provided for a dose of 300 µg nanoparticles per mouse. (**A–C**) The experiments were performed on BALB/c mice (blue and red bars) and on C57Bl/6 mice (green bars), n = 3 for each bar. Asterisks indicate significant difference from the control: Welch's t-test, *—$p < 0.05$; **—$p < 0.01$; ***—$p < 0.001$. Adapted with permission from Ref. [182]. (**D**) The degradation of magnetic particles is investigated with the MPQ approach noninvasively in a setup where the magnetic signal from the particles is measured by a magnetic coil placed under the mouse liver and spleen. (**E**) MPQ data of accumulation and degradation kinetics in the liver of 100 µg of citrate-stabilized MPs, and (**F**) representative MRI of the two-month evolution studies in the liver (cyan lines) and spleen (red). (**G**) Dependence of degradation kinetics on particle size for 50, 100, 200 nm MPs coated with carboxymethyldextran. (**D–G**). Adapted with permission from Ref. [195]. Copyright 2021 American Chemical Society.

Although the magnetic scaffolds are proven to trigger osteogenic differentiation and therefore have been mostly involved in bone tissue engineering, they have also shown utility for the repair of other tissue types [137,217–221], and in relation to specific uses other than the direct modulation of cellular processes (e.g., bioimaging, cell patterning, robotics, drug delivery) [215,222]. In particular, magnetic actuation of scaffolds can serve to localize and control drug delivery [223]. Porous and fibrous scaffolds have been widely used to deliver drugs, proteins, and living cells for both tissue regeneration and cancer therapies [224–228]. However, in conventional scaffolds the release of bioagents is typically governed by the intrinsic chemical and structural properties of the matrix, that determine the kinetics of crucial processes, such as the diffusion of dispersed molecules, the scaffold material degradation, and the cell migration. Conventional scaffolds, therefore, do not permit dynamic external regulation of the topical therapy [224]. As a consequence, different routes to release biofactors on-demand from implantable scaffolds have been investigated [223,229]. In this context, magnetically driven approaches are emerging due to the favorable properties of magnetism, such as the remote controllability of materials, the high tissue penetration depth of magnetic fields, and the versatility of magnetic actuation mechanisms [213,216,230]. As an example, scaffolds can be enriched with magnetic material and loaded with biofactors. Then, externally applied magnetic fields can modulate the release of the biofactors via different mechanisms, which depend on the method for magnetic stimulation, the properties of the scaffold matrix, and the strategy of drug incorporation. If MNPs are stably conjugated to the scaffold matrix, they can induce physical deformations of the structure, so that the included substances are extruded. By subjecting MNPs to AMF, it is possible to locally increase the temperature, thus altering the structure of the surrounding matrix and the drug release kinetics. In such systems, bioactive molecules can be directly loaded into the scaffolds and conjugated to the polymeric structure. Alternatively, drugs can be loaded onto the MNPs which are then incorporated into the scaffold. When a magnetic field is applied, the MNPs vibrate detaching the coupled drugs and accelerating their release.

Magnetic nanocomposites (especially in the form of hydrogels) are typically investigated as controllable drug delivery systems, although only a minority of them have been proposed for tissue regeneration applications. Magnetized scaffolds for tissue repair are typically loaded with factors that affect the three most crucial processes in implantology, namely tissue regeneration, graft vascularization, and fusion with the host tissue [224,231]. Growth factors are often applied to support these processes, as they play a pivotal role in directing regenerative pathways for many cell populations [223]. Magnetic nanocomposite scaffolds can be loaded with the growth factors that are necessary to stimulate cells over time, leading to complete biological and histomorphological tissue maturation. One of the first studies in this area was reported in 2011. Luo et al. loaded plasmid genes of the Vascular Endothelial Growth Factor (VEGF) onto chitosan/gelatin microspheres that were enriched with superparamagnetic magnetite crystals [206,208,224,232]. The magnetic microspheres were then incorporated into a scaffold for bone tissue repair. Oscillating and static magnetic fields induced local micro-movements of the microspheres, resulting in the release of the plasmids and transfection of the surrounding cells. The group demonstrated improved angiogenesis and osteogenesis within the scaffold, as well as enhanced scaffold vascularization and large bone defect repair. Magnetized scaffolds have been also proposed for wound dressing. In a recent study, itaconic acid was copolymerized onto starch and alginic acid in the presence of graphene sheets and magnetite nanoparticles [233]. The resulting magnetic composite hydrogel was loaded with guaifenesin to enhance wound healing. Guaifenesin could be released in a controlled manner by modulating the applied magnetic fields.

Exposing MNPs entrapped into the scaffolds' matrix to AMF causes the temperature to locally increase. The resulting thermal variations can affect the release kinetics of drugs loaded onto the scaffolds. For instance, when magnetite nanoparticles (NPs) were incorporated into PLGA scaffolds and then exposed to magnetic fields as per hyperthermia treatment protocol, heat was generated [234]. The heat varied proportionally to the concentration of MNPs and affected the release of minocycline, a drug for the treatment of glioblastoma. At temperatures close to the polymer's glass transition temperature, the polymer's chain becomes more mobile resulting in enhanced drug diffusion out of the scaffold. The same paradigm was adopted by Amini et al., who proposed magnetized nanofibers as a device to be implanted in the proximity of bone tumors for topic cancer therapy. In this system, magnetic bioactive glasses and the antitumor drug cisplatin were loaded onto chitosan-grafted PCL nanofibers [126]. These fibers were used to test the efficacy of the simultaneous application of chemotherapy and hyperthermia on MG-63 osteosarcoma cells. While no initial burst release of cisplatin occurred from the fibers, the release rate was accelerated under increased temperature (43 °C) as compared with the physiological condition.

Controlling drug release from scaffolds is crucial for tissue regeneration. In fact, many tissue engineering approaches aim to recapitulate the sequences of events that drive tissue development in nature, and that are regulated by growth factors. For example, the sequence and timing of delivery of growth factors play a pivotal role in bone regeneration [235]. Although various biomaterials that deliver multiple factors have been presented, there still exists a strong demand for strategies to control the sequential drug delivery and replicate the key steps of tissue development to improve regenerative outcomes. To address this issue, Madani et al. presented a bone tissue regenerative biomaterial that could rapidly recruit stem cells and expose them to the Bone Morphogenetic Protein 2 (BMP-2, a potent inducer of osteogenic differentiation) with programmed timing [236]. BMP-2 can inhibit cell proliferation, resulting in a poor cell population of the scaffold. Therefore, BMP-2 has to be provided only when the scaffold has intensely been invaded by cells and a robust cell population is established. To delay the BMP-2 release after implantation, the group proposed a biphasic biomaterial that consisted of an outer porous gelatin compartment and an inner ferrogel compartment. The outer layer served to harbor stem cells and was therefore loaded with a stem cell recruitment factor, the stromal cell-derived factor 1-α, (SDF-1α). The core of the system contained the BMP-2. By applying external magnetic fields (0.56 T), the BMP-2 could be released from immediate to a delay of up to 11 days after implantation.

In 2011, Zhao et al. described a porous magnetic scaffold that could macroscopically deform under the application of an external magnetic field, and they proposed that the volume variation could serve to extrude different bioagents (Figure 6C,D), including mitoxantrone, plasmid DNA, and a chemokine from the scaffold [203]. Moreover, the scaffold could serve as a depot of various cells, whose release can be magnetically controlled. More recently, hollow fibers composed of alginate and enriched with IONPs were prepared by coaxial 3D printing [130]. The fibers also encapsulated drugs, proteins, or living cells in the core, in which a lower alginate concentration was used. When an external magnetic field was applied, the resulting scaffolds deformed, causing the various biofactors to be extruded. A formulation consisting of 10 wt.% of alginate and 13% of MNPs displayed repeated on-demand release capability in vitro and in vivo under intermittent magnetic stimulation. Cells loaded onto magnetized scaffolds can also be stimulated by the magnetoactive components of the substrate (Figure 6E) [237], and be then released in the proximity of the scaffold. This perspective holds promise for future drug-free strategies to control cell therapy in vivo.

Even if intense efforts have been devoted to magnetic targeted drug delivery, fixing drugs after targeting is also extremely important in local therapies, such as tissue implantation. Research has therefore intensified on devices that guarantee the residence of delivered bioagents to the target site. A wearable drug fixation device was presented for cartilage repair via stem cell therapy [238]. An array of permanent magnets was mounted on a wearable band capable of wrapping the region of interest. The magnet configuration was determined through univariate search optimization and 3D simulation. Yielding a strong magnetic flux density (>40 mT) at the target site, the band could immobilize magnetic carriers at the cartilage defect sites in a rat model. In particular, the system's retention action was tested on a stem cell spheroid made from cells labeled with MNPs, and a micro-scaffold composed of PLGA functionalized with MNPs via amino bond formation. The next research steps in magnetic therapy fixation are expected to generate devices matching the criteria for clinical translatability.

Even in the absence of magnetic fields, doping the scaffolds with MNPs can alter their activity as a drug delivery system. For instance, the addition of MNPs reduced the degradation of silk fibroin scaffolds that have been proposed as multifunctional composite for tissue engineering [239]. In particular, the degradation rate of these sponges was strongly affected by the presence of the MNPs and correlated with the availability of active sites for proteases' binding. First, hydrogen bonds formed between fibroin and MNPs, which strengthened the matrix structure. In addition, the complexation of iron atoms and tyrosines decreased the activity of hydrolases. These two processes synergized to delay the degradation of the system, in which cell viability was preserved.

In summary, magnetized scaffolds can be used not only to control the release of pharmaceuticals via different mechanisms (such as thermo-responsive disassociation and physical extrusion via structural deformation), but they can also serve to modulate the behavior of loaded cells, and also to affect the kinetics of material degradation, thus prolonging the permanence of bioagents in the desired site.

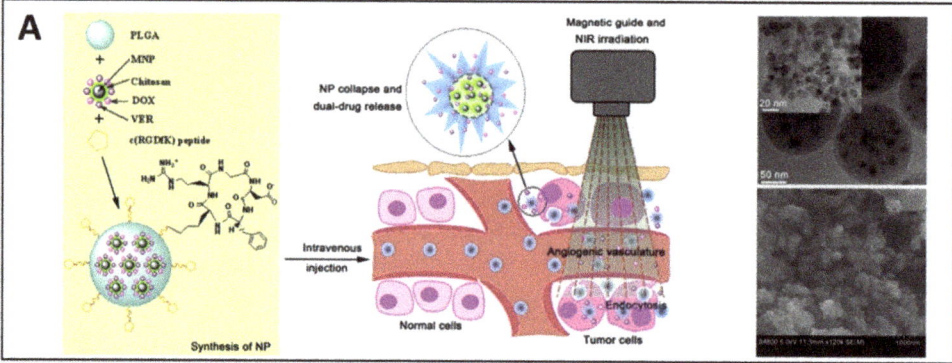

Figure 6. Cont.

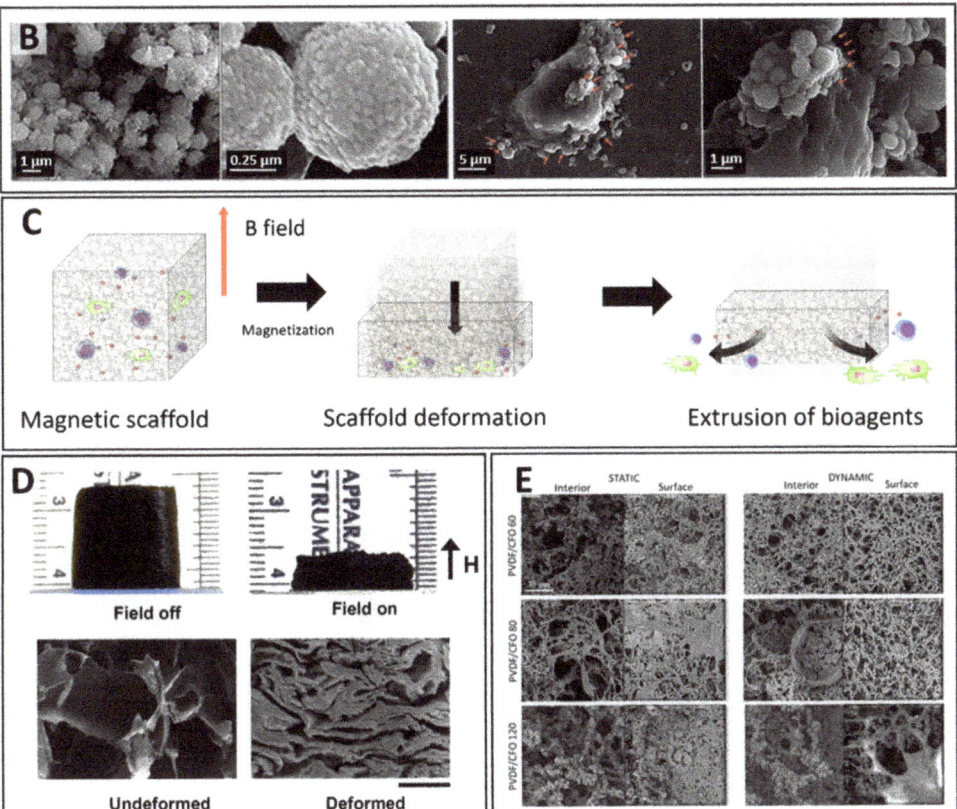

Figure 6. MCs for drug delivery as injectable agents or magnetizing agents for scaffolds. (**A**) Drug delivery via injectable MNPs: Shen et al. synthesized MNPs loaded with DOX and verapamil and coated with PLGA. Soon after the injection the mice were treated with a constant magnetic field (about 2000 Oe) laid outside the skin near the tumor site at 25 °C for 3 h. Drug release was then obtained by NIR irradiation. Right: TEM (up) and SEM (bottom) images of the developed MNPs. Reproduced with permission from Ref. [201]. (**B**) SEM images of iron oxide mesoporous microparticles (left) and cells incubated with particles (right). Microparticles (indicated by red arrows) associated with breast cancer MCF7 cells, both as attached to the periphery, and internalized inside. Adapted and Reproduced with permission from Ref. [202] (**C**) Drug release from magnetized scaffold in response to macroscopic structural deformation. (**D**) A cylinder of a macroporous ferrogel reduced its height ~70% when subjected to a vertical magnetic-field gradient of ~38 A/m^2, and SEM images of a freeze-dried sample in the undeformed and deformed states. (Scale bar: 500 μm.) Adapted and Reproduced with permission from Ref. [203] (**E**) SEM images of a magnetized scaffold made of a piezoelectric polymer, poly(vinylidene fluoride) (PVDF), and magnetostrictive particles of CoFe$_2$O$_4$. Preosteoblast cells were cultured for four days on scaffolds with different pore sizes. When magnetically stimulated, cells grew around the pores, probably guided by the magnetomechanical and magnetoelectrical stimuli. Adapted and Reproduced with permission from Ref. [240].

4. Magnetic Guidance

Magnetic materials and forces enable us to target different classes of therapeutic agents to the site of disease, feeding the interest in developing approaches that can overcome the inherent limitations in human use. One of the main challenges in the magnetic targeting of bioagents is that it limits targeting in deep tissue in human subjects, therefore making the translation into clinic arduous. Nevertheless, in recent years, novel approaches have emerged as a clinically viable solution for the site-specific delivery of small-molecule pharmaceuticals, biotherapeutics, and cells in patients. For example, the effects of deep-penetrating uniform magnetic fields can be exploited to accurately provide antirestenotic therapy to vascular targets and prevent injury-triggered re-obstruction of stented blood vessels [241]. The magnetization and guidance of cells can serve for different applications, including: magnetophoresis, cell-sorting, tissue engineering, and cell vectorization to target sites in vivo. Magnetic materials and forces can also control the motion of dynamic devices at the microscale, called micro-robots. In the following paragraphs, we will describe the use of magnetic forces to control the movement of cells and microrobots in different environments, highlighting their potential in biomedical applications.

4.1. Magnetic Guidance of Cells

4.1.1. Magnetic Cell Manipulation

Up to several million MNPs can concentrate within endosomes and lysosomes of labeled cells, rendering these cells responsive to inhomogeneous magnetic fields. MNPs can therefore mediate controlled and contactless movements of cells and subcellular structure. Cell magnetophoresis describes the motion of cells, which is caused by their loading with magnetic materials and their subsequent exposure to a nonuniform magnetic field created either by electromagnets or a permanent magnet [242]. In the last two decades, magnetophoresis of cells has been investigated for applications like cell separation, localized cell therapy, and tissue engineering. Within a unidirectional magnetic field gradient, the labeled cells experience a magnetic force, which can be defined as $M(B) \Delta B$, where $M(B)$ is the magnetic moment of the cell in the field B and ΔB is the magnetic field gradient. The magnetic moment of the cell can be approximated as the magnetic moment of one MNP multiplied by the number of MNPs per cell. The magnetic force that the cell experiences can drag it to a precise direction, but it is counteracted by several factors, like adhesivity of cells to substrates, the architecture of surroundings (e.g., porous or compact matrices), and viscosity of the medium. For cells suspended in liquids, the viscous force is defined as $6\pi\eta RV$, where R is the cell radius, V is the cell velocity and η is the viscosity of the medium. Magnetophoresis of cells within more complex 3D environments (like tissues, scaffolds, or hydrogels) is also strongly affected by the physicochemical features of the embedding matrix.

The magnetic fields generated by permanent magnets typically vary in the range of 10–50 T/m over a distance of approximately 1 cm. Labeled cells with an average iron load of 5 to 15 pg/cell therefore experience magnetic forces that range from 1 pN to a few nN. It is possible to deduce the iron load through magnetophoresis by determining the cell radius and velocity from video-microscopy. To label the cells with MCs in a quantifiable and reproducible way, the MNP formulation must be stable in the conditions selected for cell incubation. Importantly, magnetophoresis can detect potential nanoparticle aggregation and non-specific particle attachment to the cell membrane. If the cells are covered with particle aggregates, the outcome of the magnetophoretic assay will not lead to calculating the correct value of the intracellular iron load, as the obtained value will be higher due to extracellular aggregate pods. Magnetophoresis will instead reflect the cell velocity that is due to both internalized and membrane-attached nanoparticles. Therefore, as an iron dosage method, single-cell magnetophoresis allows one to identify potential artifacts that alter the quantification of iron, which is internalized by the cells [243]. Moreover, magnetophoretic cell sorting can serve multiple purposes, such as: (i) obtaining a fraction of labeled cells with a precise and homogenous iron load; (ii) separating magnetic cells from complex mixtures; or (iii) sorting them according to their endocytosis ability.

In 2009, Wilhelm et al. demonstrated the ability to control cell migration with external magnetic forces [244]. In this study, the magnetic field created by a thin magnetic tip, with forces as low as 30 pN, was sufficient to control the movements of aggregated and magnetized cells from the early developmental stage of *Dictyostelium discoideum*. The effects of magnetotaxis on cells in the presence of chemical gradients were studied in a competitive assay, where cAMP was slowly injected through a micropipette. Cells initially aggregated around the pipette, but when the injection was terminated, the cells were found to move towards the magnetic tip forming streams that were observable under the microscope.

Blümler et al. presented a system of permanent magnets in a cylindrical arrangement, which allows not only for steering the SPIONs on arbitrary trajectories but also for dragging labeled cells in suspensions along with predetermined directions (Figure 7A) [245]. Control over the movements of intracellular structures within individual cells was also demonstrated. In this work, the researchers combined a strong, homogeneous dipolar magnetic field with a constantly graded quadrupolar field, which was superimposed on the first. Whereas the homogeneous field aligned the nanoparticles' magnetic moments, the quadrupolar gradient field exerted a force on the particles, predominantly guiding them with constant force and in a single direction. In a setup with a coaxial arrangement of two Halbach cylinders, the force direction could be simply adjusted by varying the angle between quadrupole and dipole. Furthermore, the force strength can be additionally scaled by adding another quadrupole, even doubling that of a single quadrupole upon mutual field rotation. The authors showed that magnetized macrophages could form needle-shaped aggregates of cells containing intra- and extracellular SPIONs, and that this aggregation tendency increased while increasing SPION-loading per cell, and can be reversed by removing the magnetic fields. Long pole-ladder-like structures might be useful to model or engineering structures with unidirectional cellular growth, having potential for research in cellular barriers, capillaries, and neural communication pathways [246]. Cells labeled with SPIONs were placed in the magnetic arena and then guided on a square path by changing the direction of the quadrupole by 45° every minute. This caused the cells to move at twice that angle. The direction and velocity of cells were analyzed based on the recorded videos, proving that labeled cells have increased magnetically induced average velocity that depends not only on the SPION concentration but also on the cell type. Moreover, SPIONs inside the cells strictly followed the dipole's rotation, and the stirring effect could move the SPIONs-loaded endosomes inside the cells. Guryanov et al. used polyelectrolyte-stabilized IONPs to label neuronal progenitor cells and neurons differentiated from fibroblast-derived induced pluripotent stem cells. Then, they spatially guided these cells by using a permanent magnet, opening a new perspective on the future cell therapy for the treatment of neurodegenerative diseases [247]. In this regard, Marcus et al. published a thorough study on the optimal type and concentration of particles, as well as optimized incubation time, for magnetic manipulation of neuronal cells. Uncoated maghemite IONPs were efficiently internalized in neurons, without reported cytotoxicity. Furthermore, the MNP amount, which increased with the incubation time, reached a plateau after 24 h of exposure [248].

Magnetophoresis of cells can also be used to characterize and separate cells for biochemical analysis. For instance, this technique enabled the differential migration of red blood cells exposed to a high magnetic field, based on the content of deoxy and methemoglobins, which have paramagnetic properties as contrasted to the diamagnetic character of oxyhemoglobin [249]. Having a high hemoglobin concentration, human erythrocytes were ideal candidates to measure their migration velocity in cell tracking velocimetry. Under a magnetic field of 1.40 T and a gradient of 0.131 T/mm, erythrocytes with 100% deoxygenated hemoglobin or methemoglobins displayed similar magnetophoretic mobility of 3.86 and 3.66 $\times$ 10^{-6} mm^3 s/kg. In contrast, the magnetophoretic mobility of oxygenated erythrocytes (from -0.2 to 0.3×10^{-6} mm^3 s/kg) revealed a significant diamagnetic component relative to the suspension medium.

In 2009, mesenchymal stromal cells from human bone marrow were magnetically targeted to fragments of degenerated human cartilage in vitro. Briefly, cells were injected into tissue culture flasks, in which osteochondral fragments were attached to the sidewall, and a magnetic force was applied for 6 h to the direction of the cartilage, resulting in the formation of a cell layer inclusive of extracellular matrix on the degenerated cartilage, suggesting that the method has applicative potential in tissue regeneration [250].

Magnetic cell manipulation has also several other applications in tissue engineering. Engineered tissue grafts are typically generated from biocompatible porous or fibrous scaffolds that are seeded with cells. Local seeding of cells in 3D scaffolds can result in inhomogeneous cell dispersion over the construct volume, but magnetic field gradients can drag magnetized cells through the constructs to distribute them more homogeneously. In addition, magnetic attraction can serve to induce, through magnetic tweezers, a defined 3-D cellular patterning. Damien et al. labeled mesenchymal stem cells with anionic MNPs and applied magnetic micro-manipulation not only to enhance the cell seeding process into pullulan/dextran scaffolds, but also to confine the cells into specific positions to achieve a well-defined spatial cellular organization (Figure 7B) [251].

Alternatively, magnetized cells can be manipulated for scaffold-free 3D cell culture systems [252]. In this case, the process of cellular assembly in multicellular aggregates occurs via magnetic levitation of labeled cells within the culture environment. This biofabrication technique is based on the ability of the magnetic force acting on the cells to be equilibrated with the gravitational force. Optimal levitation of cells leads to the aggregation and condensation of cells, which then form complex 3D cellular structures (Figure 7C) [108]. Magnetic levitation enables the production of spheroids with higher densities and more symmetrical structures than those obtained from traditional techniques (like the centrifugation method) [108]. Moreover, by differently patterning the magnetic fields, it is possible to modulate the morphology of cell constructs, and also to improve the fusion among spheroids, opening perspectives in creating larger and more complex structures in tissue substitutes [253]. Magnetic cell levitation is typically performed by labeling the cells with MNPs, and scaffold-free tissue constructs with different morphologies have been generated in this way from endothelial cells [254], fibroblasts, preadipocytes [108], stem cells, muscle cells [255], epithelial cells [256], tumors [257], and others.

Contactless magnetic assembly of cell clusters has been primarily proposed to bypass the need for artificial scaffolds. Nevertheless, magnetic levitation can also apply to magnetized matrices. For example, Souza et al. levitated human glioblastoma cells within magnetized phage-based hydrogels consisting of gold nanoparticles, MNPs, and filamentous bacteriophage. By spatially controlling the magnetic field, the authors modulated the geometry of the mass of seeded cells and also obtained multicellular clustering of different cell types [258].

When submitted to remote magnetic forces, the cells suspended in liquids can move more freely than those that are adherent to substrates or those embedded in viscous or more solid matrices. Adherent and embedded cells cannot move if the magnetic forces do not overcome the adhesion constraints or if physical entities (like solid structures of the matrix) obstruct the movement. In such cases, the magnetic force acts on the organelles (i.e., endo-lysosomes) that have internalized the MCs. Wilhelm et al. demonstrated that magneto-manipulation of physically constrained cells allowed for the formation of vascular networks within an ECM-like matrix (Matrigel) [259]. Briefly, endothelial progenitor cells were labeled with uncoated anionic iron oxide nanoparticles, seeded on Matrigel, and magnetically manipulated at a distance by applying a magnetic field gradient that served to pilot different phases of tubulogenesis. In the first 14 h after seeding, cells migrated towards the magnetic tip to form dense tissue just around and below the tip (Figure 7D, top left). After 18 h, the magnet tip was applied, and local connections of vascular tubes started to form towards it. Lines of cells perpendicular to the direction of the magnetic tip were deformed due to the attraction force to the tip (Figure 7D, top right). In addition, the magnetic guidance of cell homing was demonstrated in vivo. Labeled cells were injected into matrigels implanted in subcutaneous pockets on the flanks of mice. A permanent magnet was placed on the opposite side of the injection site, resulting in a magnetic field gradient that guided the cells towards the magnetic pole, as shown by the elongated hypointense signal detected by MRI (Figure 7D, bottom).

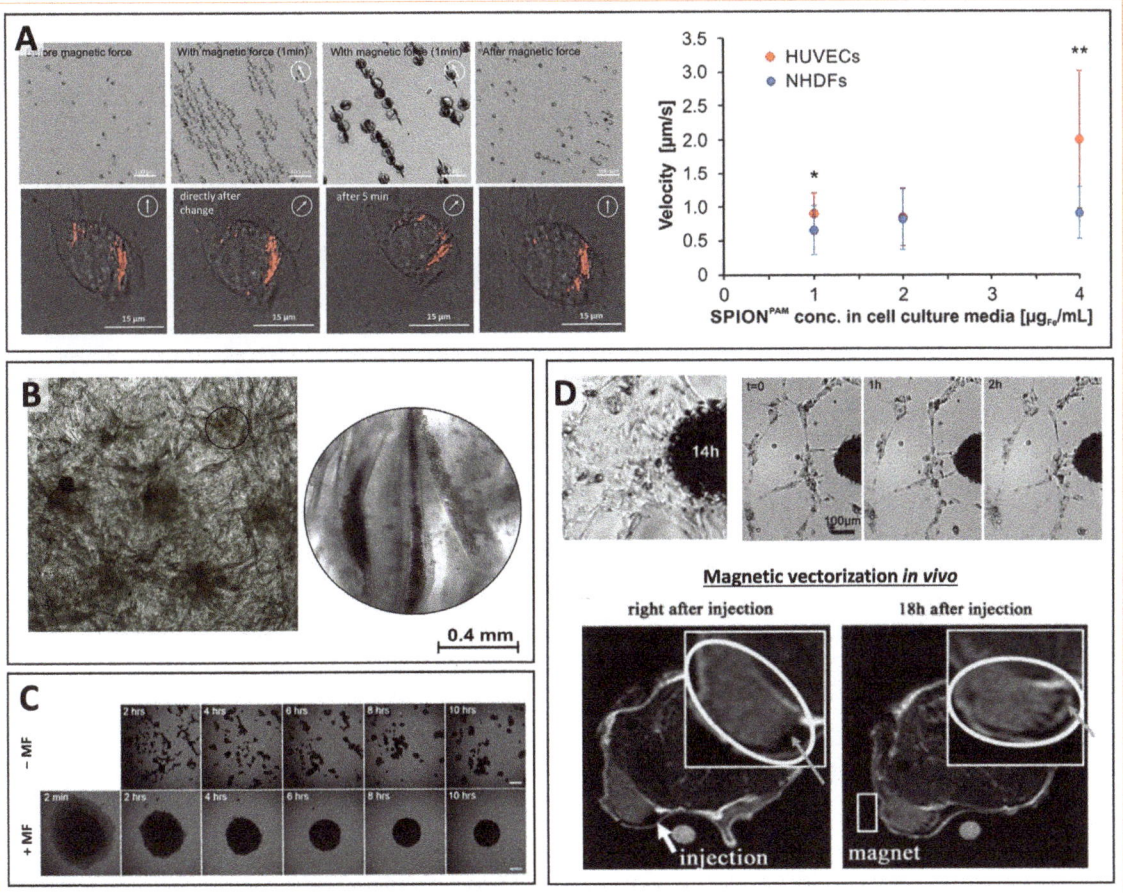

Figure 7. **Magnetic guidance of cells**. (**A**) Magnetic guidance for the formation of aligned cell clusters that are largely reversible upon removal of the magnet system. Magnetically induced velocities of SPION-labeled fibroblasts, endothelial cells, and cell clusters, guided on a square path by changing the direction of the quadrupole. Statistical significances between SPION-loaded HUVEC and NHDF cells are indicated with * and **. The respective confidential intervals are $p \leq 0.05$ and $p \leq 0.0005$, respectively, and were calculated via Student's *t*-test. Confocal laser microscopy of macrophages labeled with fluorescent SPIONs guiding the intracellular organization of SPION-loaded endosomes upon varying magnet orientations. Adapted and Reproduced with permission from Ref. [245]. (**B**) Photograph of a scaffold seeded with magnetic Mesenchymal Stem Cells (MSCs), under the influence of the magnetic fields created by the tips. Magnetic fields induced precise cell patterning creating hexagonal multicellular assemblies within the scaffold. Adapted and Reproduced with permission from Ref. [251]. (**C**) Formation of pre-adipocyte cell spheroids in the presence (+MF) and absence (-MF) of an applied magnetic field over time (scale bars: 300 μm). Adapted and Reproduced with permission from Ref. [108]. (**D**) Top: endothelial cells magnetized with MNPs form dense tissue around the magnetic tip 14 h after seeding onto Matrigel, and ex novo formation of tubular structures connecting to the magnetic tip in 2 h. Bottom: MRI of in vivo guidance of magnetized endothelial cells injected into Matrigel, implanted subcutaneously in mice. After 18 h of application of gradient magnetic field, negative enhancement is diluted on Matrigel submitted to the magnet, whereas controls still show a localized area of hypointense signal (not shown). Adapted and reproduced with permission from Ref. [259].

This experiment was an example of vectorizing magnetic cells in vivo following their topical injection within tissue or tissue substitutes. Nevertheless, in vivo cell targeting experiments are typically performed by releasing the magnetized cells into the systemic circulation, before applying a magnetic gradient that guides or retains them to the desired locations. In the following section, we will survey the use of MCs to direct cells to a specific target after their injection into the circulatory system.

4.1.2. Magnetic Targeting of Cells In Vivo

The successful outcome of cell-based therapies primarily depends on the effective cell delivery to the target site. As certain lesions are located in sites that are difficult to reach (such as the spinal cord and the joints), novel strategies to localize and maintain the position of curative cells are demanded. Magnetic vectorization of cells in vivo is an emergent approach that holds promise for cell therapy, regenerative medicine, and tumor treatment [260]. Briefly, cells with a therapeutic function are labeled with magnetic materials (like the MNPs) and injected into the circulation of living beings. Then, a static magnetic field is used to retain the cells within specific locations in the body. Magnetic targeting can promote more effective or focused cell delivery to areas that cells usually do not reach after systemic administration. Being compatible with use in biological systems and noninvasive control of cells in vivo, magnetic cell targeting is an attractive solution to improve the localization and the beneficial effects of stem cells used for tissue regeneration, as well as immune cells exploited for tumor treatment [261].

For magnetic cell targeting in vivo, cells must be loaded with MNPs following some general observations. The labeling procedure must ensure that cell viability and functions are preserved, but also that a sufficient amount of particles are internalized. Whereas MNPs with large size (>200 nm in diameter) and a neutral charge are readily internalized by macrophages, the particles with small size (<200 nm in diameter) and positive or negative charges can be efficiently taken up by nonphagocytic cells, like the stem cells [262]. Several reports of stem cell magnetic targeting have used MNPs with sizes ranging between 20 and 100 nm, which is considered as an optimal size range for uptake by stem cells. Nevertheless, it was also demonstrated that larger MNPs (with a diameter above 0.9 μm), if positively charged, can also be well incorporated by both cardiac-derived stem cells and endothelial cells, and then used for magnetic targeting in vivo [263–266].

Another fundamental requirement in magnetic cell targeting is that the MNPs are noncytotoxic at the needed concentrations. The toxicity of MNPs strongly depends on the physicochemical profile of the particles, which accounts for the type of coating, the tendency to aggregation, and the overall stability. Moreover, other factors that define the interaction of the MNPs with the biological systems contribute to the toxicity of the particle formulation (e.g., pathways of internalization and stability within the cells, as well as their interaction with membranes and cytoskeleton). It has to be noted that the suitability of MNPs for magnetic targeting is not directly related to their performance as imaging agents, namely as contrast agents for MRI or other imaging technologies. As an example, the commercial formulation "FluidMAG" performs better than Ferumoxides (Feridex in the USA, Endorem in Europe) as a cell-targeting vector, since FluidMAG is better taken up by cells and displays improved magnetic properties helping the spatial guidance [267,268].

Once the cells are labeled and injected into the living system, their localization to the desired region occurs in response to the interaction between their magnetic cargo, and the magnetic field which is applied. Most studies have used externally placed magnets, including permanent magnets [259,264,265,269–272], electromagnets [250,273], and even MRI systems [267]. Some in vivo studies of magnetic cell targeting are listed in Table 1 and categorized in relation to the target organ.

Table 1. In vivo studies of magnetic cell targeting categorized in relation to the target organ.

Target Organ	Injected/Transplanted Cells	Magnetizing Material	Recipient	Animal Model	Cell Delivery Modality	Magnetic Set Up	Magnet Position	Cell Accumulation (Increase Folds)	Time Point	Ref.
Heart	Cardiosphere-derived cells (rat)	Superparamagnetic microspheres (diameter: 0.9 μm, Bangs Laboratories)	Rat	Cardiac ischemia	Intramyocardial injections	Magnet (1.3 T) was applied during the injection and the following 10 min	External, above the apex	2–3	24 h	[265]
	Cord blood EPCs (human)	Anionic citrate-coated maghemite nanoparticles (diameter: 8 nm)	Rat		Cardiac intraventricular injection	Neodymium magnet (0.1 T, magnetic gradient 11 T/m at 4 mm from its surface) applied for 24 h	Subcutaneously implanted	10	24 h	[274]
	Cardiosphere-derived cells (rat)	Superparamagnetic microspheres	Rat	Myocardial infarction	Cardiac intraventricular injection	Circular NdFeB magnet (1.3 T)	External	4 3	24 h 3 weeks	[264]
	MSCs (rat)	Ferucarbotran/Resovist®	Rat		Intracoronary infusion	Permanent Neodymium-iron-boron (NdFeB) magnet cylinder	External	3.9	24 h	[275]
	MSCs (rat)	Ferucarbotran/Resovist® (diameter: 62 nm)	Rat		Transjugular injection into the left cardiac vein	Permanent NdFeB magnet cylinder (0.6 T; diameter: 8 mm)	External	2.7–2.9	24 h	[276]
	Cardiosphere-derived stem cells (human)	Ferumoxytol (Feraheme®)	Rat	Myocardial infarction	Intracoronary injection	A permanent NdFeB magnet (1.3 T) was applied during the injection and the following 10 min	External	3	24 h	[272]
	MSCs (pig)	Gadolinium nanotubes and Molday ION(−)®	Pig	Cardiomyoplasty	Transepicardial injection	Permanent a 1.3 T NdFeB ring magnet	Implanted (sutured onto the cardiac ventricle)	3	24 h	[277]
	MSCs (rat)	Ferucarbotran/Resovist® (diameter: 62 nm)	Rat		Cardiac intraventricular injection	Cylindrical NdFeB magnets (0.15 T, 0.3 T, and 0.6 T)	External	4	24 h	[278]

Table 1. *Cont.*

Target Organ	Injected/Transplanted Cells	Magnetizing Material	Recipient	Animal Model	Cell Delivery Modality	Magnetic Set Up	Magnet Position	Cell Accumulation (Increase Folds)	Time Point	Ref.
Vascular system	Endothelial progenitor cells (pig)	Superparamagnetic microspheres (diameter: 0.9 μm, Bangs Laboratories)	Pig		Infusion in the central lumen of an angioplastic balloon	Magnetized stent	Implanted	6–30	24 h	[263]
	Aortic endothelial cells (bovine)	Polylactide MNPs (diameter: 290 nm)	Rat		Transthoracical cardiac intraventricular injection	Magnetic field source of 500 G	Implanted (magnetized stent)	6	2 days	[279]
	Endothelial cells (human) (HUVECs)	MNPs-TransMAG, Chemicell	Mouse	Denuded carotid artery	Injection in external carotid artery	3 small NdFeB magnets (diameter: 1 mm; 10-mm length)	External	Improved (non-quantitative)	24 h	[280]
	Endothelial progenitor cells	SPIO Endorem	Rat		Infusion in the common carotid artery	Magnetic array	External	5.4	24 h	[281]
	MSCs (rabbit)	FluidMAG-D® (Chemicell)	Rabbit		Infusion in the central lumen of the over the wire angioplastic balloon	Permanent cylindrical magnet (Halbach cylinder)	External	6.2	24 h	[268]
Brain	Neural stem cells (human)	Ferumoxide	Rat	Ischemia	Intravenous injection (tail vein)	Neodymium magnet	External (attached to the skull for 7 days)	6	7 days	[282]
	Endothelial progenitor cells (mouse)	SPIONs (diameter: 50 nm)	Mouse		Intravenous injection (tail vein)	Two permanent FeNdB magnets (3 × 42 mm; 0.3 T)	Implanted (left hemisphere)	Non defined	24 h	[283]
	Embryonic stem cell (ESC)-derived spherical neural masses (human)	FIONs (60 nm)	Rat	Intracerebral hemorrhage	Intravenous injection (tail vein)	Neodymium magnet (5 × 10 × 2 mm³; 0.32 T)	External (helmet)	2	3 days	[284]
	Human nasal turbinate stem cells	SPIONs (diameter: 15–20 nm)	Mouse		Intranasal delivery	Permanent cylindrical magnet	External (attached to the restrainer)	Non quantitative	2 days	[285]

Table 1. Cont.

Target Organ	Injected/Transplanted Cells	Magnetizing Material	Recipient	Animal Model	Cell Delivery Modality	Magnetic Set Up	Magnet Position	Cell Accumulation (Increase Folds)	Time Point	Ref.
Knee joint (cartilage)	MSCs (rat, human)	Feridex (Tanabe Seiyaku)	Rabbit, pig	Osteochondral defect	Local transplantation (knee joint)	Permanent	External	Improved (non-quantitative)		[250,286]
	MSCs (rabbit)	Ferucarbotran/Resovist®	Rabbit		Local transplantation	Permanent	External			[287]
	MSCs (rat)	Poly-L-lysine-coated SPIONs	Rat	Osteochondral defect	Intrathecal transplantation	Permanent; Neodymium magnet (13 × 7 × 2 mm; 0.35 T; remnant magnetic field B_r = 1.2 T)	Implanted at 4–4.5 mm above the lesion site	3.7	1 week	[288]
Spinal cord	MSCs (rat)		Rat	Contusion-derived injury	Subarachnoidal injection	Permanent; Neodymium magnet	Implanted in paravertebral muscles, T7 level	30	24 h	[289,290]
	MSCs	SPIONs	Rat	Lesion	Intrathecal transplantation (L5–L6 level, 10 cm from the lesion)	Permanent; two cylindrical NdFeB magnets (1 cm × 5 cm; remanent magnetization: 1.2 T)	External	4	2 days	[291]
Lung	MSCs (human)	DMSA-coated maghemite nanoparticles	Mouse	Silicosis	Systemic inoculation	Permanent; Neodymium circular magnets (20 mm × 2 mm)	External, onto the thoracic region	1.5	2 days	[292]
Skeletal muscle	MSCs (human)	Ferucarbotran/Resovist®	Rat	Local damage	Local transplantation (between the fascia and scar tissue)	Electromagnet (magnetic field intensities from 0 to 5 T)	External	3–20	2 days	[293]
Bone	MSCs (rat)	Ferucarbotran/Resovist®	Rat	Non-healing fracture	Percutaneous injection	Permanent superconducting bulk magnet system (3T)	External	2–3	3 days	[271]

Table 1. Cont.

Target Organ	Injected/Transplanted Cells	Magnetizing Material	Recipient	Animal Model	Cell Delivery Modality	Magnetic Set Up	Magnet Position	Cell Accumulation (Increase Folds)	Time Point	Ref.
Retina	MSCs (rat)	FluidMAG-D® (Chemicell)	Rat	Retinal degeneration (S334ter-4 transgenic rats)	Intravitreal or tail vein injection	Gold-plated neodymium disc magnet	Implanted within the orbit, but outside the eye	37 (intravitreal injection) 10 (intravenous injection)	1 week	[294]
Tumor	Macrophages (human)	SPIONs (25 nm)	Mouse	Prostate tumor (in dorsolateral prostate) Lung tumor metastasis	Intravenous injection (tail vein)	MRI Scanner: small bore magnet (7 T, gradient of $660\,\text{mT}\,\text{m}^{-1}$)	External	Improved (non quantitative)	1 h	[295]
	CD8+ T cells (mouse)	Magnetic nanoparticles (80 nm)	Mouse	Lymphoma (from E.G7-OVA cell line), subcutaneous in the flank	Intravenous injection (tail vein)	Permanent; neodymium magnet (8×6 mm)	External magnet, applied over the tumor	Improved (non quantitative)	2 weeks	[296]
	CD8+ T cells (mouse)		Mouse	Lymphoma (from E.G7-OVA cell line) Breast cancer (from 4T1 cell line)	Intratumoral injection					[297]
	NK-92 cell (human)	CD56-conjugated magnetic nanoparticles (75 nm)	Mouse	Hepatoma (from H22 cell line) Melanoma (from B16 cell line)	Intravenous injection (tail vein)	Permanent magnet	External magnet, applied over the tumor (6h)	Undefined (non quantitative)	2 weeks	[298]

One typical application is to guide transplanted stem or progenitor cells to the injured tissues, where they are held by the magnetic fields [260]. Once in the lesion site, stem cells can restore the damaged cell population by differentiating into specific cell types or by acting as localized immune modulators [299,300]. In particular, stem cells can secrete vesicles and soluble factors with immunoregulatory or reparative effects. Kobayashi et al. investigated the opportunity to accumulate Mesenchymal Stem Cells (MSCs) in osteochondral defects of the patellae in rabbits and pigs after intra-articular injection [286]. In the rabbits, the magnetized cells localized to the osteochondral defect when exposed to an external magnetic field, as shown by macroscopic and histological observation. In the pigs, cells attached to the chondral defect were observed arthroscopically. By proving that labeled MSCs can be magnetically guided from a local injection site to the desired place in the knee joint, the authors proposed that the magnetic targeting of locally injected therapies is applicable to repair the human cartilage defects caused by osteoarthritis or trauma by means of a scarcely invasive technique [219,286,301].

Magnetic cell targeting has widely been investigated for applications within the cardiocirculatory system. For example, it was used for intracoronary delivery of cardiosphere-derived cells or endothelial progenitor cells in a rat model of acute ischemia/reperfusion [265,272]. In these studies, magnetized cells were guided through permanent magnets to the heart tissue. The magnetic setup used by Cheng et al. relied on an actual flux intensity of 0.3 Tesla that was generated by a 1.3 T magnet placed ~1 cm above the heart [264,265]. A four-times greater cell retention was observed in the hearts of animals receiving the magnetically targeted cell therapy. Chaudeurge et al. implanted a 0.1 T magnet into the subcutaneous layer, producing an actual working intensity that was far less than 0.1 T. Magnetic targeting at these conditions failed to augment the cell retention as quantified by Reverse Transcriptase-Polymerase Chain Reaction (RT-PCR) [274]. Another group showed that cell capture efficiency positively related to the magnetic flux density between 10 and 640 mT, reaching around 90% of cell retention with a magnetic flux density of 640 mT, a magnetic intensity gradient of 38.4 T/m, and a flow velocity of 0.8 mm/s [278].

Riegler et al. targeted magnetized MSCs in a rabbit model of vascular injury treated with interventional balloon angioplasty [268]. By using a clinically applicable permanent magnet, it was possible to increase cell localization up to six folds as compared to the controls. Augmented cell retention led to a reduction in restenosis three weeks after cell delivery [268]. Magnetic targeting of mesenchymal stromal cells (MSCs) in the lungs was recently investigated [302]. In this study, MSCs were made magnetically responsive by incubation with citrate functionalized magnetic nanoparticles. After being intravenously injected into murine models of experimental silicosis, the magnetized cells were retained in the lungs with the aid of a pair of circular magnets attached to the chest of each mouse through a surgical tape jacket. The higher fraction of retained cells was associated with enhanced beneficial effects for the treatment of silicosis in mice, holding promise for the cure of chronic lung disease. Magnetic cell targeting in the respiratory system has not been well explored, as most cells that are administered systemically might remain trapped in the narrow capillaries of the lungs. In this study, the technique was investigated in the perspective to rather prolong than improve the MSC retention on site. Nevertheless, the investigations were inconclusive in explaining how the cells are cleared from the capillaries, a process that remains incompletely understood.

Magnetic targeting can be followed by the imaging technologies that are typically used for the visualization of MCs in vivo, such as MRI. Intriguingly, clinical MRI scanners are able not only to track the location of magnetized cells but also to act on the process of the targeted vectorization. This integration between imaging and guidance is referred to as *magnetic resonance targeting*. In 2010, Riegler et al. applied magnetic field gradients inherent to the MRI system, to steer ferromagnetic particles to their target region [267]. This work proved the feasibility of imaging-integrated magnetic cell targeting in a vascular bifurcation phantom while varying the flow rates and testing different types of particles. In conclusion, 75% of labeled cells could be guided within the vascular bifurcation. Muthana et al. labeled macrophages with SPIONs and then apply pulsed magnetic field gradients to guide the magnetized cells to tumors and metastatic sites in vivo [295]. This way, the authors steered the circulating macrophages from the bloodstream to the tumor site and the lungs in orthotopic prostate xenograft murine models. The enhanced macrophage tumor infiltration was accompanied by a reduction in the tumor burden and metastasis to the lungs (Figure 8), being promising for future theranostic development of magnetic cell guidance.

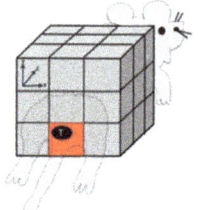 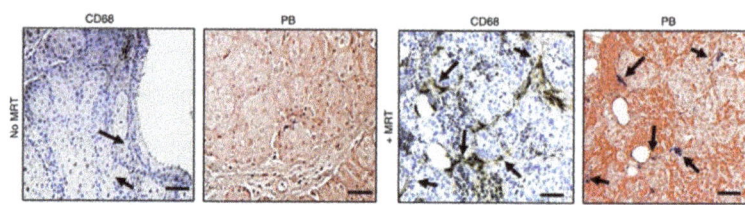

Figure 8. Magnetic Resonance Targeting technique. Magnetic resonance targeting (MRT) to vectorize circulating macrophages to prostate tumors and lung metastasis. Schematic of targeted regions using imaging gradients for MRT, where a–y gradient was equally applied across the animal to target the location of the prostate as depicted (red box). Histological staining of paraffin wax-embedded tumor sections with an anti-human CD68 antibody and Prussian blue (CD68-positive macrophages are brown and SPIO-positive macrophages are blue: see arrows). Adapted and Reproduced with permission from Ref. [295].

Other examples of immune cells that were magnetically targeted against the tumors include cytotoxic lymphocytes and natural killer cells (Table 1). These reports demonstrated that systemically administered immune cells can be retained at the tumor and metastasis sites upon magnetic attraction.

Although magnetic cell retention is very promising for cell therapy, to date it is still uncertain if magnetic forces can efficiently allow the cells to reach the target site in many applications. In fact, for successful delivery, magnetic forces then need to overcome the force that opposes the movement of cells under guidance. When cells are systemically injected, the magnetic drag has to overcome the force of the bloodstream flow. When cells are locally transplanted into the tissue, the magnetic drag must exceed the physical hindrance of the matrix surrounding the cells. Moreover, after engrafting to the target site, the cells might not persist in that location after removing the magnetic forces. Therefore, magnetically guided systems endowed with intrinsic mobility are promising to develop effective magnetic guidance of cell therapy, as well as drug release. The next section describes dynamic devices under magnetic guidance that have the potential to improve the efficacy of magnetically targeted therapy.

4.2. Magnetic Guidance of Microrobots

The MCs can be used to confer controllability to microscaled materials that are endowed with intrinsic motion abilities or designs that improve motion dynamics. Such devices referred to as microrobots, are typically involved in biomedical applications, due to their size, which renders them effective to interact with cells and tissues, and move through anatomical structures. In microrobotics, magnetic actuation provides the intrinsic advantage of using magnetic fields, which are safe for biological tissues and can penetrate them. Moreover, as microsystems have limited space for onboard energy storage, magnetic fields are also advantageous as they wirelessly transfer actuation power to the microrobots.

Microrobots based on MCs can be classified into two major classes based on their composition: purely synthetic microrobots and bio-hybrid robots. Purely synthetic microrobots are made of synthetic materials only, and their actuation results from a combination of applied magnetic forces and propellant designs. Bio-hybrid robots are robots that contain a biological living component [303,304]. In bio-hybrid robots, this component is represented by cells endowed with self-propulsion ability. In the next paragraphs, we will describe these two main categories of magnetic microrobots.

4.2.1. Synthetic Microrobots

Synthetic microrobots use the MCs as a platform to release a therapeutic cargo to a specific destination. Magnetic microrobots can be guided to the target sites through contactless magnetic control, and deliver bioagents like cells or drugs. The microrobot body can be made of hard or soft magnetic materials [305]. Microparticles made of NdFeB, SPIONs, CrO_2 powder, and FePt nanoparticles are the most typically used materials, but also other magnetic micro/nanoscale entities (such as molecules, discs, or wires) can be embedded inside the robot, absorbed, or sputtered onto the robot to form a nanofilm coating [305].

Using microrobots for biomedical applications requires locomotion in fluids. Microrobots that can controllably move in liquid environments have been designed and developed by mimicking the natural locomotion of the swimming microorganisms [306]. These propulsion strategies have been of great inspiration to realize specific types of locomotion under magnetic control [306]. As an example, micropropellers with a helical design are inspired by bacterial flagella (Figure 9A) [307,308]. To drive helical microrobots, external rotating magnetic fields are typically used in order to avoid the need for an on–board rotary micromotor. In addition, the steering of these robots is achieved by varying the axis of rotation of the magnetic field, whereas the overall speed mostly depends on their helical shape [309]. Another bioinspired design for magnetic microrobots resembles the anatomy of sperm cells and other unicellular eukaryotes swimming through flexible flagella that are at least ten times larger and longer than bacterial flagella [310]. Such flexible filaments can propagate bending waves that provide propulsion to the whole cell body. Magnetic microrobots with flexible tails have been proposed, which move similarly to eukaryotic flagellated propellers [311–314]. For example, Pak et al. demonstrated a microrobot with a magnetic head and a flexible nanowire tail that can move in liquids as actuated by a magnetic rotating field [315]. Moreover, flexible structures with distributed magnetization can be used to achieve the propagation of bending waves leading to the undulation of artificial structures under magnetic actuation. Undulatory motion enabling locomotion in fluids can also be realized in the form of oscillating sheet-like structures [316,317]. One example is represented by milliscaled magnetoelastic robots [316,318]. By applying an external rotating magnetic field, a sheet with a periodic magnetization profile can be actuated in a way that propagating bending waves run along the axis of the magnetization gradient [319]. Magnetic flexible composites can be produced at the microscale as well if SPIONs are loaded onto the elastomeric structure [317].

MCs-based microrobots can be loaded with molecules, drugs, and cells. When reaching the target site, the delivery of the bioagents can occur via a passive or an active mechanism. In fact, magnetic microrobots can spontaneously release their cargo in response to a specific environmental condition. The most typical example is the pH variation, which is a hallmark of inflamed tissue or tumors. Chitosan hydrogels for instance have been proposed for conditional localized drug delivery upon acidic conditions, as they undergo physical deformations that can induce the release of loaded molecules [320]. Fusco et al. covered their magnetic microrobots with a chitosan gel layer and demonstrated the sustained and efficient release of a model drug [321,322]. The delivery of the payload from magnetic microrobots can also be controlled and triggered in response to an external trigger or an environmental cue [306]. For instance, thermoresponsive hydrogels containing MNPs can release molecules on demand from microrobots via heating by magnetic induction [323].

Magnetic microrobots can also be used to manipulate cells or carry them to a specific destination (Figure 9B) [324,325]. Breger et al. reported Poly(N-isopropylacrylamide) (PNIPAM)-based thermoresponsive microrobots with a gripper design [326]. They embedded iron oxide NPs into the porous hydrogel layer used to fabricate the microrobots, and showed that they could command them remotely through magnetic fields to actuate some conformational change. Through these movements, their microrobots could excise and capture cells from a live cell fibroblast clump. Microrobots are increasingly studied for their potential in the delivery of cells for targeted cell therapy. Li et al. proposed a magnetic microrobot for the delivery of stem cells [324]. They fabricated magnetic skeletons with burr-like porous spherical structures to improve the magnetic driving capability and maximize the cell-carrying capacity. Jeon et al. presented a helix-shaped microrobot where hippocampal neural stem cells could attach, proliferate, and differentiate into various mature neural phenotypes [327]. These microrobots were also loaded with MSCs derived from the human nose and then manipulated in vivo, inside the intraperitoneal cavity of a nude mouse. Go et al. used magnetic microrobots to control the localization and delivery of human adipose-derived MSCs for knee cartilage regeneration in a rabbit model [328]. The robot's scaffold consisted of a spherical PLGA microstructure decorated with MNPs (Ferumoxytol), and the stem cells could grow over its surface. Following injection into the wounded-knee of the animal, the microrobots were actuated through an oscillating magnetic field. The robots accumulated in the target injured site within the knee, and then were retained in place through a permanent magnet. This way, the cells were facilitated in adhering to the surrounding tissue and proliferating there. Magnetically propelled spherical microrobots with burr-like designs have already been tested for transporting cells in different animal models [324]. For instance, MC3T3-E1 cells have been transported within the yolk of zebrafish embryos. Although the successful use of these cell-carrying machines in vivo was recently demonstrated [285,324], controlling them while they move noninvasively through the body to a target site is still an open challenge.

4.2.2. Bio-Hybrid Microrobots

Robots with a biological living component that actively contributes to the movement of the whole assembly are referred to as bio-hybrid robots [304]. Bio-hybrid microrobots use natural biological actuation for propulsion. They integrate motile cells (such as bacteria, algae, sperm cells, or leukocytes), and are typically controlled via magnetic actuation [329]. Park et al. proved that multifunctional bacteria-driven MCs enable enhanced drug transfer as compared to passive microparticles (Figure 9C) [330]. The authors prepared stochastic microswimmers by loading *Escherichia coli* (*E. coli*) bacteria with polyelectrolyte-based DOX-loaded microparticles embedded with MNPs. The microrobots moved at average speeds of up to 22.5 µm/s, and their dynamic control was demonstrated by applying a chemoattractant gradient and a magnetic field to induce biased and directional motion, respectively. In another work, bio-hybrid microrobots targeting infectious biofilms were generated by combining the magnetotactic bacteria *Magnetosopirrillum gryphiswalense* with mesoporous silica microtubes loaded with antibiotics [331]. Through magnetic guidance, the particles were delivered to the matured E. coli biofilm, where they released the drug and caused the biofilm disruption.

Microrobots based on sperm cells have been intensely investigated for their multifunctionality, actuation efficiency, and their controllability (Figure 9D) [311,312,332]. A magnetic bio-hybrid microrobot based on flagellate propulsion for drug delivery was developed by Xu et al. [333]. They created a sperm-hybrid micromotor consisting of a motile sperm cell that serves both as a propulsion source and a drug carrier. The synthetic component of the bio-hybrid was a 3D-printed magnetic tetrapod, namely a tubular microstructure made of a polymer and coated with 10 nm Fe and 2 nm Ti. The tetrapod could release the sperm cell in situ, as it consisted of four arms that could deform upon mechanical input. The arms could bend externally and open the tetrapod to liberate the sperm when the robot hit the tumor wall. This microrobotic platform was not only endowed with high drug loading capacity and self-propulsion, but it could also be magnetically guided to the target site and release its cargo upon mechanical trigger while exploiting the sperm penetration ability.

The cellular component of the bio-hybrid microrobots contributes to their biodegradability and structural compliance, which helps for safe interaction with biological structures and circulation within living environments. Moreover, the small size and swimming abilities of the integrated microorganisms and motile cells turn helpful to efficiently navigate within biological fluids and/or through intricate anatomical structures (e.g., vessel networks), which are a prerogative for most biomedical applications, like targeted delivery and cell/tissue manipulation. These abilities combine with the capability of self-repair, flexibility, and even multiple degrees of freedom [334]. Thus, bio-hybrid robots at the microscale have a great potential for biological applications.

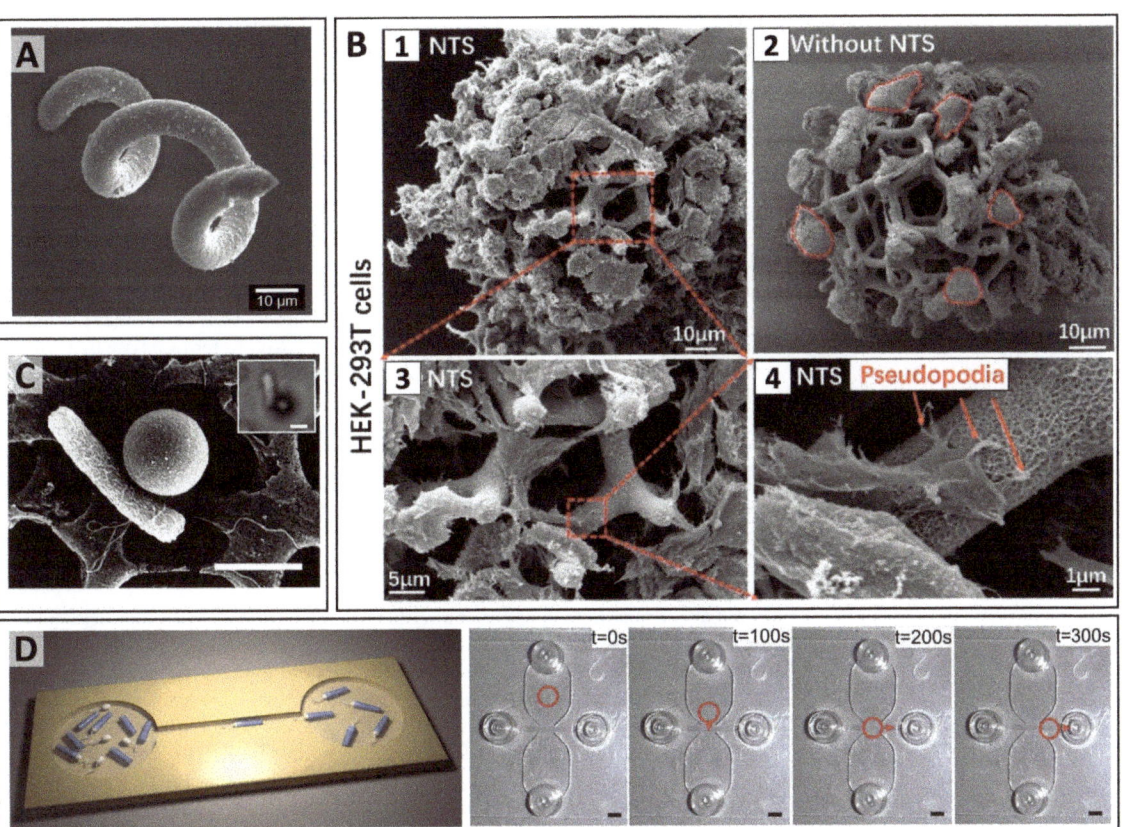

Figure 9. **Magnetic microrobots.** (**A**) SEM image of Spirulina-templated magnetic helical microswimmers. Adapted and Reproduced with permission from Ref. [308]; (**B**) Cell-carrying magnetic microbots with bioactive nanostructured titanate surface for enhanced cell adhesion. SEM image of HEK-293T cells on microrobot after 1 day of incubation with (B1) and without (B2) nanostructured titanate surface; (B3) high-magnification SEM image of HEK-293T cells on microrobot with NTS; (B4) High-magnification SEM image of single HEK-293T cell with protrusive pseudopodia on microrobot with NTS. Adapted and Reproduced with permission from Ref. [325]; (**C**) Bacteria-driven microswimmers for targeted active drug delivery. Adapted and Reproduced with permission from Ref. [330]; (**D**) Magnetic guidance of microrobots based on sperm cells from a chamber containing a mixture of microtubes and sperm cells (left chamber) into a clean chamber (right) of an ibidi-chip. Scale bar: 3 mm. Adapted and reproduced with permission from Ref. [332].

5. Imaging of Magnetic Micro/Nano-Systems

MCs can be visualized via various imaging techniques (Figure 10). This makes them extremely attractive as delivery systems, as they enable one to monitor the delivery of bioagents in real-time. In Section 5.1, the main imaging techniques allowing for the visualization of MCs will be presented, describing the basic principles and enabling technologies, as well as the properties that are needed in the MCs to be detected via a specific technique. In Section 5.2, we will discuss the opportunity to combine mechanisms of imaging and therapy within one MC and realize multifunctional platforms.

5.1. Imaging Techniques to Visualize MCs

MRI is by far the most popular imaging modality to track magnetic systems. MRI uses powerful magnets and radiofrequency pulses to visualize soft tissues within the body. In principle, during an MR scan, a living subject is exposed to an intense magnetic field (1.5 or 3 T for clinical MRI; and 1, 3, 7, 9.4, or 11.7 T for preclinical MRI) that induces the alignment of magnetically active atomic nuclei in the matter [335,336]. In biological tissues, these nuclei are mostly protons (namely hydrogen nuclei) of fat and water molecules. Nuclei can be aligned parallelly or anti-parallelly to the magnetic field direction. The net difference in the number of nuclei aligning in the opposite directions is what generates the MR signal [337]. A pulsed radiofrequency can be then used to temporarily disturb the alignment distribution of the nuclei, which subsequently returns to the original state in a process known as relaxation. Relaxation occurs through two concomitant processes, termed longitudinal and transverse relaxation and characterized by specific kinetics. In these two processes, the time required to restore the equilibrium state is referred to as relaxation time, with T_1 and T_2 indicating the longitudinal and transverse relaxation times, respectively. The relaxation rates in the two processes depend on the physico-chemical environment surrounding the nuclei [338].

In MRI, the relaxation times are therefore assessed to gain information about the investigated matter. To increase the sensitivity of the technique, the relaxation processes can be altered by administering contrast agents, namely materials with the capability to vary the local magnetism experienced by the tissue nuclei. In particular, these compounds work by augmenting the longitudinal or transverse relaxation rates (R_1 and R_2) [339,340]. For instance, the MCs based on iron oxide predominately shorten the transverse relaxation time, therefore acting as T_2 contrast agents [42,341]. Iron oxide-based MCs include: SPIOs (superparamagnetic iron oxide nanoparticles); USPIOs (ultrasmall superparamagnetic iron oxide nanoparticles); and MPIOs (microparticles of iron oxide). Although MCs are typically used to accelerate the transverse relaxation, some IONPs can also serve as T_1 contrast agents, primarily reducing the longitudinal relaxation time [61,342–344]. The most commonly used MCs in the clinics are the USPIOs, which have typical diameters ranging from 20 to 50 nm and are characterized by a long circulation half-life. In contrast, as compared to USPIOs, MPIOs display a larger size and contrast-to-noise value per particle, and higher ligand valency, but also a shorter half-life (<5 min). As such, MPIOs are extremely suitable for cell labeling procedures (including in vivo labeling) but less performant in applications that require direct systemic administration of the particles [345]. In general, MCs for MRI are composed of a metallic core surrounded by a biocompatible coating, which varies in composition, total net charge, and surface functionalization. In micro/nanocomposites, the magnetic NPs can be loaded not only in the system's core but also inside the shell. Detailed reviews about iron oxide-based MR contrast agents are available in the literature [346–348].

The main advantages of MRI are the absence of ionizing radiations, the high spatial resolution (preclinical: micrometers, clinical: ~1 mm), and unlimited tissue penetration [349,350]. The major limitation is a low sensitivity, which derives from the fact that the difference between the nuclei populations that, under the application of a magnetic field distribute in the two possible alignment states, is very small. The low sensitivity then makes it necessary to provide highly concentrated contrast agents (10^{-3} to 10^{-5} mol/L) [335,336]. In view of visualizing the bioagents delivery, an important consideration is that the MRI signal is not directly associated with the contrast agent itself, but it represents changes in environmental magnetic properties induced by the contrast agents. As such, this signal lacks information in terms of contrast agent localization at a high spatial resolution.

A more recent and likewise appealing technique is Magnetic Particle Imaging (MPI). MPI was firstly proposed in the early 2000s to track magnetic systems with an extremely high spatio-temporal resolution and without any background signal [351]. In MPI, a magnetic system (like a MC) is administered to a living subject exposed to a magnetic field similar to that used for MRI. However, in contrast to MRI, the magnetic systems themselves are detected, rather than the effects that they induce in their local environment [352]. Changing magnetic fields are applied to produce a single magnetic-field free region (the field-free point, FFP) in the imaged sample, and, if any magnetic material is located at the position of FFP, it will generate a signal. By steering this region across the sample, an image of the magnetic tracer is generated. As only information about the magnetic tracer is provided, anatomical images have to be co-acquired with a structural imaging technique (e.g., Computed Tomography (CT) or MRI) to precisely localize the MPI signal [353]. Nevertheless, MPI offers several advantages including high spatial resolution, fast acquisition times, unlimited tissue penetration depth, and high sensitivity [354]. As such, MPI is likely to become crucial for imaging MCs in the near future. MPI contrast agents are typically IONPs with size-dependent magnetic relaxation properties and designs that are tuned to maximize signal generation [355]. Available SPIOs are grossly inadequate for MPI and do not translate well in vivo. In fact, ideal MPI contrast agents should be extremely uniform in size (monodispersity of the core and hydrodynamic sizes of NPs), as for a given excitation field frequency there is a single particle diameter associated with a resonant, or perfectly in-phase, magnetization response for maximum signal intensity. For example, a diameter of ~25 nm represents the optimum core diameter for a 25 kHz excitation frequency [356]. However, nanosystems were not the only systems to be developed for MPI. In fact, certain magnetic microcarriers have also found applications as MPI agents [357]. Zahn et al. designed magnetic microspheres with a size that could be tuned between 1 and 2 µm, which displayed MNPs onto the surface [358]. Co-loading a drug onto such microsystems opens perspectives in MPI-monitored targeted drug delivery.

Another emerging technology that uses MCs as contrast agents is magnetomotive ultrasounds (MMUs) [359]. MMUs use the vibration induced in the adjacent tissues by SPIONs, or other magnetic systems, subjected to an external magnetic field. MCs can be systemically administered to patients or living beings; then electromagnets, permanent magnets, or a combination of the two, are applied on the surface of the body. The MCs are displaced within the tissue through their immediate surrounding. When the magnetic field is removed, the MCs tend to return to their initial position [360]. This movement is then detected with ultrasound imaging by using a frequency- or time-domain analysis of echo data. MMUs can also be combined with photoacoustic (PA) imaging, a biomedical imaging technique that relies on the generation of sound waves as a result of light absorption [361]. These sound waves are then detected by ultrasound transducers and analyzed to produce images. PA and MMUs can be combined by merging IONPs with nanorods in larger MCs, with the advantage of PA background noise suppression [361].

Similar to MMUs, magnetomotive optical coherence tomography (MMOCT) exploits the movement of a magnetic contrast agent induced by an external magnetic field [362]. The resulting motion is then tracked by using low-coherence NIR light and interferometry. In comparison to MMUs, the penetration depth of MMOCT is limited [360].

Molecular imaging techniques using ionizing radiations, such as Positron Emission Tomography (PET) and Single-Photon Emission Computed Tomography (SPECT), can also be employed to track the delivery of MCs and the release of bioagents, but in such a case one needs to add a radioactive isotope to the MC [363]. The hybrid systems can be employed for multimodal imaging, with a wide range of applications.

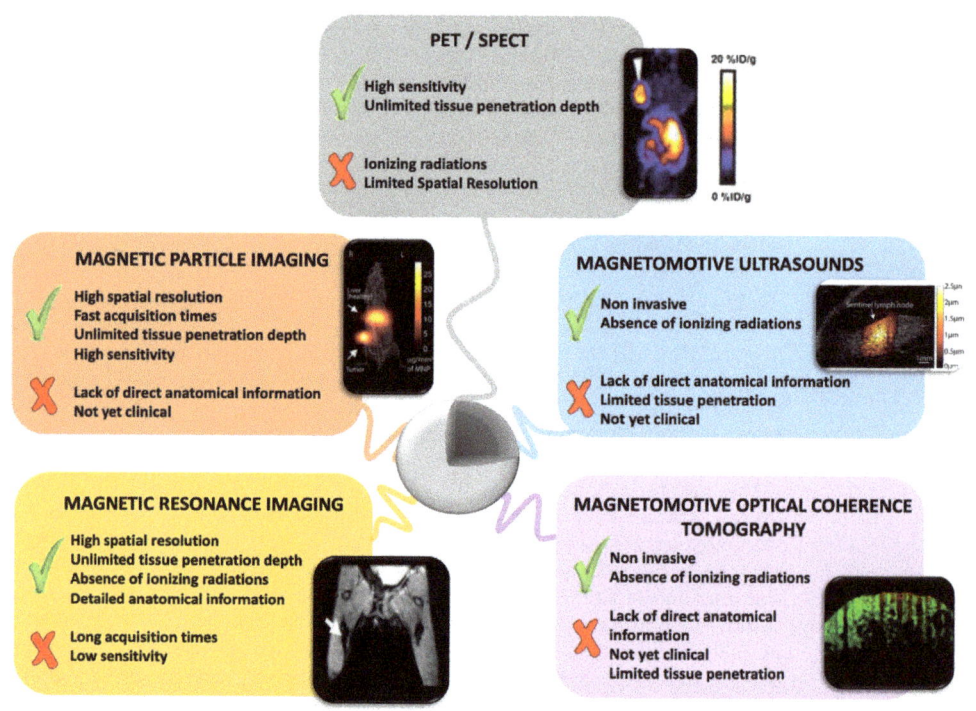

Figure 10. **Advantages and limitations of the imaging techniques for tracking of MCs.** MCs can be tracked by various imaging techniques (PET/SPECT; magnetic particle imaging, magnetomotive ultrasounds, magnetic resonance imaging, and magnetomotive optical coherence tomography). The selection of a precise technique is based on the assessment of the associated advantages and disadvantages. Adapted with permission from Refs. [364–368].

5.2. Theranostic Applications of MCs

Since its first outbreak at the end of the 1990s, the word "theranostics" has become increasingly popular among scientists. This term derives from the combination of the words "therapy" and "diagnostics" [369,370], and refers to the use of systems that deliver an imaging and a therapeutic agent at the same time [371]. Theranostic systems can be exploited in a variety of fields [372], including: drug discovery [373]; real-time assessment of drug pharmacokinetics [374]; therapy screening/selection [375]; imaging drug delivery [376]; and imaging triggered drug release [377]. For these purposes, the contrast agents should be designed to combine capabilities of diagnosis and treatment in one platform. MCs are ideal candidates for such an application, as they can be used in MRI, MPI, magnetic targeting, hyperthermia, cell labeling, and controlled drug release. The following paragraphs briefly survey the use of MCs in different fields of theranostics.

5.2.1. Magnetic Fluid Hyperthermia

Magnetic fluid hyperthermia (MFH) is a technique in which a biocompatible magnetic fluid suspension is locally or systemically administered and then heated up, above 37 °C. The temperature increase (to about 41–46 °C) causes cell apoptosis, and/or enhances the susceptibility of the target tissue to other therapies such as radiation and chemotherapy [378–380]. Heating can be induced through high intensity focused ultrasounds (HIFU), radiofrequency heating, microwave radiation, and laser photocoagulation [381], with high tissue penetration and specificity, and minimal heating of the surrounding tissues. The heating ability of the magnetic nanoparticles depends on their specific absorption rate (SAR), a property that is a function of particle size, shape, coating, surface charge, aggregation state, interaction with proteins corona, and even the dispersion medium [382]. The magnetic properties of the suspensions used for MFH allow for imaging during hyperthermia by both MRI and MPI. This enables the selective heating of pathological tissue, without affecting the organs where MCs tend to non-specifically accumulate (i.e., spleen and liver). In order to design the ideal nanoagent for simultaneous MRI and MFH, the T_1 and T_2 relaxivities and the SAR should be maximized [383]. For this purpose, iron/platinum (FePt) nanoparticles were combined with kaolinite and cetyltrimethylammonium bromide, and loaded with DOX, to serve as MFH agents. These nanoparticles presented a transversal relaxivity r_2 of 29.32 $mM^{-1} s^{-1}$ and a SAR of 36.91 W/g and were used as a multifunctional drug delivery system for MRI, magnetically guided targeting, and therapeutic treatment [384]. As compared to Fe_3O_4, metallic Fe, Au–Fe_3O_4, and FeCo nanoparticles, the FePt nanoparticles for combined MRI and MFH, demonstrated higher chemical stability, and better performance as contrast agents for MRI [385]. Following stereotactical administration of IONPs in patients bearing recurrent glioblastomas, the particle distribution could be imaged by MRI and the heating boundaries could be estimated by applying a forward model [386]. The combined MFH and radiation therapy significantly increased the overall survival of patients [386].

Zahn et al., instead, prepared PLGA microspheres loaded with magnetic multicore NPs and the anticancer drug Camptothecin to investigate the hyperthermia-induced drug release [357]. SAR measurements showed a promising heating power of the microspheres (161 W/g) that led to the release of around 80% Camptothecin at 43 °C. Similar results were obtained by Zhang et al. who used nano-in-micro thermo-responsive microspheres loaded with Methotrexate or 5-Fluorouracil [387]. Stocke et al. developed inhalable magnetic nanocomposite microparticles for targeted pulmonary delivery via spray drying [388]. This magnetic nanocomposite displayed excellent thermal properties (SAR > 200 W/g), holding promise for innovative bronchial lung cancer treatment.

More recently, MFH was co-performed with real-time MPI [389]. A combined MPI–MFH scanner was developed, in which seamless switching between imaging and therapy while the subject remains in the scanner was possible. This system is capable of selectively heating SPIO particles separated by a distance of 7 mm, and providing imaging at a high spatial resolution. As a proof of concept in vivo, mice bearing two tumor xenografts of human origin each were injected with SPIONs, which were then visualized by MPI (Figure 11A–C) [364]. One only tumor in the couples was selectively heated by MFH, while the other tumor served as the negative control. Ex vivo studies proved that heat damage was indeed localized to the target tumor, while the off-target tumor or off-target healthy clearance organs were spared.

MCs for simultaneous MPI and MFH must have a high SAR and reach magnetization saturation at a small magnetic field. Even if IONPs are promising as MFH agents, they have not yet been translated to the clinical practice of this application due to the following reasons: (i) the low heating power of the clinically approved IONPs; (ii) their limited ability to target tumor cells; and (iii) their inhomogeneous distribution in the tumor tissue. The direct injection of magnetic nanoparticles into the tumor is necessary to achieve sufficient local concentrations of particles that make the MFH treatments effective [390].

5.2.2. Imaging of Targeted Drug Delivery from Smart Systems

Magnetically targeted drug delivery has been already reviewed in paragraph three, however, the possibility to image this process has not been thoroughly treated yet. One of the great advantages of using MCs resides not only in the possibility of driving them from the external, but also in the possibility of tracking this process [391]. The basic requirement is the stability of the co-loading of the drug and the imaging agent. If the drug is released before the imaging agent, in fact, the information gained is not reliable. This does not apply to MCs employed for MFH, as the magnetic system is the therapeutic agent itself. Moreover, as mentioned in paragraph three, MCs used for drug delivery should be biocompatible and their clearance should be quite fast, in order to allow multiple drug delivery administrations in a selected period of time. Among the imaging techniques employed to follow drug delivery from MCs, MRI is the most exploited, however, also PAI, OI, and MPI can be employed. Svenskaya et al., for example, recently developed and optimized biodegradable capsules sensitive to magnet navigation, demonstrating the possibility of tracking their delivery by MRI in a breast cancer murine model. For this purpose, an external magnet was applied for one hour on the mouse tumor, then the MRI contrast was followed up 48 h post-injection [100]. Guo et al. endowed methotrexate thermosensitive magnetoliposomes with both the lipophilic fluorescent dye Cy5.5 and magnetic nanoparticles in the bilayer, and DOX in the core. By the combination of magnetic and receptor targeting, the administered nanosystem reached the tumor, and thanks to the application of an alternating magnetic field the drug can be selectively released. Drug delivery was tracked both by MRI and Optical Imaging (OI) [392]. Foy et al. demonstrated the possibility of combining magnetic targeting with OI to track drug delivery of a hydrophobic anticancer drug to breast cancer tumors, using magnetic nanoparticles [393,394]. Recently, MPI was also employed to track magnetic guided delivery, even if this field is still in its infancy. Zhang et al. proposed a real-time imaging-based guidance system for simultaneous monitoring and actuation of MCs. However, this system was limited to 1D monitoring [395,396]. Le et al. later developed a 2D monitoring system that rendered magnetic particle imaging-based navigation more feasible [397]. Meantime Yu et al. demonstrated for the first time the use of MPI for in vivo cancer imaging with systemic drug-free tracer administration [398], while Jung et al. reported the MPI tracking of SPIO-labeled therapeutic exosomes distribution in vivo in breast cancer-affected mice [399]. However, simultaneous magnetic guidance and drug targeting imaging by MPI has not been reported yet. Interestingly, while monitoring the intratumoral release of Doxorubicin from Fe_3O_4-PLGA core-shell nanocomposites, Zhu et al. found a strong correlation between the MPI signal and the % of doxorubicin release, and demonstrated that ~67% of doxorubicin molecules in nanocomposites was released in the tumor site within the first 48 h after injection [400]. Concerning PAI of magnetically controlled drug delivery, only a few papers have been published [396]. Huang et al. designed a "yolk-shell" structure, confining movable gold nanorods in iron oxide nanoshells loaded with DOX and functionalized with hyaluronic acid. The obtained system was injected intravenously and magnetically driven to a 4T1 tumor. Drug delivery was successfully imaged by both PAI and MRI (Figure 11D,E) [401].

5.2.3. Imaging of Cell Therapy In Situ

The use of cell therapies in clinical trials is constantly increasing [402]. The success of these therapies depends entirely on the accurate delivery of cells to the target organs. Therefore, having the ability of labeling cells and tracking them directly in vivo with high spatial resolution and high sensitivity is of utmost importance to predict the therapy outcome. The main cell therapies adopted employ two different kinds of cells: immune and stem cells [403]. The most diffuse and sensitive method to track them makes use of magnetic systems [346]. It consists of labeling them with SPIO and monitoring them by MRI after the administration [346,404,405]. The advantages of using MRI are represented by the high spatial resolution, the co-registration of anatomical images that allow to spatially locate administered cells, the excellent soft tissues contrast, and the absence of ionizing radiations [406]. The first effective MRI tracking of ex vivo MCs labeled cells in human patients was reported in 2005 by I. de Vries et al. [403]. In more detail, they labeled autologous dendritic cells with SPIO and tracked them after ultrasound-guided intranodal administration in melanoma patients. As a result, MRI allowed the assessment of the accuracy of dendritic cell delivery and inter- and intra-nodal cell migration patterns. They estimated that 1.5×10^5 migrated cells and ~2×10^3 cells/voxel could be visualized. In this case, the immature dendritic cells naturally endocytosed SPIO, but in the case of non-phagocytic cells, the addition of transfection agents is generally required [407]. Of course, both the imaging and the transfection agents should not affect the cell function, viability, phenotype, and receptor expression. The advantage of labeling cells with MCs resides also in the possibility of guiding them to the pathological site selectively using pulsed magnetic field gradients or applying a local external magnet [269,281,295]. Besides the possibility of controlling cells remotely, moreover, labeling cells with MCs allows for cell function modulation, by external magnetic fields application [131,137,205]. This chance is of the utmost importance, especially in the field of cell therapy for tissue regeneration, as already reviewed in Section 3.2.

Even if MRI is undoubtedly the main imaging technique employed to track MCs labeled cells, MPI, PAI, and US have also been recently proposed. MPI cell tracking displays great potential for overcoming the challenges of MRI-based cell tracking allowing for both high cellular sensitivity and high specificity and quantification of SPIO labeled cell numbers, even if the technique still needs to be fine-tuned (Figure 11F–H) [408,409]. Dhada et al. developed ROS-sensitive gold nanorods for mesenchymal stem cells tracking by PAI, with the possibility of discerning between viable and dead stem cells with high spatiotemporal resolution. They found that, after transplantation of mesenchymal stem cells into the lower limb of a mouse, a significant cell death occurred within 24 h, with an estimated 5% viability after 10 days [410]. Donnelly et al. used gold nanospheres to label MSCs and track them during and after the injection in the spinal cord by both PAI and US, detecting as few as 1×10^3 cells [411]. CT can also be employed to track MCs labeled cells [412], but with a great limit on the use of ionizing radiations, which hampers assiduous monitoring. Of course in the present paragraph, only some of most representative studies about cell tracking, combined with MCs cell labeling, were reported, just to give to the reader the idea of the great significance of this topic for future medical applications.

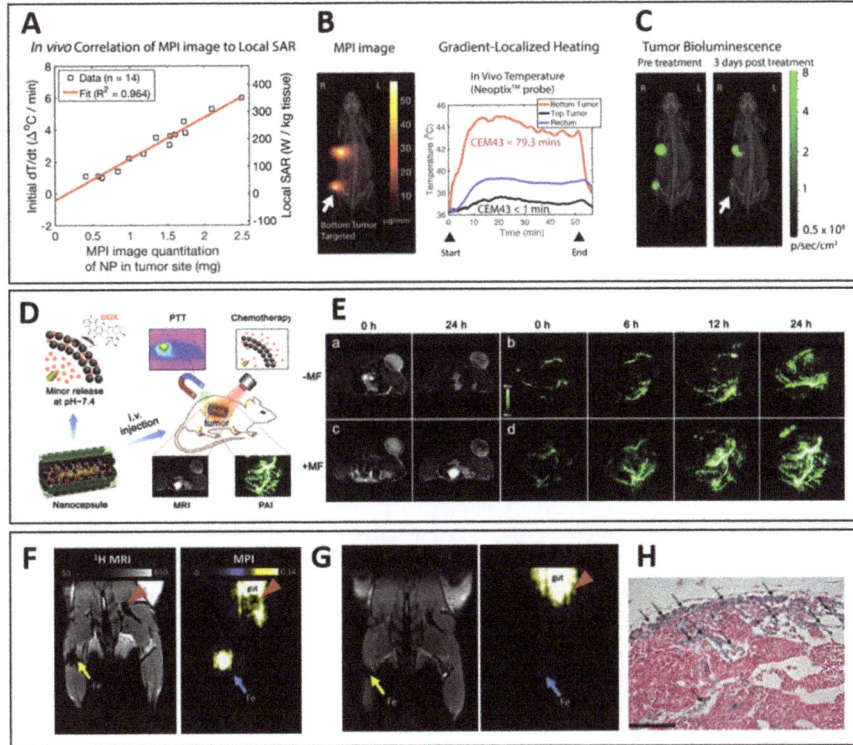

Figure 11. Imaging MCs in theranostic applications. (A–C) In vivo localized MPI guided MFH

experiment. (**A**) MPI images provided quantitative information of SAR estimate at the tumor site, (**B**) MPI image of the dual tumor MDA-MB-231-luc xenograft mice. The bottom tumor is targeted during gradient-localized heating, as described in the graph. (**C**) Luciferase bioluminescence activity of the tumor was measured before and after treatment. The results show a greater than six-fold decrease in activity for the treated bottom tumor, while the untreated tumor had almost no change in bioluminescence activity. Thus, this demonstrates the utility of MPI gradients in localizing tumor therapy. Adapted with permission from Ref. [364], Copyright 2018 American Chemical Society. (**D,E**): magnetic nanocapsules for multimodal-imaging and magnetically guided combination cancer therapy. (**D**) Schematic illustration of the gold-silica nanocapsules loaded with DOX and of the employed combination therapy. (**E**) (**a,c**) In vivo T_2-weighted MR images of 4T1 tumor-bearing mice at various time points postinjection of magnetic nanocapsules without (**a**) or with (**c**) magnetic tumor targeting. White circles indicate the positions of tumors. (**b,d**) In vivo PA images of the tumor sites at different time points, postinjection of the magnetic capsules without (**b**) or with (**d**) magnetic targeting. Reprinted with permission from Ref. [401], Copyright 2016 American Chemical Society. (**F,G**) MRI and MPI of 1×10^6 ferumoxytol-labeled MSCs implanted within the hind limb muscle of mice; imaging was performed 1 day (**F**) or 12 days (**G**) post implantation. Yellow and blue arrows indicate the transplantation site, while red triangles indicate misleading signal coming from the bone (in MRI) or from the iron in the gut of the mice (in MPI). (**H**) Prussian Blue staining with nuclear fast red counterstain confirms the presence of iron within MSCs. Adapted with permission from Ref. [409].

6. Perspectives

Targeting bioagents via magnetic forces has several key advantages. First, magnetic targeting uses a strong targeting power with high penetration depth, which can facilitate the administration of medications in the human body [153]. Second, it is safe and compliant with use in living systems, as the side effects of MCs are limited. Third, it has the potential for high therapeutic selectivity. The therapeutic selectivity refers not only to the ability to reach targets at a high molecular resolution (e.g., recognition of specific disease-associated molecules) and with patient-specificity (e.g., targeting of molecular fingerprints of tumors) but also to the opportunity to establish mechanisms for conditional and controlled release of the bioagents within the pathological site. As such, not only MCs are more promising for drug delivery as compared to other vehicles and vehiculation mechanisms, but they also are helpful in other in vivo applications, such as biosensing. For instance, MCs can serve as bioprobes to image malignancies and diseases, or to detect biomolecular interactions [158,384,396].

In the future, MCs will continue to gain medical relevance as applied to various uses, including theranostics, separation of biomolecules, sorting of diseased cells, manipulation of genetic material, and control of bioagents in vivo. Nevertheless, to render MCs successful delivery platforms, several requirements need to be addressed [121]. One first important assumption is that MCs are developed to possess sufficient magnetization making them responsive to the application of external magnetic fields. This property of MCs largely depends on their size, which in turn affects the applicability within living systems [157]. Another important feature in MC-based pharmaceutical vehicles is the long-circulation, which can be realized by implementing stealthy systems that avoid the clearance operated by the phagocyte system [184,195]. Magnetic controllability and stability in vivo are vital prerequisites for the future development of MCs as multifunctional platforms. In this regard, the most promising directions for next generation MCs are effective mechanisms to control the drug uptake and release, as well as designs that permit theranostic functions [157]. An essential aspect that still requires intense development is the interaction of the MCs with living environments. Ideal MCs should be able to efficiently move within the bodies of living systems. Depending on the application and the used MCs type, guiding magnetic materials within the body requires one to overcome several issues [325,413–415]. Among them, the anatomical structures represent a major physical barrier that hinders controlled movement at different levels (from micro to macro scale). The physical properties of injectable formulations of MCs must be compliant with the purpose of navigation within the vascular networks. As an example, not only the size and morphology of the MCs must be optimized for safe circulation within vessels, but also the systems' stability has to be maximized to reach both long circulation time and reduced risk of releasing unsafe degradation products into the circulation. Stealthy designs can dramatically extend the circulation time while masking the carriers from clearance cell effectors and protecting them from excretion [416,417]. At the same time, successful intravascular guidance of the MCs must win the forces generated by the hemodynamics [418]. To guide MCs through compact tissues, the magnetic attraction must overcome the physical barriers of the matrix and other tissue components. In addition, as cells are responsive to chemotactic cues (such as the chemokine gradients generated by other cells within pathological sites), fully controllable dragging of magnetized cells should also overcome the natural chemophoretic tendency that cells experience in their specific microenvironment [419,420]. As far as we can argue from the literature, the trend in research points out the sophistication of MCs for the targeted delivery of bioagents [157]. MCs are in fact increasingly designed to become multifunctional, while at the same time being performant in relation to their basic abilities as carriers that, as mentioned above, mostly refer to their in vivo structural stability and controllability of the bioagent liberation [157,169,202].

Even if tremendous progress was recently achieved in the MCs' development, the actual applicability of these systems needs to be discussed. MNPs have received FDA approval and have been commercialized for use as MRI contrast agents and hyperthermia [211,380]. Nevertheless, their use in magnetic targeting is not yet at a stage of clinical relevance. Some limited clinical trials (even in Phase III) have already started, but none of them has yet resulted in FDA approval or commercialization [421,422]. One example is a recent study from Kamei et al., which concluded that magnetic targeting of ferucarbotran-labeled MSCs could be performed safely in five patients with cartilage defects in the knee joint [301]. A compact magnet (one T) was placed around the knee joint of patients to allow the magnetic force to be as perpendicular to the surface of the lesion as possible. The magnet was maintained in position for 10 min after the injection of 1×10^7 magnetized cells into the knee joint. Upon magnetic targeting, the injured sites were completely refilled with cartilage-like tissues and the patients displayed improved clinical outcomes 48 weeks after treatment. Hence, magnetic cell guidance might represent a minimally invasive treatment for cartilage repair.

Magnetic cell guidance is a peculiar instance of magnetic targeting. In the last years, magnetic materials have attracted much attention for other applications in addition to specific drug release [423]. For instance, an interesting use of magnetic actuation is to control the process of the self-assembly of multicellular aggregates. Cells, cells spheroids, or matrices containing cells can be modified with magnetic particles and then, under the application of a magnetic field, can be induced to assemble into morphologies that mimic the structure of target organs (e.g., rings for vessels, sheets for skin, etc.) [110,424]. Thus, magnetic tissue engineering emerges as an important tool to enhance bioprinting functionality and realize future multiplex bioconstructs and organ replicas. Moreover, MCs are also promising for robotics as applied to drug delivery and regenerative medicine [305,323,324,327]. Thus far, microrobotic MCs have been widely studied in vitro and scarcely in animal models, with no reported use in humans [305,425–428]. Nevertheless, more and more reports validate the performance of their robot in phantoms that replicate the anatomical structures of patients as realized from imaging data [426,429]. Although this field is moving forward in the direction of clinical applicability, consistent efforts are still needed before starting experimentation in humans [426].

One of the major problems is to magnetically control the delivery of bioagents in-depth in the body as realized with the practical implementation of the magnetic fields. In-body sources of magnetic fields (like magnetized or magnetic scaffolds) have been proposed as a solution for portability [279,425,430], which might enhance the targeting efficiency when vehiculating drugs or guiding the homing of medical cells. However, the use of magnetic implants requires surgical intervention and it is not always possible. Hence, more research effort is needed to expand the applicability spectrum of magnetic targeting, as well as its implementation in real-life use.

Author Contributions: Conceptualization: F.G., Y.S., B.P. and M.F.; writing—original draft preparation, F.G., Y.S. and M.F.; writing—review and editing, F.G., Y.S. and M.F.; supervision, M.F. All authors have read and agreed to the published version of the manuscript.

Funding: Ministry of Science and Higher *Educ.* of the Russian Federation (project number FSRR-2020-0002).

Institutional Review Board Statement: Not applicable.

Informed Consent Statement: Not applicable.

Acknowledgments: The contribution of F.G. was supported by the Italian Research Program PNR 2021–2027 and PON "Ricerca e Innovazione" 2014–2020. The contribution of Y.S. on the magnetic microcapsules in biomedicine and MNPs for systemic circulation was supported by the Ministry of Science and Higher *Educ.* of the Russian Federation (project number FSRR-2020-0002).

Conflicts of Interest: The authors declare no conflict of interest.

References

1. Öztürk-Atar, K.; Eroğlu, H.; Çalış, S. Novel Advances in Targeted Drug Delivery. *J. Drug Target.* **2018**, *26*, 633–642. [CrossRef] [PubMed]
2. Zhao, X.; Ye, Y.; Ge, S.; Sun, P.; Yu, P. Cellular and Molecular Targeted Drug Delivery in Central Nervous System Cancers: Advances in Targeting Strategies. *Curr. Top. Med. Chem.* **2020**, *20*, 2762–2776. [CrossRef]
3. Kumari, P.; Ghosh, B.; Biswas, S. Nanocarriers for Cancer-Targeted Drug Delivery. *J. Drug Target.* **2016**, *24*, 179–191. [CrossRef] [PubMed]
4. Alsaggar, M.; Liu, D. Organ-Based Drug Delivery. *J. Drug Target.* **2018**, *26*, 385–397. [CrossRef]
5. Jain, K.K. An Overview of Drug Delivery Systems. *Methods Mol. Biol.* **2020**, *2059*, 1–54. [CrossRef] [PubMed]
6. Francis, D.M.; Thomas, S.N. Progress and Opportunities for Enhancing the Delivery and Efficacy of Checkpoint Inhibitors for Cancer Immunotherapy. *Adv. Drug Deliv. Rev.* **2017**, *114*, 33–42. [CrossRef]
7. Czech, T.; Lalani, R.; Oyewumi, M.O. Delivery Systems as Vital Tools in Drug Repurposing. *AAPS PharmSciTech* **2019**, *20*, 116. [CrossRef]
8. Siemer, S.; Wünsch, D.; Khamis, A.; Lu, Q.; Scherberich, A.; Filippi, M.; Krafft, M.P.; Hagemann, J.; Weiss, C.; Ding, G.-B.; et al. Nano Meets Micro-Translational Nanotechnology in Medicine: Nano-Based Applications for Early Tumor Detection and Therapy. *Nanomaterials* **2020**, *10*, 383. [CrossRef]
9. Mitchell, M.J.; Billingsley, M.M.; Haley, R.M.; Wechsler, M.E.; Peppas, N.A.; Langer, R. Engineering Precision Nanoparticles for Drug Delivery. *Nat. Rev. Drug Discov.* **2021**, *20*, 101–124. [CrossRef]
10. Lee, W.-H.; Loo, C.-Y.; Traini, D.; Young, P.M. Nano- and Micro-Based Inhaled Drug Delivery Systems for Targeting Alveolar Macrophages. *Expert Opin. Drug Deliv.* **2015**, *12*, 1009–1026. [CrossRef]
11. Ahadian, S.; Finbloom, J.A.; Mofidfar, M.; Diltemiz, S.E.; Nasrollahi, F.; Davoodi, E.; Hosseini, V.; Mylonaki, I.; Sangabathuni, S.; Montazerian, H.; et al. Micro and Nanoscale Technologies in Oral Drug Delivery. *Adv. Drug Deliv. Rev.* **2020**, *157*, 37–62. [CrossRef] [PubMed]
12. Kurmi, B.D.; Kayat, J.; Gajbhiye, V.; Tekade, R.K.; Jain, N.K. Micro- and Nanocarrier-Mediated Lung Targeting. *Expert Opin. Drug Deliv.* **2010**, *7*, 781–794. [CrossRef]
13. Sizikov, A.A.; Nikitin, P.I.; Nikitin, M.P. Magnetofection In Vivo by Nanomagnetic Carriers Systemically Administered into the Bloodstream. *Pharmaceutics* **2021**, *13*, 1927. [CrossRef] [PubMed]
14. Liu, Y.-L.; Chen, D.; Shang, P.; Yin, D.-C. A Review of Magnet Systems for Targeted Drug Delivery. *J. Control. Release Off. J. Control. Release Soc.* **2019**, *302*, 90–104. [CrossRef]
15. Moshfeghi, A.A.; Peyman, G.A. Micro- and Nanoparticulates. *Adv. Drug Deliv. Rev.* **2005**, *57*, 2047–2052. [CrossRef] [PubMed]
16. Yetisgin, A.A.; Cetinel, S.; Zuvin, M.; Kosar, A.; Kutlu, O. Therapeutic Nanoparticles and Their Targeted Delivery Applications. *Molecules* **2020**, *25*, 2193. [CrossRef] [PubMed]
17. Filippi, M.; Catanzaro, V.; Patrucco, D.; Botta, M.; Tei, L.; Terreno, E. First in Vivo MRI Study on Theranostic Dendrimersomes. *J. Control. Release Off. J. Control. Release Soc.* **2017**, *248*, 45–52. [CrossRef]
18. Filippi, M.; Patrucco, D.; Martinelli, J.; Botta, M.; Castro-Hartmann, P.; Tei, L.; Terreno, E. Novel Stable Dendrimersome Formulation for Safe Bioimaging Applications. *Nanoscale* **2015**, *7*, 12943–12954. [CrossRef]
19. Filippi, M.; Martinelli, J.; Mulas, G.; Ferraretto, M.; Teirlinck, E.; Botta, M.; Tei, L.; Terreno, E. Dendrimersomes: A New Vesicular Nano-Platform for MR-Molecular Imaging Applications. *Chem. Commun. Camb. Engl.* **2014**, *50*, 3453–3456. [CrossRef]

20. Filippi, M.; Remotti, D.; Botta, M.; Terreno, E.; Tei, L. GdDOTAGA(C18)2: An Efficient Amphiphilic Gd(Iii) Chelate for the Preparation of Self-Assembled High Relaxivity MRI Nanoprobes. *Chem. Commun. Camb. Engl.* **2015**, *51*, 17455–17458. [CrossRef]
21. Rastogi, R.; Anand, S.; Koul, V. Flexible Polymerosomes–an Alternative Vehicle for Topical Delivery. *Colloids Surf. B Biointerfaces* **2009**, *72*, 161–166. [CrossRef]
22. Shimanovich, U.; Bernardes, G.J.L.; Knowles, T.P.J.; Cavaco-Paulo, A. Protein Micro- and Nano-Capsules for Biomedical Applications. *Chem. Soc. Rev.* **2014**, *43*, 1361–1371. [CrossRef] [PubMed]
23. Jaganathan, M.; Madhumitha, D.; Dhathathreyan, A. Protein Microcapsules: Preparation and Applications. *Adv. Colloid Interface Sci.* **2014**, *209*, 1–7. [CrossRef] [PubMed]
24. Fu, P.; Zhang, J.; Li, H.; Mak, M.; Xu, W.; Tao, Z. Extracellular Vesicles as Delivery Systems at Nano-/Micro-Scale. *Adv. Drug Deliv. Rev.* **2021**, *179*, 113910. [CrossRef] [PubMed]
25. Garello, F.; Stefania, R.; Aime, S.; Terreno, E.; Delli Castelli, D. Successful Entrapping of Liposomes in Glucan Particles: An Innovative Micron-Sized Carrier to Deliver Water-Soluble Molecules. *Mol. Pharm.* **2014**, *11*, 3760–3765. [CrossRef]
26. Shao, J.; Zaro, J.; Shen, Y. Advances in Exosome-Based Drug Delivery and Tumor Targeting: From Tissue Distribution to Intracellular Fate. *Int. J. Nanomedicine* **2020**, *15*, 9355–9371. [CrossRef]
27. Batrakova, E.V.; Kim, M.S. Using Exosomes, Naturally-Equipped Nanocarriers, for Drug Delivery. *J. Control. Release Off. J. Control. Release Soc.* **2015**, *219*, 396–405. [CrossRef]
28. Delcassian, D.; Patel, A.K.; Cortinas, A.B.; Langer, R. Drug Delivery across Length Scales. *J. Drug Target.* **2019**, *27*, 229–243. [CrossRef]
29. Wong, C.Y.; Al-Salami, H.; Dass, C.R. Microparticles, Microcapsules and Microspheres: A Review of Recent Developments and Prospects for Oral Delivery of Insulin. *Int. J. Pharm.* **2018**, *537*, 223–244. [CrossRef]
30. Truong, N.P.; Whittaker, M.R.; Mak, C.W.; Davis, T.P. The Importance of Nanoparticle Shape in Cancer Drug Delivery. *Expert Opin. Drug Deliv.* **2015**, *12*, 129–142. [CrossRef]
31. Biffi, S.; Voltan, R.; Bortot, B.; Zauli, G.; Secchiero, P. Actively Targeted Nanocarriers for Drug Delivery to Cancer Cells. *Expert Opin. Drug Deliv.* **2019**, *16*, 481–496. [CrossRef] [PubMed]
32. Karimi, M.; Eslami, M.; Sahandi-Zangabad, P.; Mirab, F.; Farajisafiloo, N.; Shafaei, Z.; Ghosh, D.; Bozorgomid, M.; Dashkhaneh, F.; Hamblin, M.R. PH-Sensitive Stimulus-Responsive Nanocarriers for Targeted Delivery of Therapeutic Agents. *Wiley Interdiscip. Rev. Nanomed. Nanobiotechnol.* **2016**, *8*, 696–716. [CrossRef] [PubMed]
33. Karimi, M.; Ghasemi, A.; Sahandi Zangabad, P.; Rahighi, R.; Moosavi Basri, S.M.; Mirshekari, H.; Amiri, M.; Shafaei Pishabad, Z.; Aslani, A.; Bozorgomid, M.; et al. Smart Micro/Nanoparticles in Stimulus-Responsive Drug/Gene Delivery Systems. *Chem. Soc. Rev.* **2016**, *45*, 1457–1501. [CrossRef] [PubMed]
34. Vázquez-González, M.; Willner, I. Aptamer-Functionalized Micro- and Nanocarriers for Controlled Release. *ACS Appl. Mater. Interfaces* **2021**, *13*, 9520–9541. [CrossRef]
35. Gagliardi, M. Novel Biodegradable Nanocarriers for Enhanced Drug Delivery. *Ther. Deliv.* **2016**, *7*, 809–826. [CrossRef]
36. Katz, E. Synthesis, Properties and Applications of Magnetic Nanoparticles and Nanowires—A Brief Introduction. *Magnetochemistry* **2019**, *5*, 61. [CrossRef]
37. Kolhatkar, A.G.; Jamison, A.C.; Litvinov, D.; Willson, R.C.; Lee, T.R. Tuning the Magnetic Properties of Nanoparticles. *Int. J. Mol. Sci.* **2013**, *14*, 15977–16009. [CrossRef]
38. Akbarzadeh, A.; Samiei, M.; Davaran, S. Magnetic Nanoparticles: Preparation, Physical Properties, and Applications in Biomedicine. *Nanoscale Res. Lett.* **2012**, *7*, 144. [CrossRef]
39. Issa, B.; Obaidat, I.M.; Albiss, B.A.; Haik, Y. Magnetic Nanoparticles: Surface Effects and Properties Related to Biomedicine Applications. *Int. J. Mol. Sci.* **2013**, *14*, 21266–21305. [CrossRef]
40. Myrovali, E.; Maniotis, N.; Makridis, A.; Terzopoulou, A.; Ntomprougkidis, V.; Simeonidis, K.; Sakellari, D.; Kalogirou, O.; Samaras, T.; Salikhov, R.; et al. Arrangement at the Nanoscale: Effect on Magnetic Particle Hyperthermia. *Sci. Rep.* **2016**, *6*, 37934. [CrossRef]
41. Kouhpanji, M.R.Z.; Stadler, B.J.H. A Guideline for Effectively Synthesizing and Characterizing Magnetic Nanoparticles for Advancing Nanobiotechnology: A Review. *Sensors* **2020**, *20*, 2554. [CrossRef] [PubMed]
42. Laurent, S.; Forge, D.; Port, M.; Roch, A.; Robic, C.; Vander Elst, L.; Muller, R.N. Magnetic Iron Oxide Nanoparticles: Synthesis, Stabilization, Vectorization, Physicochemical Characterizations, and Biological Applications. *Chem. Rev.* **2008**, *108*, 2064–2110. [CrossRef]
43. Ajinkya, N.; Yu, X.; Kaithal, P.; Luo, H.; Somani, P.; Ramakrishna, S. Magnetic Iron Oxide Nanoparticle (IONP) Synthesis to Applications: Present and Future. *Mater. Basel Switz.* **2020**, *13*, 4644. [CrossRef] [PubMed]
44. Caizer, C. Nanoparticle Size Effect on Some Magnetic Properties. In *Handbook of Nanoparticles*; Aliofkhazraei, M., Ed.; Springer International Publishing: Cham, Switzerland, 2016; pp. 475–519. ISBN 978-3-319-15337-7.
45. Chen, P.; Cui, B.; Bu, Y.; Yang, Z.; Wang, Y. Synthesis and Characterization of Mesoporous and Hollow-Mesoporous $MxFe_3-XO_4$ (M=Mg, Mn, Fe, Co, Ni, Cu, Zn) Microspheres for Microwave-Triggered Controllable Drug Delivery. *J. Nanoparticle Res.* **2017**, *19*, 398. [CrossRef]
46. Farah, F.H. Magnetic Microspheres: A Novel Drug Delivery System. *J. Anal. Pharm. Res.* **2016**, *3*, 67. [CrossRef]
47. Ali, A.; Zafar, H.; Zia, M.; ul Haq, I.; Phull, A.R.; Ali, J.S.; Hussain, A. Synthesis, Characterization, Applications, and Challenges of Iron Oxide Nanoparticles. *Nanotechnol. Sci. Appl.* **2016**, *9*, 49–67. [CrossRef] [PubMed]

48. Wallyn, J.; Anton, N.; Vandamme, T.F. Synthesis, Principles, and Properties of Magnetite Nanoparticles for In Vivo Imaging Applications-A Review. *Pharmaceutics* **2019**, *11*, 601. [CrossRef]
49. Li, J.; Wang, S.; Shi, X.; Shen, M. Aqueous-Phase Synthesis of Iron Oxide Nanoparticles and Composites for Cancer Diagnosis and Therapy. *Adv. Colloid Interface Sci.* **2017**, *249*, 374–385. [CrossRef]
50. Martínez-Cabanas, M.; López-García, M.; Barriada, J.L.; Herrero, R.; Sastre de Vicente, M.E. Green Synthesis of Iron Oxide Nanoparticles. Development of Magnetic Hybrid Materials for Efficient As(V) Removal. *Chem. Eng. J.* **2016**, *301*, 83–91. [CrossRef]
51. Narayanan, K.B.; Sakthivel, N. Biological Synthesis of Metal Nanoparticles by Microbes. *Adv. Colloid Interface Sci.* **2010**, *156*, 1–13. [CrossRef]
52. Cuenya, B.R. Synthesis and Catalytic Properties of Metal Nanoparticles: Size, Shape, Support, Composition, and Oxidation State Effects. *Thin Solid Films* **2010**, *518*, 3127–3150. [CrossRef]
53. Yadwade, R.; Kirtiwar, S.; Ankamwar, B. A Review on Green Synthesis and Applications of Iron Oxide Nanoparticles. *J. Nanosci. Nanotechnol.* **2021**, *21*, 5812–5834. [CrossRef] [PubMed]
54. Jacinto, M.J.; Silva, V.C.; Valladão, D.M.S.; Souto, R.S. Biosynthesis of Magnetic Iron Oxide Nanoparticles: A Review. *Biotechnol. Lett.* **2021**, *43*, 1–12. [CrossRef] [PubMed]
55. Wu, W.; He, Q.; Jiang, C. Magnetic Iron Oxide Nanoparticles: Synthesis and Surface Functionalization Strategies. *Nanoscale Res. Lett.* **2008**, *3*, 397–415. [CrossRef] [PubMed]
56. Wu, S.; Sun, A.; Zhai, F.; Wang, J.; Xu, W.; Zhang, Q.; Volinsky, A.A. Fe3O4 Magnetic Nanoparticles Synthesis from Tailings by Ultrasonic Chemical Co-Precipitation. *Mater. Lett.* **2011**, *65*, 1882–1884. [CrossRef]
57. Benetti, F.; Bregoli, L.; Olivato, I.; Sabbioni, E. Effects of Metal(Loid)-Based Nanomaterials on Essential Element Homeostasis: The Central Role of Nanometallomics for Nanotoxicology. *Met. Integr. Biometal Sci.* **2014**, *6*, 729–747. [CrossRef]
58. Chen, J.; Rogers, S.C.; Kavdia, M. Analysis of Kinetics of Dihydroethidium Fluorescence with Superoxide Using Xanthine Oxidase and Hypoxanthine Assay. *Ann. Biomed. Eng.* **2013**, *41*, 327–337. [CrossRef]
59. Novotna, B.; Jendelova, P.; Kapcalova, M.; Rossner, P.; Turnovcova, K.; Bagryantseva, Y.; Babic, M.; Horak, D.; Sykova, E. Oxidative Damage to Biological Macromolecules in Human Bone Marrow Mesenchymal Stromal Cells Labeled with Various Types of Iron Oxide Nanoparticles. *Toxicol. Lett.* **2012**, *210*, 53–63. [CrossRef]
60. Küstermann, E.; Himmelreich, U.; Kandal, K.; Geelen, T.; Ketkar, A.; Wiedermann, D.; Strecker, C.; Esser, J.; Arnhold, S.; Hoehn, M. Efficient Stem Cell Labeling for MRI Studies. *Contrast Media Mol. Imaging* **2008**, *3*, 27–37. [CrossRef]
61. Wei, H.; Hu, Y.; Wang, J.; Gao, X.; Qian, X.; Tang, M. Superparamagnetic Iron Oxide Nanoparticles: Cytotoxicity, Metabolism, and Cellular Behavior in Biomedicine Applications. *Int. J. Nanomedicine* **2021**, *16*, 6097–6113. [CrossRef]
62. Yang, K.; Liu, Y.; Liu, Y.; Zhang, Q.; Kong, C.; Yi, C.; Zhou, Z.; Wang, Z.; Zhang, G.; Zhang, Y.; et al. Cooperative Assembly of Magneto-Nanovesicles with Tunable Wall Thickness and Permeability for MRI-Guided Drug Delivery. *J. Am. Chem. Soc.* **2018**, *140*, 4666–4677. [CrossRef] [PubMed]
63. Wang, Y.; Wei, G.; Zhang, X.; Huang, X.; Zhao, J.; Guo, X.; Zhou, S. Multistage Targeting Strategy Using Magnetic Composite Nanoparticles for Synergism of Photothermal Therapy and Chemotherapy. *Small* **2018**, *14*, 1702994. [CrossRef] [PubMed]
64. Ao, L.; Wu, C.; Liu, K.; Wang, W.; Fang, L.; Huang, L.; Su, W. Polydopamine-Derived Hierarchical Nanoplatforms for Efficient Dual-Modal Imaging-Guided Combination in Vivo Cancer Therapy | ACS Applied Materials & Interfaces. *ACS Appl. Mater. Interfaces* **2018**, *10*, 12544–12552. [CrossRef]
65. Singh, M.N.; Hemant, K.S.Y.; Ram, M.; Shivakumar, H.G. Microencapsulation: A Promising Technique for Controlled Drug Delivery. *Res. Pharm. Sci.* **2010**, *5*, 65–77. [PubMed]
66. Sinha, V.R.; Goyal, V.; Bhinge, J.R.; Mittal, B.R.; Trehan, A. Diagnostic Microspheres: An Overview. *Crit. Rev. Ther. Drug Carrier Syst.* **2003**, *20*, 431–460. [CrossRef]
67. Arshady, R. Microspheres for Biomedical Applications: Preparation of Reactive and Labelled Microspheres. *Biomaterials* **1993**, *14*, 5–15. [CrossRef]
68. Häfeli, U.O. Magnetically Modulated Therapeutic Systems. *Int. J. Pharm.* **2004**, *277*, 19–24. [CrossRef]
69. Gupta, P.K.; Hung, C.T. Magnetically Controlled Targeted Micro-Carrier Systems. *Life Sci.* **1989**, *44*, 175–186. [CrossRef]
70. Ranney, D.F.; Huffaker, H.H. Magnetic Microspheres for the Targeted Controlled Release of Drugs and Diagnostic Agents. *Ann. N. Y. Acad. Sci.* **1987**, *507*, 104–119. [CrossRef]
71. Senyei, A.E.; Widder, K.J. Drug Targeting: Magnetically Responsive Albumin Microspheres–a Review of the System to Date. *Gynecol. Oncol.* **1981**, *12*, 1–13. [CrossRef]
72. Long, S.; Xiao, Y.; Zhang, X. Progress in Preparation of Silk Fibroin Microspheres for Biomedical Applications. *Pharm. Nanotechnol.* **2020**, *8*, 358–371. [CrossRef]
73. Leong, W.; Wang, D.-A. Cell-Laden Polymeric Microspheres for Biomedical Applications. *Trends Biotechnol.* **2015**, *33*, 653–666. [CrossRef]
74. Dhamecha, D.; Movsas, R.; Sano, U.; Menon, J.U. Applications of Alginate Microspheres in Therapeutics Delivery and Cell Culture: Past, Present and Future. *Int. J. Pharm.* **2019**, *569*, 118627. [CrossRef] [PubMed]
75. Lengyel, M.; Kállai-Szabó, N.; Antal, V.; Laki, A.J.; Antal, I. Microparticles, Microspheres, and Microcapsules for Advanced Drug Delivery. *Sci. Pharm.* **2019**, *87*, 20. [CrossRef]
76. Toll, R.; Jacobi, U.; Richter, H.; Lademann, J.; Schaefer, H.; Blume-Peytavi, U. Penetration Profile of Microspheres in Follicular Targeting of Terminal Hair Follicles. *J. Investig. Dermatol.* **2004**, *123*, 168–176. [CrossRef] [PubMed]

77. Svenskaya, Y.I.; Genina, E.A.; Parakhonskiy, B.V.; Lengert, E.V.; Talnikova, E.E.; Terentyuk, G.S.; Utz, S.R.; Gorin, D.A.; Tuchin, V.V.; Sukhorukov, G.B. A Simple Non-Invasive Approach toward Efficient Transdermal Drug Delivery Based on Biodegradable Particulate System. *ACS Appl. Mater. Interfaces* **2019**, *11*, 17270–17282. [CrossRef] [PubMed]
78. Lademann, J.; Richter, H.; Schaefer, U.F.; Blume-Peytavi, U.; Teichmann, A.; Otberg, N.; Sterry, W. Hair Follicles—A Long-Term Reservoir for Drug Delivery. *Skin Pharmacol. Physiol.* **2006**, *19*, 232–236. [CrossRef]
79. Genina, E.A.; Svenskaya, Y.I.; Yanina, I.Y.; Dolotov, L.E.; Navolokin, N.A.; Bashkatov, A.N.; Terentyuk, G.S.; Bucharskaya, A.B.; Maslyakova, G.N.; Gorin, D.A.; et al. In Vivo Optical Monitoring of Transcutaneous Delivery of Calcium Carbonate Microcontainers. *Biomed. Opt. Express* **2016**, *7*, 2082–2087. [CrossRef]
80. Grigoriev, D.O.; Bukreeva, T.; Möhwald, H.; Shchukin, D.G. New Method for Fabrication of Loaded Micro- and Nanocontainers: Emulsion Encapsulation by Polyelectrolyte Layer-by-Layer Deposition on the Liquid Core. *Langmuir* **2008**, *24*, 999–1004. [CrossRef]
81. Cui, J.; Wang, Y.; Postma, A.; Hao, J.; Hosta-Rigau, L.; Caruso, F. Monodisperse Polymer Capsules: Tailoring Size, Shell Thickness, and Hydrophobic Cargo Loading via Emulsion Templating. *Adv. Funct. Mater.* **2010**, *20*, 1625–1631. [CrossRef]
82. Shchukin, D.G.; Köhler, K.; Möhwald, H.; Sukhorukov, G.B. Gas-Filled Polyelectrolyte Capsules. *Angew. Chem. Int. Ed.* **2005**, *44*, 3310–3314. [CrossRef]
83. Georgieva, R.; Moya, S.; Donath, E.; Bäumler, H. Permeability and Conductivity of Red Blood Cell Templated Polyelectrolyte Capsules Coated with Supplementary Layers. *Langmuir* **2004**, *20*, 1895–1900. [CrossRef]
84. Sukhorukov, G.B.; Rogach, A.L.; Garstka, M.; Springer, S.; Parak, W.J.; Muñoz-Javier, A.; Kreft, O.; Skirtach, A.G.; Susha, A.S.; Ramaye, Y.; et al. Multifunctionalized Polymer Microcapsules: Novel Tools for Biological and Pharmacological Applications. *Small* **2007**, *3*, 944–955. [CrossRef] [PubMed]
85. Donath, E.; Sukhorukov, G.B.; Caruso, F.; Davis, S.A.; Möhwald, H. Novel Hollow Polymer Shells by Colloid-Templated Assembly of Polyelectrolytes. *Angew. Chem. Int. Ed Engl.* **1998**, *37*, 2201–2205. [CrossRef]
86. Geest, B.G.D.; Koker, S.D.; Sukhorukov, G.B.; Kreft, O.; Parak, W.J.; Skirtach, A.G.; Demeester, J.; Smedt, S.C.D.; Hennink, W.E. Polyelectrolyte Microcapsules for Biomedical Applications. *Soft Matter* **2009**, *5*, 282–291. [CrossRef]
87. Peyratout, C.S.; Dähne, L. Tailor-Made Polyelectrolyte Microcapsules: From Multilayers to Smart Containers. *Angew. Chem. Int. Ed.* **2004**, *43*, 3762–3783. [CrossRef]
88. Zhao, S.; Caruso, F.; Dähne, L.; Decher, G.; De Geest, B.G.; Fan, J.; Feliu, N.; Gogotsi, Y.; Hammond, P.T.; Hersam, M.C.; et al. The Future of Layer-by-Layer Assembly: A Tribute to ACS Nano Associate Editor Helmuth Möhwald. *ACS Nano* **2019**, *13*, 6151–6169. [CrossRef]
89. Vikulina, A.S.; Campbell, J. Biopolymer-Based Multilayer Capsules and Beads Made via Templating: Advantages, Hurdles and Perspectives. *Nanomaterials* **2021**, *11*, 2502. [CrossRef]
90. Brazel, C.S.; Peppas, N.A. Modeling of Drug Release from Swellable Polymers. *Eur. J. Pharm. Biopharm. Off. J. Arbeitsgemeinschaft Pharm. Verfahrenstechnik EV* **2000**, *49*, 47–58. [CrossRef]
91. Berkland, C.; Kipper, M.J.; Narasimhan, B.; Kim, K.K.; Pack, D.W. Microsphere Size, Precipitation Kinetics and Drug Distribution Control Drug Release from Biodegradable Polyanhydride Microspheres. *J. Control. Release Off. J. Control. Release Soc.* **2004**, *94*, 129–141. [CrossRef]
92. Delcea, M.; Möhwald, H.; Skirtach, A.G. Stimuli-Responsive LbL Capsules and Nanoshells for Drug Delivery. *Adv. Drug Deliv. Rev.* **2011**, *63*, 730–747. [CrossRef]
93. Skirtach, A.G.; Yashchenok, A.M.; Möhwald, H. Encapsulation, Release and Applications of LbL Polyelectrolyte Multilayer Capsules. *Chem. Commun.* **2011**, *47*, 12736–12746. [CrossRef] [PubMed]
94. German, S.V.; Bratashov, D.N.; Navolokin, N.A.; Kozlova, A.A.; Lomova, M.V.; Novoselova, M.V.; Burilova, E.A.; Zyev, V.V.; Khlebtsov, B.N.; Bucharskaya, A.B.; et al. In Vitro and in Vivo MRI Visualization of Nanocomposite Biodegradable Microcapsules with Tunable Contrast. *Phys. Chem. Chem. Phys.* **2016**, *18*, 32238–32246. [CrossRef]
95. Szczęch, M.; Szczepanowicz, K. Polymeric Core-Shell Nanoparticles Prepared by Spontaneous Emulsification Solvent Evaporation and Functionalized by the Layer-by-Layer Method. *Nanomater. Basel Switz.* **2020**, *10*, 496. [CrossRef] [PubMed]
96. Valdepérez, D.; Del Pino, P.; Sánchez, L.; Parak, W.J.; Pelaz, B. Highly Active Antibody-Modified Magnetic Polyelectrolyte Capsules. *J. Colloid Interface Sci.* **2016**, *474*, 1–8. [CrossRef]
97. Read, M.; Möslinger, C.; Dipper, T.; Kengyel, D.; Hilder, J.; Thenius, R.; Tyrrell, A.; Timmis, J.; Schmickl, T. Profiling Underwater Swarm Robotic Shoaling Performance Using Simulation. In Proceedings of the Towards Autonomous Robotic Systems, Oxford, UK, 28–30 August 2013; Springer: Berlin/Heidelberg, Germany, 2013; pp. 404–416.
98. Kozlova, A.A.; German, S.V.; Atkin, V.S.; Zyev, V.V.; Astle, M.A.; Bratashov, D.N.; Svenskaya, Y.I.; Gorin, D.A. Magnetic Composite Submicron Carriers with Structure-Dependent MRI Contrast. *Inorganics* **2020**, *8*, 11. [CrossRef]
99. Yoon, H.-J.; Lim, T.G.; Kim, J.-H.; Cho, Y.M.; Kim, Y.S.; Chung, U.S.; Kim, J.H.; Choi, B.W.; Koh, W.-G.; Jang, W.-D. Fabrication of Multifunctional Layer-by-Layer Nanocapsules toward the Design of Theragnostic Nanoplatform. *Biomacromolecules* **2014**, *15*, 1382–1389. [CrossRef] [PubMed]
100. Svenskaya, Y.; Garello, F.; Lengert, E.; Kozlova, A.; Verkhovskii, R.; Bitonto, V.; Ruggiero, M.R.; German, S.; Gorin, D.; Terreno, E. Biodegradable Polyelectrolyte/Magnetite Capsules for MR Imaging and Magnetic Targeting of Tumors. *Nanotheranostics* **2021**, *5*, 362–377. [CrossRef]

101. Lepik, K.V.; Muslimov, A.R.; Timin, A.S.; Sergeev, V.S.; Romanyuk, D.S.; Moiseev, I.S.; Popova, E.V.; Radchenko, I.L.; Vilesov, A.D.; Galibin, O.V.; et al. Mesenchymal Stem Cell Magnetization: Magnetic Multilayer Microcapsule Uptake, Toxicity, Impact on Functional Properties, and Perspectives for Magnetic Delivery. *Adv. Healthc. Mater.* **2016**, *5*, 3182–3190. [CrossRef]
102. Zharkov, M.N.; Brodovskaya, E.P.; Kulikov, O.A.; Gromova, E.V.; Ageev, V.P.; Atanova, A.V.; Kozyreva, Z.V.; Tishin, A.M.; Pyatakov, A.P.; Pyataev, N.A.; et al. Enhanced Cytotoxicity Caused by AC Magnetic Field for Polymer Microcapsules Containing Packed Magnetic Nanoparticles. *Colloids Surf. B Biointerfaces* **2021**, *199*, 111548. [CrossRef]
103. Pavlov, A.M.; Gabriel, S.A.; Sukhorukov, G.B.; Gould, D.J. Improved and Targeted Delivery of Bioactive Molecules to Cells with Magnetic Layer-by-Layer Assembled Microcapsules. *Nanoscale* **2015**, *7*, 9686–9693. [CrossRef]
104. Voronin, D.V.; Sindeeva, O.A.; Kurochkin, M.A.; Mayorova, O.; Fedosov, I.V.; Semyachkina-Glushkovskaya, O.; Gorin, D.A.; Tuchin, V.V.; Sukhorukov, G.B. In Vitro and in Vivo Visualization and Trapping of Fluorescent Magnetic Microcapsules in a Bloodstream. *ACS Appl. Mater. Interfaces* **2017**, *9*, 6885–6893. [CrossRef]
105. Read, J.E.; Luo, D.; Chowdhury, T.T.; Flower, R.J.; Poston, R.N.; Sukhorukov, G.B.; Gould, D.J. Magnetically Responsive Layer-by-Layer Microcapsules Can Be Retained in Cells and under Flow Conditions to Promote Local Drug Release without Triggering ROS Production. *Nanoscale* **2020**, *12*, 7735–7748. [CrossRef] [PubMed]
106. Wilhelm, S.; Tavares, A.J.; Dai, Q.; Ohta, S.; Audet, J.; Dvorak, H.F.; Chan, W.C.W. Analysis of Nanoparticle Delivery to Tumours. *Nat. Rev. Mater.* **2016**, *1*, 1–12. [CrossRef]
107. Mayorova, O.A.; Sindeeva, O.A.; Lomova, M.V.; Gusliakova, O.I.; Tarakanchikova, Y.V.; Tyutyaev, E.V.; Pinyaev, S.I.; Kulikov, O.A.; German, S.V.; Pyataev, N.A.; et al. Endovascular Addressing Improves the Effectiveness of Magnetic Targeting of Drug Carrier. Comparison with the Conventional Administration Method. *Nanomedicine Nanotechnol. Biol. Med.* **2020**, *28*, 102184. [CrossRef]
108. Jafari, J.; Han, X.; Palmer, J.; Tran, P.A.; O'Connor, A.J. Remote Control in Formation of 3D Multicellular Assemblies Using Magnetic Forces. *ACS Biomater. Sci. Eng.* **2019**, *5*, 2532–2542. [CrossRef]
109. Armstrong, J.P.K.; Stevens, M.M. Using Remote Fields for Complex Tissue Engineering. *Trends Biotechnol.* **2020**, *38*, 254–263. [CrossRef] [PubMed]
110. Koudan, E.V.; Zharkov, M.N.; Gerasimov, M.V.; Karshieva, S.S.; Shirshova, A.D.; Chrishtop, V.V.; Kasyanov, V.A.; Levin, A.A.; Parfenov, V.A.; Karalkin, P.A.; et al. Magnetic Patterning of Tissue Spheroids Using Polymer Microcapsules Containing Iron Oxide Nanoparticles. *ACS Biomater. Sci. Eng.* **2021**, *7*, 5206–5214. [CrossRef]
111. Zyuzin, M.V.; Timin, A.S.; Sukhorukov, G.B. Multilayer Capsules Inside Biological Systems: State-of-the-Art and Open Challenges. *Langmuir* **2019**, *35*, 4747–4762. [CrossRef] [PubMed]
112. Inozemtseva, O.A.; Lomova, M.V.; Sindeeva, O.A.; Svenskaya, Y.I.; Gorin, D.A.; Sukhorukov, G.B. Remote Controlled Delivery Systems. On a Road to Medical Applications. *Rev. Adv. Chem.* **2021**, *11*, 73–84. [CrossRef]
113. Shchukin, D.G.; Gorin, D.A.; Möhwald, H. Ultrasonically Induced Opening of Polyelectrolyte Microcontainers. *Langmuir* **2006**, *22*, 7400–7404. [CrossRef]
114. Carregal-Romero, S.; Guardia, P.; Yu, X.; Hartmann, R.; Pellegrino, T.; Parak, W.J. Magnetically Triggered Release of Molecular Cargo from Iron Oxide Nanoparticle Loaded Microcapsules. *Nanoscale* **2014**, *7*, 570–576. [CrossRef] [PubMed]
115. Kozlovskaya, V.; Kharlampieva, E. Anisotropic Particles through Multilayer Assembly. *Macromol. Biosci.* **2022**, *22*, 2100328. [CrossRef] [PubMed]
116. Bhuyan, T.; Singh, A.K.; Dutta, D.; Unal, A.; Ghosh, S.S.; Bandyopadhyay, D. Magnetic Field Guided Chemotaxis of IMushbots for Targeted Anticancer Therapeutics. *ACS Biomater. Sci. Eng.* **2017**, *3*, 1627–1640. [CrossRef]
117. Langeveld, S.A.G.; Meijlink, B.; Kooiman, K. Phospholipid-Coated Targeted Microbubbles for Ultrasound Molecular Imaging and Therapy. *Curr. Opin. Chem. Biol.* **2021**, *63*, 171–179. [CrossRef] [PubMed]
118. Shi, D.; Wallyn, J.; Nguyen, D.-V.; Perton, F.; Felder-Flesch, D.; Bégin-Colin, S.; Maaloum, M.; Krafft, M.P. Microbubbles Decorated with Dendronized Magnetic Nanoparticles for Biomedical Imaging: Effective Stabilization via Fluorous Interactions. *Beilstein J. Nanotechnol.* **2019**, *10*, 2103–2115. [CrossRef] [PubMed]
119. Beguin, E.; Bau, L.; Shrivastava, S.; Stride, E. Comparing Strategies for Magnetic Functionalization of Microbubbles. *ACS Appl. Mater. Interfaces* **2019**, *11*, 1829–1840. [CrossRef] [PubMed]
120. Owen, J.; Crake, C.; Lee, J.Y.; Carugo, D.; Beguin, E.; Khrapitchev, A.A.; Browning, R.J.; Sibson, N.; Stride, E. A Versatile Method for the Preparation of Particle-Loaded Microbubbles for Multimodality Imaging and Targeted Drug Delivery. *Drug Deliv. Transl. Res.* **2018**, *8*, 342–356. [CrossRef]
121. Price, P.M.; Mahmoud, W.E.; Al-Ghamdi, A.A.; Bronstein, L.M. Magnetic Drug Delivery: Where the Field Is Going. *Front. Chem.* **2018**, *6*, 619. [CrossRef] [PubMed]
122. Beguin, E.; Gray, M.D.; Logan, K.A.; Nesbitt, H.; Sheng, Y.; Kamila, S.; Barnsley, L.C.; Bau, L.; McHale, A.P.; Callan, J.F.; et al. Magnetic Microbubble Mediated Chemo-Sonodynamic Therapy Using a Combined Magnetic-Acoustic Device. *J. Controlled Release* **2020**, *317*, 23–33. [CrossRef] [PubMed]
123. Chertok, B.; Langer, R. Circulating Magnetic Microbubbles for Localized Real-Time Control of Drug Delivery by Ultrasonography-Guided Magnetic Targeting and Ultrasound. *Theranostics* **2018**, *8*, 341–357. [CrossRef]
124. Liu, Y.; Yang, F.; Yuan, C.; Li, M.; Wang, T.; Chen, B.; Jin, J.; Zhao, P.; Tong, J.; Luo, S.; et al. Magnetic Nanoliposomes as in Situ Microbubble Bombers for Multimodality Image-Guided Cancer Theranostics. *ACS Nano* **2017**, *11*, 1509–1519. [CrossRef] [PubMed]

125. Perera, A.S.; Zhang, S.; Homer-Vanniasinkam, S.; Coppens, M.-O.; Edirisinghe, M. Polymer–Magnetic Composite Fibers for Remote-Controlled Drug Release. *ACS Appl. Mater. Interfaces* **2018**, *10*, 15524–15531. [CrossRef] [PubMed]
126. Amini, Z.; Rudsary, S.S.; Shahraeini, S.S.; Dizaji, B.F.; Goleij, P.; Bakhtiari, A.; Irani, M.; Sharifianjazi, F. Magnetic Bioactive Glasses/Cisplatin Loaded-Chitosan (CS)-Grafted- Poly (ε-Caprolactone) Nanofibers against Bone Cancer Treatment. *Carbohydr. Polym.* **2021**, *258*, 117680. [CrossRef] [PubMed]
127. Abasalta, M.; Asefnejad, A.; Khorasani, M.T.; Saadatabadi, A.R. Fabrication of Carboxymethyl Chitosan/Poly(ε-Caprolactone)/Doxorubicin/Nickel Ferrite Core-Shell Fibers for Controlled Release of Doxorubicin against Breast Cancer. *Carbohydr. Polym.* **2021**, *257*, 117631. [CrossRef]
128. Suneet, K.; De, T.; Rangarajan, A.; Jain, S. Magnetic Nanofibers Based Bandage for Skin Cancer Treatment: A Non-Invasive Hyperthermia Therapy. *Cancer Rep.* **2020**, *3*, e1281. [CrossRef]
129. Nikolaou, M.; Avraam, K.; Kolokithas-Ntoukas, A.; Bakandritsos, A.; Lizal, F.; Misik, O.; Maly, M.; Jedelsky, J.; Savva, I.; Balanean, F.; et al. Superparamagnetic Electrospun Microrods for Magnetically-Guided Pulmonary Drug Delivery with Magnetic Heating. *Mater. Sci. Eng. C Mater. Biol. Appl.* **2021**, *126*, 112117. [CrossRef]
130. Wang, Z.; Liu, C.; Chen, B.; Luo, Y. Magnetically-Driven Drug and Cell on Demand Release System Using 3D Printed Alginate Based Hollow Fiber Scaffolds. *Int. J. Biol. Macromol.* **2021**, *168*, 38–45. [CrossRef]
131. Filippi, M.; Garello, F.; Yasa, O.; Kasamkattil, J.; Scherberich, A.; Katzschmann, R.K. Engineered Magnetic Nanocomposites to Modulate Cellular Function. *Small Weinh. Bergstr. Ger.* **2021**, *18*, e2104079. [CrossRef]
132. Bramhill, J.; Ross, S.; Ross, G. Bioactive Nanocomposites for Tissue Repair and Regeneration: A Review. *Int. J. Environ. Res. Public. Health* **2017**, *14*, 66. [CrossRef]
133. Wu, S.; Hu, W.; Ze, Q.; Sitti, M.; Zhao, R. Multifunctional Magnetic Soft Composites: A Review. *Multifunct. Mater.* **2020**, *3*, 042003. [CrossRef]
134. Sapir-Lekhovitser, Y.; Rotenberg, M.Y.; Jopp, J.; Friedman, G.; Polyak, B.; Cohen, S. Magnetically Actuated Tissue Engineered Scaffold: Insights into Mechanism of Physical Stimulation. *Nanoscale* **2016**, *8*, 3386–3399. [CrossRef] [PubMed]
135. Borin, D. Targeted Patterning of Magnetic Microparticles in a Polymer Composite. *Philos. Trans. R. Soc. Math. Phys. Eng. Sci.* **2020**, *378*, 20190256. [CrossRef] [PubMed]
136. Spangenberg, J.; Kilian, D.; Czichy, C.; Ahlfeld, T.; Lode, A.; Günther, S.; Odenbach, S.; Gelinsky, M. Bioprinting of Magnetically Deformable Scaffolds. *ACS Biomater. Sci. Eng.* **2021**, *7*, 648–662. [CrossRef] [PubMed]
137. Filippi, M.; Dasen, B.; Guerrero, J.; Garello, F.; Isu, G.; Born, G.; Ehrbar, M.; Martin, I.; Scherberich, A. Magnetic Nanocomposite Hydrogels and Static Magnetic Field Stimulate the Osteoblastic and Vasculogenic Profile of Adipose-Derived Cells. *Biomaterials* **2019**, *223*, 119468. [CrossRef]
138. Pardo, A.; Gómez-Florit, M.; Barbosa, S.; Taboada, P.; Domingues, R.M.A.; Gomes, M.E. Magnetic Nanocomposite Hydrogels for Tissue Engineering: Design Concepts and Remote Actuation Strategies to Control Cell Fate. *ACS Nano* **2021**, *15*, 175–209. [CrossRef]
139. Chen, F.; Li, C.; Zhu, Y.-J.; Zhao, X.-Y.; Lu, B.-Q.; Wu, J. Magnetic Nanocomposite of Hydroxyapatite Ultrathin Nanosheets/Fe3O4 Nanoparticles: Microwave-Assisted Rapid Synthesis and Application in PH-Responsive Drug Release. *Biomater. Sci.* **2013**, *1*, 1074–1081. [CrossRef]
140. Walsh, D.; Kim, Y.-Y.; Miyamoto, A.; Meldrum, F.C. Synthesis of Macroporous Calcium Carbonate/Magnetite Nanocomposites and Their Application in Photocatalytic Water Splitting. *Small* **2011**, *7*, 2168–2172. [CrossRef]
141. Parakhonskiy, B.V.; Abalymov, A.; Ivanova, A.; Khalenkow, D.; Skirtach, A.G. Magnetic and Silver Nanoparticle Functionalized Calcium Carbonate Particles—Dual Functionality of Versatile, Movable Delivery Carriers Which Can Surface-Enhance Raman Signals. *J. Appl. Phys.* **2019**, *126*, 203102. [CrossRef]
142. German, S.V.; Novoselova, M.V.; Bratashov, D.N.; Demina, P.A.; Atkin, V.S.; Voronin, D.V.; Khlebtsov, B.N.; Parakhonskiy, B.V.; Sukhorukov, G.B.; Gorin, D.A. High-Efficiency Freezing-Induced Loading of Inorganic Nanoparticles and Proteins into Micron- and Submicron-Sized Porous Particles. *Sci. Rep.* **2018**, *8*, 17763. [CrossRef]
143. Zheltova, V.; Vlasova, A.; Bobrysheva, N.; Abdullin, I.; Semenov, V.; Osmolowsky, M.; Voznesenskiy, M.; Osmolovskaya, O. Fe3O4@HAp Core–Shell Nanoparticles as MRI Contrast Agent: Synthesis, Characterization and Theoretical and Experimental Study of Shell Impact on Magnetic Properties. *Appl. Surf. Sci.* **2020**, *531*, 147352. [CrossRef]
144. Yusoff, A.H.M.; Salimi, M.N.; Gopinath, S.C.B.; Abdullah, M.M.A.; Samsudin, E.M. Catechin Adsorption on Magnetic Hydroxyapatite Nanoparticles: A Synergistic Interaction with Calcium Ions. *Mater. Chem. Phys.* **2020**, *241*, 122337. [CrossRef]
145. Sergeeva, A.; Sergeev, R.; Lengert, E.; Zakharevich, A.; Parakhonskiy, B.; Gorin, D.; Sergeev, S.; Volodkin, D. Composite Magnetite and Protein Containing CaCO3 Crystals. External Manipulation and Vaterite → Calcite Recrystallization-Mediated Release Performance. *ACS Appl. Mater. Interfaces* **2015**, *7*, 21315–21325. [CrossRef] [PubMed]
146. Hoshiar, A.K.; Le, T.-A.; Amin, F.U.; Kim, M.O.; Yoon, J. Studies of Aggregated Nanoparticles Steering during Magnetic-Guided Drug Delivery in the Blood Vessels. *J. Magn. Magn. Mater.* **2017**, *427*, 181–187. [CrossRef]
147. Shaw, S.; Sutradhar, A.; Murthy, P. Permeability and Stress-Jump Effects on Magnetic Drug Targeting in a Permeable Microvessel Using Darcy Model. *J. Magn. Magn. Mater.* **2017**, *429*, 227–235. [CrossRef]
148. Shamsi, M.; Sedaghatkish, A.; Dejam, M.; Saghafian, M.; Mohammadi, M.; Sanati-Nezhad, A. Magnetically Assisted Intraperitoneal Drug Delivery for Cancer Chemotherapy. *Drug Deliv.* **2018**, *25*, 846–861. [CrossRef]

149. Iacobazzi, R.M.; Vischio, F.; Arduino, I.; Canepa, F.; Laquintana, V.; Notarnicola, M.; Scavo, M.P.; Bianco, G.; Fanizza, E.; Lopedota, A.A.; et al. Magnetic Implants in Vivo Guiding Sorafenib Liver Delivery by Superparamagnetic Solid Lipid Nanoparticles. *J. Colloid Interface Sci.* **2021**, *608*, 239–254. [CrossRef]
150. International Commission on Non-Ionizing Radiation Protection Guidelines on Limits of Exposure to Static Magnetic Fields. *Health Phys.* **2009**, *96*, 504–514. [CrossRef]
151. Al-Jamal, K.T.; Bai, J.; Wang, J.T.-W.; Protti, A.; Southern, P.; Bogart, L.; Heidari, H.; Li, X.; Cakebread, A.; Asker, D.; et al. Magnetic Drug Targeting: Preclinical in Vivo Studies, Mathematical Modeling, and Extrapolation to Humans. *Nano Lett.* **2016**, *16*, 5652–5660. [CrossRef]
152. Shapiro, B.; Kulkarni, S.; Nacev, A.; Muro, S.; Stepanov, P.Y.; Weinberg, I.N. Open Challenges in Magnetic Drug Targeting. *WIREs Nanomedicine Nanobiotechnology* **2015**, *7*, 446–457. [CrossRef]
153. Shapiro, B. Towards Dynamic Control of Magnetic Fields to Focus Magnetic Carriers to Targets Deep inside the Body. *J. Magn. Magn. Mater.* **2009**, *321*, 1594–1599. [CrossRef]
154. Ge, J.; Zhang, Y.; Dong, Z.; Jia, J.; Zhu, J.; Miao, X.; Yan, B. Initiation of Targeted Nanodrug Delivery in Vivo by a Multifunctional Magnetic Implant. *ACS Appl. Mater. Interfaces* **2017**, *9*, 20771–20778. [CrossRef] [PubMed]
155. Saatchi, K.; Tod, S.E.; Leung, D.; Nicholson, K.E.; Andreu, I.; Buchwalder, C.; Schmitt, V.; Häfeli, U.O.; Gray, S.L. Characterization of Alendronic- and Undecylenic Acid Coated Magnetic Nanoparticles for the Targeted Delivery of Rosiglitazone to Subcutaneous Adipose Tissue. *Nanomedicine Nanotechnol. Biol. Med.* **2017**, *13*, 559–568. [CrossRef] [PubMed]
156. Kush, P.; Kumar, P.; Singh, R.; Kaushik, A. Aspects of High-Performance and Bio-Acceptable Magnetic Nanoparticles for Biomedical Application. *Asian J. Pharm. Sci.* **2021**, *16*, 704–737. [CrossRef] [PubMed]
157. Aslam, H.; Shukrullah, S.; Naz, M.Y.; Fatima, H.; Hussain, H.; Ullah, S.; Assiri, M.A. Current and Future Perspectives of Multifunctional Magnetic Nanoparticles Based Controlled Drug Delivery Systems. *J. Drug Deliv. Sci. Technol.* **2022**, *67*, 102946. [CrossRef]
158. Li, X.; Li, W.; Wang, M.; Liao, Z. Magnetic Nanoparticles for Cancer Theranostics: Advances and Prospects. *J. Controlled Release* **2021**, *335*, 437–448. [CrossRef]
159. Gholami, A.; Mousavi, S.M.; Hashemi, S.A.; Ghasemi, Y.; Chiang, W.-H.; Parvin, N. Current Trends in Chemical Modifications of Magnetic Nanoparticles for Targeted Drug Delivery in Cancer Chemotherapy. *Drug Metab. Rev.* **2020**, *52*, 205–224. [CrossRef]
160. Liu, Y.; Cao, F.; Sun, B.; Bellanti, J.A.; Zheng, S.G. Magnetic Nanoparticles: A New Diagnostic and Treatment Platform for Rheumatoid Arthritis. *J. Leukoc. Biol.* **2021**, *109*, 415–424. [CrossRef]
161. Fisher, J.D.; Acharya, A.P.; Little, S.R. Micro and Nanoparticle Drug Delivery Systems for Preventing Allotransplant Rejection. *Clin. Immunol. Orlando Fla* **2015**, *160*, 24–35. [CrossRef]
162. Rodrigues, G.R.; López-Abarrategui, C.; de la Serna Gómez, I.; Dias, S.C.; Otero-González, A.J.; Franco, O.L. Antimicrobial Magnetic Nanoparticles Based-Therapies for Controlling Infectious Diseases. *Int. J. Pharm.* **2019**, *555*, 356–367. [CrossRef]
163. Xu, C.; Akakuru, O.U.; Zheng, J.; Wu, A. Applications of Iron Oxide-Based Magnetic Nanoparticles in the Diagnosis and Treatment of Bacterial Infections. *Front. Bioeng. Biotechnol.* **2019**, *7*, 141. [CrossRef]
164. Jat, S.K.; Gandhi, H.A.; Bhattacharya, J.; Sharma, M.K. Magnetic Nanoparticles: An Emerging Nano-Based Tool to Fight against Viral Infections. *Mater. Adv.* **2021**, *2*, 4479–4496. [CrossRef]
165. Arias, L.S.; Pessan, J.P.; Vieira, A.P.M.; de Lima, T.M.T.; Delbem, A.C.B.; Monteiro, D.R. Iron Oxide Nanoparticles for Biomedical Applications: A Perspective on Synthesis, Drugs, Antimicrobial Activity, and Toxicity. *Antibiot. Basel Switz.* **2018**, *7*, 46. [CrossRef] [PubMed]
166. Liu, J.F.; Jang, B.; Issadore, D.; Tsourkas, A. Use of Magnetic Fields and Nanoparticles to Trigger Drug Release and Improve Tumor Targeting. *WIREs Nanomedicine Nanobiotechnology* **2019**, *11*, e1571. [CrossRef] [PubMed]
167. Derfus, A.M.; von Maltzahn, G.; Harris, T.J.; Duza, T.; Vecchio, K.S.; Ruoslahti, E.; Bhatia, S.N. Remotely Triggered Release from Magnetic Nanoparticles. *Adv. Mater.* **2007**, *19*, 3932–3936. [CrossRef]
168. Lim, E.-K.; Huh, Y.-M.; Yang, J.; Lee, K.; Suh, J.-S.; Haam, S. PH-Triggered Drug-Releasing Magnetic Nanoparticles for Cancer Therapy Guided by Molecular Imaging by MRI. *Adv. Mater. Deerfield Beach Fla* **2011**, *23*, 2436–2442. [CrossRef]
169. Albinali, K.E.; Zagho, M.M.; Deng, Y.; Elzatahry, A.A. A Perspective on Magnetic Core–Shell Carriers for Responsive and Targeted Drug Delivery Systems. *Int. J. Nanomedicine* **2019**, *14*, 1707–1723. [CrossRef]
170. Mai, B.T.; Fernandes, S.; Balakrishnan, P.B.; Pellegrino, T. Nanosystems Based on Magnetic Nanoparticles and Thermo- or PH-Responsive Polymers: An Update and Future Perspectives. *Acc. Chem. Res.* **2018**, *51*, 999–1013. [CrossRef]
171. Moros, M.; Idiago-López, J.; Asín, L.; Moreno-Antolín, E.; Beola, L.; Grazú, V.; Fratila, R.M.; Gutiérrez, L.; de la Fuente, J.M. Triggering Antitumoural Drug Release and Gene Expression by Magnetic Hyperthermia. *Adv. Drug Deliv. Rev.* **2019**, *138*, 326–343. [CrossRef]
172. Matsumoto, Y.; Nichols, J.W.; Toh, K.; Nomoto, T.; Cabral, H.; Miura, Y.; Christie, R.J.; Yamada, N.; Ogura, T.; Kano, M.R.; et al. Vascular Bursts Enhance Permeability of Tumour Blood Vessels and Improve Nanoparticle Delivery. *Nat. Nanotechnol.* **2016**, *11*, 533–538. [CrossRef]
173. Subhan, M.A.; Yalamarty, S.S.K.; Filipczak, N.; Parveen, F.; Torchilin, V.P. Recent Advances in Tumor Targeting via EPR Effect for Cancer Treatment. *J. Pers. Med.* **2021**, *11*, 571. [CrossRef]
174. Torchilin, V. Tumor Delivery of Macromolecular Drugs Based on the EPR Effect. *Adv. Drug Deliv. Rev.* **2011**, *63*, 131–135. [CrossRef] [PubMed]

175. Zhi, D.; Yang, T.; Yang, J.; Fu, S.; Zhang, S. Targeting Strategies for Superparamagnetic Iron Oxide Nanoparticles in Cancer Therapy. *Acta Biomater.* 2020, *102*, 13–34. [CrossRef] [PubMed]
176. Tomitaka, A.; Kaushik, A.; Kevadiya, B.D.; Mukadam, I.; Gendelman, H.E.; Khalili, K.; Liu, G.; Nair, M. Surface-Engineered Multimodal Magnetic Nanoparticles to Manage CNS Diseases. *Drug Discov. Today* 2019, *24*, 873–882. [CrossRef]
177. Abd Elrahman, A.A.; Mansour, F.R. Targeted Magnetic Iron Oxide Nanoparticles: Preparation, Functionalization and Biomedical Application. *J. Drug Deliv. Sci. Technol.* 2019, *52*, 702–712. [CrossRef]
178. Dadfar, S.M.; Roemhild, K.; Drude, N.I.; von Stillfried, S.; Knüchel, R.; Kiessling, F.; Lammers, T. Iron Oxide Nanoparticles: Diagnostic, Therapeutic and Theranostic Applications. *Adv. Drug Deliv. Rev.* 2019, *138*, 302–325. [CrossRef] [PubMed]
179. Dai, Q.; Bertleff-Zieschang, N.; Braunger, J.A.; Björnmalm, M.; Cortez-Jugo, C.; Caruso, F. Particle Targeting in Complex Biological Media. *Adv. Healthc. Mater.* 2018, *7*, 1700575. [CrossRef]
180. Shubayev, V.I.; Pisanic, T.R.; Jin, S. Magnetic Nanoparticles for Theragnostics. *Adv. Drug Deliv. Rev.* 2009, *61*, 467–477. [CrossRef]
181. Nikitin, M.P.; Zelepukin, I.V.; Shipunova, V.O.; Sokolov, I.L.; Deyev, S.M.; Nikitin, P.I. Enhancement of the Blood-Circulation Time and Performance of Nanomedicines via the Forced Clearance of Erythrocytes. *Nat. Biomed. Eng.* 2020, *4*, 717–731. [CrossRef]
182. Zelepukin, I.V.; Yaremenko, A.V.; Yuryev, M.V.; Mirkasymov, A.B.; Sokolov, I.L.; Deyev, S.M.; Nikitin, P.I.; Nikitin, M.P. Fast Processes of Nanoparticle Blood Clearance: Comprehensive Study. *J. Control. Release Off. J. Control. Release Soc.* 2020, *326*, 181–191. [CrossRef]
183. Muthiah, M.; Park, I.-K.; Cho, C.-S. Surface Modification of Iron Oxide Nanoparticles by Biocompatible Polymers for Tissue Imaging and Targeting. *Biotechnol. Adv.* 2013, *31*, 1224–1236. [CrossRef]
184. Keselman, P.; Yu, E.Y.; Zhou, X.Y.; Goodwill, P.W.; Chandrasekharan, P.; Ferguson, R.M.; Khandhar, A.P.; Kemp, S.J.; Krishnan, K.M.; Zheng, B.; et al. Tracking Short-Term Biodistribution and Long-Term Clearance of SPIO Tracers in Magnetic Particle Imaging. *Phys. Med. Biol.* 2017, *62*, 3440–3453. [CrossRef] [PubMed]
185. Liu, S.; Chiu-Lam, A.; Rivera-Rodriguez, A.; DeGroff, R.; Savliwala, S.; Sarna, N.; Rinaldi-Ramos, C.M. Long Circulating Tracer Tailored for Magnetic Particle Imaging. *Nanotheranostics* 2021, *5*, 348–361. [CrossRef] [PubMed]
186. Roohi, F.; Lohrke, J.; Ide, A.; Schütz, G.; Dassler, K. Studying the Effect of Particle Size and Coating Type on the Blood Kinetics of Superparamagnetic Iron Oxide Nanoparticles. *Int. J. Nanomedicine* 2012, *7*, 4447–4458. [CrossRef] [PubMed]
187. Gamucci, O.; Bertero, A.; Gagliardi, M.; Bardi, G. Biomedical Nanoparticles: Overview of Their Surface Immune-Compatibility. *Coatings* 2014, *4*, 139–159. [CrossRef]
188. Degors, I.M.S.; Wang, C.; Rehman, Z.U.; Zuhorn, I.S. Carriers Break Barriers in Drug Delivery: Endocytosis and Endosomal Escape of Gene Delivery Vectors. *Acc. Chem. Res.* 2019, *52*, 1750–1760. [CrossRef]
189. Hatakeyama, H.; Akita, H.; Harashima, H. The Polyethyleneglycol Dilemma: Advantage and Disadvantage of PEGylation of Liposomes for Systemic Genes and Nucleic Acids Delivery to Tumors. *Biol. Pharm. Bull.* 2013, *36*, 892–899. [CrossRef]
190. Brenner, J.S.; Pan, D.C.; Myerson, J.W.; Marcos-Contreras, O.A.; Villa, C.H.; Patel, P.; Hekierski, H.; Chatterjee, S.; Tao, J.-Q.; Parhiz, H.; et al. Red Blood Cell-Hitchhiking Boosts Delivery of Nanocarriers to Chosen Organs by Orders of Magnitude. *Nat. Commun.* 2018, *9*, 2684. [CrossRef]
191. Liu, C.; Choi, H.; Zhou, R.; Chen, I.-W. RES Blockade: A Strategy for Boosting Efficiency of Nanoparticle Drug. *Nano Today* 2015, *10*, 11–21. [CrossRef]
192. Garello, F.; Boido, M.; Miglietti, M.; Bitonto, V.; Zenzola, M.; Filippi, M.; Arena, F.; Consolino, L.; Ghibaudi, M.; Terreno, E. Imaging of Inflammation in Spinal Cord Injury: Novel Insights on the Usage of PFC-Based Contrast Agents. *Biomedicines* 2021, *9*, 379. [CrossRef]
193. Ackun-Farmmer, M.A.; Xiao, B.; Newman, M.R.; Benoit, D.S.W. Macrophage Depletion Increases Target Specificity of Bone-Targeted Nanoparticles. *J. Biomed. Mater. Res. A* 2022, *110*, 229–238. [CrossRef]
194. Mirkasymov, A.B.; Zelepukin, I.V.; Nikitin, M.P.; Deyev, S.M. In Vivo Blockade of Mononuclear Phagocyte System with Solid Nanoparticles: Efficiency and Affecting Factors. *J. Controlled Release* 2021, *330*, 111–118. [CrossRef] [PubMed]
195. Zelepukin, I.V.; Yaremenko, A.V.; Ivanov, I.N.; Yuryev, M.V.; Cherkasov, V.R.; Deyev, S.M.; Nikitin, P.I.; Nikitin, M.P. Long-Term Fate of Magnetic Particles in Mice: A Comprehensive Study. *ACS Nano* 2021, *15*, 11341–11357. [CrossRef] [PubMed]
196. Arami, H.; Khandhar, A.; Liggitt, D.; Krishnan, K.M. In Vivo Delivery, Pharmacokinetics, Biodistribution and Toxicity of Iron Oxide Nanoparticles. *Chem. Soc. Rev.* 2015, *44*, 8576–8607. [CrossRef]
197. Zoppellaro, G. Iron Oxide Magnetic Nanoparticles (NPs) Tailored for Biomedical Applications. In *Magnetic Nanoheterostructures: Diagnostic, Imaging and Treatment*; Nanomedicine and Nanotoxicology; Sharma, S.K., Javed, Y., Eds.; Springer International Publishing: Cham, Switzerland, 2020; pp. 57–102. ISBN 978-3-030-39923-8.
198. Stepien, G.; Moros, M.; Pérez-Hernández, M.; Monge, M.; Gutiérrez, L.; Fratila, R.M.; de las Heras, M.; Menao Guillén, S.; Puente Lanzarote, J.J.; Solans, C.; et al. Effect of Surface Chemistry and Associated Protein Corona on the Long-Term Biodegradation of Iron Oxide Nanoparticles In Vivo. *ACS Appl. Mater. Interfaces* 2018, *10*, 4548–4560. [CrossRef]
199. Zhang, T.; Li, G.; Miao, Y.; Lu, J.; Gong, N.; Zhang, Y.; Sun, Y.; He, Y.; Peng, M.; Liu, X.; et al. Magnetothermal Regulation of in Vivo Protein Corona Formation on Magnetic Nanoparticles for Improved Cancer Nanotherapy. *Biomaterials* 2021, *276*, 121021. [CrossRef] [PubMed]
200. Rojas, J.M.; Gavilán, H.; Del Dedo, V.; Lorente-Sorolla, E.; Sanz-Ortega, L.; da Silva, G.B.; Costo, R.; Perez-Yagüe, S.; Talelli, M.; Marciello, M.; et al. Time-Course Assessment of the Aggregation and Metabolization of Magnetic Nanoparticles. *Acta Biomater.* 2017, *58*, 181–195. [CrossRef]

201. Shen, J.-M.; Gao, F.-Y.; Yin, T.; Zhang, H.-X.; Ma, M.; Yang, Y.-J.; Yue, F. CRGD-Functionalized Polymeric Magnetic Nanoparticles as a Dual-Drug Delivery System for Safe Targeted Cancer Therapy. *Pharmacol. Res.* **2013**, *70*, 102–115. [CrossRef]
202. El-Boubbou, K.; Ali, R.; Al-Zahrani, H.; Trivilegio, T.; Alanazi, A.H.; Khan, A.L.; Boudjelal, M.; AlKushi, A. Preparation of Iron Oxide Mesoporous Magnetic Microparticles as Novel Multidrug Carriers for Synergistic Anticancer Therapy and Deep Tumor Penetration. *Sci. Rep.* **2019**, *9*, 9481. [CrossRef]
203. Zhao, X.; Kim, J.; Cezar, C.A.; Huebsch, N.; Lee, K.; Bouhadir, K.; Mooney, D.J. Active Scaffolds for On-Demand Drug and Cell Delivery. *Proc. Natl. Acad. Sci. USA* **2011**, *108*, 67–72. [CrossRef]
204. Cardoso, V.F.; Francesko, A.; Ribeiro, C.; Bañobre-López, M.; Martins, P.; Lanceros-Mendez, S. Advances in Magnetic Nanoparticles for Biomedical Applications. *Adv. Healthc. Mater.* **2018**, *7*, 1700845. [CrossRef]
205. Filippi, M.; Born, G.; Felder-Flesch, D.; Scherberich, A. Use of Nanoparticles in Skeletal Tissue Regeneration and Engineering. *Histol. Histopathol.* **2020**, *35*, 331–350. [CrossRef] [PubMed]
206. Chen, X.; Fan, H.; Deng, X.; Wu, L.; Yi, T.; Gu, L.; Zhou, C.; Fan, Y.; Zhang, X. Scaffold Structural Microenvironmental Cues to Guide Tissue Regeneration in Bone Tissue Applications. *Nanomater. Basel Switz.* **2018**, *8*, 960. [CrossRef] [PubMed]
207. Singla, R.; Abidi, S.M.S.; Dar, A.I.; Acharya, A. Nanomaterials as Potential and Versatile Platform for next Generation Tissue Engineering Applications. *J. Biomed. Mater. Res. B Appl. Biomater.* **2019**, *107*, 2433–2449. [CrossRef] [PubMed]
208. Villanueva-Flores, F.; Castro-Lugo, A.; Ramírez, O.T.; Palomares, L.A. Understanding Cellular Interactions with Nanomaterials: Towards a Rational Design of Medical Nanodevices. *Nanotechnology* **2020**, *31*, 132002. [CrossRef] [PubMed]
209. Cun, X.; Hosta-Rigau, L. Topography: A Biophysical Approach to Direct the Fate of Mesenchymal Stem Cells in Tissue Engineering Applications. *Nanomater. Basel Switz.* **2020**, *10*, 2070. [CrossRef] [PubMed]
210. Rivera-Rodriguez, A.; Rinaldi-Ramos, C.M. Emerging Biomedical Applications Based on the Response of Magnetic Nanoparticles to Time-Varying Magnetic Fields. *Annu. Rev. Chem. Biomol. Eng.* **2021**, *12*, 163–185. [CrossRef]
211. Baki, A.; Wiekhorst, F.; Bleul, R. Advances in Magnetic Nanoparticles Engineering for Biomedical Applications-A Review. *Bioeng. Basel Switz.* **2021**, *8*, 134. [CrossRef]
212. Bramson, M.T.K.; Van Houten, S.K.; Corr, D.T. Mechanobiology in Tendon, Ligament, and Skeletal Muscle Tissue Engineering. *J. Biomech. Eng.* **2021**, *143*. [CrossRef]
213. Li, S.; Wei, C.; Lv, Y. Preparation and Application of Magnetic Responsive Materials in Bone Tissue Engineering. *Curr. Stem Cell Res. Ther.* **2020**, *15*, 428–440. [CrossRef]
214. Farr, A.C.; Hogan, K.J.; Mikos, A.G. Nanomaterial Additives for Fabrication of Stimuli-Responsive Skeletal Muscle Tissue Engineering Constructs. *Adv. Healthc. Mater.* **2020**, *9*, e2000730. [CrossRef]
215. Fan, D.; Wang, Q.; Zhu, T.; Wang, H.; Liu, B.; Wang, Y.; Liu, Z.; Liu, X.; Fan, D.; Wang, X. Recent Advances of Magnetic Nanomaterials in Bone Tissue Repair. *Front. Chem.* **2020**, *8*, 745. [CrossRef] [PubMed]
216. Xia, Y.; Sun, J.; Zhao, L.; Zhang, F.; Liang, X.-J.; Guo, Y.; Weir, M.D.; Reynolds, M.A.; Gu, N.; Xu, H.H.K. Magnetic Field and Nano-Scaffolds with Stem Cells to Enhance Bone Regeneration. *Biomaterials* **2018**, *183*, 151–170. [CrossRef]
217. Sapir, Y.; Ruvinov, E.; Polyak, B.; Cohen, S. Magnetically Actuated Alginate Scaffold: A Novel Platform for Promoting Tissue Organization and Vascularization. *Methods Mol. Biol.* **2014**, *1181*, 83–95. [CrossRef] [PubMed]
218. Sapir, Y.; Cohen, S.; Friedman, G.; Polyak, B. The Promotion of in Vitro Vessel-like Organization of Endothelial Cells in Magnetically Responsive Alginate Scaffolds. *Biomaterials* **2012**, *33*, 4100–4109. [CrossRef] [PubMed]
219. Zhang, C.; Cai, Y.-Z.; Lin, X.-J.; Wang, Y. Magnetically Actuated Manipulation and Its Applications for Cartilage Defects: Characteristics and Advanced Therapeutic Strategies. *Front. Cell Dev. Biol.* **2020**, *8*, 526. [CrossRef]
220. Sun, T.; Shi, Q.; Huang, Q.; Wang, H.; Xiong, X.; Hu, C.; Fukuda, T. Magnetic Alginate Microfibers as Scaffolding Elements for the Fabrication of Microvascular-like Structures. *Acta Biomater.* **2018**, *66*, 272–281. [CrossRef]
221. Manjua, A.C.; Cabral, J.M.S.; Portugal, C.A.M.; Ferreira, F.C. Magnetic Stimulation of the Angiogenic Potential of Mesenchymal Stromal Cells in Vascular Tissue Engineering. *Sci. Technol. Adv. Mater.* **2021**, *22*, 461–480. [CrossRef]
222. Friedrich, R.P.; Cicha, I.; Alexiou, C. Iron Oxide Nanoparticles in Regenerative Medicine and Tissue Engineering. *Nanomater. Basel Switz.* **2021**, *11*, 2337. [CrossRef]
223. Cheah, E.; Wu, Z.; Thakur, S.S.; O'Carroll, S.J.; Svirskis, D. Externally Triggered Release of Growth Factors—A Tissue Regeneration Approach. *J. Control. Release Off. J. Control. Release Soc.* **2021**, *332*, 74–95. [CrossRef]
224. Rambhia, K.J.; Ma, P.X. Controlled Drug Release for Tissue Engineering. *J. Control. Release Off. J. Control. Release Soc.* **2015**, *219*, 119–128. [CrossRef]
225. Mohammadian, F.; Eatemadi, A. Drug Loading and Delivery Using Nanofibers Scaffolds. *Artif. Cells Nanomedicine Biotechnol.* **2017**, *45*, 881–888. [CrossRef] [PubMed]
226. Doostmohammadi, M.; Forootanfar, H.; Ramakrishna, S. Regenerative Medicine and Drug Delivery: Progress via Electrospun Biomaterials. *Mater. Sci. Eng. C Mater. Biol. Appl.* **2020**, *109*, 110521. [CrossRef] [PubMed]
227. Limongi, T.; Susa, F.; Allione, M.; di Fabrizio, E. Drug Delivery Applications of Three-Dimensional Printed (3DP) Mesoporous Scaffolds. *Pharmaceutics* **2020**, *12*, 851. [CrossRef] [PubMed]
228. Mei, Y.; He, C.; Gao, C.; Zhu, P.; Lu, G.; Li, H. 3D-Printed Degradable Anti-Tumor Scaffolds for Controllable Drug Delivery. *Int. J. Bioprinting* **2021**, *7*, 418. [CrossRef] [PubMed]
229. Gelmi, A.; Schutt, C.E. Stimuli-Responsive Biomaterials: Scaffolds for Stem Cell Control. *Adv. Healthc. Mater.* **2021**, *10*, 2001125. [CrossRef] [PubMed]

230. Sensenig, R.; Sapir, Y.; MacDonald, C.; Cohen, S.; Polyak, B. Magnetic Nanoparticle-Based Approaches to Locally Target Therapy and Enhance Tissue Regeneration in Vivo. *Nanomed.* **2012**, *7*, 1425–1442. [CrossRef] [PubMed]
231. Ekenseair, A.K.; Kasper, F.K.; Mikos, A.G. Perspectives on the Interface of Drug Delivery and Tissue Engineering. *Adv. Drug Deliv. Rev.* **2013**, *65*, 89–92. [CrossRef]
232. Luo, C.; Yang, X.; Li, M.; Huang, H.; Kang, Q.; Zhang, X.; Hui, H.; Zhang, X.; Cen, C.; Luo, Y.; et al. A Novel Strategy for in Vivo Angiogenesis and Osteogenesis: Magnetic Micro-Movement in a Bone Scaffold. *Artif. Cells Nanomedicine Biotechnol.* **2018**, *46*, 636–645. [CrossRef]
233. Forouzandehdel, S.; Forouzandehdel, S.; Rezghi Rami, M. Synthesis of a Novel Magnetic Starch-Alginic Acid-Based Biomaterial for Drug Delivery. *Carbohydr. Res.* **2020**, *487*, 107889. [CrossRef]
234. Arriaga, M.A.; Enriquez, D.M.; Salinas, A.D.; Garcia, R.; Trevino De Leo, C.; Lopez, S.A.; Martirosyan, K.S.; Chew, S.A. Application of Iron Oxide Nanoparticles to Control the Release of Minocycline for the Treatment of Glioblastoma. *Future Med. Chem.* **2021**, *13*, 1833–1843. [CrossRef]
235. Dosier, C.R.; Uhrig, B.A.; Willett, N.J.; Krishnan, L.; Li, M.-T.A.; Stevens, H.Y.; Schwartz, Z.; Boyan, B.D.; Guldberg, R.E. Effect of Cell Origin and Timing of Delivery for Stem Cell-Based Bone Tissue Engineering Using Biologically Functionalized Hydrogels. *Tissue Eng. Part A* **2015**, *21*, 156–165. [CrossRef] [PubMed]
236. Madani, S.Z.M.; Reisch, A.; Roxbury, D.; Kennedy, S.M. A Magnetically Responsive Hydrogel System for Controlling the Timing of Bone Progenitor Recruitment and Differentiation Factor Deliveries. *ACS Biomater. Sci. Eng.* **2020**, *6*, 1522–1534. [CrossRef] [PubMed]
237. Fernandes, M.M.; Correia, D.M.; Ribeiro, C.; Castro, N.; Correia, V.; Lanceros-Mendez, S. Bioinspired Three-Dimensional Magnetoactive Scaffolds for Bone Tissue Engineering. *ACS Appl. Mater. Interfaces* **2019**, *11*, 45265–45275. [CrossRef] [PubMed]
238. Lee, K.; Go, G.; Yoo, A.; Kang, B.; Choi, E.; Park, J.-O.; Kim, C.-S. Wearable Fixation Device for a Magnetically Controllable Therapeutic Agent Carrier: Application to Cartilage Repair. *Pharmaceutics* **2020**, *12*, 593. [CrossRef] [PubMed]
239. Liu, X.; Sun, Y.; Chen, B.; Li, Y.; Zhu, P.; Wang, P.; Yan, S.; Li, Y.; Yang, F.; Gu, N. Novel Magnetic Silk Fibroin Scaffolds with Delayed Degradation for Potential Long-Distance Vascular Repair. *Bioact. Mater.* **2022**, *7*, 126–143. [CrossRef]
240. Fernandes, D.C.; Canadas, R.F.; Reis, R.L.; Oliveira, J.M. Dynamic Culture Systems and 3D Interfaces Models for Cancer Drugs Testing. In *Biomaterials- and Microfluidics-Based Tissue Engineered 3D Models*; Advances in Experimental Medicine and Biology; Oliveira, J.M., Reis, R.L., Eds.; Springer International Publishing: Cham, Switzerland, 2020; pp. 137–159. ISBN 978-3-030-36588-2.
241. Chorny, M.; Fishbein, I.; Forbes, S.; Alferiev, I. Magnetic Nanoparticles for Targeted Vascular Delivery. *IUBMB Life* **2011**, *63*, 613–620. [CrossRef]
242. Kolosnjaj-Tabi, J.; Wilhelm, C.; Clément, O.; Gazeau, F. Cell Labeling with Magnetic Nanoparticles: Opportunity for Magnetic Cell Imaging and Cell Manipulation. *J. Nanobiotechnology* **2013**, *11*, S7. [CrossRef]
243. Wilhelm, C.; Gazeau, F.; Bacri, J.-C. Magnetophoresis and Ferromagnetic Resonance of Magnetically Labeled Cells. *Eur. Biophys. J. EBJ* **2002**, *31*, 118–125. [CrossRef]
244. Wilhelm, C.; Rivière, C.; Biais, N. Magnetic Control of Dictyostelium Aggregation. *Phys. Rev. E Stat. Nonlin. Soft Matter Phys.* **2007**, *75*, 041906. [CrossRef]
245. Blümler, P.; Friedrich, R.P.; Pereira, J.; Baun, O.; Alexiou, C.; Mailänder, V. Contactless Nanoparticle-Based Guiding of Cells by Controllable Magnetic Fields. *Nanotechnol. Sci. Appl.* **2021**, *14*, 91–100. [CrossRef]
246. Falconieri, A.; De Vincentiis, S.; Raffa, V. Recent Advances in the Use of Magnetic Nanoparticles to Promote Neuroregeneration. *Nanomed.* **2019**, *14*, 1073–1076. [CrossRef]
247. Guryanov, I.; Naumenko, E.; Konnova, S.; Lagarkova, M.; Kiselev, S.; Fakhrullin, R. Spatial Manipulation of Magnetically-Responsive Nanoparticle Engineered Human Neuronal Progenitor Cells. *Nanomedicine Nanotechnol. Biol. Med.* **2019**, *20*, 102038. [CrossRef] [PubMed]
248. Marcus, M.; Karni, M.; Baranes, K.; Levy, I.; Alon, N.; Margel, S.; Shefi, O. Iron Oxide Nanoparticles for Neuronal Cell Applications: Uptake Study and Magnetic Manipulations. *J. Nanobiotechnology* **2016**, *14*, 37. [CrossRef] [PubMed]
249. Zborowski, M.; Ostera, G.R.; Moore, L.R.; Milliron, S.; Chalmers, J.J.; Schechter, A.N. Red Blood Cell Magnetophoresis. *Biophys. J.* **2003**, *84*, 2638–2645. [CrossRef]
250. Kobayashi, T.; Ochi, M.; Yanada, S.; Ishikawa, M.; Adachi, N.; Deie, M.; Arihiro, K. Augmentation of Degenerated Human Cartilage In Vitro Using Magnetically Labeled Mesenchymal Stem Cells and an External Magnetic Device. *Arthrosc. J. Arthrosc. Relat. Surg.* **2009**, *25*, 1435–1441. [CrossRef] [PubMed]
251. Robert, D.; Fayol, D.; Le Visage, C.; Frasca, G.; Brulé, S.; Ménager, C.; Gazeau, F.; Letourneur, D.; Wilhelm, C. Magnetic Micro-Manipulations to Probe the Local Physical Properties of Porous Scaffolds and to Confine Stem Cells. *Biomaterials* **2010**, *31*, 1586–1595. [CrossRef]
252. Naumenko, E.A.; Dzamukova, M.R.; Fakhrullina, G.I.; Akhatova, F.S.; Fakhrullin, R.F. Nano-Labelled Cells—a Functional Tool in Biomedical Applications. *Curr. Opin. Pharmacol.* **2014**, *18*, 84–90. [CrossRef]
253. Byun, H.; Lee, S.; Jang, G.N.; Lee, H.; Park, S.; Shin, H. Magnetism-Controlled Assembly of Composite Stem Cell Spheroids for the Biofabrication of Contraction-Modulatory 3D Tissue. *Biofabrication* **2021**, *14*, 015007. [CrossRef]
254. Whatley, B.R.; Li, X.; Zhang, N.; Wen, X. Magnetic-Directed Patterning of Cell Spheroids. *J. Biomed. Mater. Res. A* **2014**, *102*, 1537–1547. [CrossRef]

255. Mattix, B.M.; Olsen, T.R.; Casco, M.; Reese, L.; Poole, J.T.; Zhang, J.; Visconti, R.P.; Simionescu, A.; Simionescu, D.T.; Alexis, F. Janus Magnetic Cellular Spheroids for Vascular Tissue Engineering. *Biomaterials* **2014**, *35*, 949–960. [CrossRef]
256. Ghosh, S.; Kumar, S.R.P.; Puri, I.K.; Elankumaran, S. Magnetic Assembly of 3D Cell Clusters: Visualizing the Formation of an Engineered Tissue. *Cell Prolif.* **2016**, *49*, 134–144. [CrossRef] [PubMed]
257. Ho, V.H.B.; Müller, K.H.; Barcza, A.; Chen, R.; Slater, N.K.H. Generation and Manipulation of Magnetic Multicellular Spheroids. *Biomaterials* **2010**, *31*, 3095–3102. [CrossRef] [PubMed]
258. Souza, G.R.; Molina, J.R.; Raphael, R.M.; Ozawa, M.G.; Stark, D.J.; Levin, C.S.; Bronk, L.F.; Ananta, J.S.; Mandelin, J.; Georgescu, M.-M.; et al. Three-Dimensional Tissue Culture Based on Magnetic Cell Levitation. *Nat. Nanotechnol.* **2010**, *5*, 291–296. [CrossRef] [PubMed]
259. Wilhelm, C.; Bal, L.; Smirnov, P.; Galy-Fauroux, I.; Clément, O.; Gazeau, F.; Emmerich, J. Magnetic Control of Vascular Network Formation with Magnetically Labeled Endothelial Progenitor Cells. *Biomaterials* **2007**, *28*, 3797–3806. [CrossRef] [PubMed]
260. Silva, L.H.A.; Cruz, F.F.; Morales, M.M.; Weiss, D.J.; Rocco, P.R.M. Magnetic Targeting as a Strategy to Enhance Therapeutic Effects of Mesenchymal Stromal Cells. *Stem Cell Res. Ther.* **2017**, *8*, 58. [CrossRef]
261. Connell, J.J.; Patrick, P.S.; Yu, Y.; Lythgoe, M.F.; Kalber, T.L. Advanced Cell Therapies: Targeting, Tracking and Actuation of Cells with Magnetic Particles. *Regen. Med.* **2015**, *10*, 757–772. [CrossRef]
262. Thorek, D.L.J.; Tsourkas, A. Size, Charge and Concentration Dependent Uptake of Iron Oxide Particles by Non-Phagocytic Cells. *Biomaterials* **2008**, *29*, 3583–3590. [CrossRef]
263. Pislaru, S.V.; Harbuzariu, A.; Gulati, R.; Witt, T.; Sandhu, N.P.; Simari, R.D.; Sandhu, G.S. Magnetically Targeted Endothelial Cell Localization in Stented Vessels. *J. Am. Coll. Cardiol.* **2006**, *48*, 1839–1845. [CrossRef]
264. Cheng, K.; Malliaras, K.; Li, T.-S.; Sun, B.; Houde, C.; Galang, G.; Smith, J.; Matsushita, N.; Marbán, E. Magnetic Enhancement of Cell Retention, Engraftment, and Functional Benefit after Intracoronary Delivery of Cardiac-Derived Stem Cells in a Rat Model of Ischemia/Reperfusion. *Cell Transplant.* **2012**, *21*, 1121–1135. [CrossRef]
265. Cheng, K.; Li, T.-S.; Malliaras, K.; Davis, D.R.; Zhang, Y.; Marbán, E. Magnetic Targeting Enhances Engraftment and Functional Benefit of Iron-Labeled Cardiosphere-Derived Cells in Myocardial Infarction. *Circ. Res.* **2010**, *106*, 1570–1581. [CrossRef]
266. Foged, C.; Brodin, B.; Frokjaer, S.; Sundblad, A. Particle Size and Surface Charge Affect Particle Uptake by Human Dendritic Cells in an in Vitro Model. *Int. J. Pharm.* **2005**, *298*, 315–322. [CrossRef]
267. Riegler, J.; Wells, J.A.; Kyrtatos, P.G.; Price, A.N.; Pankhurst, Q.A.; Lythgoe, M.F. Targeted Magnetic Delivery and Tracking of Cells Using a Magnetic Resonance Imaging System. *Biomaterials* **2010**, *31*, 5366–5371. [CrossRef] [PubMed]
268. Riegler, J.; Liew, A.; Hynes, S.O.; Ortega, D.; O'Brien, T.; Day, R.M.; Richards, T.; Sharif, F.; Pankhurst, Q.A.; Lythgoe, M.F. Superparamagnetic Iron Oxide Nanoparticle Targeting of MSCs in Vascular Injury. *Biomaterials* **2013**, *34*, 1987–1994. [CrossRef] [PubMed]
269. Landázuri, N.; Tong, S.; Suo, J.; Joseph, G.; Weiss, D.; Sutcliffe, D.J.; Giddens, D.P.; Bao, G.; Taylor, W.R. Magnetic Targeting of Human Mesenchymal Stem Cells with Internalized Superparamagnetic Iron Oxide Nanoparticles. *Small Weinh. Bergstr. Ger.* **2013**, *9*, 4017–4026. [CrossRef] [PubMed]
270. Kang, H.-J.; Kim, J.-Y.; Lee, H.-J.; Kim, K.-H.; Kim, T.-Y.; Lee, C.-S.; Lee, H.-C.; Park, T.H.; Kim, H.-S.; Park, Y.-B. Magnetic Bionanoparticle Enhances Homing of Endothelial Progenitor Cells in Mouse Hindlimb Ischemia. *Korean Circ. J.* **2012**, *42*, 390–396. [CrossRef] [PubMed]
271. Kodama, A.; Kamei, N.; Kamei, G.; Kongcharoensombat, W.; Ohkawa, S.; Nakabayashi, A.; Ochi, M. In Vivo Bioluminescence Imaging of Transplanted Bone Marrow Mesenchymal Stromal Cells Using a Magnetic Delivery System in a Rat Fracture Model. *J. Bone Joint Surg. Br.* **2012**, *94-B*, 998–1006. [CrossRef]
272. Vandergriff, A.C.; Hensley, T.M.; Henry, E.T.; Shen, D.; Anthony, S.; Zhang, J.; Cheng, K. Magnetic Targeting of Cardiosphere-Derived Stem Cells with Ferumoxytol Nanoparticles for Treating Rats with Myocardial Infarction. *Biomaterials* **2014**, *35*, 8528–8539. [CrossRef] [PubMed]
273. Kobayashi, A.; Yamakoshi, K.; Yajima, Y.; Utoh, R.; Yamada, M.; Seki, M. Preparation of Stripe-Patterned Heterogeneous Hydrogel Sheets Using Microfluidic Devices for High-Density Coculture of Hepatocytes and Fibroblasts. *J. Biosci. Bioeng.* **2013**, *116*, 761–767. [CrossRef]
274. Chaudeurge, A.; Wilhelm, C.; Chen-Tournoux, A.; Farahmand, P.; Bellamy, V.; Autret, G.; Ménager, C.; Hagège, A.; Larghéro, J.; Gazeau, F.; et al. Can Magnetic Targeting of Magnetically Labeled Circulating Cells Optimize Intramyocardial Cell Retention? *Cell Transplant.* **2012**, *21*, 679–691. [CrossRef]
275. Huang, Z.; Shen, Y.; Pei, N.; Sun, A.; Xu, J.; Song, Y.; Huang, G.; Sun, X.; Zhang, S.; Qin, Q.; et al. The Effect of Nonuniform Magnetic Targeting of Intracoronary-Delivering Mesenchymal Stem Cells on Coronary Embolisation. *Biomaterials* **2013**, *34*, 9905–9916. [CrossRef]
276. Huang, Z.; Shen, Y.; Sun, A.; Huang, G.; Zhu, H.; Huang, B.; Xu, J.; Song, Y.; Pei, N.; Ma, J.; et al. Magnetic Targeting Enhances Retrograde Cell Retention in a Rat Model of Myocardial Infarction. *Stem Cell Res. Ther.* **2013**, *4*, 149. [CrossRef]
277. Tran, L.A.; Hernández-Rivera, M.; Berlin, A.N.; Zheng, Y.; Sampaio, L.; Bové, C.; Cabreira-Hansen, M.d.G.; Willerson, J.T.; Perin, E.C.; Wilson, L.J. The Use of Gadolinium-Carbon Nanostructures to Magnetically Enhance Stem Cell Retention for Cellular Cardiomyoplasty. *Biomaterials* **2014**, *35*, 720–726. [CrossRef] [PubMed]

278. Shen, Y.; Liu, X.; Huang, Z.; Pei, N.; Xu, J.; Li, Z.; Wang, Y.; Qian, J.; Ge, J. Comparison of Magnetic Intensities for Mesenchymal Stem Cell Targeting Therapy on Ischemic Myocardial Repair: High Magnetic Intensity Improves Cell Retention but Has No Additional Functional Benefit. *Cell Transplant.* **2015**, *24*, 1981–1997. [CrossRef] [PubMed]
279. Polyak, B.; Fishbein, I.; Chorny, M.; Alferiev, I.; Williams, D.; Yellen, B.; Friedman, G.; Levy, R.J. High Field Gradient Targeting of Magnetic Nanoparticle-Loaded Endothelial Cells to the Surfaces of Steel Stents. *Proc. Natl. Acad. Sci. USA* **2008**, *105*, 698–703. [CrossRef] [PubMed]
280. Hofmann, A.; Wenzel, D.; Becher, U.M.; Freitag, D.F.; Klein, A.M.; Eberbeck, D.; Schulte, M.; Zimmermann, K.; Bergemann, C.; Gleich, B.; et al. Combined Targeting of Lentiviral Vectors and Positioning of Transduced Cells by Magnetic Nanoparticles. *Proc. Natl. Acad. Sci. USA* **2009**, *106*, 44–49. [CrossRef]
281. Kyrtatos, P.G.; Lehtolainen, P.; Junemann-Ramirez, M.; Garcia-Prieto, A.; Price, A.N.; Martin, J.F.; Gadian, D.G.; Pankhurst, Q.A.; Lythgoe, M.F. Magnetic Tagging Increases Delivery of Circulating Progenitors in Vascular Injury. *JACC Cardiovasc. Interv.* **2009**, *2*, 794–802. [CrossRef]
282. Song, M.; Kim, Y.-J.; Kim, Y.; Roh, J.; Kim, S.U.; Yoon, B.-W. Using a Neodymium Magnet to Target Delivery of Ferumoxide-Labeled Human Neural Stem Cells in a Rat Model of Focal Cerebral Ischemia. *Hum. Gene Ther.* **2010**, *21*, 603–610. [CrossRef]
283. Carenza, E.; Barceló, V.; Morancho, A.; Levander, L.; Boada, C.; Laromaine, A.; Roig, A.; Montaner, J.; Rosell, A. In Vitro Angiogenic Performance and in Vivo Brain Targeting of Magnetized Endothelial Progenitor Cells for Neurorepair Therapies. *Nanomedicine Nanotechnol. Biol. Med.* **2014**, *10*, 225–234. [CrossRef]
284. Kang, M.K.; Kim, T.J.; Kim, Y.-J.; Kang, L.; Kim, J.; Lee, N.; Hyeon, T.; Lim, M.; Mo, H.J.; Shin, J.H.; et al. Targeted Delivery of Iron Oxide Nanoparticle-Loaded Human Embryonic Stem Cell-Derived Spherical Neural Masses for Treating Intracerebral Hemorrhage. *Int. J. Mol. Sci.* **2020**, *21*, 3658. [CrossRef]
285. Jeon, S.; Park, S.H.; Kim, E.; Kim, J.-Y.; Kim, S.W.; Choi, H. A Magnetically Powered Stem Cell-Based Microrobot for Minimally Invasive Stem Cell Delivery via the Intranasal Pathway in a Mouse Brain. *Adv. Healthc. Mater.* **2021**, *10*, e2100801. [CrossRef]
286. Kobayashi, T.; Ochi, M.; Yanada, S.; Ishikawa, M.; Adachi, N.; Deie, M.; Arihiro, K. A Novel Cell Delivery System Using Magnetically Labeled Mesenchymal Stem Cells and an External Magnetic Device for Clinical Cartilage Repair. *Arthrosc. J. Arthrosc. Relat. Surg.* **2008**, *24*, 69–76. [CrossRef]
287. Mahmoud, E.E.; Kamei, G.; Harada, Y.; Shimizu, R.; Kamei, N.; Adachi, N.; Misk, N.A.; Ochi, M. Cell Magnetic Targeting System for Repair of Severe Chronic Osteochondral Defect in a Rabbit Model. *Cell Transplant.* **2016**, *25*, 1073–1083. [CrossRef] [PubMed]
288. Vaněček, V.; Zablotskii, V.; Forostyak, S.; Růžička, J.; Herynek, V.; Babič, M.; Jendelová, P.; Kubinová, S.; Dejneka, A.; Syková, E. Highly Efficient Magnetic Targeting of Mesenchymal Stem Cells in Spinal Cord Injury. *Int. J. Nanomedicine* **2012**, *7*, 3719–3730. [CrossRef] [PubMed]
289. Nishida, K.; Tanaka, N.; Nakanishi, K.; Kamei, N.; Hamasaki, T.; Yanada, S.; Mochizuki, Y.; Ochi, M. Magnetic Targeting of Bone Marrow Stromal Cells into Spinal Cord: Through Cerebrospinal Fluid. *NeuroReport* **2006**, *17*, 1269–1272. [CrossRef] [PubMed]
290. Sasaki, H.; Tanaka, N.; Nakanishi, K.; Nishida, K.; Hamasaki, T.; Yamada, N.; Ochi, M. Therapeutic Effects with Magnetic Targeting of Bone Marrow Stromal Cells in a Rat Spinal Cord Injury Model. *Spine* **2011**, *36*, 933–938. [CrossRef]
291. Tukmachev, D.; Lunov, O.; Zablotskii, V.; Dejneka, A.; Babic, M.; Syková, E.; Kubinová, Š. An Effective Strategy of Magnetic Stem Cell Delivery for Spinal Cord Injury Therapy. *Nanoscale* **2015**, *7*, 3954–3958. [CrossRef]
292. Silva, L.H.A.; da Silva, J.R.; Ferreira, G.A.; Silva, R.C.; Lima, E.C.D.; Azevedo, R.B.; Oliveira, D.M. Labeling Mesenchymal Cells with DMSA-Coated Gold and Iron Oxide Nanoparticles: Assessment of Biocompatibility and Potential Applications. *J. Nanobiotechnology* **2016**, *14*, 59. [CrossRef]
293. Oshima, S.; Kamei, N.; Nakasa, T.; Ochi, M.; Yasunaga, Y. Enhancement of Muscle Repair Using Human Mesenchymal Stem Cells with a Magnetic Targeting System in a Subchronic Muscle Injury Model. *J. Orthop. Sci.* **2014**, *19*, 478–488. [CrossRef]
294. Yanai, A.; Häfeli, U.O.; Metcalfe, A.L.; Soema, P.; Addo, L.; Gregory-Evans, C.Y.; Po, K.; Shan, X.; Moritz, O.L.; Gregory-Evans, K. Focused Magnetic Stem Cell Targeting to the Retina Using Superparamagnetic Iron Oxide Nanoparticles. *Cell Transplant.* **2012**, *21*, 1137–1148. [CrossRef]
295. Muthana, M.; Kennerley, A.J.; Hughes, R.; Fagnano, E.; Richardson, J.; Paul, M.; Murdoch, C.; Wright, F.; Payne, C.; Lythgoe, M.F.; et al. Directing Cell Therapy to Anatomic Target Sites in Vivo with Magnetic Resonance Targeting. *Nat. Commun.* **2015**, *6*, 1–11. [CrossRef]
296. Sanz-Ortega, L.; Portilla, Y.; Pérez-Yagüe, S.; Barber, D.F. Magnetic Targeting of Adoptively Transferred Tumour-Specific Nanoparticle-Loaded CD8+ T Cells Does Not Improve Their Tumour Infiltration in a Mouse Model of Cancer but Promotes the Retention of These Cells in Tumour-Draining Lymph Nodes. *J. Nanobiotechnology* **2019**, *17*, 87. [CrossRef]
297. Nie, W.; Wei, W.; Zuo, L.; Lv, C.; Zhang, F.; Lu, G.-H.; Li, F.; Wu, G.; Huang, L.; Xi, X.; et al. Magnetic Nanoclusters Armed with Responsive PD-1 Antibody Synergistically Improved Adoptive T-Cell Therapy for Solid Tumors. *ACS Nano* **2019**, *13*, 1469–1478. [CrossRef]
298. Zhao, S.; Duan, J.; Lou, Y.; Gao, R.; Yang, S.; Wang, P.; Wang, C.; Han, L.; Li, M.; Ma, C.; et al. Surface Specifically Modified NK-92 Cells with CD56 Antibody Conjugated Superparamagnetic Fe3O4 Nanoparticles for Magnetic Targeting Immunotherapy of Solid Tumors. *Nanoscale* **2021**, *13*, 19109–19122. [CrossRef]
299. Wu, X.; Jiang, J.; Gu, Z.; Zhang, J.; Chen, Y.; Liu, X. Mesenchymal Stromal Cell Therapies: Immunomodulatory Properties and Clinical Progress. *Stem Cell Res. Ther.* **2020**, *11*, 345. [CrossRef] [PubMed]

300. Zheng, Q.; Zhang, S.; Guo, W.-Z.; Li, X.-K. The Unique Immunomodulatory Properties of MSC-Derived Exosomes in Organ Transplantation. *Front. Immunol.* **2021**, *12*, 659621. [CrossRef] [PubMed]
301. Kamei, N.; Ochi, M.; Adachi, N.; Ishikawa, M.; Yanada, S.; Levin, L.S.; Kamei, G.; Kobayashi, T. The Safety and Efficacy of Magnetic Targeting Using Autologous Mesenchymal Stem Cells for Cartilage Repair. *Knee Surg. Sports Traumatol. Arthrosc. Off. J. ESSKA* **2018**, *26*, 3626–3635. [CrossRef] [PubMed]
302. Silva, L.H.A.; Silva, M.C.; Vieira, J.B.; Lima, E.C.D.; Silva, R.C.; Weiss, D.J.; Morales, M.M.; Cruz, F.F.; Rocco, P.R.M. Magnetic Targeting Increases Mesenchymal Stromal Cell Retention in Lungs and Enhances Beneficial Effects on Pulmonary Damage in Experimental Silicosis. *STEM CELLS Transl. Med.* **2020**, *9*, 1244–1256. [CrossRef]
303. Ricotti, L.; Trimmer, B.; Feinberg, A.W.; Raman, R.; Parker, K.K.; Bashir, R.; Sitti, M.; Martel, S.; Dario, P.; Menciassi, A. Biohybrid Actuators for Robotics: A Review of Devices Actuated by Living Cells. *Sci. Robot.* **2017**, *2*. [CrossRef]
304. Filippi, M.; Buchner, T.; Yasa, O.; Weirich, S.; Katzschmann, R.K. Microfluidic Tissue Engineering and Bio-Actuation. *Adv. Mater.* **2022**, 2108427. [CrossRef]
305. Sitti, M.; Wiersma, D.S. Pros and Cons: Magnetic versus Optical Microrobots. *Adv. Mater.* **2020**, *32*, 1906766. [CrossRef]
306. Palagi, S.; Fischer, P. Bioinspired Microrobots. *Nat. Rev. Mater.* **2018**, *3*, 113–124. [CrossRef]
307. Jarrell, K.F.; McBride, M.J. The Surprisingly Diverse Ways That Prokaryotes Move. *Nat. Rev. Microbiol.* **2008**, *6*, 466–476. [CrossRef] [PubMed]
308. Gong, D.; Cai, J.; Celi, N.; Feng, L.; Jiang, Y.; Zhang, D. Bio-Inspired Magnetic Helical Microswimmers Made of Nickel-Plated Spirulina with Enhanced Propulsion Velocity. *J. Magn. Magn. Mater.* **2018**, *468*, 148–154. [CrossRef]
309. Walker, D.; Kübler, M.; Morozov, K.I.; Fischer, P.; Leshansky, A.M. Optimal Length of Low Reynolds Number Nanopropellers. *Nano Lett.* **2015**, *15*, 4412–4416. [CrossRef] [PubMed]
310. Cell Movements: From Molecules to Motility. Available online: https://www.routledge.com/Cell-Movements-From-Molecules-to-Motility/Bray/p/book/9780815332824 (accessed on 28 February 2022).
311. Magdanz, V.; Vivaldi, J.; Mohanty, S.; Klingner, A.; Vendittelli, M.; Simmchen, J.; Misra, S.; Khalil, I.S.M. Impact of Segmented Magnetization on the Flagellar Propulsion of Sperm-Templated Microrobots. *Adv. Sci. Weinh. Baden-Wurtt. Ger.* **2021**, *8*, 2004037. [CrossRef]
312. Magdanz, V.; Khalil, I.S.M.; Simmchen, J.; Furtado, G.P.; Mohanty, S.; Gebauer, J.; Xu, H.; Klingner, A.; Aziz, A.; Medina-Sánchez, M.; et al. IRONSperm: Sperm-Templated Soft Magnetic Microrobots. *Sci. Adv.* **2020**, *6*, eaba5855. [CrossRef]
313. Peyer, K.E.; Siringil, E.; Zhang, L.; Nelson, B.J. Magnetic Polymer Composite Artificial Bacterial Flagella. *Bioinspir. Biomim.* **2014**, *9*, 046014. [CrossRef]
314. Huang, H.-W.; Chao, Q.; Sakar, M.S.; Nelson, B.J. Optimization of Tail Geometry for the Propulsion of Soft Microrobots. *IEEE Robot. Autom. Lett.* **2017**, *2*, 727–732. [CrossRef]
315. Pak, O.S.; Gao, W.; Wang, J.; Lauga, E. High-Speed Propulsion of Flexible Nanowire Motors: Theory and Experiments. *Soft Matter* **2011**, *7*, 8169–8181. [CrossRef]
316. Diller, E.; Zhuang, J.; Zhan Lum, G.; Edwards, M.R.; Sitti, M. Continuously Distributed Magnetization Profile for Millimeter-Scale Elastomeric Undulatory Swimming. *Appl. Phys. Lett.* **2014**, *104*, 174101. [CrossRef]
317. Evans, B.A.; Shields, A.R.; Carroll, R.L.; Washburn, S.; Falvo, M.R.; Superfine, R. Magnetically Actuated Nanorod Arrays as Biomimetic Cilia. *Nano Lett.* **2007**, *7*, 1428–1434. [CrossRef]
318. Xu, T.; Zhang, J.; Salehizadeh, M.; Onaizah, O.; Diller, E. Millimeter-Scale Flexible Robots with Programmable Three-Dimensional Magnetization and Motions. *Sci. Robot.* **2019**, *4*, eaav4494. [CrossRef] [PubMed]
319. Hu, W.; Lum, G.Z.; Mastrangeli, M.; Sitti, M. Small-Scale Soft-Bodied Robot with Multimodal Locomotion. *Nature* **2018**, *554*, 81–85. [CrossRef] [PubMed]
320. Bhattarai, N.; Gunn, J.; Zhang, M. Chitosan-Based Hydrogels for Controlled, Localized Drug Delivery. *Adv. Drug Deliv. Rev.* **2010**, *62*, 83–99. [CrossRef]
321. Fusco, S.; Chatzipirpiridis, G.; Sivaraman, K.M.; Ergeneman, O.; Nelson, B.J.; Pané, S. Chitosan Electrodeposition for Microrobotic Drug Delivery. *Adv. Healthc. Mater.* **2013**, *2*, 1037–1044. [CrossRef]
322. Chatzipirpiridis, G.; Ergeneman, O.; Pokki, J.; Ullrich, F.; Fusco, S.; Ortega, J.A.; Sivaraman, K.M.; Nelson, B.J.; Pané, S. Electroforming of Implantable Tubular Magnetic Microrobots for Wireless Ophthalmologic Applications. *Adv. Healthc. Mater.* **2015**, *4*, 209–214. [CrossRef] [PubMed]
323. Tabatabaei, S.N.; Lapointe, J.; Martel, S. Shrinkable Hydrogel-Based Magnetic Microrobots for Interventions in the Vascular Network. *Adv. Robot.* **2011**, *25*, 1049–1067. [CrossRef]
324. Li, J.; Li, X.; Luo, T.; Wang, R.; Liu, C.; Chen, S.; Li, D.; Yue, J.; Cheng, S.; Sun, D. Development of a Magnetic Microrobot for Carrying and Delivering Targeted Cells. *Sci. Robot.* **2018**, *3*, eaat8829. [CrossRef]
325. Li, J.; Fan, L.; Li, Y.; Wei, T.; Wang, C.; Li, F.; Tian, H.; Sun, D. Development of Cell-Carrying Magnetic Microrobots with Bioactive Nanostructured Titanate Surface for Enhanced Cell Adhesion. *Micromachines* **2021**, *12*, 1572. [CrossRef]
326. Breger, J.C.; Yoon, C.; Xiao, R.; Kwag, H.R.; Wang, M.O.; Fisher, J.P.; Nguyen, T.D.; Gracias, D.H. Self-Folding Thermo-Magnetically Responsive Soft Microgrippers. *ACS Appl. Mater. Interfaces* **2015**, *7*, 3398–3405. [CrossRef]
327. Jeon, S.; Kim, S.; Ha, S.; Lee, S.; Kim, E.; Kim, S.Y.; Park, S.H.; Jeon, J.H.; Kim, S.W.; Moon, C.; et al. Magnetically Actuated Microrobots as a Platform for Stem Cell Transplantation. *Sci. Robot.* **2019**, *4*, eaav4317. [CrossRef]

328. Go, G.; Jeong, S.-G.; Yoo, A.; Han, J.; Kang, B.; Kim, S.; Nguyen, K.T.; Jin, Z.; Kim, C.-S.; Seo, Y.R.; et al. Human Adipose–Derived Mesenchymal Stem Cell–Based Medical Microrobot System for Knee Cartilage Regeneration in Vivo. *Sci. Robot.* **2020**, *5*, eaay6626. [CrossRef]
329. Li, J.; Dekanovsky, L.; Khezri, B.; Wu, B.; Zhou, H.; Sofer, Z. Biohybrid Micro- and Nanorobots for Intelligent Drug Delivery. *Cyborg Bionic Syst.* **2022**, *2022*. [CrossRef]
330. Park, B.-W.; Zhuang, J.; Yasa, O.; Sitti, M. Multifunctional Bacteria-Driven Microswimmers for Targeted Active Drug Delivery. *ACS Nano* **2017**, *11*, 8910–8923. [CrossRef] [PubMed]
331. Stanton, M.M.; Park, B.-W.; Vilela, D.; Bente, K.; Faivre, D.; Sitti, M.; Sánchez, S. Magnetotactic Bacteria Powered Biohybrids Target E. Coli Biofilms. *ACS Nano* **2017**, *11*, 9968–9978. [CrossRef]
332. Magdanz, V.; Sanchez, S.; Schmidt, O.G. Development of a Sperm-Flagella Driven Micro-Bio-Robot. *Adv. Mater.* **2013**, *25*, 6581–6588. [CrossRef] [PubMed]
333. Xu, H.; Medina-Sánchez, M.; Magdanz, V.; Schwarz, L.; Hebenstreit, F.; Schmidt, O.G. Sperm-Hybrid Micromotor for Targeted Drug Delivery. *ACS Nano* **2018**, *12*, 327–337. [CrossRef] [PubMed]
334. Lin, Z.; Jiang, T.; Shang, J. The Emerging Technology of Biohybrid Micro-Robots: A Review. *Bio-Des. Manuf.* **2022**, *5*, 107–132. [CrossRef]
335. James, M.L.; Gambhir, S.S. A Molecular Imaging Primer: Modalities, Imaging Agents, and Applications. *Physiol. Rev.* **2012**, *92*, 897–965. [CrossRef]
336. Kircher, M.F.; Willmann, J.K. Molecular Body Imaging: MR Imaging, CT, and US. Part I. Principles. *Radiology* **2012**, *263*, 633–643. [CrossRef]
337. Pooley, R.A. Fundamental Physics of MR Imaging. *RadioGraphics* **2005**, *25*, 1087–1099. [CrossRef]
338. Strijkers, G.J.; Mulder, W.J.M.; van Tilborg, G.A.F.; Nicolay, K. MRI Contrast Agents: Current Status and Future Perspectives. *Anticancer Agents Med. Chem.* **2007**, *7*, 291–305. [CrossRef] [PubMed]
339. Wahsner, J.; Gale, E.M.; Rodríguez-Rodríguez, A.; Caravan, P. Chemistry of MRI Contrast Agents: Current Challenges and New Frontiers. *Chem. Rev.* **2019**, *119*, 957–1057. [CrossRef] [PubMed]
340. Aime, S.; Botta, M.; Terreno, E. Gd(III)-BASED CONTRAST AGENTS FOR MRI. In *Advances in Inorganic Chemistry*; Academic Press: Cambridge, MA, USA, 2005; Volume 57, pp. 173–237.
341. Nisticò, R.; Cesano, F.; Garello, F. Magnetic Materials and Systems: Domain Structure Visualization and Other Characterization Techniques for the Application in the Materials Science and Biomedicine. *Inorganics* **2020**, *8*, 6. [CrossRef]
342. Jeon, M.; Halbert, M.V.; Stephen, Z.R.; Zhang, M. Iron Oxide Nanoparticles as T1 Contrast Agents for Magnetic Resonance Imaging: Fundamentals, Challenges, Applications, and Prospectives. *Adv. Mater.* **2021**, *33*, 1906539. [CrossRef]
343. Wei, H.; Bruns, O.T.; Kaul, M.G.; Hansen, E.C.; Barch, M.; Wiśniowska, A.; Chen, O.; Chen, Y.; Li, N.; Okada, S.; et al. Exceedingly Small Iron Oxide Nanoparticles as Positive MRI Contrast Agents. *Proc. Natl. Acad. Sci. USA* **2017**, *114*, 2325–2330. [CrossRef]
344. Na, H.B.; Song, I.C.; Hyeon, T. Inorganic Nanoparticles for MRI Contrast Agents. *Adv. Mater.* **2009**, *21*, 2133–2148. [CrossRef]
345. Perez-Balderas, F.; van Kasteren, S.I.; Aljabali, A.A.A.; Wals, K.; Serres, S.; Jefferson, A.; Sarmiento Soto, M.; Khrapitchev, A.A.; Larkin, J.R.; Bristow, C.; et al. Covalent Assembly of Nanoparticles as a Peptidase-Degradable Platform for Molecular MRI. *Nat. Commun.* **2017**, *8*, 14254. [CrossRef]
346. Bulte, J.W.M.; Kraitchman, D.L. Iron Oxide MR Contrast Agents for Molecular and Cellular Imaging. *NMR Biomed.* **2004**, *17*, 484–499. [CrossRef]
347. Laurent, S.; Boutry, S.; Mahieu, I.; Vander Elst, L.; Muller, R.N. Iron Oxide Based MR Contrast Agents: From Chemistry to Cell Labeling. *Curr. Med. Chem.* **2009**, *16*, 4712–4727. [CrossRef]
348. Gauberti, M.; Martinez de Lizarrondo, S. Molecular MRI of Neuroinflammation: Time to Overcome the Translational Roadblock. *Neuroscience* **2021**, *474*, 30–36. [CrossRef]
349. Garello, F.; Terreno, E. Sonosensitive MRI Nanosystems as Cancer Theranostics: A Recent Update. *Front. Chem.* **2018**, *6*, 157. [CrossRef] [PubMed]
350. Grover, V.P.B.; Tognarelli, J.M.; Crossey, M.M.E.; Cox, I.J.; Taylor-Robinson, S.D.; McPhail, M.J.W. Magnetic Resonance Imaging: Principles and Techniques: Lessons for Clinicians. *J. Clin. Exp. Hepatol.* **2015**, *5*, 246–255. [CrossRef] [PubMed]
351. Gleich, B.; Weizenecker, J. Tomographic Imaging Using the Nonlinear Response of Magnetic Particles. *Nature* **2005**, *435*, 1214–1217. [CrossRef] [PubMed]
352. Bulte, J.W.M. Superparamagnetic Iron Oxides as MPI Tracers: A Primer and Review of Early Applications. *Adv. Drug Deliv. Rev.* **2019**, *138*, 293–301. [CrossRef]
353. Wu, L.C.; Zhang, Y.; Steinberg, G.; Qu, H.; Huang, S.; Cheng, M.; Bliss, T.; Du, F.; Rao, J.; Song, G.; et al. A Review of Magnetic Particle Imaging and Perspectives on Neuroimaging. *AJNR Am. J. Neuroradiol.* **2019**, *40*, 206–212. [CrossRef]
354. Saritas, E.U.; Goodwill, P.W.; Croft, L.R.; Konkle, J.J.; Lu, K.; Zheng, B.; Conolly, S.M. Magnetic Particle Imaging (MPI) for NMR and MRI Researchers. *J. Magn. Reson. San Diego Calif 1997* **2013**, *229*, 116–126. [CrossRef]
355. Arami, H.; Khandhar, A.P.; Tomitaka, A.; Yu, E.; Goodwill, P.W.; Conolly, S.M.; Krishnan, K.M. In Vivo Multimodal Magnetic Particle Imaging (MPI) with Tailored Magneto/Optical Contrast Agents. *Biomaterials* **2015**, *52*, 251–261. [CrossRef]
356. Khandhar, A.P.; Ferguson, R.M.; Arami, H.; Krishnan, K.M. Monodisperse Magnetite Nanoparticle Tracers for in Vivo Magnetic Particle Imaging. *Biomaterials* **2013**, *34*, 3837–3845. [CrossRef]

357. Zahn, D.; Weidner, A.; Saatchi, K.; Häfeli, U.O.; Dutz, S. Biodegradable Magnetic Microspheres for Drug Targeting, Temperature Controlled Drug Release, and Hyperthermia. *Curr. Dir. Biomed. Eng.* **2019**, *5*, 161–164. [CrossRef]
358. Zahn, D.; Ackers, J.; Dutz, S.; Buzug, T.; Graeser, M. Magnetic Microspheres for MPI and Magnetic Actuation. *Int. J. Magn. Part. Imaging* **2022**, *8*. [CrossRef]
359. Oh, J.; Feldman, M.D.; Kim, J.; Condit, C.; Emelianov, S.; Milner, T.E. Detection of Magnetic Nanoparticles in Tissue Using Magneto-Motive Ultrasound. *Nanotechnology* **2006**, *17*, 4183–4190. [CrossRef]
360. Sjöstrand, S.; Evertsson, M.; Jansson, T. Magnetomotive Ultrasound Imaging Systems: Basic Principles and First Applications. *Ultrasound Med. Biol.* **2020**, *46*, 2636–2650. [CrossRef] [PubMed]
361. Qu, M.; Mehrmohammadi, M.; Truby, R.; Graf, I.; Homan, K.; Emelianov, S. Contrast-Enhanced Magneto-Photo-Acoustic Imaging in Vivo Using Dual-Contrast Nanoparticles. *Photoacoustics* **2014**, *2*, 55–62. [CrossRef] [PubMed]
362. Oldenburg, A.L.; Toublan, F.J.-J.; Suslick, K.S.; Wei, A.; Boppart, S.A. Magnetomotive Contrast for in Vivo Optical Coherence Tomography. *Opt. Express* **2005**, *13*, 6597–6614. [CrossRef] [PubMed]
363. Yang, C.-T.; Ghosh, K.K.; Padmanabhan, P.; Langer, O.; Liu, J.; Eng, D.N.C.; Halldin, C.; Gulyás, B. PET-MR and SPECT-MR Multimodality Probes: Development and Challenges. *Theranostics* **2018**, *8*, 6210–6232. [CrossRef]
364. Tay, Z.W.; Chandrasekharan, P.; Chiu-Lam, A.; Hensley, D.W.; Dhavalikar, R.; Zhou, X.Y.; Yu, E.Y.; Goodwill, P.W.; Zheng, B.; Rinaldi, C.; et al. Magnetic Particle Imaging-Guided Heating in Vivo Using Gradient Fields for Arbitrary Localization of Magnetic Hyperthermia Therapy. *ACS Nano* **2018**, *12*, 3699–3713. [CrossRef]
365. Torres Martin de Rosales, R.; Tavaré, R.; Paul, R.L.; Jauregui-Osoro, M.; Protti, A.; Glaria, A.; Varma, G.; Szanda, I.; Blower, P.J. Synthesis of 64Cu(II)-Bis(Dithiocarbamatebisphosphonate) and Its Conjugation with Superparamagnetic Iron Oxide Nanoparticles: In Vivo Evaluation as Dual-Modality PET-MRI Agent. *Angew. Chem. Int. Ed Engl.* **2011**, *50*, 5509–5513. [CrossRef] [PubMed]
366. Xing, Y.; Zhao, J.; Conti, P.S.; Chen, K. Radiolabeled Nanoparticles for Multimodality Tumor Imaging. *Theranostics* **2014**, *4*, 290–306. [CrossRef]
367. Evertsson, M.; Kjellman, P.; Cinthio, M.; Andersson, R.; Tran, T.A.; in't Zandt, R.; Grafström, G.; Toftevall, H.; Fredriksson, S.; Ingvar, C.; et al. Combined Magnetomotive Ultrasound, PET/CT, and MR Imaging of 68Ga-Labelled Superparamagnetic Iron Oxide Nanoparticles in Rat Sentinel Lymph Nodes in Vivo. *Sci. Rep.* **2017**, *7*, 4824. [CrossRef]
368. John, R.; Rezaeipoor, R.; Adie, S.G.; Chaney, E.J.; Oldenburg, A.L.; Marjanovic, M.; Haldar, J.P.; Sutton, B.P.; Boppart, S.A. In Vivo Magnetomotive Optical Molecular Imaging Using Targeted Magnetic Nanoprobes. *Proc. Natl. Acad. Sci. USA* **2010**, *107*, 8085–8090. [CrossRef]
369. Jeelani, S.; Reddy, R.; Maheswaran, T.; Asokan, G.; Dany, A.; Anand, B. Theranostics: A Treasured Tailor for Tomorrow. *J. Pharm. Bioallied Sci.* **2014**, *6*, S6–S8. [CrossRef] [PubMed]
370. Idée, J.-M.; Louguet, S.; Ballet, S.; Corot, C. Theranostics and Contrast-Agents for Medical Imaging: A Pharmaceutical Company Viewpoint. *Quant. Imaging Med. Surg.* **2013**, *3*, 292–297.
371. Caldorera-Moore, M.E.; Liechty, W.B.; Peppas, N.A. Responsive Theranostic Systems: Integration of Diagnostic Imaging Agents and Responsive Controlled Release Drug Delivery Carriers. *Acc. Chem. Res.* **2011**, *44*, 1061–1070. [CrossRef] [PubMed]
372. Lammers, T.; Aime, S.; Hennink, W.E.; Storm, G.; Kiessling, F. Theranostic Nanomedicine. *Acc. Chem. Res.* **2011**, *44*, 1029–1038. [CrossRef]
373. Ahn, B.-C. Personalized Medicine Based on Theranostic Radioiodine Molecular Imaging for Differentiated Thyroid Cancer. *BioMed Res. Int.* **2016**, *2016*, 1680464. [CrossRef]
374. Theek, B.; Rizzo, L.Y.; Ehling, J.; Kiessling, F.; Lammers, T. The Theranostic Path to Personalized Nanomedicine. *Clin. Transl. Imaging* **2014**, *2*, 66–76. [CrossRef]
375. Fleuren, E.D.G.; Versleijen-Jonkers, Y.M.H.; Heskamp, S.; van Herpen, C.M.L.; Oyen, W.J.G.; van der Graaf, W.T.A.; Boerman, O.C. Theranostic Applications of Antibodies in Oncology. *Mol. Oncol.* **2014**, *8*, 799–812. [CrossRef]
376. Hapuarachchige, S.; Artemov, D. Theranostic Pretargeting Drug Delivery and Imaging Platforms in Cancer Precision Medicine. *Front. Oncol.* **2020**, *10*, 1131. [CrossRef]
377. Patrucco, D.; Terreno, E. MR-Guided Drug Release From Liposomes Triggered by Thermal and Mechanical Ultrasound-Induced Effects. *Front. Phys.* **2020**, *8*, 325. [CrossRef]
378. Jordan, A.; Scholz, R.; Wust, P.; Fähling, H.; Felix, R. Magnetic Fluid Hyperthermia (MFH): Cancer Treatment with AC Magnetic Field Induced Excitation of Biocompatible Superparamagnetic Nanoparticles. *J. Magn. Magn. Mater.* **1999**, *201*, 413–419. [CrossRef]
379. Jordan, A.; Wust, P.; Fähling, H.; John, W.; Hinz, A.; Felix, R. Inductive Heating of Ferrimagnetic Particles and Magnetic Fluids: Physical Evaluation of Their Potential for Hyperthermia. *Int. J. Hyperth. Off. J. Eur. Soc. Hyperthermic Oncol. North Am. Hyperth. Group* **1993**, *9*, 51–68. [CrossRef] [PubMed]
380. Thiesen, B.; Jordan, A. Clinical Applications of Magnetic Nanoparticles for Hyperthermia. *Int. J. Hyperth. Off. J. Eur. Soc. Hyperthermic Oncol. North Am. Hyperth. Group* **2008**, *24*, 467–474. [CrossRef] [PubMed]
381. Suriyanto; Ng, E.Y.K.; Kumar, S.D. Physical Mechanism and Modeling of Heat Generation and Transfer in Magnetic Fluid Hyperthermia through Néelian and Brownian Relaxation: A Review. *Biomed. Eng. Online* **2017**, *16*, 36. [CrossRef] [PubMed]
382. Fatima, H.; Charinpanitkul, T.; Kim, K.-S. Fundamentals to Apply Magnetic Nanoparticles for Hyperthermia Therapy. *Nanomater. Basel Switz.* **2021**, *11*, 1203. [CrossRef]
383. Liu, X.L.; Fan, H.M. Innovative Magnetic Nanoparticle Platform for Magnetic Resonance Imaging and Magnetic Fluid Hyperthermia Applications. *Curr. Opin. Chem. Eng.* **2014**, *4*, 38–46. [CrossRef]

384. Chan, M.-H.; Hsieh, M.-R.; Liu, R.-S.; Wei, D.-H.; Hsiao, M. Magnetically Guided Theranostics: Optimizing Magnetic Resonance Imaging with Sandwich-Like Kaolinite-Based Iron/Platinum Nanoparticles for Magnetic Fluid Hyperthermia and Chemotherapy. *Chem. Mater.* **2020**, *32*, 697–708. [CrossRef]
385. Ho, D.; Sun, X.; Sun, S. Monodisperse Magnetic Nanoparticles for Theranostic Applications. *Acc. Chem. Res.* **2011**, *44*, 875–882. [CrossRef]
386. Maier-Hauff, K.; Ulrich, F.; Nestler, D.; Niehoff, H.; Wust, P.; Thiesen, B.; Orawa, H.; Budach, V.; Jordan, A. Efficacy and Safety of Intratumoral Thermotherapy Using Magnetic Iron-Oxide Nanoparticles Combined with External Beam Radiotherapy on Patients with Recurrent Glioblastoma Multiforme. *J. Neurooncol.* **2011**, *103*, 317–324. [CrossRef]
387. Zhang, Z.; Wang, Y.; Rizk, M.M.I.; Liang, R.; Wells, C.J.R.; Gurnani, P.; Zhou, F.; Davies, G.-L.; Williams, G.R. Thermo-Responsive Nano-in-Micro Particles for MRI-Guided Chemotherapy. *Mater. Sci. Eng. C* **2022**, 112716. [CrossRef]
388. Stocke, N.A.; Meenach, S.A.; Arnold, S.M.; Mansour, H.M.; Hilt, J.Z. Formulation and Characterization of Inhalable Magnetic Nanocomposite Microparticles (MnMs) for Targeted Pulmonary Delivery via Spray Drying. *Int. J. Pharm.* **2015**, *479*, 320–328. [CrossRef]
389. Chandrasekharan, P.; Tay, Z.W.; Hensley, D.; Zhou, X.Y.; Fung, B.K.; Colson, C.; Lu, Y.; Fellows, B.D.; Huynh, Q.; Saayujya, C.; et al. Using Magnetic Particle Imaging Systems to Localize and Guide Magnetic Hyperthermia Treatment: Tracers, Hardware, and Future Medical Applications. *Theranostics* **2020**, *10*, 2965–2981. [CrossRef]
390. Ruggiero, M.R.; Crich, S.G.; Sieni, E.; Sgarbossa, P.; Forzan, M.; Cavallari, E.; Stefania, R.; Dughiero, F.; Aime, S. Magnetic Hyperthermia Efficiency And1H-NMR Relaxation Properties of Iron Oxide/Paclitaxel-Loaded PLGA Nanoparticles. *Nanotechnology* **2016**, *27*, 285104. [CrossRef]
391. Laurent, S.; Saei, A.A.; Behzadi, S.; Panahifar, A.; Mahmoudi, M. Superparamagnetic Iron Oxide Nanoparticles for Delivery of Therapeutic Agents: Opportunities and Challenges. *Expert Opin. Drug Deliv.* **2014**, *11*, 1449–1470. [CrossRef] [PubMed]
392. Guo, Y.; Zhang, Y.; Ma, J.; Li, Q.; Li, Y.; Zhou, X.; Zhao, D.; Song, H.; Chen, Q.; Zhu, X. Light/Magnetic Hyperthermia Triggered Drug Released from Multi-Functional Thermo-Sensitive Magnetoliposomes for Precise Cancer Synergetic Theranostics. *J. Control. Release Off. J. Control. Release Soc.* **2018**, *272*, 145–158. [CrossRef] [PubMed]
393. Foy, S.P.; Manthe, R.L.; Foy, S.T.; Dimitrijevic, S.; Krishnamurthy, N.; Labhasetwar, V. Optical Imaging and Magnetic Field Targeting of Magnetic Nanoparticles in Tumors. *ACS Nano* **2010**, *4*, 5217–5224. [CrossRef] [PubMed]
394. Jain, T.K.; Morales, M.A.; Sahoo, S.K.; Leslie-Pelecky, D.L.; Labhasetwar, V. Iron Oxide Nanoparticles for Sustained Delivery of Anticancer Agents. *Mol. Pharm.* **2005**, *2*, 194–205. [CrossRef] [PubMed]
395. Zhang, C.; Wang, J.; Wang, W.; Xi, N.; Wang, Y.; Liu, L. Modeling and Analysis of Bio-Syncretic Micro-Swimmers for Cardiomyocyte-Based Actuation. *Bioinspir. Biomim.* **2016**, *11*, 056006. [CrossRef]
396. Zhang, X.; Le, T.-A.; Yoon, J. Development of a Real Time Imaging-Based Guidance System of Magnetic Nanoparticles for Targeted Drug Delivery. *J. Magn. Magn. Mater.* **2017**, *C*, 345–351. [CrossRef]
397. Le, T.-A.; Zhang, X.; Hoshiar, A.K.; Yoon, J. Real-Time Two-Dimensional Magnetic Particle Imaging for Electromagnetic Navigation in Targeted Drug Delivery. *Sensors* **2017**, *17*, 2050. [CrossRef]
398. Yu, E.Y.; Bishop, M.; Zheng, B.; Ferguson, R.M.; Khandhar, A.P.; Kemp, S.J.; Krishnan, K.M.; Goodwill, P.W.; Conolly, S.M. Magnetic Particle Imaging: A Novel in Vivo Imaging Platform for Cancer Detection. *Nano Lett.* **2017**, *17*, 1648–1654. [CrossRef]
399. Jung, K.O.; Jo, H.; Yu, J.H.; Gambhir, S.S.; Pratx, G. Development and MPI Tracking of Novel Hypoxia-Targeted Theranostic Exosomes. *Biomaterials* **2018**, *177*, 139–148. [CrossRef] [PubMed]
400. Zhu, X.; Li, J.; Peng, P.; Hosseini Nassab, N.; Smith, B.R. Quantitative Drug Release Monitoring in Tumors of Living Subjects by Magnetic Particle Imaging Nanocomposite. *Nano Lett.* **2019**, *19*, 6725–6733. [CrossRef] [PubMed]
401. Huang, L.; Ao, L.; Hu, D.; Wang, W.; Sheng, Z.; Su, W. Magneto-Plasmonic Nanocapsules for Multimodal-Imaging and Magnetically Guided Combination Cancer Therapy. *Chem. Mater.* **2016**, *28*, 5896–5904. [CrossRef]
402. Deinsberger, J.; Reisinger, D.; Weber, B. Global Trends in Clinical Trials Involving Pluripotent Stem Cells: A Systematic Multi-Database Analysis. *NPJ Regen. Med.* **2020**, *5*, 15. [CrossRef]
403. De Vries, I.J.M.; Lesterhuis, W.J.; Barentsz, J.O.; Verdijk, P.; van Krieken, J.H.; Boerman, O.C.; Oyen, W.J.G.; Bonenkamp, J.J.; Boezeman, J.B.; Adema, G.J.; et al. Magnetic Resonance Tracking of Dendritic Cells in Melanoma Patients for Monitoring of Cellular Therapy. *Nat. Biotechnol.* **2005**, *23*, 1407–1413. [CrossRef] [PubMed]
404. Arbab, A.S.; Frank, J.A. Cellular MRI and Its Role in Stem Cell Therapy. *Regen. Med.* **2008**, *3*, 199–215. [CrossRef] [PubMed]
405. Bull, E.; Madani, S.Y.; Sheth, R.; Seifalian, A.; Green, M.; Seifalian, A.M. Stem Cell Tracking Using Iron Oxide Nanoparticles. *Int. J. Nanomedicine* **2014**, *9*, 1641–1653. [CrossRef]
406. Ngen, E.J.; Artemov, D. Advances in Monitoring Cell-Based Therapies with Magnetic Resonance Imaging: Future Perspectives. *Int. J. Mol. Sci.* **2017**, *18*, 198. [CrossRef]
407. Bulte, J.W.M. In Vivo MRI Cell Tracking: Clinical Studies. *AJR Am. J. Roentgenol.* **2009**, *193*, 314–325. [CrossRef]
408. Sehl, O.C.; Gevaert, J.J.; Melo, K.P.; Knier, N.N.; Foster, P.J. A Perspective on Cell Tracking with Magnetic Particle Imaging. *Tomography* **2020**, *6*, 315–324. [CrossRef]
409. Sehl, O.C.; Makela, A.V.; Hamilton, A.M.; Foster, P.J. Trimodal Cell Tracking In Vivo: Combining Iron- and Fluorine-Based Magnetic Resonance Imaging with Magnetic Particle Imaging to Monitor the Delivery of Mesenchymal Stem Cells and the Ensuing Inflammation. *Tomography* **2019**, *5*, 367–376. [CrossRef]

410. Dhada, K.S.; Hernandez, D.S.; Suggs, L.J. In Vivo Photoacoustic Tracking of Mesenchymal Stem Cell Viability. *ACS Nano* **2019**, *13*, 7791–7799. [CrossRef] [PubMed]
411. Donnelly, E.M.; Kubelick, K.P.; Dumani, D.S.; Emelianov, S.Y. Photoacoustic Image-Guided Delivery of Plasmonic-Nanoparticle-Labeled Mesenchymal Stem Cells to the Spinal Cord. *Nano Lett.* **2018**, *18*, 6625–6632. [CrossRef] [PubMed]
412. Nafiujjaman, M.; Kim, T. Gold Nanoparticles as a Computed Tomography Marker for Stem Cell Tracking. *Methods Mol. Biol. Clifton NJ* **2020**, *2126*, 155–166. [CrossRef]
413. Yang, L.; Zhang, L. Motion Control in Magnetic Microrobotics: From Individual and Multiple Robots to Swarms. *Annu. Rev. Control Robot. Auton. Syst.* **2021**, *4*, 509–534. [CrossRef]
414. Yesin, K.B.; Vollmers, K.; Nelson, B.J. Modeling and Control of Untethered Biomicrorobots in a Fluidic Environment Using Electromagnetic Fields. *Int. J. Robot. Res.* **2006**, *25*, 527–536. [CrossRef]
415. Shao, Y.; Fahmy, A.; Li, M.; Li, C.; Zhao, W.; Sienz, J. Study on Magnetic Control Systems of Micro-Robots. *Front. Neurosci.* **2021**, *15*, 736730. [CrossRef] [PubMed]
416. Salmaso, S.; Caliceti, P. Stealth Properties to Improve Therapeutic Efficacy of Drug Nanocarriers. *J. Drug Deliv.* **2013**, *2013*, e374252. [CrossRef]
417. Friedl, J.D.; Nele, V.; De Rosa, G.; Bernkop-Schnürch, A. Bioinert, Stealth or Interactive: How Surface Chemistry of Nanocarriers Determines Their Fate In Vivo. *Adv. Funct. Mater.* **2021**, *31*, 2103347. [CrossRef]
418. Manshadi, M.K.D.; Saadat, M.; Mohammadi, M.; Shamsi, M.; Dejam, M.; Kamali, R.; Sanati-Nezhad, A. Delivery of Magnetic Micro/Nanoparticles and Magnetic-Based Drug/Cargo into Arterial Flow for Targeted Therapy. *Drug Deliv.* **2018**, *25*, 1963–1973. [CrossRef]
419. Hocking, A.M. The Role of Chemokines in Mesenchymal Stem Cell Homing to Wounds. *Adv. Wound Care* **2015**, *4*, 623–630. [CrossRef] [PubMed]
420. Vanden Berg-Foels, W.S. In Situ Tissue Regeneration: Chemoattractants for Endogenous Stem Cell Recruitment. *Tissue Eng. Part B Rev.* **2014**, *20*, 28–39. [CrossRef] [PubMed]
421. Wang, R.; Billone, P.S.; Mullett, W.M. Nanomedicine in Action: An Overview of Cancer Nanomedicine on the Market and in Clinical Trials. *J. Nanomater.* **2013**, *2013*, e629681. [CrossRef]
422. Min, Y.; Caster, J.M.; Eblan, M.J.; Wang, A.Z. Clinical Translation of Nanomedicine. *Chem. Rev.* **2015**, *115*, 11147–11190. [CrossRef] [PubMed]
423. Grillone, A.; Ciofani, G. Magnetic Nanotransducers in Biomedicine. *Chem. Eur. J.* **2017**, *23*, 16109–16114. [CrossRef]
424. Goranov, V.; Shelyakova, T.; De Santis, R.; Haranava, Y.; Makhaniok, A.; Gloria, A.; Tampieri, A.; Russo, A.; Kon, E.; Marcacci, M.; et al. 3D Patterning of Cells in Magnetic Scaffolds for Tissue Engineering. *Sci. Rep.* **2020**, *10*, 2289. [CrossRef]
425. Ceylan, H.; Giltinan, J.; Kozielski, K.; Sitti, M. Mobile Microrobots for Bioengineering Applications. *Lab. Chip* **2017**, *17*, 1705–1724. [CrossRef]
426. Ceylan, H.; Yasa, I.C.; Kilic, U.; Hu, W.; Sitti, M. Translational Prospects of Untethered Medical Microrobots. *Prog. Biomed. Eng.* **2019**, *1*, 012002. [CrossRef]
427. Sitti, M.; Ceylan, H.; Hu, W.; Giltinan, J.; Turan, M.; Yim, S.; Diller, E. Biomedical Applications of Untethered Mobile Milli/Microrobots. *Proc. IEEE Inst. Electr. Electron. Eng.* **2015**, *103*, 205–224. [CrossRef]
428. Li, J.; Esteban-Fernández de Ávila, B.; Gao, W.; Zhang, L.; Wang, J. Micro/Nanorobots for Biomedicine: Delivery, Surgery, Sensing, and Detoxification. *Sci. Robot.* **2017**, *2*, eaam6431. [CrossRef]
429. Bakenecker, A.C.; von Gladiss, A.; Schwenke, H.; Behrends, A.; Friedrich, T.; Lüdtke-Buzug, K.; Neumann, A.; Barkhausen, J.; Wegner, F.; Buzug, T.M. Navigation of a Magnetic Micro-Robot through a Cerebral Aneurysm Phantom with Magnetic Particle Imaging. *Sci. Rep.* **2021**, *11*, 14082. [CrossRef] [PubMed]
430. Tefft, B.J.; Gooden, J.Y.; Uthamaraj, S.; Harburn, J.J.; Klabusay, M.; Holmes, D.R.; Simari, R.D.; Dragomir-Daescu, D.; Sandhu, G.S. Magnetizable Duplex Steel Stents Enable Endothelial Cell Capture. *IEEE Trans. Magn.* **2013**, *49*, 463–466. [CrossRef]

MDPI
St. Alban-Anlage 66
4052 Basel
Switzerland
Tel. +41 61 683 77 34
Fax +41 61 302 89 18
www.mdpi.com

Pharmaceutics Editorial Office
E-mail: pharmaceutics@mdpi.com
www.mdpi.com/journal/pharmaceutics

www.ingramcontent.com/pod-product-compliance
Lightning Source LLC
LaVergne TN
LVHW070049120526
838202LV00101B/1847